Gardens and Academies in Early Modern Italy and Beyond

Intersections

INTERDISCIPLINARY STUDIES IN EARLY MODERN CULTURE

VOLUME 94 – 2025

The titles published in this series are listed at *brill.com/inte*

Gardens and Academies in Early Modern Italy and Beyond

Edited by

Denis Ribouillault

BRILL

LEIDEN | BOSTON

Cover illustrations: (central image) Louis Rouhier (engraver), View of the Villa Giulia Nymphaeum, *c.*1653–1691. Etching, 25.1 × 28.2 cm. From: *Le fontane ne' Palazzi e ne' giardini di Roma* [...] *parte terza* (Rome, Giovanni Giacomo de' Rossi: *c.*1653–1691). Amsterdam, Rijksmuseum, inv. BI-1893-A39-58. Image © Rijksmuseum. See also figure 6.3; (background image) Detail of Giovanni Battista Derchi (1879–1912), The fishpond seen from the Belvedere in the garden of the villa Imperiale Scassi a Genova Sampierdarena, pencil drawing, 21.5 × 23 cm. Private collection. Image © Lauro Magnani. See also figure 12.12.

The Library of Congress Cataloging-in-Publication Data is available online at https://catalog.loc.gov
LC record available at https://lccn.loc.gov/2024035731

Typeface for the Latin, Greek, and Cyrillic scripts: "Brill". See and download: brill.com/brill-typeface.

ISSN 1568-1181
ISBN 978-90-04-51753-0 (hardback)
ISBN 978-90-04-51754-7 (e-book)
DOI 10.1163/9789004517547

Contents

Acknowledgements

This book grew out of three sessions on "Gardens and Academies in Early Modern Italy", organised by Denis Ribouillault and Ginette Vagenheim at the Renaissance Society of America conference in 2021. The editor is very grateful to Ginette Vagenheim for her invaluable contribution to this conference and for her ongoing support of the publication project. Essays by Katrina Grant, Christiane Lauterbach, Laurent Paya and Denis Ribouillault have been added to the original contributions.

The editor received financial support from the research project "The Shared Languages of Art and Science in the Early Modern Period" (Social Sciences and Humanities Research Council (SSHRC) Insight Grant, Canada), directed by Denis Ribouillault at Université de Montréal.

At Université de Montréal, Mariah O'Brien provided valuable help with editing papers and streamlining translation issues. Thanks are also due to Dr. Victoria Addona for her painstaking work in editing and proofreading the volume. Some Latin translations have been revised by Alin Mocanu, whom we would also like to thank.

Chapter 6 is a translation of a paper already published in French: Denis Ribouillault, "La fortune poétique du nymphée de la Villa Giulia", in Jakob M. – Berchtold J. (eds.), *Jardins en images. Stratégie de representation au fil des siècles* (Geneva: 2020) 98–157. We are thankful to the editors for permission to publish it in English in this volume. This second revised and expanded edition of the text includes new and previously unpublished material.

Figures

Note on the Editor

Denis Ribouillault
is Professor of early modern art history at Université de Montréal. His research focuses on Renaissance villa culture, cultural landscapes and gardens and the intersection of art, science and literature. He is the recipient of numerous grants including a Florence J. Gould Fellowship at Villa I Tatti (2008–9), a French Academy in Rome Fellowship (2009–2011), a Dumbarton Oaks Fellowship (2017). In 2024, he was awarded the Panofsky Professorship at the Zentralinstitut für Kunstgeschichte in Munich and the Montaigne Professorship at the Université de Bordeaux. His recent books include *Sacred Landscape. Landscape as Exegesis in Early Modern Europe* (2011, with Michel Weemans), *Rome en ses jardins. Paysage et pouvoir au XVIe siècle* (2013), *De la peinture au jardin* (2016, with Hervé Brunon), *Jardiner/Gardening* (*Intermediality* 35 (2020)), and *The Villa Barbaro at Maser. Science, Philosophy and the Family in Venetian Renaissance Art* (2023).

Notes on the Contributors

George Brocklehurst
is a Ph.D. student at the Warburg Institute, University of London. His thesis examines the reception of the classical art of the symposium in Italian humanism, with a focus on the academy of Giovanni Pontano at Naples. His work has received the 2021 *Society for Neo-Latin Studies* Early Career Prize and the 2022 Gaetano Cozzi Prize for studies in the history of play. He holds degrees in classics from the University of Saint Andrews and the University of London.

Tiziana Checchi
holds a degree in Humanities, History of Art, from the University of Rome La Sapienza, where she also received a Postgraduate Diploma in History of Medieval and Modern Art and a Ph.D. in History of Art (2011). Since 2012 she has been the director of the Ecclesiastical Cultural Heritage Office at the Territorial Abbey of Subiaco and a member of the Lazio Regional Council for Ecclesiastical Cultural Heritage. She has participated in numerous national and international conferences and research projects, collaborating, among others, with the University of Rome La Sapienza, the Vatican Museums, the University of Southampton, the Academy of Architecture University of Italian Switzerland (USI), the University of Salerno and the Galleria Colonna in Rome.

Alessia Dessì
holds a Ph.D. in Early Modern Art History and is a graduate archivist of the Vatican School of Palaeography, Diplomatics and Archivistics and the State Archives in Rome. She is the author of several publications on the relationship between art, literature and antiquarian culture in the second half of the fifteenth century and has presented in and organised numerous international conferences. She is currently a lecturer in Museology and Criticism of Art and Restoration at the Sapienza University of Rome and works as an archivist at the Historical Archives of Fondazione Camillo Caetani in Rome.

Katrina Grant
is a Research Fellow at the Power Institute for Arts and Visual Culture, The University of Sydney. She is an art historian and digital humanities expert with a focus on early modern Italy. Her research focuses on the history of early modern Italian gardens and landscapes, the visual culture of performance in the same period and the application of digital technologies to art history and art collections (digital mapping in particular). Her publications include articles

and book chapters on the garden history of Italy, history of emotions and set design, digitisation in the gallery and museum sector, gaming and European history, the Arcadian Academy and culture in early eighteenth-century Rome and artistic relationships between Britain and Italy in the eighteenth century. She is the author of *Landscape in Early Modern Italy: Theatre, Garden and Visual Culture* (2022), which explores how the idea of gardens as theatres emerged at the same time as opera was developed in Italian courts around the turn of the seventeenth century.

Christiane Lauterbach
is a researcher and deputy head librarian at the Germanisches Nationalmuseum, Nuremberg. She studied art history, German philology and history at the University of Cologne and has published on German and Dutch garden history in the sixteenth and seventeenth centuries.

Maria Cristina Loi
is Associate Professor at the Faculty of Civil Architecture at the Politecnico di Milano. She graduated in Architecture from the Sapienza University in Rome in 1986. In 1989 she obtained a Master of Arts at the Department of Art History and Archaeology of Columbia University, New York, and in 1995 a Ph.D. at Sapienza University. Her main areas of research are American architecture (focusing on the process of defining American architecture after the Revolution, with particular emphasis on the work of Thomas Jefferson and the phenomenon of Palladianism), modern and contemporary Italian architecture (especially Milanese) and architectural drawing. She has published widely on these subjects and has organised several conferences and exhibitions with the Politecnico di Milano and other institutions such as the Accademia di San Luca in Rome and the Archivio del Moderno in Mendrisio.

Lauro Magnani
was Professor of the History of Modern Art at the University of Genoa. His research has focused on pictorial and sculptural production between the sixteenth and eighteenth centuries, the relationship between artists, patrons and the public, religious iconography, and the interpretation of aristocratic interiors. He has published over two hundred books and articles and curated numerous exhibitions. Magnani has published widely on historic gardens and his monograph *The Temple of Venus* is now in its third edition. A summer fellow at Dumbarton Oaks in Washington D.C. (1997), he returned in 2001, where he initiated various collaborations. Most recently, he curated the exhibition *Grottoes and Gardens in the Time of Rubens. Delight and Wonder in Genoa at the*

Beginning of the Seventeenth Century and produced the accompanying catalogue (Genoa 2022).

Giorgio Mangani
teaches intercultural heritage geography at the University of Bologna, Department of Cultural Heritage, Ravenna Campus. He is a historian of geographical thought and cartography. He has published books and essays on Abraham Ortelius, Gerard Mercator, Cyriac of Ancona, the painter and botanist Gherardo Cibo, humanist views of cities, the persuasive power of maps, the representation of landscape, and sixteenth-century Dutch and Italian maps and atlases. His work is available at www.giorgiomangani.it

Sergio Monferrini
is an archivist at Dal Pozzo d'Annone Archive and F. Crespi Archive and an independent scholar. His main interests are the history of Milan, Lombardy and the Province of Novara in the seventeenth and eighteenth centuries, with particular attention paid to noble families, especially the Borromeo and Arese. He has published widely on these subjects as well as on baroque art, music and theatre in Northern Italy.

Laurent Paya
holds a Master's in Landscape Architecture from the Agrocampus Ouest (Angers, France) and a Ph.D. in Art History from the Centre d'Études Supérieures de la Renaissance (CESR, Tours). As Chief Engineer at the Ministry of Agriculture, he teaches landscape architecture and garden design. He is an associate researcher in art history at the CESR and the Centre de Recherches Interdisciplinaires en Sciences humaines et sociales (Université Montpellier III). His research focuses on the aesthetics of the garden, ornament, decoration, and the city in the early modern period.

Alessandro Spila
is Research Fellow and Professor of the History of Early Modern Architecture at the Politecnico di Torino, Faculty of Architecture. Previously, he was Adjunct Professor at the Sapienza University of Rome (2018–2019) and Research Fellow at the Humboldt University of Berlin, Census of Antique Works of Art and Architecture known in the Renaissance (2015–2017). His research interests include the history of architecture in Rome between the seventeenth and eighteenth centuries, with a focus on the history of gardens, the history of festivals and ephemeral representations, and the reception of antiquity. He is the author *Palazzo Colonna nel Settecento. Architettura e potere nella Roma del secolo dei Lumi* (2020).

Note on the Translations

The essays by Laurent Paya and Christiane Lauterbach have been translated from French and German by Denis Ribouillault and Mariah O'Brien. Denis Ribouillault's essay on the Nymphaeum of Villa Giulia has been translated from the French by Mariah O'Brien, Victoria Addona and Denis Ribouillault. All other translations are by the authors but have been checked for consistency by Mariah O'Brien, Victoria Addona and Denis Ribouillault.

CHAPTER 1

Introduction: "Atque inter silvas Academi quaerere verum" – Early Modern Gardens and the Academic Ideal

Denis Ribouillault

This collection of essays aims to explore the role of gardens in early modern academies for the first time and, conversely, the place of what might be called 'academic culture' in early modern gardens. Research on academies has developed considerably in recent decades, emphasising the importance of Italian and European academic networks and their members' publications.[1] However, little has been written about the sites where they met, including gardens. How did an academy choose its location and why? How did literary and scientific activities and debates influence the chosen site's architectural, artistic and/or horticultural qualities? In other words, can the design and iconography of gardens be linked to academic activities? While studies of botanical gardens have often focused on their links with a research institution, our intent here is deliberately broader, seeking to examine the interactions and the intersections between the built environment of the early modern garden and the more or less organised social and intellectual life it sustained.

The ambivalent definition of the word "academy" justifies the study of its diverse places of practice. The academy refers both to a gathering of people – a kind of intellectual and polymorphous collective – and the place where these gatherings occur. The Vocabulario della Crusca's entry for the word *accademia*

1 This introduction is partly based on an earlier paper: Ribouillault D., "Hortus academicus: les académies de la Renaissance et le jardin", in Jakob M. (ed.), *Des jardins et des livres* (Geneva: 2018) 23–34. Some important publications on Italian academies are Maylender M., *Storia delle accademie d'Italia*, 5 vols. (Bologna: 1926–1930); Boehm L. – Raimondi E. (eds.), *Università, accademie e società scientifiche in Italia e in Germania dal Cinquecento al Settecento* (Bologna: 1981); Quondam A., "L'accademia," in *Letteratura italiana. Il letterato e le istituzioni* (Turin: 1982) 823–898; Chambers D.S. – Quiviger F. (eds.), *Italian Academies of the Sixteenth Century* (London: 1995); Barbe J.-P. – Pigeaud J. (eds.), *Les Académies (Antiquité–XIXe siècle). Sixièmes "Entretiens" de la Garenne Lemot* (Québec: 2005); Demaraix M. et al., (eds)., *Les académies dans l'Europe humaniste. Idéaux et pratiques* (Geneva: 2008); Testa S., *Italian Academies and Their Networks, 1525–1700. From Local to Global* (New York: 2015); Everson J.E. – Reidy D.V. – Simpson L. (eds.), *The Italian Academies (1520–1700). Networks of Culture, Innovation and Dissent* (Cambridge & New York: 2016).

 | DOI:10.1163/9789004517547_002

makes this clear: 'I. Oggi comunemente per Accademia s'intende un' Adunanza d'uomini, instituita con autorità pubblica, o per privato consenso, sotto particolari denominazioni e statuti, a fine di dare opera all'accrescimento delle lettere, delle scienze e d'altre discipline. [...] II. Ed anche il luogo dove s'adunano gli Accademici. [...]' ('I. An academy is now generally understood to be an assembly of men, constituted by public authority or by private agreement, under certain names and statutes, for the purpose of working for the advancement of literature, science, and other disciplines. [...] II. And also the place where academicians meet. [...]').[2] Consequently, academies occupied an ambivalent space, at the same time real and ideal – an ambivalence that historians cannot easily dismiss. Some of the questions posed in this book are located in this indeterminate space. What are the textual and visual sources that refer to academic gardens? Are they purely literary formulations suggesting a symbolic dimension, or do they refer to tangible realities? In other words, are these academic gardens purely fictional places, or are they rooted in the physical world? How do we approach the constant hybridisation of literary imaginaries and the concrete realities of place from a cultural, historical and art historical perspective?

The various debates on the reality or fiction of the Platonic Academy of Florence are a good example of the fascinating paradox of this topic and the difficulties that can be encountered when studying it. The founding of the *Accademia Platonica* by Marsilio Ficino in the gardens of the Medici villa of Careggi in 1462, instigated by Cosimo de' Medici (the Elder), has long dominated histories of Florentine academies. Generations of historians, first among them Arnaldo della Torre, have vivaciously documented and imagined the role played by 'Marsilio, the new Plato, who had received as a gift from Cosimo the new Academy, which, like the old one, was situated not far from the city and, like the old one, in a pleasant and charming environment'. There, as della Torre writes, '[Ficino] could have brooded over the speculations that the divine Plato had conceived in the depth of his intellect and the genius of his soul, or in the shade of the tall plane trees in Academos's garden'.[3] In a series of seminal articles published in the 1990s, James Hankins threw a spanner into studies that perpetuate this kind of romantic narrative, one that was very popular from the nineteenth century, by arguing that the Academy referenced by Marsilio Ficino in several crucial documents primarily comprised Plato's Greek manuscripts – manuscripts that he was busy translating – rather than a specific place. Hankins rightly stresses the unstable meaning of the

2 *Vocabolario degli Accademici della Crusca*, vol. 1 (Florence: 1863) 73–74.
3 della Torre A., *Storia dell'Accademia platonica di Firenze* (Florence: 1902) 562. My translation.

word *accademia* in Renaissance Italy and the fact that it was rarely used to designate real places, but rather schools, universities or informal gatherings of scholars (*sodalitates, studia, gymnasia, coetus letteratorum*, etc.). For Hankins, then, the villa of Careggi and the modest house at Monte Vecchio that Ficino owned nearby were never the seats of the new Platonic Academy, as advanced in the historiography on the subject. Is the Platonic Academy thus a pure myth?[4] Indeed, in his poetic description of the villa of Careggi (*Descriptio horti Laurentii Medicis, c.*1480), Alessandro Braccesi never specifically mentions that it might have played such a role. Moreover, Ficino's only use of the term "academiola Saturnii Montis" (small academy) to refer to his modest residence in a letter to Lorenzo de' Medici in 1483 could be understood, according to Hankins, as a purely literary allusion derived by a humanist who considered himself an *alter Plato*.[5] Reviewing available evidence, Christophe Poncet has pointed out other interesting passages that support Hankins's claim. Ficino, for example, mentions a painting in his *gymnasium* with 'the globe of the world and on either side Democritus and Heraclitus' in direct emulation of Plato's Academy, which seems to describe his small villa where, as other documents confirm, the humanists in his circle occasionally gathered.[6] Whereas this reference again shows a willingness to evoke the prestigious memories of the ancient Athenian academy, it does not describe a philosophical school in any way. Scholarly confusion regarding the form of the Platonic Academy may originate in the fact that, when using words such as *academia* or *gymnasium* in these few cryptic passages, Ficino describes a place of calm and solitude rather than an organised school for teaching. In this way, he interprets the fragmentary memories of Plato's academy through the lens of the widespread medieval concept of *solitudo*, notably embraced by Petrarch.[7] For Poncet, 'the *academiola* was more

4 Hankins J., "The Myth of the Platonic Academy of Florence", *Renaissance Quarterly* 44.3 (1991) 429–475; Idem, "Humanist Academies and the 'Platonic Academy of Florence'", *Analecta Romana Instituti Danici* 42 (2011) 31–46. On the villa Medici at Careggi, see also Fabiani Giannetto R., *Medici Gardens. From Making to Design* (Philadelphia: 2008) 31–53.

5 Hankins, "The Myth" 456.

6 Poncet C., "Ficino's Little Academy of Careggi", *Bruniana & Campanelliana* 24.1 (2013) 67–76. See also Arthur Field, "The Platonic Academy of Florence", in Allen M.J.B. – Rees V. – Davies M. (eds.), *Marsilio Ficino: His Theology, His Legacy* (Leiden: 2002) 359–376.

7 On this topic see Enenkel K.A.E. – Göttler C. (eds), *Solitudo. Spaces, Places, and Times of Solitude in Late Medieval and Early Modern Cultures* (Leiden: 2018); and on Petrarch, see Brunon H., "*Locus secretus*: topique et topophilie", in Luciani D. – Mosser M. (eds.), *Petrarca e i suoi luoghi. Spazi reali e paesaggi poetici alle origini del moderno senso della natura* (Treviso: 2010) 41–56.

a hermitage than a school, and its *gymnasium* was probably used more often as his working cabinet or library than as a meeting place or classroom'.[8]

The case of the Platonic Academy in Florence underlines a need to think about the topographical reality of early modern academies within a much broader cultural framework, which Ingrid Rowland, referring to the activities of Giulio Pomponio Leto's Roman Academy, has termed 'creative anachronism', or what I have previously called 'topographical anachronism'.[9] In sum, Ficino, like Petrarch and many humanists before him, projected his knowledge of an ancient site that had eventually been mythologised onto a contemporary location; an operation that, far from being "fantastic", resulted in an almost spiritual belief in the transfer of the "aura" of the ancient place to the new one.

•••

The subject of this book has generated surprisingly scant scholarship, with the possible exception of literature on the Platonic Academy of Florence, mentioned above, and other canonical sites associated with academies, such as the eighteenth-century Bosco Parrasio in Rome, the seat of the Accademia degli Arcadi.[10] In contrast, numerous literary studies have stressed the importance of gardens as places of philosophical dialogue and poetic composition, though they have paid little interest to actual gardens and their histories. Similarly, studies on academies rarely dwell on the form, history and function of their meeting places. For this reason, it may be useful to give a broad overview of the subject, without aiming to be exhaustive, as research on this topic is still in its infancy. To do this, we must start with Plato's Academy, the model for academies *par excellence* [Fig. 1.1].

Located on the outskirts of Athens, the Academy, founded in 387 BCE, took its name from the presumed tomb of the hero Academus. Planted with olive and plane trees, lined with paths, and equipped with a gymnasium and several altars, this vast sacred enclosure was dedicated to Athena, the goddess of wisdom. When he wasn't teaching in one of the classrooms adjoining the gymnasium, Plato gathered his most advanced students in an exedra he had built in the garden of his modest home nearby. Here, he placed a small shrine dedicated to the Muses and a statue of Apollo. The only room in Plato's house

---

8 Poncet, "Ficino's Little Academy" 76.

9 Rowland I., *The Culture of the High Renaissance* (Cambridge: 1998), 7–41; Ribouillault D., "Landscape *all'antica* and Topographical Anachronism in Roman Fresco Painting of the Sixteenth Century", *Journal of the Warburg and Courtauld Institutes* 71 (2008) 211–237.

10 See Katrina Grant's essay in this volume.

FIGURE 1.1 *Plato's Academy*, 1st century AD, Roman mosaic (House of T. Siminius Stephanus, Pompei), Naples, Museo Nazionale Archeologico, inv. 124545
IMAGE © WIKIMEDIA COMMONS

was decorated with two paintings illustrating the life of his teacher Socrates. It also contained his scrolls, an alarm clock of his design, celestial and terrestrial globes, maps, and an armillary sphere.[11]

Philosophical research at the Academy involved the study of a wide range of disciplines including mathematics and astronomy. The Earth's spherical shape was first proven in Plato's academic garden. Some contemporary comedies even mocked the research carried out by Plato and his students on the classification of animal and plant species – for many, it was inappropriate to

11 Diogene Laertius, iv, 19. See Baltes M., "Plato's School, the Academy", *Hermathena*, no. 155 (1993) 5–26.

philosophise about pumpkins! The range of these activities, to which we can add rhetoric, poetry, geography, history, grammar and linguistics, mechanics, music, medicine and meteorology, also animated lively social events rooted in friendship, which often took the form of banquets held in honour of the gods. Many scenes in Plato's dialogues evoke the rural setting of the academy, such as the discussions on love in *Symposium* or the praise of the ideal landscape (*locus amoenus*) in *Phaedrus* (230 BCE), which was, with its plane tree, spring, sanctuary and soft grass, a place conducive to the relaxation of body and mind and to philosophical dialogue. To philosophise in a garden was to recognise not only the essential link between man, nature, and the cosmos but also the links between individuals (*polis*).[12] Philosophy in the garden meant nourishing the body as well as the mind. Initially, students could practise in the gymnasium, but the adjoining garden also provided a place to walk, as walking was considered conducive to philosophical reflection. Later, the students of Aristotle's Lyceum were called peripatetics, from *peripatetikós*, meaning "those who like to walk and talk", referring to the paths in the gymnasium garden where classes were held.[13]

The cosmological dimension of an academy garden allowed it to be transposed easily to other geographical and philosophical domains. Although later gardens would evoke the diverse *topoi* associated with Plato's garden (Apollo and the Muses, the spring, the plane trees, conversation, the pleasantness of the place, etc.), the philosophical ideal it embodied cemented its widespread popularity in Western history. Theophrastus, a pupil of Plato and Aristotle, author of *Inquiry into Plants*, and the so-called "father of biology", founded a school modelled on Plato's Academy featuring a sanctuary of the Muses and a portico decorated with maps. He bequeathed it to his friends 'who wish to devote themselves to study and philosophy'.[14] Epicurus did the same with his garden, which gave its name to the school. Far from the city, men, women and enslaved people worked together to advance philosophical writing and self-perfection, in addition to cultivating the garden in a spirit of self-sufficiency, well-being and friendship. In this case as well, the garden embodied an ethical principle: a source of pleasure that healed both body and soul. The enclosed, sheltering and confining garden represented the tranquillity of the soul detached from worldly contingencies, which Epicurus believed the philosopher should

12 Harrison R.P., *Gardens. An Essay on the Human Condition* (Chicago: 2008).

13 Delorme J., *Gymnasion: étude sur les monuments consacrés à l'éducation en Grèce* (Paris: 1960) 335–336. On this topic in the context of Rome, see O'Sullivan T.M., *Walking in Roman Culture* (Cambridge: 2011) 94–96.

14 Diogene Laertius, v, 2, 51–55.

strive to achieve (*ataraxia*).[15] Like Plato's garden, Epicurus's was used as an ideal model for several early modern gardens, notably those of Justus Lipsius, Carolus Clusius and Peter Paul Rubens or, in England, the gardens of John Evelyn and William Temple.[16]

However, Cicero provided the strongest point of contact between the traditions of the Greek and Renaissance academies, having written extensively about them, especially in the last chapter of *De finibus*.[17] In this text, Cicero's descriptions of his visit to Athens and its emblematic sites – including the ancient Academy and Aristotle's – become opportunities to evoke the memory of the great figures of Hellenic culture. The *auctoritas* of the ancients left its mark on his literary and philosophical works, as well as on the architecture of the villa he bought and renovated around 68 BCE in Tusculum, about twenty kilometres from Rome near modern Frascati. We know from Cicero's correspondence with his friend Atticus that he paid close attention to the design of two garden gymnasiums on two superimposed terraces of his villa, built on the slopes of Mount Albanus. He called the first lower garden the Academy and the second upper gymnasium the Lyceum. By decorating his garden gymnasiums with statues imported from Greece, in particular the *hermae* found in the philosophical parks of Athens, Cicero paid tribute to Greek culture and sought to establish for himself, his friends and his young students places conducive to the practice of *otium*, based on the model of the Athenian schools.[18]

Cicero's philhellenism left an enduring mark on Roman culture. A century and a half after Cicero's time, the emperor Hadrian brought about a true Athenian renaissance; his immense villa at Tibur never failed to evoke the mythical places, real or imagined, that were most dear to him. The *Lives of Augustus* recounts how he decorated the villa with admirable buildings representing famous places such as the Lyceum, the Academy, the Prytaneum,

15 Campbell G., "Epicurus, the Garden, and the Golden Age", in O'Brien D. (ed.), *Gardening. Philosophy for Everyone* (Oxford: 2010) 220–231; Harrison, *Gardens*.

16 Morford M., "The Stoic Garden", *The Journal of Garden History* 7.2 (1987) 151–175; Small A. – Small C., "John Evelyn and the Garden of Epicurus", *Journal of the Warburg and Courtauld Institutes* 60 (1997) 194–214.

17 On this aspect, see also Chambers D.S., "The Earlier 'Academies' in Italy", in Chambers – Quiviger (eds.), *Italian Academies of the Sixteenth Century*, 1–14; Carlo Tosco, "Giardini platonici: una lettura del 'Fedro'", in *Architettura del giardino in Europa* (Naples: 2020) 96–100.

18 Cicero, *De diuinatione*, i, 8, 2; *Ad Att.*, i, 9, 2. On Cicero and gardens, see Grimal P., *Les jardins romains* (Paris: 1969) 358–363.

Canopus, the Poecile and Tempe. Not wanting to leave anything out, he even included Hell among its monuments and landscapes.[19]

The first references to the term "academy" in the Renaissance clearly stem from knowledge of Cicero's texts. The humanist Poggio Bracciolini is credited with first using the term in a letter to the great Hellenist Niccolò Niccoli. Like Cicero to his friend Atticus, whom Poggio explicitly quotes, Poggio describes the modest villa he had bought in 1427 in Terranuova in the Val d'Arno, where he intended to retire. He mentions a room full of busts and statues and other objects he wanted to acquire for his garden, and which would decorate his "academy".[20] The same Ciceronian academic ideal animated one of the circles that Poggio attended in Rome, the Accademia Bessarionea, which met in the garden of the Byzantine émigré Basil Bessarion near the Basilica of the Santi Apostoli. While the garden contained numerous inscriptions in Greek and Latin, the literary imagination of its members rather than the modest decoration of the garden made this place a new Parnassus and the domain of Apollo and the Muses, as some poems suggest.[21] Moreover, the academy's interests weren't specifically Platonic but rooted in a broader attempt to recover the Greek heritage that had recently fallen into Ottoman hands with the fall of Constantinople in 1453.

Although the antiquarian dimension of the garden was not yet fully formed and the collection of ancient fragments not yet clearly established, the literary trope linking the garden to friendship and the quest for knowledge and philosophical dialogue persisted in the Middle Ages. In his monumental *Storia dell'Accademia platonica*, published in 1902, della Torre gives several examples of early pseudo-academic gatherings, such as the *Consistoire du Gai Savoir*

19 Historia Augusta, *Life of Hadrian*, 24. Other famous gardens of Roman antiquity were undoubtedly the setting for philosophical and literary meetings. Varro was the first author to describe an architectural structure specific to the garden, which was to prove extremely popular, namely the famous "Varro's Aviary". Some researchers believe that it served as a meeting place for a cenacle of Academicians *ante litteram*, friends of Varro. Ginette Vagenheim has shown that the purpose of the description was above all an expression of the refined luxury of Roman villas at the end of the Republic. See, among several contributions on the topic, Ginette Vagenheim, "Comment nourrissait-on les canards 'convives' dans la volière de Varron (3,5,16)? *Risus* et *inventio* dans les *Res rusticae*", *Rivista di filologia e di istruzione classica* 148, 2 (2020), 390–418.

20 Bracciolini P., *Epistolae*, iv, 18 in Fubini R. (ed.), *Opera omnia* (Turin: 1964). See Franzoni C., "'Rimembranze d'infinite cose'. Le collezioni rinascimentali d'antichità", in Settis S. (ed.), *Memoria dell'antico nell'arte italiana*: *L'uso dei classici* (Turin: 1984) 309, 316–317.

21 Christian K.W., *Empire Without End. Antiquities Collections in Renaissance Rome, c. 1350–1527* (New Haven – London: 2010) 124. Callimaco Esperiente (Filippo Buonaccorsi), Giannantonio Campano and Martino Filetico wrote poems about the *Hortum Bissarionis*.

FIGURE 1.2 Giovanni Boccaccio, *Decamerone* (Venice: Per Giovanni & Gregorio de Gregorii, 1492) IMAGE © WIKIMEDIA COMMONS

(later known as the *Académie des Jeux floraux*), founded in Toulouse around 1323. Every Sunday in the garden of the Augustinian nuns, this circle brought together notable citizens to discuss and recite their poetic compositions in the hope of winning a prize symbolised by a flower.[22] Giovanni Boccaccio provides another essential source for the study of the philosophical garden typology, especially in the *Decameron*.[23] To this end, illustrations of the *Decameron* are illuminating, as they show gatherings of men and women – since love is a central theme of the garden – almost always seated in a circle or semicircle in an enclosed garden. A famous edition printed in Venice in 1492 shows such a gathering surmounted by an *all'antica* arch, at the top of which two books frame the figure of Apollo [Fig. 1.2]. This iconographic model, featuring a fountain, pergolas, and musical instruments, seems to be directly inspired by Francesco

22 della Torre A., *Storia dell'Accademia platonica di Firenze* (Florence: 1902) 120.

23 Raja M.E., *Le muse in giardino: il paesaggio ameno nelle opere di Giovanni Boccaccio* (Alessandria: 2003); Harrison, *Gardens*.

Colonna, the author of the *Hypnerotomachia Poliphilii*, and his illustrator, since the representation of the garden of Venus in this famous work, published in 1499, follows a similar typology. It also corresponds to a very traditional type of aristocratic garden already described, for example, in Corniolo Della Cornia's *La Divina Villa* in the late fourteenth and early fifteenth centuries.[24]

A specific allusion to the Athenian school is also found in *Filocolo*, composed in 1336, in an episode recounting how an elegant party invited the hero, shipwrecked in the Gulf of Naples, to a beautiful garden. Taken to a meadow 'very beautiful and covered with sweet herbs and flowers, surrounded by beautiful young trees that guarded it and defended it with their branches and green leaves from the malice of the sun', the young man sat down with the company 'around a beautiful and clear fountain'. A long discussion about love began, which Queen Fiametta ended by deferring the unanswered questions to the 'philosophers of Athens'.[25]

Continuing the tradition begun by Boccaccio, Giovanni di Gherardo da Prato's *Paradiso degli Alberti*, written around 1426, once again makes the garden the main stage for the intellectual life of a villa. The text not only develops literary tropes of the medieval garden of love and the *locus amoenus* of Latin poetry but also celebrates the garden of a real villa, the Villa del Paradiso in Gavinana located south of Florence. The villa belonged to the humanist Antonio Alberti who, in 1389, set up an Academy of Letters and Sciences on its grounds.[26] Leon Battista Alberti mentions this academy in his *Trattato della famiglia*: 'Master Antonio Alberti, a great scholar [...] as our father Lorenzo Alberti used to say, used to converse with his learned friends in your splendid gardens about the great loss both of our immense ancient empire and our ancient and splendid Latin language'.[27] Here Alberti points to a fundamental characteristic of Florentine proto-academies: their interest in language and the ancient heritage that they sought to revive.

Even if the tangible link that Cicero established between Plato's Academy in Athens and his villa in Tusculum seemed to have faded by the Renaissance, villas and their gardens remained essential places for Italian intellectual circles. In Fiesole, the Villa Medici served as a retreat for Poliziano, who composed

24 Segre A., "Untangling the Knot: Garden Design in Francesco Colonna's *Hypnerotomachia Poliphili*", *Word & Image* 14.1–2 (1998), 82–108.

25 Boccaccio Giovanni, *Le Philocope de Messire Iehan Boccace Florentin*, trad. André Sevin (Paris, Denys Janot: 1542) 100, 123. My translation.

26 Lanza A., "Il giardino tardogotico del Paradiso degli Alberti", *Italies* 8 (2004) [https://journals.openedition.org/italies/1060].

27 Alberti L.B., *I libri della famiglia*, Romano R. – Tenenti A. – Furlan F. (eds.) (Turin: 1994) 161. My translation.

several of his works there, while frequent meetings of scholars in the area are well documented. The location of the villa and its extraordinary relationship to the landscape – an important innovation of the architecture of Renaissance villas – intersected with the literary interests of these poets. The physical reality of the garden and landscape and the literary imaginary inspired by the ancients mutually fed each other, as is confirmed, for example, in a letter from Poliziano to Lorenzo the Magnificent in 1479, in which the poet compared the garden culture of Fiesole with the culture of letters.[28] We have observed the same process of merging one landscape with another, a physical and imaginary one, in Ficino's prose.

Bernardo Rucellai's garden, known as the Orti Oricellari, was another famous Florentine Garden where philosophical and literary meetings took place. Here, in the last years of the fifteenth century, some of the greatest scholars of the time gathered, including members of Ficino's and Poliziano's circles, as well as Machiavelli, Jacopo Nardi and later Pope Leo X. Although no documents specifically refer to these meetings as academies, they certainly played a decisive role in the cultural and political life of Italy in the period.[29] Banquets, recitals and plays were regularly held in the garden. The garden, which was said to be cool and shady, is mainly referred to in the sources as *selva*, a term that denotes a natural park planted with trees. The term also had an especially strong literary connotation, alluding to Dante and Petrarch in particular, who associated it with shade, silence and contemplation. Machiavelli, for his part, spoke of the seats placed in the 'most secret and darkest part' of the garden. *Selva* could also refer to its Latin equivalent, *silva*, and to Horace's famous expression 'inter silvas Academi quaerere uerum' ('to seek the truth among the groves of Academus').[30] The park also contained many animals: rabbits,

28 Poliziano A., *Opera omnia*, vol. 3, ed. Ida Maïer (Turin: 1971), 550–551. See Hankins, "Cosimo de' Medici and the 'Platonic Academy'", note 37, 155–156. On this topic, see also Simonin M., "Poësie est un pré', 'Poëme est une fleur': métaphore horticole et imaginaire du texte à la Renaissance", in *La letteratura e i giardini* (Florence: 1987), 45–56.

29 Gilbert F., "Bernardo Rucellai and the Orti Oricellari: A Study on the Origin of Modern Political Thought", *Journal of the Warburg and Courtauld Institutes* 12 (1949), 101–131; De Gaetano A.L., "The Florentine Academy and the Advancement of Learning Through the Vernacular: The Orti Oricellari and the Sacra Accademia", *Bibliothèque d'Humanisme et Renaissance* 30.1 (1968), 19–52; Morolli G., "L'accademia, la selva, le grotte, il famedio", in *Il gigante degli Orti Oricellari* (Rome: 1993): 9–68; Rinaldi A., "Ideologia e tipologia del giardino urbano a Firenze tra XV e XVI secolo", in *Stanze per un giardino: Il paesaggio e il giardino nella cultura umanistica. V centenario della morte di Agnolo Ambrogini detto il Poliziano* (Siena: 1994) 1–27; Comanducci R.M., "Gli Orti Oricellari", *Interpres. Rivista di studi quattrocenteschi* 15 (1995–1996) 302–358.

30 Horace, *Epistles*, ii, 2, 45.

fawns, goats and even porcupines, as well as rare plants, which, according to Niccolò Machiavelli, evoked all the species mentioned in classical literature. Rucellai possibly found inspiration for his Florentine garden in the splendid Barchi of Naples since it is known that he attended the Academy of Giovanni Pontano 'in a garden at the foot of Vesuvius' in 1495. Diarists long claimed that the paths of this garden were lined with busts of famous men from antiquity, but this information is not confirmed. However, ancient fragments and statues, which Rucellai passionately collected, were certainly fundamental decorative elements in the garden. Finally, a summer pavilion and benches were placed under the trees, inviting visitors to read, chat or hold conferences and other festive events. This example demonstrates that the Renaissance garden was not conceived in slavish imitation of the gardens of the Athenian philosophers. Instead, it tended to combine spaces and artistic collections to stimulate intellectual research and conversation, as well as to provide an embodied way of living the lessons of the poets and wise men of antiquity.

In Rome as well, humanists designed gardens to closely evoke their literary interests and academic activities. Pomponio Leto and his famous Roman academy, which was strongly influenced by the Ciceronian model, met in a modest garden on the Quirinal acquired in 1474 and situated among the ruins of the Baths of Constantine. Jacopo Galli's garden, close to the Castel Sant'Angelo, was used for philosophical meetings aimed at reviving Bacchic mysteries, as Cardinal Jacopo Sadoleto poetically described it in his dialogue *Il Fedro* (1538) – a double homage to the nickname of the humanist Tommaso Inghirami and Plato's famous dialogue. In addition to containing Michelangelo's *Bacchus*, a statue of a nymph, and many other ancient fragments, the garden featured a portico decorated with paintings, a shady courtyard with an exedra, and an area known as the Hippodrome – a design that recalled Cicero's gymnasium gardens and those of ancient Greece [Fig. 1.3].[31] In a letter written in 1529, Sadoleto enthusiastically remembers the intellectual life of the Roman gardens: 'Oh, how often I recall those colloquies, those banquets that we used to hold, when, either in your gardens (*orti suburbani*) or in mine on the Quirinal, or in the Circus Maximus, or on the banks of the Tiber at the Temple of Hercules, or elsewhere, we gathered together learned men, respectable no less for virtue than for reputation: where, after the customary banquets, seasoned less with gluttony than with wit, we would recite poems and declaim orations

31 Wind E., *Pagan Mysteries of the Renaissance* (London: 1958); Coffin D., *Gardens and Gardening in Papal Rome* (Princeton: 1991) 235–236. On Roman academies and gardens in Rome in the fifteenth and early sixteenth century, see especially Christian, *Empire Without End.* On Jacopo Galli's garden in Rome, see Christian K.W., *Raffaele Riario, Jacopo Galli, and Michelangelo's 'Bacchus', 1471–1572* (Turnhout: 2024).

FIGURE 1.3 Maarten van Heemskerck, *Garden of Casa Galli with Michelangelo's Bacchus*, c.1533–36, drawing, Kupferstichkabinett, Staatliche Museen, Berlin, 79 d 2, fol. 72 recto
IMAGE © STAATLICHE MUSEEN

to the great satisfaction of all, because they revealed a greatness of spirit (*altezza d'ingegno*), yet were tinged with joy and grace'.[32] In these gardens, musical concerts took on a neo-Platonic character, leading the soul towards the eternal music of the spheres.[33] The ancient statues placed in the space were appreciated not only for their aesthetic qualities but also because they entered into a genuine dialogue with poetry. As Kathleen Wren Christian has described Leto's Roman academy, '[v]erses recited in the presence of sculpture, written in the "voice" of ancient sculptures and physically attached to these works drew out the inspirited quality that antique sculpture was thought to share with poetry'.[34] In short, sculptures began to speak and converse, a practice strongly linked to the decoration of Renaissance gardens – poems written by humanists were often attached to sculptures or even on trees in the garden! On St. Anne's Day, for example, Johann Goritz's companions wrote verses

32 Quoted in Gnoli D., "Orti letterari nella Roma di Leone X", *Nuova antologia* (1930) 15. For this letter, see also Giorgio Mangani's essay in this book.

33 Cummings A., "Informal Academies and Music in Pope Leo X's Rome", *Italica* 86.4 (2009) 583–601.

34 Christian K.W., "Poetry and Spirited Ancient Sculpture in Renaissance Rome", in Kenda B. (ed.), *Aeolian Winds and the Spirit in Renaissance Architecture* (London: 2006) 104.

that were pinned to the trees in his garden near the Basilica of Maxentius and Constantine. Some of these epigrams were published in 1524 in a collection by the poet Blosio Palladio titled *Coryciana*. Highlighting a creative anachronism typical of Renaissance *all'antica* gardens, Palladio places Goritz Academy in fifth-century Athens rather than in the ruins of Rome and moves the Muses of the mythical Helicon to the Quirinal Hill and the Tarpeian Rock.[35]

The model of Plato's Academy certainly played a part in this kind of rhetorical ritual. Diogenes Laertius recalls how, after Plato's death, his disciples buried him in the Academy garden and composed epigrams in his honour, which they placed on his tomb.[36] Some of these verses were well known in Roman humanist circles, in large part due to the 1495 publication of the Byzantine Maximus Planudes's version of the *Greek Anthology*, a collection of epigrams written in memory of illustrious poets and philosophers of great interest in the Renaissance. This fashion for poetic epigrams also developed around a new type of fountain, that of the sleeping nymph, which was associated with an ancient cult of the Muses. Many examples of this practice can be traced in the histories of Renaissance gardens. To this end, in Rome, we should mention Goritz's garden, but also that of his friend Angelo Colocci near the Trevi Fountain, who was named as the head of the new Greek Academy commissioned by Leo X, sometimes called the Gymnasium of Monte Cavallo. According to Elizabeth MacDougall, 'for the contemporaries of Colocci and Goritz, the Fountain of the Sleeping Nymph was a symbol of the presence of the Muses, who presided over the new academies of learning and the renaissance of poetic art'.[37] The two Nymphaea by Colocci and Goritz were inspired by the one installed in 1512 in the courtyard of the Belvedere in the Vatican, at the centre of which was placed an ancient statue identified as Ariadne or Cleopatra. In the Vatican, as in humanist gardens, the themes of the ancient Greek Academy, Parnassus and the Muses were deeply intertwined. It is no coincidence, then, that Raphael's *Parnassus*, next to his famous *School of Athens* in the Stanza della Segnatura (*c.*1509–1511), was placed above the window opening directly onto the great courtyard garden [Fig. 1.4].[38]

35 Bober P.P., "The Coryciana and the Nymph Corycia", *Journal of the Warburg and Courtauld Institutes* 40 (1977) 237; IJsewijn J., "Poetry in a Roman Garden: The Coryciana", in Godman P. – Murray O. (eds.), *Latin Poetry and the Classical Tradition: Essays in Medieval and Renaissance Literature* (Oxford: 1990) 211–231.

36 Diogenes Laertius, iii, 43–45.

37 MacDougall E.B., "The Sleeping Nymph: Origins of a Humanist Fountain Type", *The Art Bulletin* 57.3 (1975) 363.

38 Watson P.W., "On a Window in Parnassus", *Artibus et Historiae* 8.16 (1987) 142.

FIGURE 1.4 Marcantonio Raimondi, *Parnassus* (after Raphael's fresco in the Vatican), *c.*1517–20, engraving, 35.6 × 46.9 cm, New York, The Metropolitan Museum of Art
IMAGE © MET

Other important examples of nymphaea from the mid-sixteenth century can be found in the courtyards of the Villa d'Este in Tivoli and Villa Giulia built for Pope Julius III, as is discussed in this book. Several academies had links with important gardens in the sixteenth century. To name a few, the Accademia degli Agevoli (1571) had its headquarters at Villa d'Este in Tivoli and its famous gardens. The Accademia delle Notti Vaticane, founded in 1562 under the patronage of Carlo Borromeo, was in the Casino di Pio IV in the Vatican gardens, today the seat of the Pontifical Academy of Sciences [Fig. 1.5]. As Pirro Ligorio, the architect of the Casino, recalls, the iconography of the Casino, dedicated to the Muses, directly related to the ancient academies: 'Et Musaeo anchora si diceva il tempio delle Muse che si edificano come quelli delle Accademie, nellj Gymnasij et presso delli Fori e piazze delle città illustre et schuole de scientie et colonie de philosophia, come furono dedicati in Athene, in Corinto, in Mitylene. Nella villa Hadriana Tiburtina nel luogo della sua Accademia' ('The temples of the Muses built in the academies, gymnasiums, forums and squares of famous cities, the scientific schools and philosophical colonies,

FIGURE 1.5 *Casina di Pio IV, Vatican Gardens, the seat of the Accademia delle Notti Vaticane, founded in 1562 under the patronage of Carlo Borromeo*
IMAGE © WIKIMEDIA COMMONS

were then called Musaeums, like those dedicated to Athens, Corinth, Mitylene and Hadrian's villa at Tibur, in the place called Academy').[39] There, operating under the allegorical name "Chaos", Borromeo presided over Latin oratorical jousts between his friends Silvio Antoniano ("Il Risoluto"), Agostino Valier ("L'Obbediente"), Sperone Speroni ("Nestor") and the future Pope Gregory XIII, Ugo Boncompagni, among others. The nickname "Chaos", according to one of the academicians, referred to the collaborative work of scientific research: 'al principio confuso di tutte le cose secondo le dottrine mitologiche, dal quale uscirono poi distinte le singole parti dell'universo, prendendo la forma conveniente [...] in quel caos accademico comprendevansi tutte le doti eminenti di quel nobilissimo ceto e venivano poi sviluppate nei singoli membri' ('to the confused beginning of all things, according to mythological doctrines, from which the individual parts of the universe then clearly emerged, taking the

39 On the links between the Casino of Pius IV and the ancient academy, see especially Fagiolo M. – Madonna M.L., "La casina di Pio IV in Vaticano. Pirro Ligorio e l'architettura come geroglifico", *Storia dell'arte* 15–16 (1972) 237–281, here 242–243.

appropriate form [...] in this academic chaos were contained all the eminent qualities of this most noble class, and were then developed in the individual members').[40] Maria Losito has clearly shown how the stucco decorations based on ancient and pagan themes and designed by Pirro Ligorio on the outside of the Casino, as well as the frescoes of sacred themes that adorn the interior, closely reflected speeches delivered by Carlo Borromeo and the members of the academy, which were inspired directly by Aulus Gellius's famous *Attic Nights*. The antithetical iconography of the Casino – "pagan" on the outside and "Christian" on the inside – can be understood as reflecting Borromeo's thinking before and after the Council of Trent. As Marc Fumaroli has pointed out, '[a]fter 1563, [Borromeo's] ordination, and the end of the Council, the themes of the Academy changed: the subject of the speeches was no longer Cicero, Livy or Lucretius, but the Beatitudes and the theological virtues'.[41] It is therefore irrelevant to consider the duality of this iconography as resulting from an ideological split, as has often been written. Rather, it should be seen as the fruit of a specific intellectual movement linked to the activities of this Academy.

The function of the garden as an ideal place for teaching and inspiring poetry and philosophy also influenced artists, as illustrated by a late sixteenth-century engraving by Jan van der Straet [Fig. 1.6]. Titled *Litterae*, it depicts the god Apollo as the protector of the arts and sciences, with a series of objects and instruments relating to music, astronomy, literature, mathematics and art collected in the foreground. The setting is the garden and landscape of a large Italian-style villa. In the main avenue, a group of people discuss the many sculptures and fountains in the garden. In the background, an assembly gathers around a large circular table on which musical instruments are placed. Not far away, labourers till and plant the garden. On the left, a large building houses a conference room, a library on the first floor, and a laboratory in the loggia. Pegasus and the Muses, perched on a hill at the bottom of the garden, preside over these studious activities. Finally, on the right, a sculptor, who bears a striking resemblance to the statue of the river god behind him, is at work, while painters busily decorate the portico of the villa.

The artistic activities that took place in this ideal garden typology are documented as having occurred in several known Renaissance gardens. For example, the antiquities collected in the garden of Cardinal Della Valle in Rome

40 Losito M., "L'iconografia dell'Accademia delle Noctes Vaticanae e la cultura umanistica di San Carlo Borromeo", *Arte Cristiana* 98 (2010) 421–439, 424.

41 Fumaroli M., *L'âge de l'éloquence, rhétorique et « res literaria » de la Renaissance au seuil de l'époque classique* (Geneva: 2002), 136. My translation.

FIGURE 1.6 Johannes Sadeler, after Jan van der Straet, *Apollo as Patron of the Arts and Sciences*, 1597, engraving, 21.4 × 28 cm, Rijksmuseum, Amsterdam
IMAGE © RIJKSMUSEUM

were intended to inspire painters and poets alike, according to an inscription at its entrance.[42] Later, the Florentine sculptor Baccio Bandinelli founded a true academy of art in the gardens of the Belvedere in Rome. Maarten van Heemskerck's drawings of the Cortile delle Statue, dated 1532–1533, and Hendrick van Cleve's large views of the Villa Belvedere, which include draughtsmen copying antiques, show how the garden developed into an open-air studio.

Like the poetic academies, Bandinelli's Roman academy had its roots in Florence: it was modelled on the artistic academy set up in Lorenzo the Magnificent's famous garden 'sulla piazza di San Marco' ('in piazza San Marco') where Michelangelo cut his teeth. According to Giorgio Vasari and Ascanio Condivi, this academy acted as a kind of school (*scuola*) for young artists who were encouraged to copy the ancient statues in the garden. A tapestry based on a design by Jan van der Straet dating to 1571 and a passage from the eulogy of Michelangelo, who died in 1564, evoke this idea of an artistic academy,

42 Coffin, *Gardens and Gardening* 246.

which preceded the creation of the Accademia delle Arti del Disegno founded by Cosimo I de' Medici in 1563 by almost one hundred years: '[Lorenzo the Magnificent] had, among other things, a studio built for talented painters and sculptors in his garden by the piazza San Marco, and those prevented by poverty from attending were supported by him with salaries and other gifts. And in order that they should not miss any occasion to learn, over and above that virtuous and noble endeavour to excel which springs up between young men competing with each other, there were experienced masters who guided the youths along the right path. Among those selected was Michelangelo, and it was there that he laid the foundation for his greatness. Thus, if we were to look for a single cause of his greatness, we should say the garden of San Marco was in a sense the origin of it'.[43] While for many art historians like Ernst Gombrich, 'the seed of a new conception of art was planted in the garden of Lorenzo the Magnificent',[44] others have stressed the highly constructed nature of such an account, which has been attributed to the skilful courtier Vasari and inextricably aligns Michelangelo's precocious artistic genius with the Medici's enlightened patronage.[45] Yet several documents clearly show that artists, including Leonardo da Vinci, had access to the garden and its collections. Rather than disentangling historical truth from myth, it is worth noting that this phenomenon of rewriting history is similar to the above-mentioned elision of myth and reality in the case of Ficino's Platonic Academy and the Villa di Careggi. In addition to the role played by ancient sculpture in artistic learning – considered suitable for teaching young artists the perfection of art –, the garden, more than the traditional studio, embodied the humanistic intellectual values that Florentine artists wished to promote. In other words, the garden brought together two archetypal domains of artistic creation and inspiration

43 Wittkower R. – Wittkower M. (eds. and trans.), *The Divine Michelangelo. The Florentine Academy's Homage on his Death in 1564* (London: 1964) 90–93. On the garden, see Elam C. "Lorenzo de' Medici Sculpture Garden", *Mitteilungen des Kunsthistorischen Institutes in Florenz* 36. 1–2 (1992) 41–84. On the artists who frequented it, Baldini N. "'Quasi Adonidos hortum'. Il giovane Michelangelo al giardino mediceo delle sculture", in Weil-Garris Brandt K. et al. (eds.), *Giovinezza di Michelangelo* (Florence: 1999) 49–56.

44 Gombrich E. "The Early Medici as Patrons of Art", in *Norm and Form* (London: 1966), 56–57.

45 See Chastel A. "Vasari et la légende médicéenne: L'École du Jardin de Saint-Marc", *Studi* (1950), 159–167; Pommier É., "Notes sur le jardin dans la littérature artistique de la Renaissance italienne", in Pigeaud J. – Barbe J.-P. (eds.), *Histoires de jardins. Lieux et imaginaire* (Paris: 2001) 134–139; Cox-Rearick J. "Michelangelo's Mask of the Head of a Faun: Myth and the Mythic in the Sculpture Garden of Lorenzo il Magnifico", in Israëls M. – Waldman L.A. (eds.), *Renaissance Studies in Honor of Joseph Connors* I (Milan: 2013) 322–333.

in a single place: the studio, on the one hand, and nature, on the other, where the innate talents of Renaissance forebearers such as Giotto were supposedly discovered.

The gardens of Rome and their collections of antiquities also played an important role in the training of foreign artists, who often occupied marginal roles in the workshops of Italian masters. The Dutchman Michael Sweerts, who lived in Rome from 1646 to 1655, is an interesting example in this respect since several of his works deal with the artistic academy and are undoubtedly linked to his ambition to create an academy of life drawing. In at least two paintings, Sweerts shows an artist drawing in a Roman garden, identifying it as a privileged place of learning.[46]

Many academies were directly attached to botanical gardens, such as the botanical garden at Leiden in Holland (*c.*1590), which was affiliated with the university and called *hortus academicus*.[47] In Italy, university students used the Botanical Gardens of Padua, while the Società Botanica Fiorentina and the Accademia dei Georgofili also had gardens in the eighteenth and nineteenth centuries. However, their exact role in the life of scientific proto-academies in the Renaissance remains unclear at the present stage of research. The first academy dedicated to natural philosophy, the Accademia dei Segreti (or *Academia secretorum naturae*) founded in Naples in the 1560s by Giambattista della Porta, was modelled after his summer palace and its garden was used to grow exotic plants to complement his collection of rare animals, plants, and minerals.[48]

The Lycium philosophal Sanmarcellin run by Jacques Gohory in the Faubourg Saint-Marceau-les-Prés in Paris and discussed by Laurent Paya in this book was a noteworthy French example of the Renaissance academic garden. Of Florentine origin and familiar to the court of Catherine de Médicis, Gohory was an alchemist, magician, physician, follower of Paracelsus and translator of Machiavelli. His so-called "Happy Academy" was organised in a garden where

46 Yeager-Crasselt L., *Michael Sweerts (1618–1664): Shaping the Artist and the Academy in Rome and Brussels* (Turnhout: 2015). See especially Michael Sweerts, *An Artist Drawing in the Garden of Villa Montalto in Rome*, *c.*1646–1648, oil on canvas, 85.5 × 120 cm, Rotterdam, Museum Boymans-van Beuningen, Inv. Nr. 2358.

47 On this topic, see Egmond F., "Sixteenth-Century University Gardens in a Medical and Botanical Context", in Baldassari F. – Zampieri F. (eds.), *Scientiae in the History of Medicine* (Rome: 2021) 89–120.

48 A vast literature exists on botanical gardens. Important contributions include those by Tongiorgi Tomasi L. See, for instance, "Gardens of Knowledge and the République des Gens de Sciences", in Michel Conan (ed.), *Baroque Garden Cultures: Emulation, Sublimation, Subversion* (Washington D.C.: 2005) 85–129.

alchemy was practised, and remedies and talismans were concocted. Learned visitors went there to admire its rare trees and plants and to enjoy the 'music of voices and instruments in the historic gallery'. A labyrinth and floral sundial were included among the many ornaments in the garden.[49]

Duke Federico Cesi undoubtedly conceived the idea of creating his famous Accademia dei Lincei in Naples, where he was in close contact with Della Porta. Founded in 1603, the academy held several meetings in the Cesi Garden in Rome; a tamed leopard or lynx served as the group's mascot.[50] In the *Lynceographum*, which outlined the Academy's statutes (1604), Cesi described the various structures necessary for scientific meetings, including a library, a museum, a scientific laboratory and a kind of botanical garden.[51] A dense network of botanists who were close to or members of the Accademia dei Lincei, such as Enrico Corvino, Teofilo Muller, Ferrante Imperato, Giovanni Battista Ferrari and Pietro Castelli worked in Rome's many magnificent gardens, including those of the Quirinal, Francesco Barberini's, and Cardinal Odoardo Farnese's on the Palatine (Horti farnesiani).[52] In this book, Alessandro Spila shows that Cardinal Francesco Maria Del Monte also employed famous botanists. In short, gardens played a fundamental role in the scientific life of the period, although they cannot be considered the permanent headquarters of such proto-academies.

The essentially peripatetic nature of the academies continued well into the eighteenth century in Italy and the rest of Europe. As Katrina Grant's essay explains, the famous Academia Arcadica in Rome, established in 1690 following Giovanni Mario Crescimbeni's transformation of the Accademia Regia founded by Queen Christina of Sweden, met in many of Rome's gardens: The Villa of the Duke of Paganica in San Pietro in Vincoli, the Villa of the Princes Mattei Orsini on the Esquiline Hill, the gardens of Palazzo Riario alla Lungara (former residence of Queen Christina of Sweden), the garden of Vincenzo Giustiniani, the Orti Farnesiani on the Palatine Hill (supposedly the residence of Evander, King of Arcadia), and many others. The garden, known as Bosco

49 Gohory Jacques, *Instruction sur l'herbe Petum* (Paris, Galio du Pré: 1572). See also Gorris Camos R., "L'Hysope et la rose: le Lycium philosophal de Jacques Gohory", in Deramaix M. et al. (eds.), *Les Académies dans l'Europe humaniste* (Geneva, 2008) 553–594.

50 Bentz K., "The Afterlife of the Cesi Garden: Family Identity, Politics, and Memory in Early Modern Rome", *Journal of the Society of Architectural Historians* 72.2 (2013) 161.

51 "Aerolam conficiendo hortulo Botanici studij, et domesticae deambulationis gratia exhibeat", in Nicolò A. (ed.), *Lynceographum: Quo norma studiosae vitae lynceorum philosophorum exponitur* (Rome: 2001), f. 73v, 88.

52 Campitelli A., *Gli horti dei papi. I giardini vaticani dal medioevo al Novecento* (Vatican City: 2009), 161–169.

Parrasio, which was very close to San Pietro in Montorio and the great fountain of the Acqua Paola, was only laid out in 1725–1726 by the architects Antonio Canevari and Nicola Salvi, following a design by Crescimbeni.[53] The structure, built around a system of twin staircases that ended at the top of the hill, provided a panoramic view of Rome and, through the arrangement of the statuary and various motifs, evoked the riches of the pastoral literature that the neo-Arcadians wanted to bring back to life on the slopes of the Janiculum. Like many Renaissance academic gardens, the Bosco Parrasio was conceived both as a theatre for meetings and as an emblematic image of the literary pastoral celebrated on the site through recitations and other Olympiads.[54] The metaphorical function of such a garden is apparent, for example, in the title of Crescimbeni's history of the Academy, *Storia dell'accademia degli arcadi, istituita in Roma l'anno 1690 per la coltivazione delle scienze, delle lettere umane e della poesia*.[55] The *impresa* of the Accademia degli Incolti – another Roman academy founded in 1658 in the Collegio Nazareno with close links to the Accademia degli Arcadi – depicts an uncultivated and a cultivated garden with the motto *inculti prosperabuntur*, foregrounding the garden as the place par excellence of an ideal order of science, beauty and wisdom to which the human soul aspires.

•••

This brief and necessarily fragmentary overview of some links between the garden and early modern academies brings to the fore several topics and specific case studies that form the subject of the essays presented in this book, which are organised in chronological order. Before briefly describing each of these chapters, some general remarks of a methodological nature are offered as a conclusion.

First, the historian of the academic garden is confronted with many literary sources and thus needs to engage with the vast field of research that concerns both gardens and literature. As Hervé Brunon recalls, 'The art of gardening reveals a 'literary imagination', in the sense that literature supplies the garden with both archetypes and images, two levels that are always intertwined, but

53 Coffin, *Gardens and Gardening*, 239–243; Hyde Minor V., *The Death of the Baroque and the Rhetoric of Good Taste* (Cambridge: 2006). See further Katrina Grant's essay in this book.

54 On the allegorical nature of academic gardens, see Marc Fumaroli's stimulating discussion in "Académie, Arcadie, Parnasse: trois lieux allégoriques du loisir lettré", *L'école du silence. Le sentiment des images au XVIIe siècle* (Paris: 1994) 19–36.

55 Crescimbeni G.M., *Storia dell'Accademia degli Arcadi* [...] (London: 1804).

which we can try to distinguish for the purposes of analysis'.[56] Furthermore, research on gardens and their links to academies also encourages scholars to consider a wide variety of representations, not only of real gardens but also idealised images of gardens or images of ideal gardens. For the garden historian, the use of such images does not so much lead to physical knowledge of historical gardens but rather contributes to sketching a more general portrait of garden culture, which in turn may illuminate aspects of its materiality or reception over time. Ultimately, these images do less to determine the shape of the flowerbeds or the layout of the statues and paths at a particular moment in the history of a specific garden than they do to illuminate the function that gardens played within a wider literary, philosophical, and scientific culture. The engraving by van der Straet mentioned above is a good example of such an idealised vision of the garden as the seat of academic activities [Fig. 1.6]. The frontispieces of theses produced within the ambit of various academies are also interesting in this regard. For instance, at the beginning of the seventeenth century, the Jesuit seminary in Rome (Seminario Romano) chose as its *impresa* a garden with the motto: 'Pomis sua nomina servant' – inspired by Virgil.[57] The *impresa* and its motto gave rise to an extensive and erudite emblematic iconography. In an engraving used as a thesis frontispiece by Johann Friedrich Greuter after Gian Francesco Romanelli, the seminary garden is depicted as the sacred wood of Academus, featuring Socrates teaching his students [Fig. 1.7]. He is shown under the plane tree and at the foot of the spring described in *Phaedrus*.[58] Two secondary episodes complete the scene. On the left, Cimon has transformed the 'bare and arid' wood into 'a beautiful park, watered by several fountains and planted with wide paths for walking', which contains the statue of Athena on whose shield the motto of the seminary is inscribed.

56 On this topic, see the many contributions by Venturi G., "Il giardino e la letteratura. Saggi d'interpretazione e problemi metodologici", in Ragioneri G. (ed.), *Il giardino storico italiano* (Florence: 1981) 251–277; idem, "'Picta poësis': ricerche sulla poesia e il giardino dalle origini al Seicento", in De Seta C. (ed.), *Storia d'Italia. Annali 5: Il paesaggio* (Turin: 1982) 663–749; *La letteratura e i giardini, Atti del Convegno Internazionale di Studi di Verona* (Florence: 1987); Fariello A., *I giardini nella letteratura: dal giardino classico al giardino paesistico* (Rome: 1998); Duport D., *Le jardin et la nature. Ordre et variété dans la littérature de la Renaissance* (Geneva: 2002); Brunon H., "De la littérature au jardin", in Venturi G. – Ceccarelli F. (eds.), *Delizie in villa. Il giardino rinascimentale e i suoi committenti* (Florence: 2008) 27 [My translation]; Jakob M. (ed.), *Des jardins et des livres* (Geneva: 2018).

57 Virgil, *Georgics*, 2, 239–40.

58 Plato, *Phaedrus*, 230 b–c.

FIGURE 1.7 Giovanni Francesco Romanelli, *Socrates teaching his disciples in the garden of the Academy*, c.1630–1640, drawing, 27.4 × 37.3 cm, San Francisco, Fine Arts Museums
IMAGE © SAN FRANCISCO, FINE ARTS MUSEUMS

On the right, the artist depicted the destruction of the wood during the siege of Athens under the reign of Sylla.[59]

The frontispiece by Giovenale Boetto for Paulo Pasta's philosophical dissertation, which he defended in 1654 upon his admission to the Accademia dei Solenghi (founded in Rome around 1626), features another remarkable *hortus academicus* [Fig. 1.8]. This garden is directly connected to the garden of the villa known as Ludovica, then Della Regina, belonging to Cardinal-Prince Maurice of Savoy and located near Turin, the seat of the academy. The engraving illustrates the concept of the garden as an academy and an encyclopaedic system of knowledge. It shows a theatre at the top of the garden made up of fifteen arcades, each bearing the name of a science, depicting activities and historical figures associated with the scientific discipline below each arch. In the centre, a maze of grass is reflected in a conical mirror, revealing the Academy's motto 'OMNIS IN UNUM', inspired by one of Virgil's verses. In the centre of the

59 Plutarch, *Cimon*, 13.8; *Sylla*, 12.13. See Rice L., "*Pomis sua nomina servant*: The Emblematic Thesis Prints of the Roman Seminary", *Journal of the Warburg and Courtauld Institutes* 70 (2007) 222–223.

FIGURE 1.8 Giovenale Boetto, *Horti academi. The garden of Prince Maurice of Savoy, seat of the Accademia dei Solenghi*, frontispiece to Paolo Pasta's thesis, 1654, Turin, Galleria sabauda, inv. Stampe 2432
IMAGE © MUSEI REALI TORINO

theatre, Apollo and the Muses frame the bust of Prince Maurice of Savoy.[60] This garden and its anamorphic perspective were probably invented by the famous rhetorician Emanuele Tesauro, who mentions the device and the garden in his *Cannocchiale aristotelico* dedicated to the Cardinal (1654), which features the motif of the conical mirror in the centre of the frontispiece. Tesauro describes and explains Boetto's engraving in detail in his *Inscriptiones*, first published in 1666.[61] In this image, the Accademia dei Solenghi is compared to Plato's Academy. According to Tesauro, the anamorphosis in the image was a

60 On the engraving, see de Vesme A., *Peintre-graveur italien* (Milan: 1906), 43–44; Carboneri N. – Griseri A., *Giovenale Boetto* (Fossano: 1966), 55, 78–79 (cat. 55); Oberli M., *Magnificentia Principis: Das Mäzenatentum des Prinzen und Kardinals Maurizio von Savoyen (1593–1657)* (Weimar: 1999), 168; Arnaldi di Balme C. – Varallo F. (eds.), *Feste Barocche. Ceremonie e spettacoli alla corte dei Savoia tra Cinque e Settecento* (Milan: 2009) 26 (cat. 11).

61 Tesauro E., *Il cannocchiale aristotelico* (Turin, Gio Sinibaldo: 1654) 677–678; idem, *Inscriptiones* (Turin, Bartolomeo Zapata: 1670) 469–477.

FIGURE 1.9 Hans Tröschel, after Simon Vouet, *Satyrs looking at an anamorphosis of an elephant*, *c.*1627, engraving, 24.3 × 34.5 cm, New York, The Metropolitan Museum of Art, 45.97(78)
IMAGE © MET

particularly ingenious way of demonstrating the scientific and literary ingenuity of the Solenghi and the enlightened patronage of Maurice of Savoy. It signified the joint research of the members of the Academy, as well as the prince's ability to distinguish between reality and appearance. A similar theme is found in a much more famous engraving by Hans Tröschel, based on a drawing by Simon Vouet from 1627 [Fig. 1.9]. The image depicts satyrs watching an anamorphosis of an elephant in a garden. It has recently been suggested that this engraving may also have been conceived for a thesis frontispiece dedicated to Cardinal Maurice of Savoy. In such a case, the satyrs may represent ignorance, while the optical device associated with the elephant refers to intelligence, insight bringing order to chaos, as suggested by the motto 'format et illustrat' and the ordered geometry of the formal garden.[62]

Many images of academic gardens also contributed to the invention of a fictional past by mobilising certain myths largely for ideological purposes. This

62 See Marr A., "A Mirror of Wisdom: Simon Vouet's *Satyrs Admiring the Anamorphosis of an Elephant* and Its Afterlives", in Berger S. – Garber D. (eds.), *Teaching Philosophy in Early Modern Europe* (Cham: 2022) 265–301.

FIGURE 1.10 Luigi Mussini, *The Birthday of Plato Celebrated in the Villa di Careggi by Lorenzo the Magnificent*, 1867, oil on canvas, 103 × 137.5 cm, Lincoln College, University of Oxford
IMAGE © DENIS RIBOUILLAULT

is the case with Jan van der Straet's tapestry depicting Lorenzo de' Medici in the garden of San Marco, mentioned above, or the fresco of the same subject painted by Ottavio Vannini in Palazzo Pitti around 1635, with Michelangelo's *Faun* in the foreground. Luigi Mussini's mid-nineteenth-century paintings also reveal an ideological recuperation of the prestige of Renaissance academies, in particular his *Celebration of Plato's Birthday at the Villa Careggi* and its companion piece *The Medicean Gardens. The School of Painting and Drawing*, both now in France with copies in Turin and Siena [Fig. 1.10].[63] These paintings likely contributed to propagating the romantic vision of the Platonic Academy

63 Luigi Mussini, *The Celebration of Plato's Birthday at the Villa of Careggi*, oil on canvas, 130.3 × 97.7 cm, Turin, Galleria d'Arte Moderna; idem, *The Medicean Gardens*, oil on canvas, 36.5 × 46 cm, Siena, Palazzo Pubblico, Archivio di Stato. The original paintings are respectively in the Musée du Monastère royal de Brou in Bourg-en-Bresse (1852; inv. 864.1.E) and the Musée municipal de l'Avranchin in Avranches (1857), France. On Mussini, see the entry by Marco Pierini in *Dizionario biografico degli italiani* 77 (2012), available online.

of Florence and the garden of San Marco, which, as has been discussed, historians have struggled to reconcile with historical truths. Likewise, the great mural by Jean-Paul Laurens in the Capitole of Toulouse, dating to the end of the nineteenth century, endowed the medieval *Académie des jeux floraux* with a new grandeur that it probably had never possessed.[64]

The theatricality of the garden is another key aspect that made it an ideal place for academic gatherings. What better stage existed for groups of people who disguised themselves in togas or shepherds' garb and invented fictitious names? Moreover, the names of the academies reflected emblematic places in their literary or scientific imaginary, which often linked directly to the world of gardens: Accademia degli Ortolani, Accademia dei Vignaiuoli, Accademia del Viridario, Accademia del Platano, and so on. As Amedeo Quondam rightly observes, 'the economy of the name(s), the *impresa*, the laws [of the academy] not only makes possible the identity of the academy, but above all allows us to identify its constituent elements as a special place'.[65]

For Renaissance academicians, the garden as theatre possessed two other essential qualities. First, it was a tool to organise knowledge, as the engraving by Boetto mentioned above demonstrates so eloquently. In other words, the academic garden was a memory garden. The *Songe énigmatique sur la peinture universelle* by the Toulouse painter and poet Hilaire Pader, published in Toulouse in 1658, is a fascinating and little-known illustration of this mnemonic function of the garden. A unique work, it describes a kind of academy of artists in an extraordinary garden and is based on the treatise *L'Idea del Tempio della pittura* published in 1590 by Giovanni Paolo Lomazzo, whose famous self-portrait as a shepherd, or more precisely as an abbott of the Accademia della Val di Blenio, is worth mentioning. Pader's interest in Lomazzo can be linked to his association with the court of Cardinal-Prince Maurice of Savoy, for whom he worked in Rome and to whom he dedicated his translation of one of Lomazzo's treatises on proportions.[66] The *Songe énigmatique* takes the form of a walk through a palace of paintings in the heart of a mystical garden.

64 Jones K.A., "Art and Regional Identity: Jean-Paul Laurens and the Murals of the Capitole of Toulouse, 1892–1915", *Studies in the Decorative Arts* 12.2 (2005) 96–128.

65 Quondam, "L'Accademia", 825. My translation. On early modern gardens and the theatre, see especially Cazzato V. – Fagiolo M. – Giusti M.A., *Lo specchio del paradiso. Giardino e teatro dall'Antico al Novecento* (Cinisello Balsamo: 1997); Grant K., *Landscape and the Arts in Early Modern Italy. Theatre, Gardens and Visual Culture* (Amsterdam: 2022).

66 Pader Hilaire, *Traicté de la proportion naturelle et artificielle des choses par Jean-Pol Lomazzo, peintre milanois. Ouvrage nécessaire aux Peintres, Sculpteurs, Graveurs et à tous ceux qui prétendent à la perfection du dessin, traduit d'Italien en Français par Hilaire Pader Tolosain, peintre de l'Altesse du serenissime Prince Maurice de Savoye* (Toulouse, Arnaud Colomiez: 1649).

It is inspired not only by the *Hypnerotomachia Poliphili*, but also by Giulio Camillo's famous *Idea del theatro*, which in turn inspired Lomazzo's *Idea del tempio della pittura*. In Pader's garden, trees in each corner symbolise the most accomplished artist in the six parts of the painting. Pader sums up the organisation of such a mnemonic garden in his address to the reader: 'Painting [...] is represented by the statue I have placed at the top of her palace, just as the lady who introduces me to the Mystic Garden represents nothing other than Mother Nature, her skates the earth, her first skirt water, the second air & the rest of her clothes, up to the top of her head, the rest of the universe. The seven trees planted in the six corners of the garden, the six different categories of painters who excel in only one. And the great one, planted in the centre, the universal painter who possesses them all. The steps leading up to the Palace of Painting, Proportion; the clumsy staircase, Movement. The first floor, colour. The second, the Science of Illumination; & the third, Perspective. The Secret Cabinet, the practice of all these parts'.[67]

The garden as theatre was also a place of dialogue, civil conversation, and conviviality, as well as a privileged location for entertainment and banqueting. These features of the garden, discussed in this book by George Brocklehurst with reference to Giovanni Pontano, are essential for understanding the rituals of early modern academies. Finally, because of its theatrical dimension, the garden is a space of utopia, or a heterotopia according to Michel Foucault.[68] It is a space where deviation from the norm is somehow tolerated, at least to a greater extent than in public spaces outside the academic garden. Ezio Raimondi writes: '[The academy] wants to be [...] a voluntary and in its own way a sovereign "gathering", in a sphere removed from the arbitrariness of everyday relations, where an "ideal state" of scholars is reconstituted, equal among equals, under the tranquil star of a pacified utopia, in the form of a scenic rite or an erudite melodrama: and whoever enters the academic universe accepts to undergo an initiation, a concordant act of rebirth, transforming himself into the character whose name and enterprise he chooses in the rational order of a destiny freed from the threat of contingency'.[69] Finally, as

67 Pader Hilaire, *Songe énigmatique sur la peinture universelle* (Toulouse, Arnaud Colomiez: 1658). My translation. See Trouvé S., "Lomazzo and France: Hilaire Pader's Translation: Theoretical and Artistic Issues", in Heck M.-C. (ed.), *Lexicographie artistique: formes, usages et enjeux dans l'Europe moderne* (Montpellier: 2018) 103–112.

68 For Foucault, the theatre, as well as the garden, is a perfect example of a heterotopia.

69 Raimondi E. "Introduzione", in Boehm – Raimondi (eds.), *Università, accademie e società scientifiche*, 17. My translation. On sociability in early modern gardens, see Ribouillault D., "Hermeneutics and the Early Modern Gardens: Ingenuity, Sociability, Education", in Melion W. – Enenkel K.A.E. (eds.), *Landscape and the Visual Hermeneutics of Place, 1500–1700* (Leiden – Boston: 2020) 291–325.

a hybrid between art and nature, the heterotopic garden offers its visitors the opportunity to encounter the higher and divine forces of nature, a necessary condition for the *furor poeticus*. Not surprisingly, as several of the essays in the book show, some academic gardens were linked to astrological determinism.

The academic garden and its long, complex history as a space for dialogue, inspiration, and democracy remains a fascinating model that we should foreground, study, and remember at a time when our academic campuses are increasingly deserted and collegiality within universities and schools is under attack by models of performance and profitability incompatible with the ideal of shared human happiness and exchange. This book would also like to be a contribution towards the revival of such an ideal.

•••

George Brocklehurst recalls the academy's essential dimension of conviviality in his essay on the Accademia Pontaniana, the earliest and among the most widely imitated humanist academies of the Quattrocento. It was also the cradle of Sannazaro's *Arcadia*, an allegory of the literary life of the Neapolitan elite that inspired a vast European literature of pastoral prose and poetry. This chapter explores the real green spaces of the Accademia Pontaniana as evoked in the works of its *maestro*, Giovanni Pontano, and other contemporary humanists. The Neapolitan circle first began hosting symposia, concerts, and poetic recitals in the gardens of the Villa Pliniana, the estate belonging to Antonio Beccadelli on the coast of Resina. After Beccadelli's death, gatherings continued to take place in Pontano's Villa Antiniana in the suburb of Vomero and his urban villa on the Via dei Tribunali. This chapter considers how these gardens operated as spaces of leisure, dialogue, and literary performance and, when possible, attempts to reconstruct their topography and horticultural programs.

Alessia Dessì offers new evidence on the location of Giulio Pomponio Leto's house on the Quirinal Hill, the seat of the most important humanist circle in Rome in the Quattrocento. There, Leto displayed his collection of epigraphs and ancient marbles, which he used in his teaching. Dessì's analysis of the elements of the collection, known from a contemporary anthology, reveals their archaeological origins and explains to what extent the location of Leto's Academy influenced the development of Roman cultural and artistic environments.

Giorgio Mangani explores another famous Roman academic garden, the Horti Colotiani on the Pincian Hill, where the poets and scholars of Colocci's circle of friends gathered for symposia and lectures. The poetic compositions pronounced in the garden harmonised with the *Sleeping Nymph* sculpture in the garden's collection, as well as with Polidoro da Caravaggio and Maturino's *Histories of Perseus and Andromeda*, painted in the 1520s. This subject, previously treated by Colocci's master, Giovanni Pontano, combined Ovid's *Metamorphoses* with astrological semi-determinism and carried considerable political symbolism. The Perseus constellation, which reached its zenith on 21 April (as the *paranatellonta* of the zodiac sign of Taurus), represented the birthday of Rome: an ancient celebration restored by Pomponio Leto and the Roman Academy, and later headed by Colocci. Mangani suggests that Colocci was the first owner of the famous Farnese Atlas (known in his circle as Hercules), which depicted the Perseus constellations in detail and introduced Polidoro's cycle together with the Sleeping Nymph.

The next two papers are concerned with the production of poetry within the gardens and landscapes of Rome and the Campagna. Tiziana Checchi's contribution analyses the relationship between some of the cardinals of the Colonna family and humanist culture. She explores the hitherto unknown "places of pleasure" that these cardinals, as commendatory abbots, enjoyed at the Abbey of Subiaco between the fifteenth and sixteenth centuries. Here, surrounded by the writers of their entourage, who dedicated poetic and celebratory compositions to them, the Colonna cardinals enjoyed themselves in joyful gatherings. The analysis of these poems and related ancient descriptions found in archival documents that have not been considered in the historiography on these texts not only make it possible to understand the nature of these places, which have now disappeared, and how they were used but also to rediscover their representations in a cycle of frescoes that decorate the so-called Studiolo di Marcantonio Colonna in the Abbey of Subiaco.

Next, Denis Ribouillault brings together for the first time a collection of Latin poems written to celebrate the famous Nymphaeum of the Villa Giulia, the villa of Pope Julius III in Rome. Some of the verses were written by Giovan Francesco Commendone, which the Pope ordered to be engraved in cartouches in the Nymphaeum. An examination of these poems demonstrates their continuity with a well-known tradition of poetic composition about statues of sleeping nymphs established decades prior, which can be seen in Colocci's, Goritz's and the Vatican gardens.

The iconological method pioneered by Erwin Panofsky and the intellectual circle indebted to Aby Warburg was first applied to the history of gardens in David Coffin's famous monograph on the Villa d'Este at Tivoli in 1960. However, this approach still largely focuses on architecture and sculpture, while the study of plants, flowers and trees is generally treated as separate from the literary and poetic culture that provides a framework to understand the meaning of the Renaissance pleasure garden – as if the natural sciences and literary culture were totally independent of each other. Here, Denis Ribouillault examines the plane tree (Platanus Orientalis *L.*), which was considered a true Platonic emblem and a 'living antique'. In doing so, he considers how its use and location in several Renaissance gardens relate to ancient literature and its prestigious association with Plato's Academy.

The following paper is concerned with Parisian gardens. In the sixteenth century, the Faubourg Saint-Marcel, which extended southwards along the Grand-Rue Saint-Marcel beyond the fortified walls of Paris, was an area between town and country, home to activities and populations that were not tolerated within the city walls for reasons of space, health, or morality. In this area, adjacent to the part of the city known as the University or the Latin Quarter, three successive and different academies were developed, for which gardens played an important role. The first was the Academy of Music and Poetry founded by Jean-Antoine de Baïf; the second was the garden of the Lycium philosophicum, a scientific academy founded by Gohory; and the third was the garden of the Maison de la Charité chrétienne, the philanthropic academy founded by the apothecary Nicolas Houel.

Moving from Paris to Breslau, Germany (today Wrocław in Poland), Christiane Lauterbach sheds new light on the botanical garden of the city physician Laurentius Scholz von Rosenau, the best-documented Renaissance city garden in late sixteenth-century Germany. Numerous neo-Latin sources, which are particularly revealing of the social gatherings in the garden, describe the design and the use of the garden by its owner, friends, and guests. The Hortus Scholzianus was the meeting place of the circle of Breslau scholars gathered around Scholz and Jakob Monau, whom contemporaries called the *collegium illustrium ingeniorum*. At the same time, it functioned as the venue of the Floralia Wratislaviensia, a festival in honour of the goddess Flora.

Next, Alessandro Spila's chapter draws on several unpublished archival documents to explore the development and possible significance of Cardinal Del Monte's gardens in Rome at the beginning of the seventeenth century for the important, symbolic role that alchemy (and, closely related, botany) played in his patronage. He also discusses the role of esoteric knowledge in shaping the key concepts of the academy that the Cardinal wished to establish. Although the gardens no longer exist or have been significantly modified, Spila examines their links with architectural, antiquarian, and naturalistic models present in more formal, institutional settings of the period, such as Federico Cesi's garden in Rome and his palace and gardens in Acquasparta, the seat of the newly established Accademia dei Lincei.

Consulting largely unpublished documents and letters, Sergio Monferrini offers a new analysis of the famous garden of Isola Bella, one of the islands in northern Italy's Lake Maggiore. For over forty years, Vitaliano VI Borromeo transformed a 'bare rock' into a 'miracle of art and nature'. On his garden island, Vitaliano welcomed distinguished guests such as the writer Carlo Maria Maggi, the mathematician Pietro Paolo Caravaggio and the Jesuit fathers Barella and Edera, giving life to an academy particularly interested in moral philosophy. Although not formally institutionalised, this Accademia dell'Isola met in the gardens, where musical performances highlighted themes of pleasure and usefulness aligned with ethical debates raised by the garden participants, and which found a similar echo in the Accademia dei Faticosi in Milan. The decoration of the garden was also conceived alongside its philosophical and academic function, which this article describes and explains.

Lauro Magnani outlines the links between the academic life of important literary figures in seventeenth-century Genoa and their gardens. *Lo stato rustico,* first published in Genoa in 1607 by Giovan Vincenzo Imperiale, an important member of many Italian academies, describes a charming visit to the gardens of his villa. It also marked the peak of the Genoese aristocrat, politician, writer, patron of the arts and collector's fame. In other Genoese villas, such as Villa Di Negro, the literary and scientific interests of academicians influenced the iconography and organisation of their gardens and vice versa.

Katrina Grant's chapter focuses on the Arcadian Academy founded in Rome in 1690. For the first thirty years after its founding, the Academy's meeting place – the Bosco Parrasio (Parrhasian Wood) – had many different locations throughout Rome, from the gardens of the Padri Minori Riformati on the Janiculum Hill to the Palatine Gardens owned by the powerful Farnese family, until a permanent site was built on the Janiculum Hill in the 1720s. These were small spaces, often sections of larger gardens, adapted to the needs of the Academy for their use over months or years. Apart from the latter garden, none of the meeting places survive in their original form, but their design can be pieced together from fragments of drawings and written descriptions. Arcadian writings constantly evoke themes of nature, landscape and gardens, which are often used metaphorically to explore ideas of beauty, friendship and good taste. The chapter considers how the physical sites of the Bosco Parrasio reflected Arcadian ideals of simplicity, egalitarianism, memorialisation, and a nostalgic longing for the classical landscape.

Finally, Maria Cristina Loi examines the Milanese colony of the Arcadian Academy, established in the city in 1704 at the behest of Father Giovanni Antonio Mezzabarba, and which found an ideal site for its meetings in the garden of Count Carlo Pertusati's residence. At the beginning of the eighteenth century, Count Pertusati, a member of an important family from Alessandria and a prominent figure of the Lombard aristocracy, bought the vast green area behind his palace and transformed it into an Italian garden – which soon became the new seat of Arcadia. The Garden of Arcadia is the most important example of eighteenth-century gardening in Milan and the first Lombard example of a winter garden, a kind of garden that became widespread in the second half of the century. Loi reconstructs the life of the Academy and its headquarters, and thus the role it played in contemporary Milan.

References

Alberti L.B., *I libri della famiglia*, Romano R. – Tenenti A. – Furlan F. (eds.) (Turin: 1994).

Arnaldi di Balme C. – Varallo F. (eds.), *Feste Barocche. Ceremonie e spettacoli alla corte dei Savoia tra Cinque e Settecento* (Milan: 2009).

Baldini N., "'Quasi Adonidos hortum'. Il giovane Michelangelo al giardino mediceo delle sculture", in Weil-Garris Brandt K. et al. (eds.), *Giovinezza di Michelangelo* (Florence: 1999) 49–56.

Baltes M., "Plato's School, the Academy", *Hermathena* 155 (1993) 5–26.

Barbe J.-P. – Pigeaud J. (eds.), *Les Académies (Antiquité–XIXe siècle). Sixièmes "Entretiens" de la Garenne Lemot* (Québec: 2005).

Bentz K., "The Afterlife of the Cesi Garden: Family Identity, Politics, and Memory in Early Modern Rome", *Journal of the Society of Architectural Historians* 72.2 (2013): 134–165.

Bober P.B., "The Coryciana and the Nymph Corycia", *Journal of the Warburg and Courtauld Institutes* 40 (1977) 223–239.

Boccaccio Giovanni, *Le Philocope de Messire Iehan Boccace Florentin*, trad. André Sevin (Paris, Denys Janot: 1542).

Boehm L. – Raimondi E. (eds.), *Università, accademie e società scientifiche in Italia e in Germania dal Cinquecento al Settecento* (Bologna: 1981).

Bracciolini Poggio, *Opera omnia*, ed. R. Fubini (Turin: 1964).

Brunon H., "De la littérature au jardin", in Venturi G. – Ceccarelli F. (eds.), *Delizie in villa. Il giardino rinascimentale e i suoi committenti* (Florence: 2008) 5–31.

Brunon H., "*Locus secretus*: topique et topophilie", in Luciani D. – Mosser M. (eds.), *Petrarca e i suoi luoghi. Spazi reali e paesaggi poetici alle origini del moderno senso della natura* (Treviso: 2010) 41–56.

Campbell G., "Epicurus, the Garden, and the Golden Age", in O'Brien D. (ed.), *Gardening. Philosophy for Everyone* (Oxford: 2010) 220–231.

Campitelli A., *Gli horti dei papi. I giardini vaticani dal medioevo al Novecento* (Vatican City: 2009).

Carboneri N. – Griseri A., *Giovenale Boetto* (Fossano: 1966).

Cazzato V. – Fagiolo M. – Giusti M.A. (eds.), *Lo specchio del paradiso. Giardino e teatro dall'Antico al Novecento* (Cinisello Balsamo: 1997).

Chambers D.S. – Quiviger F. (eds.), *Italian Academies of the Sixteenth Century* (London: 1995).

Chambers D.S., "The Earlier 'Academies' in Italy", in Chambers D.S. – Quiviger F. (eds.), *Italian Academies of the Sixteenth Century* (London: 1995) 1–14.

Chastel A., "Vasari et la légende Médicéenne: L'École du Jardin de Saint-Marc", *Studi* (1950) 159–167.

Christian K.W., *Empire Without End. Antiquities Collections in Renaissance Rome, c. 1350–1527* (New Haven – London: 2010).

Christian K.W., "Poetry and Spirited Ancient Sculpture in Renaissance Rome", in Kenda B. (ed.), *Aeolian Winds and the Spirit in Renaissance Architecture* (London: 2006) 103–124.

Christian K.W., *Raffaele Riario, Jacopo Galli, and Michelangelo's 'Bacchus', 1471–1572* (Turnhout: 2024).

Cicero, *Epistulae ad Atticum*.

Cicero, *De divinatione*.

Coffin D., *Gardens and Gardening in Papal Rome* (Princeton: 1991).

Comanducci R.M., "Gli Orti Oricellari", *Interpres. Rivista di studi quattrocenteschi* 15 (1995–1996) 302–358.

Cox-Rearick J., "Michelangelo's Mask of the Head of a Faun: Myth and the Mythic in the Sculpture Garden of Lorenzo il Magnifico", in Israëls M. – Waldman L. (eds.), *Renaissance Studies in Honor of Joseph Connors*, vol. 1 (Milan: 2013) 322–333.

Crescimbeni G.M., *Storia dell'Accademia degli Arcadi* [...] (London: 1804).

Cummings A., "Informal Academies and Music in Pope Leo X's Rome", *Italica* 86.4 (2009) 583–601.

De Gaetano A.L., "The Florentine Academy and the Advancement of Learning Through the Vernacular: The Orti Oricellari and the Sacra Accademia", *Bibliothèque d'Humanisme et Renaissance* 30.1 (1968) 19–52.

De Vesme A., *Peintre-graveur italien* (Milan: 1906).

Della Torre A., *Storia dell'Accademia platonica di Firenze* (Florence: 1902).

Delorme J., *Gymnasion: étude sur les monuments consacrés à l'éducation en Grèce* (Paris: 1960).

Demaraix M. et al. (eds.), *Les académies dans l'Europe humaniste. Idéaux et pratiques* (Geneva: 2008).

Diogene Laertius, *Lives and Opinions of Eminent Philosophers*.

Duport D., *Le jardin et la nature. Ordre et variété dans la littérature de la Renaissance* (Geneva: 2002).

Egmond F., "Sixteenth-Century University Gardens in a Medical and Botanical Context", in Fabrizio Baldassari – Fabio Zampieri (eds.), *Scientiae in the History of Medicine* (Rome: 2021) 89–120.

Elam C., "Lorenzo de' Medici Sculpture Garden", *Mitteilungen des Kunsthistorischen Institutes in Florenz* 36.1–2 (1992) 41–84.

Enenkel K.A.E. – Göttler C. (eds.), *Solitudo. Spaces, Places, and Times of Solitude in Late Medieval and Early Modern Cultures* (Leiden: 2018).

Everson J.E. – Reidy D.V. – Simpson L. (eds.), *The Italian Academies (1520–1700). Networks of Culture, Innovation and Dissent* (Cambridge & New York: 2016).

Fabiani Giannetto R., *Medici Gardens. From Making to Design* (Philadelphia: 2008).

Fagiolo M. – Madonna M.L., "La Casina di Pio IV in Vaticano. Pirro Ligorio e l'architettura come geroglifico", *Storia dell'arte* 15–16 (1972) 237–281.

Fariello A., *I giardini nella letteratura: dal giardino classico al giardino paesistico* (Rome: 1998).

Field A., "The Platonic Academy of Florence", in Allen M.J.B. – Rees V. – Davies M. (eds.), *Marsilio Ficino: His Theology, his Legacy* (Leiden: 2002) 359–376.

Franzoni C., "'Rimembranze d'infinite cose'. Le collezioni rinascimentali d'antichità", in Settis S. (ed.), *Memoria dell'antico nell'arte italiana: L'uso dei classici* (Turin: 1984) 304–360.

Fumaroli M., "Académie, Arcadie, Parnasse: trois lieux allégoriques du loisir lettré", *L'école du silence. Le sentiment des images au XVIIe siècle* (Paris: 1994), 19–36.

Gilbert F., "Bernardo Rucellai and the Orti Oricellari: A Study on the Origin of Modern Political Thought", *Journal of the Warburg and Courtauld Institutes* 12 (1949): 101–131.

Gnoli D., "Orti letterari nella Roma di Leone X", *Nuova antologia* (1930) 3–19.

Gohory J., *Instruction sur l'herbe Petum* (Paris, Galio du Pré: 1572).

Gombrich E., "The Early Medici as Patrons of Art", in *Norm and Form* (London: 1966) 35–57.

Gorris Camos R., "L'Hysope et la rose: le Lycium philosophal de Jacques Gohory", in Deramaix M. et al. (eds.), *Les Académies dans l'Europe humaniste* (Geneva: 2008) 553–594.

Grant K., *Landscape and the Arts in Early Modern Italy. Theatre, Gardens and Visual Culture* (Amsterdam: 2022).

Grimal P., *Les jardins romains* (Paris: 1969).

Hankins J., "Humanist Academies and the 'Platonic Academy of Florence'", *Analecta Romana Instituti Danici* 42 (2011) 31–46.

Hankins J., "The Myth of the Platonic Academy of Florence", *Renaissance Quarterly* 44.3 (1991) 429–475.

Harrison R.P., *Gardens. An Essay on the Human Condition* (Chicago: 2008).

Historia Augusta, *Life of Hadrian*.

Hyde Minor V., *The Death of the Baroque and the Rhetoric of Good Taste* (Cambridge: 2006).

IJsewijn J., "Poetry in a Roman Garden: The Coryciana", in Godman P. – Murray O. (eds.), *Latin Poetry and the Classical Tradition: Essays in Medieval and Renaissance Literature* (Oxford: 1990) 211–231.

Jakob M. (ed.), *Des jardins et des livres* (Geneva: 2018).

Jones K.A., "Art and Regional Identity: Jean-Paul Laurens and the Murals of the Capitole of Toulouse, 1892–1915", *Studies in the Decorative Arts* 12.2 (2005): 96–128.

La letteratura e i giardini, Atti del Convegno Internazionale di Studi di Verona (Florence: 1987).

Lanza A., "Il giardino tardogotico del Paradiso degli Alberti", *Italies* 8 (2004).

Losito M., "L'iconografia dell'Accademia delle Noctes Vaticanae e la cultura umanistica di San Carlo Borromeo", *Arte Cristiana* 98 (2010) 421–439.

Macdougall E.B., "The Sleeping Nymph: Origins of a Humanist Fountain Type", *The Art Bulletin* 57.3 (1975) 357–365.

Marr A., "A Mirror of Wisdom: Simon Vouet's *Satyrs Admiring the Anamorphosis of an Elephant* and Its Afterlives", in Berger S. – Garber D. (eds.), *Teaching Philosophy in Early Modern Europe* (Cham: 2002) 265–301.

Maylender M., *Storia delle Accademie d'Italia*, 5 vols. (Bologna: 1926–1930).

Morford M., "The Stoic Garden", *The Journal of Garden History* 7.2 (1987) 151–175.

Morolli G., "L'accademia, la selva, le grotte, il famedio", in *Il gigante degli Orti Oricellari* (Rome: 1993) 9–63.

Nicolò A. (ed.), *Lynceographum: Quo norma studiosae vitae lynceorum philosophorum exponitur* (Rome: 2001).

O'Sullivan T.M., *Walking in Roman Culture* (Cambridge: 2011).

Oberli M., *Magnificentia Principis: Das Mäzenatentum des Prinzen und Kardinals Maurizio von Savoyen (1593–1657)* (Weimar: 1999).

Pader Hilaire, *Songe énigmatique sur la peinture universelle* (Toulouse, Arnaud Colomiez: 1658).

Pader Hilaire, *Traicté de la proportion naturelle et artificielle des choses par Jean-Pol Lomazzo, peintre milanois. Ouvrage nécessaire aux Peintres, Sculpteurs, Graveurs et à tous ceux qui prétendent à la perfection du dessin, traduit d'Italien en Français par Hilaire Pader Tolosain, peintre de l'Altesse du serenissime Prince Maurice de Savoye* (Toulouse, Arnaud Colomiez: 1649).

Poliziano Angelo, *Opera omnia*, ed. I. Maïer, 4 vols. (Turin: 1971).

Pommier É., "Notes sur le jardin dans la littérature artistique de la Renaissance italienne", in Pigeaud J. – Barbe J.-P. (eds.), *Histoires de jardins. Lieux et imaginaire* (Paris: 2001) 127–140.

Poncet C., "Ficino's Little Academy of Careggi", *Bruniana & Campanelliana* 24.1 (2013) 67–76.

Quondam A., "L'accademia," in *Letteratura italiana. Il letterato e le istituzioni* (Turin: 1982) 823–898.

Raimondi E., "Introduzione", in Boehm L. – Raimondi E. (eds.), *Università, accademie e società scientifiche in Italia e in Germania dal Cinquecento al Settecento* (Bologna: 1981) 7–19.

Raja M.E., *Le muse in giardino: il paesaggio ameno nelle opere di Giovanni Boccaccio* (Alessandria: 2003).

Ribouillault D., "Hermeneutics and the Early Modern Gardens: Ingenuity, Sociability, Education", in Melion W. – Enenkel K.A.E., *Landscape and the Visual Hermeneutics of Place, 1500–1700* (Leiden – Boston: 2020) 291–325.

Ribouillault D., "Hortus academicus: les académies de la Renaissance et le jardin", in Jakob M. (ed.), *Des jardins et des livres* (Geneva: 2018) 23–34.

Ribouillault D., "Landscape *all'antica* and Topographical Anachronism in Roman Fresco Painting of the Sixteenth Century", *Journal of the Warburg and Courtauld Institutes* 71 (2008) 211–237.

Rice L., "*Pomis sua nomina servant*: The Emblematic Thesis Prints of the Roman Seminary", *Journal of the Warburg and Courtauld Institutes* 70 (2007) 195–246.

Rinaldi A., "Ideologia e tipologia del giardino urbano a Firenze tra XV e XVI secolo", in *Stanze per un giardino: Il paesaggio e il giardino nella cultura umanistica. V centenario della morte di Agnolo Ambrogini detto il Poliziano* (Siena: 1994) 1–27.

Rowland I., *The Culture of the High Renaissance* (Cambridge: 1998).

Segre A., "Untangling the Knot: Garden Design in Francesco Colonna's *Hypnerotomachia Poliphili*", *Word & Image* 14.1–2 (1998) 82–108.

Simonin M., "'Poësie est un pré', 'Poëme est une fleur': métaphore horticole et imaginaire du texte à la Renaissance", in *La letteratura e i giardini* (Florence: 1987) 45–56.

Small A. – Small C., "John Evelyn and the Garden of Epicurus", *Journal of the Warburg and Courtauld Institutes* 60 (1997) 194–214.

Tesauro Emanuele, *Il cannocchiale aristotelico* (Turin, Gio Sinibaldo: 1654).

Tesauro Emanuele, *Inscriptiones* (Turin, Bartolomeo Zapata: 1670).

Testa S., *Italian Academies and Their Networks, 1525–1700. From Local to Global* (New York: 2015).

Tongiorgi Tomasi L., "Gardens of Knowledge and the République des Gens de Sciences", in Conan M. (ed.), *Baroque Garden Cultures: Emulation, Sublimation, Subversion* (Washington DC: 2005) 85–129.

Tosco C., "Giardini platonici: una lettura del 'Fedro'", in *Architettura del giardino in Europa* (Naples: 2020) 96–100.

Trouvé S., "Lomazzo and France: Hilaire Pader's Translation: Theoretical and Artistic Issues", in Heck M.-C. (ed.), *Lexicographie artistique: formes, usages et enjeux dans l'Europe moderne* (Montpellier: 2018) 103–112.

Ginette Vagenheim, "Comment nourrissait-on les canards 'convives' dans la volière de Varron (3,5,16)? *Risus* et *inventio* dans les *Res rusticae*", *Rivista di filologia e di istruzione classica* 148, 2 (2020) 390–418.

Venturi G., "'Picta poësis': ricerche sulla poesia e il giardino dalle origini al Seicento", in De Seta C., *Storia d'Italia. Annali 5: Il paesaggio* (Turin: 1982) 663–749.

Venturi G., "Il giardino e la letteratura. Saggi d'interpretazione e problemi metodologici", in Ragioneri G. (ed.), *Il giardino storico italiano* (Florence: 1981) 251–277.

Vocabolario degli Accademici della Crusca, vol. I (Florence: 1863).

Watson P.W., "On a Window in Parnassus", *Artibus et Historiae* 8.16 (1987) 127–148.

Wind E., *Pagan Mysteries of the Renaissance* (London: 1958).

Wittkower R. – Wittkower M. (eds. and trans.), *The Divine Michelangelo. The Florentine Academy's Homage on his Death in 1564* (London: 1964).

Yeager-Crasselt L., *Michael Sweerts (1618–1664): Shaping the Artist and the Academy in Rome and Brussels* (Turnhout: 2015).

CHAPTER 2

The Green World of the Accademia Pontaniana: Conviviality, Horticulture, and Arcadia

George Brocklehurst

Abstract

The Accademia Pontaniana was the earliest and among the most widely imitated humanist academies of the Quattrocento. It was also the cradle of Sannazaro's *Arcadia*, an allegory of the literary life of the Neapolitan elite which inspired a vast European literature of pastoral prose and poetry. This chapter explores the real green spaces of the Accademia Pontaniana as depicted in the works of its *maestro*, Giovanni Pontano (1426-1503), and other contemporary humanists. The Neapolitan academy first began hosting symposia, concerts, and poetic recitals in the gardens of the Villa Pliniana, the estate belonging to Antonio Beccadelli (1394–1471) on the coast of Resina. After Beccadelli's death, banquets continued to be held in the gardens of Pontano's Villa Antiniana, and at his villa on the Via dei Tribunali. This chapter analyses these gardens as spaces of leisure, dialogue, and literary performance and, where possible, attempts to reconstruct their topography and horticultural programmes.

Keywords

Naples – Humanism – Convivium – Arcadia – Orangeries

Giovanni Pontano's Neapolitan academy (1429–1503) was one of the most influential humanist literary societies of the fifteenth century.[1] Founded in the 1440s at the court of Alfonso the Magnanimous, it was the centre of the Aragonese humanist movement, occasionally playing host to such figures as

1 On the Accademia Pontaniana and its legacies see Furstenberg-Levi S., *The Accademia Pontaniana: A Model of a Humanist Network* (Leiden: 2016); Hankins J., "Humanist Academies and the 'Platonic Academy of Florence'", in Jensen R. – Pade M. (eds.), *On Renaissance Academies: Proceedings of the Conference 'From the Roman Academy to the Danish Academy in Rome'* (Copenhagen: Odense University Press, 2011) 31–46.

 | DOI:10.1163/9789004517547_003

Lorenzo Valla, Lorenzo de' Medici and Aeneas Sylvio Piccolomini.[2] Both during its early period and in the turbulent years following Pontano's death, the academy of Pontano maintained close networks with intellectual centres throughout the Neapolitan peninsula, including Pomponio Leto's and Johann Goritz's academies in Rome, Bernardo Rucellai's Orti Oricellari in Florence, and the Ischian circle of Vittoria Colonna.[3] Through these networks, the Accademia Pontaniana helped to establish a new paradigm of humanist sociability, one which shaped the history of the Renaissance academy in a profound way. The modern myth of Arcadia, which originated in Sannazaro's *Arcadia* as a fantastic reimagining of the academy's literary and musical salons, was among its most important cultural legacies. In this chapter, I will survey the real green spaces that inspired Sannazaro's *Arcadia*, focusing on Pontano's Villa Antiniana and interpreting the social, horticultural and symbolic messages that Neapolitan humanist gardens were designed to convey.

Antonio Beccadelli (1394–1471), called Il Panormita, dominated the early decades of the Neapolitan academy. Beccadelli hosted the academy's first evening salons in the library of Castelnuovo, where passages of classical literature were recited and discussed over wine, fruit and desserts according to the example of ancient *convivia*.[4] After Alfonso's death in 1458, the academy began to hold its banquets elsewhere, sometimes gathering under a public arcade in Naples or at Beccadelli's Villa Pliniana on the coast of Resina. Concerning this villa, we only have the testimony of Pontano, who, in *De principe* (1490), describes a gathering in its forecourt where members of the academy were holding an ethical debate (*esset de virtute quaestio*) when their host invited his groundsman to join the conversation.[5] This anecdote suggests that the academy was already, in this early period, using the villa garden as a venue for literary dialogue. During the Aragonese period, it was common for urban courtiers to acquire estates through service to the crown, giving men like Beccadelli the opportunity to pursue a lifestyle of learned leisure according to ancient example.[6] The Neapolitan experience parallels that of other Quattrocento literary associations that met in the gardens of a villa property, such as the

2 Nicolini F., "Cenni storici", in *Annuario dell'Accademia Pontaniana* (Naples: 1957).

3 Furstenberg-Levi, "The Accademia Pontaniana".

4 For the banquets of the "Accademia Alfonsina", see Minieri-Riccio C., *Cenno storico della Accademia alfonsina: istituita nella città di Napoli nel 1442* (Naples: 1875).

5 Pontano Giovanni, *De principe*, in Garin E. (ed.), *Prosatori latini del Quattrocento* (Milan: 1952) 1044.

6 Tilly G., *Un manifeste posthume de l'humanisme aragonais: le* De hortis Hesperidum *de Giovanni Pontano* (Ph.D. dissertation, Normandie Université, Università degli Studi di Napoli Federico II: 2020).

Platonic Academy of the Villa Careggi in Florence, Poggio Bracciolini's *Bugiale* in the Val d'Arno and the numerous Roman *orti letterari* of the period.[7]

After Beccadelli's death in 1471, Pontano became the head of the academy that would posthumously bear his name. Pontano documented some of the gatherings of his academy in Latin dialogues, of which the *Asinus* (1488–1491) and the *Aegidius* (1501) are both set in gardens, the former on the grounds of Pontano's Villa Antiniana and the latter in the small urban garden abutting his "Portico" in the centre of Naples. Apart from these two Neapolitan gardens, Pontano also owned an estate on Ischia which later sources refer to as his *cenacolo*, though no contemporary description of this villa or the gatherings which might have taken place there has survived.[8] Of all his gardens, Pontano was especially devoted to those of his villa in the suburb of Vomero, dedicating numerous poems to them and their tutelary nymph, Antiniana. According to a document cited by Erasmo Pèrcopo, Pontano acquired this property on 9 May, 1472.[9] A leasehold for the villa dated 10 May, 1494, also discovered by Pèrcopo, records the size, contents, and surroundings of the villa:

> [...] quondam peciam terrae modiorum decem plus seu minus, cum certis arboribus et cum certis casalenis et hedeficiis ruynatis, cum cisterna et cortilio, sitam et positam in pertinentiis civitatis Neapolis, in loco ubi dicitur *Ad Antignano*, iuxta bona ecclesiae Sancti Jannarelli, iuxta bona dominae Carmosinae Cicinellae, iuxta bona ecclesiae Sancti Salvatoris ad Forum, vias publicas et alios confines.
>
> A plot of land measuring more or less ten hectares with some trees, some sheds, and ruined buildings, with a well and an internal courtyard, situated in the environs of Naples in a place called *Ad Antignano*, nearby to the Church of S. Gennariello, the home of Lady Carmosina Cicinello, the Church of St Salvatore overlooking the piazza, some public roads, and other adjoining properties.[10]

---

7 On the "garden academy" in the Quattrocento see Hankins, "Humanist Academies" 5.

8 On Pontano's villa at Ischia see Furstenberg-Levi S., "The D'Avalos-Colonna Literary Circle: A 'Renewed Parnassus'", in Cox V. – McHugh S. (eds.), *Vittoria Colonna: Poetry, Religion, Art, Impact* (Amsterdam: 2022) 45; Belli V., "La Torre Del Pontano e la sua villa a Ischia", *La Rassegna d'Ischia* 2 (2017) 42–49; Iacono A., "Gli umanisti e le acque di Ischia", in Valenti R. (ed.), *Intorno ai Campi Flegrei. Memorie dell'acqua e della terra* (Naples: 2011) 72; Iacono A., *La guerra d'Ischia nel* De Bello Neapolitano *di Giovanni Pontano* (Naples: 1996) 63–77.

9 Pèrcopo E., "La villa del Pontano ad Antignano", *Atti dell'Accademia Pontaniana* 56 (1926) 221–239.

10 The size of the measurement (*modius*) referred to cannot be determined exactly. Pèrcopo, "La Villa del Pontano" 223–224.

The size of the plot, which encompassed the ruins of the former chapel of San Pietro ad Arco, was commensurate with Pontano's senior rank in the Aragonese government. Pietro Summonte, Pontano's literary executor, later described the neighbourhood of the Villa Antiniana as fashionable (*villarum frequentia nobilis*).[11] More than just a monument to its owner's status, Pontano's villa was also the centre of his academy serving as a meeting place for scholars, poets, and musicians for the greater part of the golden age of Neapolitan humanism. In this chapter, I will investigate how the Villa Antiniana is represented as a site of academic sociability in Pontano's works and other works by members of the Accademia Pontaniana. After examining some literary descriptions of *convivia* that took place at the villa, I will attempt to decode its horticultural programme insofar as this reflects the cultural ideals and ambitions of Pontano's academy. Finally, I will consider how the academy's use of garden spaces as venues of literary and musical performance inspired the myth of Arcadia as a scholar's paradise, which was significant for the later imaginary of academic culture in Europe.

1 The Villa Antiniana as a Convivial Space

In his treatise *De splendore* (1498) concerning how the rich man should spend his wealth, Pontano emphasises the importance of gardens as places to host social functions:

> [Virum splendidum] etiam hortos habere volumus, in quibus exerceri deambulationes et convivia fieri pro tempore possint. Erunt autem horti hi ex peregrinis et egregiis arbusculis artificiose decenterque dispositi. In quibus e myrto, buxo, citrio, rore marino topiarium opus potissimum commendatur; neque enim quae parco ac lucroso patri familiae, eadem splendidi homini hortorum ratio esse debet. Conferunt etiam mirum in modum ad splendorem ipsum villae, non illae quidem rusticano ex opera, sed urbano et magnifico. Oportet enim non in urbe solum, verum etiam rure splendescere nitidi hominis supelectilem, ne, si quando seccedendum fuerit ab urbanis negociis, tanquam e luce in tenebras migrasse videatur. Quoniam autem etiam in secessu oblectamenta quaeruntur animique relaxationes tum sibi, tum comitibus et iis, qui visendi gratia aut alia de causa ad eum diverterint […] Idem familiam suam non

11 Pèrcopo, "La Villa del Pontano" 222–223.

> solum bene et laute pascet, sed multos, ut dici solet, cives peregrinosque participes quadrae suae habebit.[12]
>
> We also want [the splendid man] to have gardens in which to take the occasional stroll and to hold banquets. These gardens will be arranged artfully and becomingly from fine and exotic plantings. A work of topiary made from myrtle, boxwood, citrus, and rosemary should be deemed most important. The science of gardening is appropriate to the splendid, not to the thrifty or luxurious man. Villas made not by rustic, but by urban and magnificent workmanship confer splendour in a marvellous way. For a gentleman's furnishings should not only be splendid in the city, but also in the country lest he seem to disappear from the light into the darkness whenever the time comes to take a vacation from urban business. Since amusements and relaxation are sought during one's holiday, these should also be entertaining for [the splendid man] himself, his companions, and also those who come to visit him or for whatever other reason [...] He should not only feed his own household well and bountifully but also, as they say, the fellow citizens and foreigners with whom he will share his table.

Pontano echoes the words of Pliny the Younger, who wrote that a villa garden should be an exceedingly "urban" (*urbanus*) artwork, meaning one with beautifully arranged ornamental plantings.[13] Through the splendour of its plantings, the garden reflects the *urbanitas* and *splendor* of the property owner and becomes a means of conspicuous consumption. Pontano speaks as an experienced horticulturist who not only designed the gardens of his own properties but also helped to plan the famous gardens at the Villa Poggioreale which were used to host court banquets and festivals.[14] Pontano's love of hosting banquets – both intimate *convivia* and lavish court celebrations – is also alluded to in the subsequent essay on the virtue of 'conviviality' (*De conviventia*) in the collection of *Treatises on the Social Virtues*. Denis Ribouillault has underlined the importance of reading Renaissance gardens

12 Pontano Giovanni, "De splendore", in Tateo F. (ed.), *I trattati delle virtù sociali: De liberalitate, De beneficentia, De magnificentia, De splendore, De conviventia* (Rome: 1965) 136–138.

13 Pliny, *Letters* 5.6.35.

14 See Modesti, P., *Le delizie ritrovate: Poggioreale e la villa del rinascimento nella Napoli aragonese* (Florence: 2014) 178.

through the prism of social experience.[15] Indeed Pontano's recommendations for how the splendid man should use his villa shows that it existed primarily to serve a social purpose. The splendid man's garden, he says, is not only for its owner's enjoyment or for his family's exclusive use, but is also a space to host dinners with both 'fellow citizens and foreigners'.[16] In this regard, Pontano disagrees with Alberti, who believed that gardens should serve the pleasure of the household alone and not of outsiders.[17]

David Marsh has attributed the widespread Quattrocento humanist custom of gathering for literary *convivia* to the recovery of Cicero's *De oratore* in 1421 since this dialogue describes a banquet held in a villa garden. For Cicero and his emulators, learned leisure (*otium litteratum, otium doctum*) was fundamentally a social and discursive pursuit.[18] The circle of Antonio Beccadelli in Milan was the earliest humanist community to revive the ancient practice of the *convivium*. In the 1430s, this circle began hosting learned drinking parties *all'antica*, where literary discussions were accompanied by banter, party games and musical performances to the lyre. Francesco Filelfo, who succeeded Beccadelli as symposiarch of this circle, records one such banquet in his *Convivia mediolanensia* (1443), a dialogue that was responsible for popularising this custom among humanist communities.[19] When Beccadelli moved to the court of Aragon in the 1440s, he brought with him the tradition of hosting literary *convivia* with readings, debates and musical performances. The Neapolitan academy was born from these symposia.

In his treatise on conviviality, Pontano remarks that of all things that bring people together, the most important are shared pursuits (*societas studiorum*) and the custom of sharing a table (*consuetudo convivendi*). The greatest of human pursuits, he says, are the liberal arts (*bonae artes*) and philosophy, both

15 Ribouillault D., "Hermeneutics and the Early Modern Garden: Ingenuity, Sociability, Education", in Enenkel A.E. – Melion W. (eds.), *Landscape and the Visual Hermeneutics of Place, 1500–1700* (Leiden – Boston: 2020) 291–325.

16 Pontano, "De splendore" 136–138.

17 Coffin D.R., "The 'Lex Hortorum' and Access to Gardens of Latium during the Renaissance", *The Journal of Garden History* 2 (1982) 201–232.

18 Marsh D., *The Quattrocento Dialogue: Classical Tradition and Humanist Innovation* (Cambridge: 1980). On literary conviviality in humanism, see also Hankins J., "Vocal Music at Literary Banquets in the Renaissance", *Basler Jahrbuch für Historische Musikpraxis* 41 (2021) 227–243; Pajorin K., *Italia e Ungheria all'epoca dell'umanesimo corviniano* (Florence: 1994) 179–228. On this topic, see also Denis Ribouillault's essay in this book, *The "Academic" Plane Tree in Early Modern Gardens*.

19 Hankins, "Vocal Music"; Pajorin, "Italia e Ungheria" 179.

of which have 'a most delightful affinity' with banqueting.[20] The practice of discussing classics over wine and food appears to have shaped the academy's intellectual life in a profound way. Pontano's poetry often portrays the Villa Antiniana as a convivial space. In the first of two elegies in the *De amore coniugali* (1490–1500) dedicated to the villa and its gods, Pontano, on return from military campaign, calls for wine:

> [...] Tu prome Falernum / leniat ut curas ebria cura novas; / pone merum, pone et talos; post vina iocumque / ita sopor, hic curas diluet graves.
>
> Pour the Falernian so that drunkenness can lighten my cares. Give me the wine and give me the dice; after wine and play my stupor will be so great it will soothe my grave anxiety.[21]

Here, the Villa Antiniana becomes the setting of a Horatian *convivium*.[22] In the following poem in the sequence, Pontano, sitting alone in his garden, imagines a pastoral *fantasia* unfolding before his eyes, with streams of milk and honey swelling up from the ground and 'men young and old dining with the garden gods' ('coelitibus simul immixti iuvenesque senesque / prandebant').[23] Pontano's vision of the golden age as a rustic bacchanal mirrors the association of his gardens with conviviality. In his *De hortis Hesperidum sive de cultu citriorum* (1500), the nymph of the gardens of Antiniana and her maidens are said to 'frequent banquets' and to 'lay birthday tables', while in the *Lepidina*, the same nymph leads a train of gods bearing food and wine for the wedding of Parthenope and Sebethus.[24] In Pontano's poems, the Villa Antiniana's status as a destination for banquets and festivals seems to parallel the academy's use of these gardens as a venue for its convivial gatherings.

In an elegy in his *Eridanus*, Pontano invites his friend Girolamo Carbone to a dinner in the gardens of the Villa Antiniana:

> Fictilibus si coena placet tibi, candide Carbo, / coena parata est tibi est ruraque nostra patent / et penus et virides horti genialis et umbra.

20 Pontano Giovanni, "De conviventia", in Tateo (ed.) *I trattati delle virtù sociali* 142.
21 Pontano, *De amore coniugali* 2.2, "Villam salutat a militia regressus" 23–26.
22 Pontano, *De amore coniugali* 2.4, "Ad Bacchum consecratio" 5, 33.
23 Pontano, *De amore coniugali* 2.3, "Laetatur in villa et hortis suis constitutes" 45–46.
24 Pontano, *De hortis Hesperidum* 1.576, 2.121; *Lepidina* 6 *passim*. On the nymph Antiniana in Pontano's works, see Pèrcopo, "La villa del Pontano" 226–227.

> If, dear Carbone, you are partial to a table laden with china, then dinner is prepared and my estate, my provisions, my verdant gardens and pleasant shade lie open to you.[25]

The form of the dinner invitation is familiar from classical verse epistolography, with Pontano evoking the intimate, rustic conviviality of Horace's Sabine Farm.[26] The meal on Pontano's 'rustic table' is served by the 'hand of Antiniana' herself, with a first course of autumnal fruits from the garden followed by capons, goose and sweetmeats flavoured with citron, myrtle berry and thyme: 'rura dabunt quicquid nostra culina coquit' ('my kitchen cooks whatever the field will grow').[27] This boast calls to mind the *De splendore*, in which Pontano recommends keeping an aviary or enclosure (*vivarium*) with peacocks, pheasant, partridges and other fowl to supply the kitchen.[28] The rich, sensual description of Pontano and Carbone's 'rustic dinner' also evokes the *pompae*, or pageants of Pontano's *Lepidina*, where throngs of rustic gods and goddesses carry trays loaded with vegetables, sweetmeats and wine to the wedding table of the nymph Parthenope. The Villa Antiniana was evidently well-equipped to feed the many guests who Pontano hosted in his garden. A passage from *De hortis Hesperidum* similarly seems to refer to preparations for a birthday celebration in the gardens of Antiniana, with gilded tablets hung as decoration from the vines; birthdays being a classic occasion for academic celebration.[29]

One of Pontano's most enigmatic works, the *Asinus*, concludes with a dinner in the gardens of the Villa Antiniana. In this dialogue, Pontano imagines that he has lost his mind and fallen in love with an ass, spending his vacation at the Villa Antiniana grooming, singing, and bathing the animal in perfume. The message and intention of the *Asinus*, recently dubbed the 'impossible dialogue', is much contested.[30] The first printed edition of the work (1507) features the subtitle *De ingratitudine* (On Ingratitude). However, Francesco Tateo views this subtitle as the result of a later editorial intervention by Summonte, pointing

25 Pontano, *Eridanus* 1.40, "Ad Carbonem" 1–3. On this poem, see Casanova-Robin H., "L'adresse de Giovanni Pontano à Girolamo Carbone dans l'élégie I, 40 de l'*Eridanus*. Un idéal d'*humanitas*?" in Julhe J.C. (ed.), *Pratiques latines de la dédicace – Permanence et mutations, de l'Antiquité à la Renaissance* (Paris: 2014).

26 Horace, *Epistles* 1.5.

27 Pontano, *Eridanus* 1.40, "Ad Carbonem" 24.

28 Pontano, "De splendore" 137.

29 Pontano, *De hortis Hesperidum* 2.119–123.

30 Bistagne F., "Giovanni Pontano's *Asinus* or the Impossible Dialogue", paper presented at *The Afterlife of Apuleius* conference (London: 2016).

out that the dialogue in fact has little to do with ingratitude.[31] More recent theories attribute Pontano's strange behaviour in this dialogue to senility[32] or to his overzealous study of astrology.[33] I would like to propose an alternative interpretation that reflects the status of Villa Antiniana as a convivial space. In the final scene of the dialogue, Pontano's academic colleagues, who have been watching him play with his ass from behind a rosemary bush, come to greet him. Suddenly, Pontano returns to sanity, bidding 'farewell to asses for good'.[34] One speaker, the scholar Giovanni Pardo, compliments Pontano, remarking with palpable dramatic irony that he must have spent his time at the villa composing poetry and studying the stars, since he has never been known to pass a spell of leisure 'without activity of the mind'. The dialogue ends with the scholars sitting down to a *convivium*:

> Pont. Ego vero amicissimos homines ac Musarum nostrarum alumnos ea voluptate his in hortis accipio amplectorque, qua coelestis res diebus his in hac ipsa solitudine vel magis secessu sum contemplatus; siquidem contemplationis ipsius communicatio cum studiosis rerum earundem viris is profecto fructus est etiam suavissimus.
>
> Pontano: I welcome and embrace you, my most beloved friends and students of our Muses with the same pleasure with which I have lately contemplated astrology in this solitude or rather, this retreat, since sharing this contemplation with men who also pursue such things is no doubt a pleasure; in fact, it is the sweetest pleasure of all.[35]

We recall Pontano's remark in the treatise "On Conviviality" that dining is one of life's greatest pleasures and a boon to friendship and exchange. Liliana Monti-Sabia has suggested that *Asinus* was published in an unfinished state, with Pontano intending to continue the dialogue beyond this passage. She further suggests that the *De luna*, a coda appended by Summonte to Pontano's didactic poem *De rebus coelestibus* (1494), was formed from a fragment of the

31 Tateo F., *Umanesimo etico di Giovanni Pontano* (Lecce: 1972).

32 Casanova-Robin H., "La déraison du sage, dans l'*Asinus* de Pontano", in Casanova-Robin H. – Furlan F. – Wulfram H., (eds.) *Serio ludere: D'Alberti à Pontano et d'Érasme à Thomas More* (Paris: 2020) 151–171.

33 Haywood E., "'Iter asinarium': per una nuova interpretazione dell'*Asinus* pontaniano", in De Nichilo M. – Distaso G. – Iurilli A. (eds.), *Confini dell'umanesimo letterario, studi in onore di Francesco Tateo* (Rome: 2003) 733–746.

34 Pontano, *Asinus* 26.

35 Pontano, *Asinus* 32.

abandoned supplement to the *Asinus*, which, she suggests, featured a conversation between the scholars on astrology.[36] On the contrary, I believe that the abrupt ending of the *Asinus* is significant for our understanding of the dialogue as a work that is ultimately concerned with the importance of conviviality. In his treatise on this virtue, Pontano describes conviviality as a mean between the extremes of revelry (*comessabilitas*) and unconviviality (*inconviventia*). The first act of the *Asinus* takes place in a country inn where Pontano had previously spent a night on return from a diplomatic mission to Rome. Upon receiving news of Pontano's success, the innkeeper calls for barrels of wine, food and prostitutes. This section of the dialogue exemplifies the vice of revelry, which, according to Pontano's treatise on conviviality, causes the individual to 'drink heartily, fill his stomach, and indulge his palate, oblivious to his fellow diners'.[37] Correspondingly, the second act of the dialogue, in which the mad Pontano, home alone at his villa, plays with his donkey and tries to bargain for a night with his groundsman's wife, seems to exemplify the vice of 'unconviviality':

> Defectum igitur ipsum numquam convivendi apellabimus inconviventiam. [...] quod a convictu ipso prorsus abhorreat. [...] Nonnunquam etiam a quadam animi tristitia et studio solitudinis vitium ipsum gignitur.
>
> Therefore, we shall call the defect of never banqueting "unconviviality" [...] Because it abhors the banquet ["living together"] entirely [...] Sometimes too this vice is produced by melancholy, by a sadness of the mind, and by a zeal for solitude.[38]

Pontano's strange behaviour in the *Asinus* might be diagnosed as a symptom of melancholy (*animi tristitia*), which descriptions of the humoral type generally associated with manic behaviour and a strong libido. Seen in this light, the final scene of the *Asinus*, when the scholars sit down to dinner in the gardens of the Villa Antiniana, represents a return to conviviality and, with this, the triumph of sanity over madness. As in Pontano's social treatises, the banquet in the villa garden in the *Asinus* serves to promote harmony between members of a socio-intellectual group, something that Pontano regarded as a condition for a scholar's personal wellbeing. Not unlike the green spaces of Sannazaro's

---

36 Monti-Sabia L. – Caracciolo T. – Fido C., *Un profilo moderno e due vitae antiche di Giovanni Pontano* (Naples: 1998) 20.

37 Pontano, "De conviventia" 143.

38 Pontano, "De conviventia" 143.

Arcadia, the Villa Antiniana is portrayed in Pontano's works as a space of excess leisure and sociability – a landscape of scholars at play.

2 The Design of the Villa Antiniana

I will now attempt briefly to describe the gardens in which the above *convivia* took place. In one scene of the *Asinus*, the scholars observe Pontano taking a walk around the grounds of Antiniana with his groundsman; besides the *convivium*, the stroll around the garden (*deambulatio*) was identified by Pontano in the *De splendore* as one of the chief pleasures of owning a villa. This was true also of Roman villas, with both Cicero and the younger Pliny mentioning the promenade (*ambulatio*) as a praiseworthy pastime and as a good opportunity for post-prandial conversation.[39] Pontano uses his walk around the gardens to check in on the progress of his plantings, occasioning his display of botanical erudition as he describes how and when to graft, pollard, sow and irrigate citrus trees in harmony with the cycles of the zodiac. Giovanni Pardo, spying him from behind the rosemary bush, remarks:

> Ratio haec colendarumque citriorum cura ingenii haudquaquam delirantis videri potest. Non male igitur nobiscum agitur, nec est quod Musis gratias non agamus.
>
> The thought and attention that he gives to cultivating citrus trees may by no means be seen as a sign of manic character. So, things are not so bad with us, and we have every reason to thank the Muses.[40]

Citrus culture was considered as an art worthy of the Muses, and so Pontano's care in maintaining his garden is taken as a sure sign of sanity. Georges Tilly has recently demonstrated that Pontano's interest in agriculture, evidenced in the *Asinus* and his late didactic-botanical poem on citrus culture *De hortis Hesperidum*, was typical of the Aragonese urban aristocracy.[41] Through their loyalty to the crown, many of King Ferdinand's courtiers had received estates expropriated from the feudal nobility in the wake of the Second Baron's War (1485–1486). Inspired by humanistic ideals, the cosmopolitan urban nobility

---

39 Cicero, *Ad familiares* 2.12; Pliny, *Letters* 5.6.17. See O'Sullivan T.M., *Walking in Roman Culture* (Cambridge 2011).

40 Pontano, *Asinus* 21.

41 Tilly, *Un manifeste posthume* 15.

managed their estates following the example of Cato, who, in the prologue to his treatise on agriculture, wrote that the proprietor of a villa should always surpass his stewards in knowledge of farming and horticulture.[42] Pontano had a deep respect for agricultural knowledge; in his *Aegidius*, he says that Virgil himself learned from the farmers of Latium, while one poem in *Tumuli* is dedicated to a groundsman of his who was especially learned in agricultural matters.[43] Pontano's personal interest in maintaining his gardens leads us to question: what did he grow on the grounds of the Villa Antiniana?

In the *De splendore*, Pontano argues that particular attention should be paid to the art of ornamental gardening (*opus topiarium*) and that the splendid man should take care to have 'rare and exotic plantings' ('ex peregrinis et egregiis arbusculis'), especially myrtle, boxwood, citrus and rosemary. Alberti similarly recommends citrus, myrtle and boxwood, together with juniper, ivy and laurel, as suitable plants for an ornamental garden.[44] These plants, especially myrtle and citrus, which frequently appear in literary portraits of the Villa Antiniana, would later be counted among the typical plants to be included in a *Hortus Hesperidum*, or orangery.[45] Though citron plants had been cultivated in the Mezzogiorno since at least the thirteenth century, Spanish gardeners employed by Alfonso I first introduced the art of propagating orange groves to Italy, with significant consequences for the development of citrus culture in the peninsula more broadly.[46] Beginning at the famous orangery at the Villa Poggioreale, the humanistic tradition of citrus cultivation spread throughout Europe, inspiring the orangeries at Blois and Amboise, which were designed by the same gardener, Pacello di Mercogliano.[47] In the later years of his life,

42 Cato, *De re rustica* 1.1.

43 Pontano, *Tumuli* 1.22.5.

44 Alberti, *De re architectura* 9.4.

45 The following passages also allude to the presence of a citrus grove at the Villa Antiniana: Pontano, *Asinus* 152–153; "De splendore" 137; *Eridanus* 1.40.4. Myrtle trees were often associated with the Villa Antiniana and its guardian nymph: Pontano, "De splendore" 137; *Lepidina* 55; *De amore coniugali* 2.2.12, 3.1.31; *Eridanus* 1.40.35; *De tumulis* 1.1.5; *Hendecasyllabi* 2.27.14. Also mentioned are thyme (*Lepidina* 55; *Eridanus* 1.40.35–36); mulberry trees (*De amore coniugali* 2.4.9); pear trees (*De amore coniugali* 2.4.9); palm trees (*De hortis Hesperidum* 479); a hazel grove (*Coryle* 2.50.159–160); roses (*Eridanus* 2.31.38; *De Tumulis* 1.1.6, 1.18.7–8; *Hendecasyllabi* 2.27.13; *Lyra* 3.21.29); violets (*Lyra* 3.21), lilies (*Lyra* 3.31), and hyacinths (*Lyra* 3.32).

46 See Paulus H.E., "The Humanistic Tradition of Citrus Culture in Central Europe from the 15th to the 18th Century", in Fiorentino G. – Zech-Matterne V. (eds.), *AGRUMED: Archaeology and History of Citrus Fruit in the Mediterranean: Acclimatization, Diversifications, Uses* (Naples: 2018).

47 Paulus, "The Humanistic Tradition of Citrus Culture".

Pontano composed a book of didactic-botanical poems on citrus culture, *De hortis Hesperidum*, which also imply the presence of a citrus grove at the Villa Antiniana. According to Helmut-Eberhard Paulus, citron, pomegranate, orange and the white-flowered lemon myrtle (*citrus myrtifolia*) were among the most common trees to be cultivated in citrus groves.

These evergreen, blossoming plants were symbolic of human passion, with Pontano in *De hortis Hesperidum* describing how Venus transformed Adonis into an evergreen orange tree. The same is true of myrtle, which was said to have been used by Venus to cover her nakedness when she was born from the ocean.[48] Citrus culture held a broad range of cultural and political references, and it is not solely owing to their erotic connotation that myrtle and citrus trees, both typical orangery plants, feature so often in Pontano's poetry. Citrus and myrtle trees also allude to the myth of Hesperides and to Virgil's Golden Age, hinting at the humanist yearning to return to the primitive pleasures of the classical pastoral, which would later be celebrated by Sannazaro in his *Arcadia*. Moreover, orangeries also served as patriotic symbols of the Regno, representing Spanish culture and botanical science united with the climate and natural fecundity of Campania. In addition to conferring 'splendour' and yielding fruit for Pontano's kitchen, the presence of orangery plants in the ornamental garden of the Villa Antiniana thus also had a symbolic function, representing the cultural ideals and ambitions of the Accademia Pontaniana.

3 Music and Arcadianism

Recent scholarship has shed light on the musical accomplishments of the Accademia Pontaniana, which was one of the liveliest centres for the humanist art of poetry sung to the lyre (*cantare ad lyram*).[49] Some scholars have suggested that the art of solo singing to the lyre, along with other cultural imports of Arabic origin such as citrus culture, penetrated Italy via the Spanish court of Aragon, though some recent studies have emphasised continuities between medieval and Quattrocento Italian monophonic song traditions. In any case, the Neapolitan academy certainly made a significant contribution to the development of this art, hosting a number of celebrity *improvisatori* including

48 Ovid, *Fasti* 4.141–143.

49 Coleman J., *A Sudden Frenzy: Improvisation, Orality, and Power in Renaissance Italy* (Toronto: 2022); Wilson B., *Singing to the Lyre in Renaissance Italy* (New York: 2019) 303–336.

Cariteo, Serafino Aquilano and the Brandolini brothers.[50] The academy's banquets often centred on the performance and critique of lyric song. In Pontano's dialogue *Antonius* (1487), a banquet climaxes with the performance of some Latin lyrics by a lutenist (*lyricen*) and a carnival show led by a masked poet (*poeta personatus*) and jester (*istrio personatus*).[51] Music-making was considered appropriate to the correct practise of the symposium, with Guarino da Verona writing in 1424 that 'one reads about hardly any feasts of ancient times in which singers were not involved'.[52] One member of the Accademia Pontaniana, Johannes Tinctoris (1435–1511), who served as royal cantor at the Aragonese court, writes that music 'increases the joyfulness of banquets' and begins his *Complexus effectuum musices* (1470–1475) with the following quote from Aristotle's *Politics*: 'Musaeus says that singing is most delightful to man, for which reason men rightly believe it to be a source of delight in gatherings and parties'. Raffaele Brandolini, another Neapolitan *improvisatore* of the period and member of the Accademia Pontaniana, similarly writes in his dialogue *De musica et poesia* (1513):

> And no one should dare to impeach the use of the lyre at banquets, no one the singing and reciting of a song there, unless he be bereft of spirit and intellect.[53]

Given the high esteem in which members of the Accademia Pontaniana held the art of convivial music making, it is appropriate to wonder whether the garden banquets at the Villa Antiniana also featured musical entertainments. The location of the Villa Antiniana was perhaps chosen for its proximity to the hill of Posillipo, which hosted Virgil's tomb. This site features in the *Asinus* in an episode recounting the scholars' journey to rescue the mad Pontano, with Sannazaro remarking of how he and the poet Lucio Crasso used to sit by Virgin's grave and cap verses together:

> Nam et saepicule considebat sub arbusculis, aut muscoso aliquo conquiescebat in lapide, veterum poetarum aut siquos ipse lucubrasset

50 Haraszti E., "La technique des improvisateurs de langue vulgaire et de latin au quattrocento", in *Revue belge de Musicologie* (1955) 18.

51 Pontano, *Antonius* 102ff.

52 Cummings A., "Music and Feasts in the Fifteenth Century", in Busse Berger A.M. – Rodin J. (eds.), *The Cambridge History of Fifteenth Century Music* (Cambridge: 2015) 367.

53 Brandolini R., *On Music and Poetry* (*De musica et poesia*), A. Moyer (ed.) (Arizona: 2001) 81–83.

versiculos interim referens, multa etiam saepe aut ipse docens, aut nos qui eum sequebamur percunctatus.

For quite often he would sit under the trees or lay down on some mossy rock, sometimes singing some songs by the ancient poets or occasionally relaying some of his own verses, often teaching us many things or questioning those of us who were following him.[54]

This scene gives us an impression of the way in which green spaces served the scholars of the Neapolitan academy as a space of poetic production; the custom of reciting and singing poetry together clearly evoked the singing competitions of Virgil's shepherds. Members of the academy were christened with Arcadian names and were sometimes known to don Arcadian costume. Rustico Romano, one *improvisatore* and member of the academy, was famous for performing dressed in shepherd's clothing.[55] Antonello de Petruciis seems to refer to this custom in the following lyric, recalling happy times spent with his friends at the academy:

Or dove sono andati mo, o Barone,
Li nostri risi con li iochi e feste
Tante allegricze con mutar de veste,
Tante diverse et varie canzone,
El docto disputar le questione
De omne doctrina, e mai de cose meste?

Now where have they gone, O Baron, our laughter during games and parties, our delight when we changed our costume, all those different and varied songs, the learned disputation on questions touching upon every subject, and never upon sad things?[56]

An understanding of Sannazaro's fiction is incomplete without accounting for the pastoral games that the Neapolitan academicians played during their hours of leisure in the gardens of the Villa Antiniana, at Sannazaro's own Villa Mergellina, and in other green spaces. The Arcadian fiction experienced a rich afterlife in sixteenth-century Italian academies, thriving in the artistic production, nomenclature, music, costume, and *imprese* of societies like the Accademia degli Alterati, the Accademia della Crusca and the Accademia

54 Pontano, *Asinus* 16.
55 Bosisio M., *Il teatro di corte pre-classicista: Geografia e storia* (Milan: 2014) 179.
56 De Petruciis, *Sonetti* 25.1–6.

dell'Arcadia.[57] Though Petrarch was the first to call the life of learned leisure an "Arcadia" of the mind, immune to the seductions of city life, in *De vita solitaria*,[58] the scholars of the Accademia Pontaniana were the first to bring Arcadia to life in their rustic retreats. Virgil had referred to Parthenope as the land of 'ignoble leisure' (*ignobile otium*), where he sang his own shepherds' songs. In doing so, he left behind a legacy that led the scholars of Pontano's academy to imagine themselves the inhabitants of ancient Arcadia.[59]

4 Conclusions

Reflecting on the golden age of the Accademia Pontaniana the jurist Alessandro d'Alessandri recalled once being welcomed into the gardens of the Villa Antiniana, where the academy had gathered to celebrate Pontano's birthday over the course of several evenings.[60] For the generation of *accademici* living under the Spanish viceregency who could remember attending Pontano's banquets, the gardens at the Villa Antiniana were a space potent with nostalgia. In the *Arcadia*, Sannazaro laments:

> Antiniana, e tu perché degeneri? / Perché ruschi pungenti in te diventano / quei mirti, che fur già si molli e teneri?
>
> Antiniana, why have you fallen? Why have your myrtle trees, which were once so soft and tender, become so rough and prickly?[61]

Similarly, the humanist Giano Anisio wrote a lament in which he imagines himself exploring the ruins of the Villa Antiniana, which had been abandoned and left to rot after Pontano's death. He is greeted by the guardian nymph who takes the poet's hand:

> Et, nosti Meliseum, inquit, quem Phoebus amavit, / cui cara ante omnes Antiniana fuit? / Antiniana sum ego, coeli concessit ad astra / ille sua, et

57 See Fumaroli M., "Academia, Arcadia, Parnassus: troix lieux allégoriques de l'éloge du loisir lettré", in Chambers D.S. – Quiviger F., *Italian Academies of the Sixteenth Century* (London: 1995) 15–36.

58 Petrarch, *De vita solitaria* 1.

59 Virgil, *Georgics* 4.564.

60 d'Alessandri Alessandro, *Alexandri ab Alexandro iurisperiti Neapolitani Genialium dierum libri sex, varia ac recondita eruditione referti* (Cologne, Iaspes Gennepaeus: 1551).

61 Sannazaro, *Arcadia* 12.91–3.

> divum colloquiis fruitur. / Haec rura incoluit, haec saevit citria, et illa / dum plantat laurum, carmina curit amans: / "Ne mea, ne pereat moriens Ariadna, perenne / hoc sacro laurus cortice nomen habe." / Vidi ego, videre et multi, quum carmina falce / scripsit in hac lauro maximus ille senex. / Et crevit Laurus, crevere et carmina, et una / crevit amor, quo vel mortua corda calent.[62]
>
> And she asked, do you know Meliseus, whom Phoebus loved, and to whom Antiniana was dearest of all? I am that Antiniana. He has vanished to the stars in heaven and enjoys discourses with the divine. He had left his land uncultivated, his citrus trees were overgrown while he planted his laurel and lovingly tended to his songs: "Laurel, lest my Ariadne die and perish, have her name forever on your sacred trunk". I saw and many others have seen the songs that great old man wrote on this laurel with his sickle. The laurel tree grows, the song grows, and so does the love to which his dead heart tends.

In this passage, Pontano the great conversationalist has taken leave of his gardens, only to enjoy 'discourses with the divine'. In his absence, the gardens have returned to wilderness, leaving only an inscription on a laurel tree – the epigram carved on a tree being a commonplace of classical pastoral and amatory elegy.[63] In these lines, Anisio transforms the ruins of Antiniana into a monument to the Aragonese humanist legacy: a tree growing in a savage garden, haunted by the memory of Pontano.

The Accademia Pontaniana was the cradle of many humanistic traditions that would continue to shape European academic culture over the following century. Under both Beccadelli and Pontano, the academy revived the classical art of the literary-philosophical banquet, elevating conviviality – the fundamental trait of all academic societies – to the rank of a virtue. Beccadelli's Plinian villa at Ressina and Pontano's various gardens served as venues for *convivia* – places to socialise, exchange ideas, and play music. The horticultural programme of Pontano's Villa Antiniana, which contained one of the first orangeries in Italy, symbolised the academy's dream of the golden age restored. Through its association with the golden age, the custom of the garden

---

62 Anisio G., "De Antiniano colle", in *Varia poemata et satyrae* (Naples, Johannes Sultzbach: 1531) 37.

63 See Theocritus, *Idylls*, 2.47–48; Virgil, *Eclogues* 10; Ovid, *Heroides* 5. See also Rensselaer W.L., *Names on Trees: Ariosto into Art* (Princeton: 1977).

convivium gave rise to the legend of Arcadia, which later came to permeate the pastimes, costume, musical and poetic productions of the Renaissance academy. The famous garden banquets of the Accademia Pontaniana thereby established a paradigm and a mythology that was to enjoy a rich afterlife among the European academies of the sixteenth century.

References

d'Alessandri, Alessandro, *Alexandri ab Alexandro iurisperiti Neapolitani Genialium dierum libri sex, varia ac recondita eruditione referti* (Cologne, Iaspes Gennepaeus: 1551).

Anisio G., "De Antiniano colle", in *Varia poemata et satyrae* (Naples, Johannes Sultzbach: 1531) 37.

Belli V., "La Torre del Pontano e la sua villa a Ischia," *La Rassegna d'Ischia* 2 (2017) 42–49.

Iacono A., "Gli umanisti e le acque di Ischia", in Valenti R. (ed.), *Memorie dell'acqua e della terra* (Naples: 2011) 63–77.

Bosisio M., *Il teatro di corte pre-classicista: Geografia e storia* (Milan: 2014).

Casanova-Robin H., "La déraison du sage, dans l'*Asinus* de Pontano", in Casanova-Robin H. – Furlan F. – Wulfram H. (eds.), *Serio ludere: D'Alberti à Pontano et d'Éras me à Thomas More* (Paris: 2020) 151–171.

Casanova-Robin H., "L'adresse de Giovanni Pontano à Girolamo Carbone dans l'élégie I, 40 de l'*Eridanus*. Un idéal d'*humanitas*?", in Julhe J-C. (ed.), *Pratiques latines de la Dédicace – Permanence et mutations, de l'Antiquité à la Renaissance* (Paris: 2014) 445–463.

Coleman J., *A Sudden Frenzy: Improvisation, Orality, and Power in Renaissance Italy* (Toronto: 2022).

Fumaroli M., "*Academia, Arcadia, Parnassus*: troix lieux allégoriques de l'éloge du loisir lettré", in Chambers D.S. – Quiviger F., *Italian Academies of the Sixteenth Century* (London: 1995) 15–36.

Furstenberg-Levi S., "The D'Avalos-Colonna Literary Circle: A 'Renewed Parnassus'", in Cox V. – McHugh S. (eds.), *Vittoria Colonna: Poetry, Religion, Art, Impact* (Amsterdam: 2022) 37–54.

Furstenberg-Levi S., *The Accademia Pontaniana: A Model of a Humanist Network* (Leiden: 2016).

Hankins J., "Vocal Music at Literary Banquets in the Renaissance," *Basler Jahrbuch für Historische Musikpraxis* 41 (2021) 227–243.

Hankins J., "Humanist Academies and the 'Platonic Academy of Florence'", in Jensen R. – Pade M. (eds.), *On Renaissance Academies: Proceedings of the*

Conference 'From the Roman Academy to the Danish Academy in Rome' (Copenhagen, Odense University Press: 2011) 31–46.
Haraszti E., "La technique des improvisateurs de langue vulgaire et de latin au quattrocento", *Revue belge de Musicologie* (1955) 12–31.
Haywood E., "'Iter asinarium': per una nuova interpretazione dell'*Asinus* pontaniano", in De Nichilo M. – Distaso G. – Iurilli A. (eds.), *Confini dell'umanesimo letterario, studi in onore di Francesco Tateo* (Rome: 2003) 733–746.
Iacono A., *La guerra d'Ischia nel* De Bello Neapolitano *di Giovanni Pontano* (Naples: 1996).
Jenkyns R., "Virgil and Arcadia", *The Journal of Roman Studies* 79 (1989) 26–39.
Lee R.W., *Names on Trees: Ariosto into Art* (Princeton: 1977).
Marsh D., *The Quattrocento Dialogue: Classical Tradition and Humanist Innovation* (Cambridge: 1980).
Minieri-Riccio C., *Cenno storico della Accademia alfonsina: istituita nella città di Napoli nel 1442* (Naples: 1875).
Modesti P., *Le delizie ritrovate: Poggioreale e la villa del rinascimento nella Napoli aragonese* (Florence: 2014).
Monti-Sabia L. – Caracciolo T. – Fido C., *Un profilo moderno e due vitae antiche di Giovanni Pontano* (Naples: 1998).
Nicolini F., "Cenni storici", in *Annuario dell'Accademia Pontaniana* (Naples: 1957).
O'Sullivan T.M., *Walking in Roman Culture* (Cambridge: 2011).
Paulus H.E., "The Humanistic Tradition of Citrus Culture in Central Europe from the 15th to the 18th Century", in Fiorentino G. – Zech-Matterne V. (eds.), *AGRUMED: Archaeology and History of Citrus Fruit in the Mediterranean: Acclimatization, Diversifications, Uses* (Naples: 2018).
Pajorin K., *Italia e Ungheria all'epoca dell'umanesimo corviniano* (Florence: 1994) 179–228.
Pontano G., "De splendore," in Tateo F. (ed.), *I trattati delle virtù sociali: De liberalitate, De beneficentia, De magnificentia, De splendore, De conviventia* (Rome: 1965) 125–138.
Pontano G., "De conviventia," in Tateo F. (ed.), *I trattati delle virtù sociali: De liberalitate, De beneficentia, De magnificentia, De splendore, De conviventia* (Rome: 1965) 141–155.
Pèrcopo E., "La villa del Pontano ad Antignano", *Atti dell'Accademia Pontaniana* 56 (1926) 221–239.
Ribouillault D., "Hermeneutics and the Early Modern Garden: Ingenuity, Sociability, Education", in Enenkel A.E. – Melion W. (eds.), *Landscape and the Visual Hermeneutics of Place, 1500–1700* (Leiden – Boston: 2020) 291–325.
Soranzo M., *Poetry and Identity in Quattrocento Naples* (Farnham – Burlington: 2014).

Tateo F., *Umanesimo etico di Giovanni Pontano* (Lecce: 1972).

Tilly G., *Un manifeste posthume de l'humanisme aragonais: le* De hortis Hesperidum *de Giovanni Pontano* (Ph.D. dissertation, Normandie Université, Università degli studi di Napoli Federico II: 2020).

Wilson B., *Singing to the Lyre in Renaissance Italy* (New York: 2019).

CHAPTER 3

Pomponio Leto and His Garden: New Evidence on Its Archaeological Context

Alessia Dessì

Abstract

In the second half of the fifteenth century, the Academy founded in Rome by Pomponio Leto longed to rediscover the glories of the past with the help of tangible Roman ruins. The location of the Academy and Pomponio's house in the area called Montecavallo, near the ruins of Constantine's Baths and the statues of the Dioscuri, was crucial. Pomponio's passion for antiquity was also evident in his private garden, where he displayed his collection of epigraphs and ancient marbles, creating a link with history that he eagerly shared with his students. This chapter shows how, in organizing his collection and designing his garden, Leto sought to realize an ethical and philosophical project aimed at reviving ancient passions and virtues in the present.

Keywords

Pomponio Leto – humanism – Quirinal – antiquarianism – classicism

In an engraving made in 1575, Étienne Dupérac depicted the remains of the Baths of Constantine on the Quirinal Hill at the time of his visit [Fig. 3.1]. In this image, we observe the ruined arches of the imposing Roman structure and their natural incorporation into the contemporary Renaissance dwellings and gardens.[1] It fixes a glimpse of the Quirinal area known as Montecavallo

1 The caption of the engraving reads 'Vestigij delle Terme di Constantino nel monte quirinale dalla parte che guarda verso Libecchio [= southwest] qualli per esser molto ruinati non vi si vede adornamenti ma solo grandissime muraglie et stantie. masimamente nel giardino del Ill.mo Car.le de Vercello et da poi che io designai questa parte vi si sono fabricate case et granarij di modo che al di doggi non si puol piu vedere per esser occupata di dette fabriche' ('Vestiges of the Baths of Constantine on the Quirinal Hill on the side facing southwest which are very ruined and there is no decoration on them but only great walls. It is in the garden of cardinal de Vercelli and since I drew this part, houses and granaries have been built here, so that today it is no longer visible, because it is occupied by these buildings').

 | DOI:10.1163/9789004517547_004

FIGURE 3.1 Etienne Dupérac, *Ruins of Constantine's baths*, engraving, 37.5 × 21.5 cm. From *I Vestigi dell'antichità di Roma raccolti e ritratti in perspettiva da Stefano Du Perac parisino* (Rome, Lorenzo della Vacheria: 1575)
IMAGE © BNF / GALLICA

prior to the total alteration of its urban and architectural layout through the construction of grandiose palaces in the early seventeenth century, many of which remain to this day.[2] Thanks to Dupérac's engraving, it becomes possible to understand how Renaissance and classical environments were seamlessly blended in the fifteenth and sixteenth century. This mélange that Romans interacted with daily had a significant impact on popular interests and general culture. The stately and nostalgic aspects of the Quirinal meant that in this part of Rome more than in other areas of the Urbe, the vitality of classical Roman civilization in the modern world remained present. The interpenetration of these two cultures – the traditionally classical and the novel Renaissance classicism, understood as a conscious, retrospective gaze on the legacy of the former – was essential to their promotion and continuation.[3]

Antiquarianism became a vital piece in the puzzle of how to approach knowledge in the Renaissance. In the late fifteenth century, antiquarian interests were not exclusively directed toward objects. The study of classical texts was unquestionably the primary point of access to classical knowledge, but

2 The name Montecavallo is derived from the presence of the two statues of the Dioscuri.
3 Settis S., *Futuro del "classico"* (Turin: 2004) 18.

this was supplemented by complementary avenues of inquiry such as the study of ancient epigraphs and topography and the revival of a cultural lifestyle that emulated the past in an attempt to draw closer to it.[4] The origins of this antiquarian movement can be traced to fourteenth-century proto humanists, among whom Giovanni Colonna (1298–1343) and Zanobi da Strada (1312–1361) were the first collectors of manuscripts and epigraphs. The subsequent period of Italian humanism was led by Cola di Rienzo (1313–1354) and the 'founding father of Renaissance antiquarianism' Francesco Petrarca (1304–1374), who developed interests in the study of ancient artifacts parallel to the study of the ancient manuscripts that he had discovered. Humanists' regard for classical architecture helped foster a literary revival, as well as a new approach to figurative art; such intermingling interpretations of history and cultural heritage formed the basis for the re-emergence of antiquity. Driven by desires to create and forged through repeated attempts to reproduce classical antiquity, a new modern classicism was born.

The importance of physical places and the experience of these sites in humanist activities favoured the establishment of cultural models that were shared by educated people. Informal groups, better known as *sodalitas*, emerged in the fifteenth century because of these ongoing cultural processes. Their community offered an alternative to the institutional setting of the university. By considering the *sodalitas*, it is possible for us to trace a desire for self-individuation and a novel mode of knowledge transmission in early modern intellectual circles. Their significance is further underscored by their ideological-cultural principles, which drove 'intellectuals imbued with the myth of the paradigmatic nature of classical experience to try to conform new cultural institutions to the models offered by ancient tradition'; these principles merit further review.[5] Fifteenth-century *sodalitas* looked to the tradition of the Platonic academies as outlined by Cicero, who spoke about them extensively in his writings.[6] To this end, the significance of their meeting places and the philosophical ideals embodied by the academies should not be undervalued. In Renaissance *sodalitas*, the garden once again emerged as a place of research, one where the love of wisdom and knowledge could be taught and shared.

---

4 Acciarino D., "The Nature of Renaissance Antiquarianism: History, Methodology, Definition", *Acta Antiquia Academiae Scientiarum Hungaricae* 57 (2017) 4.

5 De Caprio V., "I cenacoli umanistici", in Asor R.A. (ed.), *La letteratura italiana, I, Il letterato e le istituzioni* (Turin: 1982) 802.

6 This is of note in the last chapter of *De finibus*, in which Aristotle's visit to Athens and the ancient Academy and Lyceum becomes an occasion to evoke the memory of the great figures of Hellenic culture. See Denis Ribouillault's introductory essay.

I have chosen Dupérac's image as the point of departure for my analysis because, since the mid-fifteenth century, the Quirinal has functioned as a privileged site for the re-establishment of ancient traditions of learning, largely due to its appearance. The vestiges of Roman customs at the site, combined with its bucolic features as a *locus amoenus*, made it a favourite seat of cardinals, particularly those with collecting and humanistic ambitions, as well as humanists.[7] The site seemed to foreground the fundamental relationship between man and nature, as well as that between man and history as mediated by ancient monuments, which together fostered a bond between the individuals who frequented this place. In this regard, Rodolfo Lanciani's description of the early modern Quirinal in his seminal work on Roman topography is striking:

> Chi avesse dunque guardato i ruderi delle Terme, tenendo le spalle rivolte al Tempio del Sole, pressappoco dal sito dell'attuale cancello di Villa Colonna, avrebbe notato da sinistra a destra primieramente la Chiesa del Salvatore e i suoi annessi, quindi la casa di Pomponio Leto e della sua Accademia: in terzo luogo quella di Bartolomeo Platina, da ultimo quella di Andreozza di Giovan di Navarra.[8]
>
> Anyone looking at the ruins of the Baths with their back to the Temple of the Sun, more or less from where the gate of Villa Colonna stood, would have seen, from left to right, first the Church of the Redeemer and its

---

7 The Neapolitan Oliviero Carafa, a patron of arts and letters who became a cardinal in 1467 and settled in the Quirinale, was one of the most important residents of the area. His house was located where the Quirinale Palace now stands, on the corner of today's Via della Dataria and Via XX Settembre. Samperi R., "La città delle vigne, dei giardini e delle ville (fine XV–XVI secolo)", in Simoncini G. (ed.), *Roma. Le trasformazioni urbane nel Cinquecento*, vol. 2 (Florence: 2011) 111. Another resident, Cardinal Francesco Gonzaga, was protector of artists and humanists such as Bartolomeo Sacchi (known as Platina) and owned a house near the Torre delle Milizie and church of Sant'Agata dei Goti as early as 1463. His garden, which extended as far as the ruins of the Baths of Constantine, was decorated with mythological scenes and a labyrinth. On the influence of cardinals' residences on the demography of a given area, see: Aurigemma M.G., "Case di fiorentini a Roma nell'ultimo decennio del '400'", in Chiabò M. – Maddalo S. – Miglio M. – Oliva A.M. (eds.), *Roma di fronte all'Europa al tempo di Alessandro VI*. Proceedings of the conference (Città del Vaticano – Roma, 1–4 dicembre 1999), Pubblicazioni degli archivi di Stato. Saggi 68 (Rome: 2021) 495–520; Aurigemma M.G., "Residenze cardinalizie tra inizio e fine del '400'", in Simoncini G. (ed.), *Roma. Le trasformazioni urbane nel Quattrocento*, vol. 2 (Florence: 2004), 117–136; Frommel C.L., *Der Römische Palastbau der Hochrenaissance* (Tübingen: 1973); Frommel C.L., *Architettura e committenza da Alberti a Bramante* (Firenze: 2006).

8 Lanciani R., *Il Panorama di Roma* (Rome: 1895) 24.

> annexes, then the house of Pomponius Leto and his Academy, then that of Bartolomeo Platina, and finally that of Andreozza di Giovan di Navarra.

Two names stand out in Lanciani's account as key contributors to the humanist history of Rome: Bartolomeo Sacchi, known as Platina, and, above all, Pomponio Leto. As two of the most important humanists in the second half of the fifteenth century, both participated in the *sodalitas* founded by Leto.

From the middle of the fifteenth century until his death in 1498, Leto was a driving force behind the classical revival. A chief exponent of Roman antiquarianism, he founded the first Roman humanist academy and occupied the role of professor at the *Studium Urbis*.[9] Pomponio gathered friends, colleagues and students who shared his love of antiquity and, through the study of the classics and the discovery of *antiquitates*, championed a living recovery of the classical past. Colleagues and pupils who were among the members or *sodales* that congregated to enjoy this learned setting included Bartolomeo Platina, the Vatican librarian under Pope Sixtus IV, Marcantonio Sabellico, his biographer, Alessandro Farnese, the future Pope Paul III, Agostino Maffei, Andrea Fulvio, the author of *Antiquaria Urbis* (1513) and many others. Leto's *sodalitas* featured an open-air meeting place that allowed participants to enjoy the ruins of the

9 Della Torre A., *Paolo Marsi da Pescina. Contributo alla storia dell'Accademia Pomponiana* (Rocca San Casciano: 1903); Zabughin V., *Giulio Pomponio Leto*, vols. 1–2 (Rome: 1909–1912); Zabughin V., *Giulio Pomponio Leto. Saggio critico*, vol. 2 (Grottaferrata: 1910–1912); Medioli Masotti P., "L'Accademia Romana e la congiura del 1468", *Italia medioevale e umanistica* 25 (1982) 89–204; Medioli Masotti P., "L'Accademia Romana e la congiura del 1468", in *Callimaco Esperiente poeta e politico del '400* (Florence: 1987) 169–179; Medioli Masotti P., "Codici scritti dagli Accademici romani nel carcere di Castel S. Angelo (1468–1469)", in Avesani R. – Ferrari M. – Foffano T. – Frasso G. – Sottili A. (eds), *Vestigia. Studi in onore di Giuseppe Billanovich* (Rome: 1984) 451–459; Miglio M., *Scritture, scrittori e storia, II, Città e corte a Roma nel Quattrocento* (Manziana: 1993) 15, 174, 195–196, 200; Viti P., "L'Umanesimo nell'Italia settentrionale e mediana", in *Storia della Letteratura italiana, III, Il Quattrocento* (Rome: 1996) 615–620; Vasoli C., "Le 'Accademie'. L''Accademia Platonica', l''Accademia Romana'", in *Storia della letteratura italiana, III, Il Quattrocento*, 78–82; Cassiani C. – Chiabò M. (eds), *Pomponio Leto e la prima accademia romana. Giornata di studio Roma 2 dicembre 2005* (Rome: 2007); Bianca C., "Pomponio Leto e l'invenzione dell'Accademia Romana", in *Les Académies dans l'Europe humaniste* 25–56; Hankins J., "Humanist Academies and the 'Platonic Academy of Florence'", in Pade M. (ed), *On Renaissance Academies. Proceedings of the International Conference "From the Roman Academy to the Danish Academy in Rome"* (The Danish Academy in Rome, 11–13 october 2006) (Rome: 2011) 31–35; Bianca C., "Le accademie a Roma nel Quattrocento", in Jensen – Pade (ed.), *On Renaissance Academies* 47–59; Stok F., "Perotti e l'Accademia Romana", in Jensen – Pade (ed.), *On Renaissance Academies* 77–90; Modigliani A. – Osmond P. – Pade M. – Ramminger J. (eds), *Pomponio Leto tra identità locale e cultura internazionale*, Proceedings of the international conference (Teggiano, 3–5 October 2008) (Rome: 2011) 1–16.

Baths of Constantine as well as his collection of epigraphs for both pleasure and study. Although the fifteenth-century Roman *sodalitas* garden had not yet taken on the formal, defined properties that would characterize the gardens in the following eras, it certainly presented itself as a place to share common ideals.

1 Pomponio Leto, Founder of the Roman *sodalitas*

Leto was the illegitimate son of Giovanni Sanseverino and brother of Roberto Sanseverino, Prince of Salerno. He arrived in Rome in the 1450s when the city was undergoing a substantial period of cultural and social development, fashioning itself in the combined image of ancient Rome and curial splendour. In this climate of cultural upheaval, Leto immediately found his place in the city, first becoming a disciple of Pietro Odo da Montopoli and then of Lorenzo Valla, although he owed the greatest intellectual debt to the former. Around the mid-1460s, Leto began teaching at the *Studium Urbis Romae*, moving to Venice in 1466. These years are best known for the series of grave events involving him and his pupils, notably an accusation by the Venetian Council of Ten of sodomy against two young men to whom he gave private lessons.[10] The trial was resolved in February 1468 but only because Leto was called back to Rome to appear before another tribunal. Pope Paul II Barbo, learning of a conspiracy for his murder apparently hatched by the Roman humanist *sodales*, ordered the imprisonment of the Pomponians in Castel Sant'Angelo. The imprisoned were to answer not only to the primary accusation of conspiring to commit murder, but also for their general behaviour, which was considered a manifestation of impiety. After two trials, Pomponio and his pupils were jailed for about a year before they were freed and cleared of their charges in the spring of 1469. An aura of uncertainty is still attached to the history of the conspiracy, mainly due to the loss of trial documents that could clarify the role of figures such as Cardinal Marco Barbo, a counsellor and trusted collaborator of the

10 Pomponio incurred high danger in Venice because of the alleged *librum inhonestum, sua manu scriptum, et per suam loquelam variatam*. In fact, concern for the defense of morality – a morality that was both stubborn and unflappable – had led to the imposition of capital punishment for sodomy under the law. Zippel G. (ed.), *Le vite di Paolo II di Gaspare da Verona e Michele Canensi* (Castello: 1904–1911) 184–185; Carini I., "La 'Difesa' di Pomponio Leto", in *Nozze Cian – Sappa – Flandinet (23 ottobre 1893)* (Bergamo: 1894) 186–187; Miglio M., "Homo totus simplex. Mitografie di un personaggio", in Modigliani – Osmond – Pade – Ramminger (eds.), *Pomponio Leto tra identità locale e cultura internazionale* 4–5.

pontiff who remained close to the Roman humanists, who sent him intercessory letters from prison.[11]

In the years that followed, Pomponio lived in Rome; his personal and professional life revolved around the Urbe. Having regained his professorship at the Studium Urbis, Pomponio enthusiastically dedicated his life to teaching about ancient authors and their works through reading, commentary and exegesis. His passion for this craft shone to such an extent that many young people who attended his lectures joined him at unorthodox hours to study. Almost like passing a test of virtue, faithful students scrupulously fixed his words in their notes (the so-called *dictata*), which they passed from hand to hand; today, these notes are crucial sources for the study of Leto's teachings.[12]

---

11 Bartolomeo Platina was the most active member in writing letters, addressing one to Barbo to invoke his protection: 'per illam amicitiam, quae est tibi cum optimo viro Omnibono Leoniceno, praeceptore meo, qui etsi absens est, hoc idem ut facias, te cum multis lacrymis et precibus orat' ('for that friendship you have with Onnibono Leoniceno, my mentor, who, though absent, begs you with many tears and prayers to do likewise'). See Della Torre, *Paolo Marsi da Pescina* 117 n. 1. Platina also sent other letters during Holy Week (11–18 April 1469) to various people in hopes that they would intercede for his release. Zabughin recalls two letters to Paul II, one to Cardinal Bessarion, one to Cardinal Borgia, the future Alexander VI, one to Francesco Gonzaga, his protector, one to Ammannati, one to Piero Morosini, a Venetian orator who came to Rome after his arrest, and, finally, one to Valescar, one of the diners recalled in the culinary splendours of the Academy and a relative of the Cardinal Vice-Chancellor. Zabughin, *Giulio Pomponio Leto* I, 127–128.

12 Michele Ferno, "Iulii Pomponii Laeti Elogium historicum", in Mansi G.D. – Fabricius J.A. (eds.), *Bibliotheca Latina mediae et infimae aetatis* 6 (Florence: 1859) 631: 'Legebat in Gallicinio sempre; non imbres, non desaeviens frigus, non immani sulla tempestas hominem coercebat, quominus praecipitem se ex illo Quirinali iugo ad Gymnasium daret; et ruebant agminati illa vel hora ad illum auditores, quantum schola ipsa vix poterat capi. Interdum pro ostio sub die longissimam velut caudam, quae testudine recipi nequiret, ducerent [...] Quamquam delectabatur magis docere quam scribere, ut essent Pomponiani plurimi discipuli, in quorum excellentia cerneret nomen suum inclarere, sicut Socrates, et factitavit Christus: qui cum ipsi (sed longe aliter hic, et aliter ille) doctissimi consummatissimi essent, in discipulis et ministrorum splendore voluerunt nomen et famam suam; ingenii et dvinitatis maiestatem elucere' ('He always read at dawn; not the rain, not the cold, not the heavy storm prevented him from rushing down from the top of the Quirinal to the gymnasium; even at that hour the crowds came to see him, so many that the gymnasium itself could hardly accommodate them, so that sometimes they formed a kind of very long queue outside the door, in the open air, which could not be accommodated by the vaulted construction of the portico [...] As for the rest, he was enjoying teaching more than writing, to the point that the Pomponiani were his most numerous disciples; he understood that as a result of their excellence, his name would be illuminated as well, as Socrates and Christ did. They too, although themselves (the case of the latter is really different, and the case of the former is distinct) were the most learned and the most accomplished, wanted their name, fame, and the majesty of their genius and divinity to shine out for their disciples and for the splendor of their ministers').

It is no coincidence that Leto studied Varro and his works the most – an author praised by Cicero as the light and guide of Romanity. Leto's pupil Pietro Marso described him in similar terms in a funeral oration honouring his master.[13] This comparison between Leto and Varro, previously highlighted in Maria Accame's studies, offers a tangible idea of the humanist's wide-ranging interests, which not only touched on the history of language and grammar, literary antiquities, geography, and the customs of distant peoples, but also concerned antiquarianism, topography, archaeology, agriculture and animals. Leto pursued some of these interests more privately, as revealed on the walls of his home. This site emerged as the hearth and seat of his *sodalitas*; it was a place that gathered those who approached antiquity as a model that could be revived.

After the 1468 accusations, Pomponio and his followers abandoned the *fratres academici* appellation that they had adopted in the early years of their assembly for the more congruous and politically correct *sodales*.[14] This choice of nomenclature carried a strong symbolic value in more than one aspect, and it was certainly influenced by a desire to align themselves with a more humanistic tenor. Pope Sixtus IV Della Rovere shared a different attitude toward the Pomponian *sodalitas* than his predecessor. By transforming the group into a lay confraternity named after patron saints and overseen by a cardinal, Pope Sixtus IV carried out what Della Torre calls an 'action of Christianisation' of the *sodalitas*. The Pomponians welcomed this decision, as they would have benefited from this official recognition to dispel any doubts or misunderstandings about their study and veneration of antiquity. Despite this shift in titular affiliation, an underlying duality remained in Pomponio's *sodalitas*, evident in the saints chosen as its Catholic protectors: St. Victor, St. Fortunatus and St. Genesius, all of whom were associated with 20 April. According to the calculations of the members, this was the day of Rome's Christmas and had become a cultic holiday for the group.[15] The revised fashioning of the Roman *sodalitas* was also reflected in Leto's change of conduct. As Della Torre argues, Leto learned that by presenting himself and the humanists close to him more

13 On Varro's study, see especially Zabughin, *Giulio Pomponio Leto* II, 112–129; Accame Lanzillotta M., "Il commento varroniano di Pomponio Leto", *Miscela Greca e Romana* 15 (1990) 309–345; Accame Lanzillotta M., "L'insegnamento di Pomponio Leto nello Studium Urbis", in Capo L. – Di Simone M.R. (eds.), *Storia della Facoltà di Lettere e Filosofia* (Rome: 2000) 71–91; Accame M., *Pomponio Leto vita e insegnamento* (Roma: 2006) 138–146.

14 Around 1468, during his months of imprisonment, Paolo Marsi dedicated a few verses 'ad frateres academicos Romae captivos' ('To imprisoned Roman academic friends').

15 The names of these saints are symbols of the city's origins, victories and good fortune. On this topic see Giorgio Mangani's chapter in this book.

formally, in a manner keeping with Christian virtues, he could protect his personal convictions and the passionate fervour for classical antiquity he shared with many of his companions.[16] This should not, however, be misunderstood as false religiosity nor as a comedic ruse staged by Leto and his collaborators with a view to pursuing paganistic aims secretly. Rather, as his students recalled, this rebranding was a compromise aligned with Leto's faith, although their testaments to his devotion may have been an attempt to rehabilitate the image of their master some thirty years after the accusations that initially marked his career.

The semantic shift from *fratres academici* to *sodales* also carried a symbolic meaning that referenced changes in the definition of academies. Associated in Rome with the *Studium*, from the end of the fifteenth century the concept of an academy as a public institution was increasingly recognized.[17] By adopting the term *sodalitas*, Leto's pupils distanced themselves from the official culture that perhaps did not fully satisfy their interests, despite the participation of some members in academies. Leto and, consequently, his pupils sought a non-standardized space to focus on the humanistic concept of the individual who could express himself in any manner and who was not bound to a singular area of studies.

Leto did not only occupy a metaphorical place; his principles were reflected in a physical place – his home – which, as I have already highlighted, was the venue for the *sodalitas* meetings. One of Leto's students, Pietro Marso, employed Cicero's words to speak of the house, defining it as a *ludus* (school) and an *officina dicendi* (workshop of eloquence) from which more masters emerged than leaders from the Trojan Horse.[18] Leto's home was located on the slopes of the Quirinal Hill on what is now via Ventiquattro Maggio, amid the ruins of the Baths of Constantine and other ancient sculptural relics. He moved from a house in the Botteghe Oscure to the house on the Quirinal in 1479, as documented by the notary Camillo Beninbene.[19] He purchased the

16 Della Torre, *Paolo Marsi da Pescina* 241.

17 Bianca, *Pomponio Leto e l'invenzione dell'Accademia romana* 27–56; Bianca, *Le accademie a Roma nel Quattrocento* 47–59.

18 Accame, *Pomponio Leto vita e insegnamento* 78.

19 Dessì A., "Pomponio Leto e la topografia del Quirinale nei protocolli notarili", *Rivista dell'Istituto di Storia dell'Europa Mediterranea* 9/I n.s. (December 2021), Special Issue: Colesanti G.T. – Piñol D. – Sakellariou E. (eds.), *Il Notaio nella società dell'Europa mediterranea (secc. XIV–XIX), The Notary in the Mediterranean European Society (14th–19th centuries)* 287–310. Della Torre believed that Pomponio's first house was near the Tiber. Della Torre, *Paolo Marsi da Pescina* 248. However, Magister argues that it was at the Botteghe Oscure, according to Fra Giocondo's epigraphic sylloge found in Biblioteca Apostolica Vaticana

house from Margherita di Novate who had sold him another bordering estate a few years earlier. During the clashes between the Orsini and Colonna families, the homes in the Montecavallo area, including Leto's, were damaged and looted. With the help of his pupils, his home was fortunately soon restored.[20] In addition to the aforementioned 1479 document, several sources point to the location. An anonymous Spaniard, for example, reported in 1568: 'Templum S. Salvatoris Cornelii, ita vocatum [...] erat haec domus academica Pomponii Laeti: apparent ad huc vestigia ruinarum antiquarum' ('The so-called temple of San Salvatore de Corneliis, [...], this was the academic house of Pomponio Leto: here are visible traces of ancient ruins'), thus confirming that the house was located opposite to the church of San Silvestro al Quirinale.[21] Two scholars tasked by Federico Borromeo in the early seventeenth century to investigate the Pomponian Academy confirmed the same location. One, Brother Giacomo Grimaldi, recalls that:

> la casa di Pomponio Leto dove egli habitava era nelle terme costantiniane nel Quirinale appresso la chiesa di S. Girolamo [cioè la chiesa di San Salvatore cornutorum] che l'anno 1615 insieme con dette terme fu spianata à terra.

(BAV), Borg. lat. 336, f. 38v, which reads 'Romae in domo D. Bartholomei del Cambio apud apothecas obscuras ubi Pomponio habitabat' ('In Rome, in the house of D. Bartolomeo del Cambio, near the *Botteghe Oscure* where Pomponius lived'). Magister S., "Pomponio Leto collezionista di antichità", in Miglio M. (ed.), *Antiquaria a Roma. Intorno a Pomponio Leto e Paolo II* (Roma: 2003) 194 n. 22. Archivio di Stato di Roma (ASR), Collegio Notai Capitolini, not. Camillus Benimbene, vol. 175, f. 130r and 140rv from 17 April 1479. A copy of this document is in BAV, Borg. lat. 1572, ff. 32–33v. The document was already mentioned in *Repertori di famiglie* in BAV, Ottob. Lat. 2550, part V; Corvisieri Costantino, "Il trionfo romano di Eleonora d'Aragona", *Archivio della R. Società Romana di Storia Patria* 10 (1887), 635–636; Lanciani R., *Storia degli scavi di Roma e notizie intorno le collezioni romane di antichità (c. 1000–1530)*, vol. 1 (Rome: 1902) 115; Zabughin, *Giulio Pomponio Leto* 194.

20 Infessura S., *Diario della città di Roma di Stefano Infessura scribasenato*, Tommasini O. (ed.), Istituto Storico Italiano. Fonti per la storia d'Italia, vol. 5 (1890) 118–122. Sabellico also shares this information. See also Carini, "La 'difesa' di Pomponio Leto" 14; Della Torre, *Paolo Marsi da Pescina* 250, 263.

21 See also De Rossi G.B., "L'Accademia di Pomponio Leto e le sue memorie scritte sulle pareti delle catacombe romane", *Bullettino di Archeologia Cristiana* 1 (1890) 87 n. 1. For the identification of the "Anonimo Spagnolo" with Ciaccioni, see Buonocore M. "Iter epigraphicum Vaticanum. Una guida ai principali testimoni della tradizione manoscritta dell'epigrafia cristiana nei codici della Biblioteca Apostolica Vaticana", in Di Stefano Manzella I. (ed.), *Le iscrizioni dei cristiani in Vaticano. Materiali e contributi per una mostra epigrafica* (Vatican City: 1997) 57.

> Pomponius Leto's house where he lived was in the Baths of Constantine in the Quirinal near the church of St. Jerome [that is the church of San Salvatore cornutorum], which was destroyed in 1615 together with the baths.[22]

A letter by Leto addressed to Angelo Poliziano inviting him to return the Lucretius codex he had previously borrowed is another source pointing to the location of the house on the Quirinal. In the letter, we read 'effice ut ad suum Quirinum redeant', i.e., 'see to it that they return to their Quirinal', thus confirming the Quirinal as his area of residence at the time of writing.[23]

More detailed knowledge of the exact location of Leto's house can be derived from a notarial document drawn up by Christophorus Antonii Paulii in 1491.[24] This document specifies a donation *inter vivos* of a house, an *aruntineto* (reed), and a *vinea* (vineyard) outside Porta Viminale. This is perhaps the place where, according to Sabellico, Leto went to do some gardening, following the precepts of ancient authors such as Columella and Varro. The donation is to three brothers from the Ceci family: Mattia, Giuliano, and their younger brother Pomponio. It is plausible to assume that the house they received as a gift corresponds to the one expanded through the 1479 purchase, and that was large enough to accommodate the meetings of the *sodalitas*. As stated in the 1491 document, the residence comprised two storeys, a hall and several rooms.

Etienne Dupérac's 1577 plan and Antonio Tempesta's 1593 plan of the Quirinal, both made before the demolitions of the seventeenth century, show two-storey buildings in the area where Leto's house still stood a century earlier.[25] More

22 The reports are in Biblioteca Ambrosiana di Milano, G 285 inf., folios 3 and 20: 'Mi ricordo anco haver visto, in Roma a Montecavallo, vicino al palagio Pontif. dove ha fabricato il suo palazzo il sig. card. Borghese, una piccola casa, dov'era scritto il nome di Pomponio Leto, et intesi dire da alcuni [...] che nella casa suddetta si faceva quest'Accademia' ('I also remember having seen, in Rome in Montecavallo, near the Papal Palace where Mr Cardinal Borghese has built his palace, a small house with the name Pomponio Leto written on it, and I heard from someone [...] that in the aforesaid house this Academy was being held').

23 Zabughin, *Giulio Pomponio Leto* II, 177, 361 n. 36.

24 ASR, Collegio Notai Capitolini, not. Christoforus Antonii Paulii, vol. 131, ff. 106r–107r. This document was drawn up in the presence of several monks on 29 August 1491 in the monastery of San Lorenzo in Panisperna. Dessì, "Pomponio Leto e la topografia del Quirinale nei protocolli notarili" 296.

25 Even Federico Borromeo spoke with regret about the seventeenth-century demolitions that distorted the Renaissance appearance of Monte Cavallo: 'pavum illus hospitium non sine moltorum dolore excidio datum est' ('This little shelter was destroyed. Not without causing resentment on the part of many'). Accame, *Pomponio Leto vita e insegnamento* 76.

details about the house can be gleaned thanks to pupils who recalled its characteristic elements: Michele Ferno spoke of a cage featuring birds of various species including peacocks, while Pietro Marso described a *sacellum* dedicated to the *Vergine Panaghia* which contained a 'sacratissimae virginis matris veneranda imago quam ipse graeco nomine quirinalem panagia appellabat' (a venerable image of the Holy Virgin Mother, whom he called by the Greek name Quirinal Panagia).[26] Marcantonio Sabellico even claims that Leto possessed some kind of *acquario* (fishpond).[27] The 1491 document also mentions an *ortum* that might correspond to the garden where Leto displayed his collection of antiquities. Indeed, Leto's house is important not only as the place where he lived and where he gathered with his *sodales*, but also as the place where he kept his large collection of epigraphs and ancient marbles – his lapidary museum.

2 Leto's Collection and Garden

Rodolfo Lanciani was the first to study Leto's collection in *Storia degli scavi di Roma* published in 1902, while Vladimir Zabughin, who dedicated the largest monograph to the humanist, is still considered essential reading on the topic today. More recently, Sara Magister has dealt with Leto's collection, revising previous scholarship with new information. A series of epigraphic syllogisms written toward the end of the fifteenth century is a key element of Leto's collection, particularly those by Fra Giocondo, Pietro Sabino, and a student of Leto's known as 'anonimo pomponiano' ('the anonymous Pompiano follower').[28] Other information about the content of the collection can be gathered from

26 Zabughin, *Giulio Pomponio Leto* I, 201. Della Torre, *Paolo Marsi da Pescina* 243. The *sacellum* is also mentioned in Pietro Sabino's sylloge as 'prope Panagiam Pomponii': BAV, Ottob. lat. 2015, f. 60v.

27 Zabughin, *Giulio Pomponio Leto* I, 200.

28 Fra Giocondo's sylloge was compiled between *c.*1478 and 1489 with a dedication to Lorenzo de' Medici and later additions up to 1492. Various copies exist, the main one being Ms. 270 in the Biblioteca Capitolare in Verona and Borg. lat. 336, which is an almost identical copy. Fra Giocondo makes a distinction between two groups in Pomponio's collection, one constituted before 1489 and another between 1489 and 1492. Magister S., "Pomponio Leto collezionista di antichità. Note sulla tradizione manoscritta di una raccolta epigrafica nella Roma del tardo Quattrocento", *Xenia Antiqua* 7 (1998) 170. Pietro Sabino compiled the sylloge between 1495 and 1498 with a dedication to Emperor Charles V. Several drafts of it exist and the main one is preserved in the Marciana library in Venice (Marc. lat. X 195, fols. 94–251). As a pupil of Leto's, Pietro Sabino also mentions the Quirinal Hill, highlighting the distinction Leto made between the Quirinal Hill and the Pincio Romossa. The sylloge

the biographies on Leto written after his death by his pupils such as Sabellico and Ferno, as well as from Federico Borromeo's seventeenth-century testimony.

Leto assembled his collection around 1479, the year when he enlarged his house on the Quirinal, but a notable portion of his initial collection was perhaps lost in the looting that followed the 1484 conflicts between Orsini and Colonna. His impressive collection was one of the largest in Rome at the time, comprising about eighty pieces, primarily ancient inscriptions.[29] The original collection contained seventeen inscriptions, while a second set acquired between 1480 and 1488 contained forty-one. A third group of nineteen inscriptions amassed between 1488 and 1498, the year of the master's death, may have served to replace some of the lost pieces. It is highly probable that Leto did not buy ancient epigraphs but rather discovered them, searching his land or the numerous sites where he ventured to study the topography of the city.

Most of the inscriptions he collected were of a pagan, sepulchral character; his collection featured a homogenous nucleus of epigraphs *proenia* from the Sallusti columbarium, which Leto presumably explored in the 1480s. The collection also included two important Christian inscriptions, one in Greek, while another focus comprised sacred pagan epigraphs: those of the Senate and the Roman People, the *tituli* of public priests, and the military. The breadth of Leto's collection demonstrates his interest in accumulating samples across epigraphic categories. Despite this variety, his choice of artifacts was neither random nor dictated by aesthetics. Rather, Leto's collection aligned with the philological, linguistic, historical and topographical interests of his *sodalitas*. In Leto's collecting, one can trace the true essence of the antiquarian: the systematic collection of artifacts from the past.[30]

In this regard, Leto's collecting practices differ from those of his contemporaries. Roman collecting was very much linked to the political sphere, as the collection of ancient Roman relics provided a means through which to ennoble one's family or to celebrate its lineage.[31] Exhibiting collections of antiquities offered an immediate way to convey political or self-celebratory messages,

is preserved in the Biblioteca Marciana in Venice, Marc. lat. X 195, fols. 1–93 and was written between 1484–1488. Ibid. 170–171.

29 Federico Borromeo informs us that small marbles were also part of the collection, perhaps fragments of architectural decorations from the monuments surrounding his house.

30 Momigliano A., "Ancient History and the Antiquarian", *Journal of the Warburg and Courtauld Institutes* 3.4 (1950) 285–315; Acciarino, "The Nature of Renaissance Antiquarianism: History, Methodology, Definition" 1–2.

31 Settis S., *Sopravvivenza dell'antichità*, Quaderni del Dipartimento di scienze dell'antichità, Università di Padova 5 (Padua: 2001) 32.

a trend that reached its peak under the pontificate of Sixtus IV della Rovere in the 1480s. Notably, this practice aligns with the period in which Leto also started acquiring epigraphs.[32] However, unlike his politically minded contemporaries, Leto was a conscious collector.[33] The pieces in his collection were objects of study and their primary purpose was to stimulate the erudition of the students who frequented his home daily. His collection was indeed public, but it did not perform a self-fashioning function. Rather, it served to display antiquity as a nucleus around which to gather and celebrate with those who shared common ideals.

From the above sources it is clear that some inscriptions were displayed in the interior rooms, but mainly in the garden courtyard of the house.[34] There, they were found leaning against the walls, scattered among the trees or simply placed on the ground.[35] The garden was thus the seat of the *sodalitas* – a repository of ancient marbles and a milieu in which to treasure the heritage of the ancients. In keeping with Cicero's conception of the academy, Leto's garden was famous for its olive trees, its collection of ancient marbles and its distinctly bucolic atmosphere; it was an ideal place to nourish the body and mind. The collection was obviously gathered with the intent to create an environment conducive to a life of contemplation and scholarship while the setting was conceived as an ancient Roman domus. This latter evocation was intentional and strengthened by the imposing remains of Constantine's baths, which were an almost integral part of the dwelling. Together with the inscriptions, the ruins offered a repertoire of ancient knowledge from which scholars could draw. Although Leto's garden had not yet adopted the formal and defined characteristics of later humanist gardens, it presented itself as a physical and living place where antiquity could not only be studied but also revived. The presence of epigraphs outside the house served as a further hyphen between the classical and the contemporary and directed the attention of those who lived in or frequented the house to antiquity. The most famous of these epigraphs carried the title of the Roman Academy, 'Pomponii Leati et Sodalitatis Escuilinalis' (Of Pomponio Leto and the Esquiline Society), emphasising how Leto's house

---

32 Cavallaro A. (ed.), *Collezioni di antichità a Roma tra '400 e '500* (Rome: 2007).

33 Magister, "Pomponio Leto collezionista di antichità" (2003) 53.

34 We find no information about where Pomponio had placed the pieces of his collection from the sylloges, apart from the indication that some of the pieces were placed in the 'cubicolo Mathie', which means 'in the room of Mattia'. See Ibid. 65. According to Michele Ferno, Mattia was Pomponio's favourite disciple and his heir. This information is confirmed by the donation of 1491.

35 Fra Giocondo, 'Romae in pariete cuisdam domunculae Pomponii sitae in colle Quirinali sub aede Sancti Silestri'. Ibid.

was increasingly becoming public property.[36] Congregating outside was a natural activity for the Romans who longed for a return to the distant past that surrounded them. Inserting themselves among the ruins of the period they wished to restore, they fashioned an idea of Rome in continuity with the present. In organising his collection and designing his garden, Leto sought to realise an ethical and philosophical project to revive ancient passions and virtues in the present.

References

Accame Lanzillotta M., "Il commento varroniano di Pomponio Leto", *Miscellanea Greca e Romana* 15 (1990) 309–345.

Accame Lanzillotta M., "L'insegnamento di Pomponio Leto nello Studium Urbis", in Capo L. – Di Simone M.R. (eds.), *Storia della Facoltà di Lettere e Filosofia* (Rome: 2000) 71–91.

Accame M., *Pomponio Leto vita e insegnamento* (Rome: 2006).

Acciarino D., "The Nature of Renaissance Antiquarianism: History, Methodology, Definition", *Acta Antiquia Academiae Scientiarum Hungaricae* 57 (2017) 485–502.

Aurigemma M.G., "Case di fiorentini a Roma nell'ultimo decennio del '400'", in Chiabò M. – Maddalo S. – Miglio M. – Oliva A.M. (eds.), *Roma di fronte all'Europa al tempo di Alessandro VI*. Proceedings of the conference (Città del Vaticano – Roma, 1–4 dicembre 1999), Pubblicazioni degli archivi di Stato. Saggi 68 (Rome: 2021) 495–520.

Bianca C., "Pomponio Leto e l'invenzione dell'Accademia Romana", in *Les académies dans l'Europe humaniste: idéaux* 25–56.

Bianca C., "Le accademie a Roma nel Quattrocento", in Jensen R. – Pade M. (eds.), *On Renaissance Academies*, Proceedings of the International Conference "From the Roman Academy to the Danish Academy in Rome" (The Danish Academy in Rome, 11–13 october 2006) (Rome: 2011) 47–59.

Buonocore M., "Iter epigraphicum Vaticanum. Una guida ai principali testimoni della tradizione manoscritta dell'epigrafia cristiana nei codici della Biblioteca Apostolica Vaticana", in Di Stefano Manzella I. (ed.), *Le iscrizioni dei Cristiani in Vaticano. Materiali e contributi per una mostra epigrafica* (Vatican City: 1997) 57–63.

Cassiani C. – Chiabò M. (eds.), *Pomponio Leto e la prima accademia romana. Giornata di studio Roma 2 dicembre 2005* (Rome: 2007).

Carini I., "La 'Difesa' di Pomponio Leto", in *Nozze Cian – Sappa – Flandinet (23 ottobre 1893)* (Bergamo: 1894) 186–187.

36 There are several versions of this inscription, with the cited one by Aldo Manuzio. The confusion between Esquiline and Quirinal is still evident.

Cavallaro A. (ed.), *Collezioni di antichità a Roma tra '400 e '500* (Rome: 2007).

Corvisieri Costantino, "Il trionfo romano di Eleonora d'Aragona", *Archivio della R. Società Romana di Storia Patria* 10 (1887) 629–687.

De Caprio V., "I cenacoli umanistici", in Asor R.A. (ed.), *La letteratura italiana, I, Il letterato e le istituzioni* (Turin: 1982) 799–822.

Della Torre A., *Paolo Marsi da Pescina. Contributo alla storia dell'Accademia Pomponiana* (Rocca San Casciano: 1903).

Dessì A., "Pomponio Leto e la topografia del Quirinale nei protocolli notarili", *Rivista dell'Istituto di Storia dell'Europa Mediterranea* 9/I n.s. (December 2021), Special Issue: Colesanti G.T. – Piñol D. – Sakellariou E. (eds.), *Il Notaio nella società dell'Europa mediterranea (secc. XIV–XIX). The Notary in the Mediterranean European Society (14th–19th centuries)* 278–310.

De Rossi G.B., "L'Accademia di Pomponio Leto e le sue memorie scritte sulle pareti delle catacombe romane", *Bullettino di Archeologia Cristiana* 1 (1890) 49–87.

Frommel C.L., *Der Römische Palastbau der Hochrenaissance* (Tübingen: 1973).

Frommel C.L., *Architettura e committenza da Alberti a Bramante* (Florence: 2006).

Hankins J., "Humanist Academies and the 'Platonic Academy of Florence'", in Jensen R. – Pade M. (eds.), *On Renaissance Academies.* Proceedings of the International Conference "From the Roman Academy to the Danish Academy in Rome" (The Danish Academy in Rome, 11–13 October 2006) (Rome: 2011) 31–35.

Infessura Stefano, *Diario della città di Roma di Stefano Infessura scribasenato*, Tommasini O. (ed), Istituto Storico Italiano. Fonti per la storia d'Italia 5 (1890) 118–122.

Lanciani R., *Il Panorama di Roma* (Rome: 1895).

Lanciani R., *Storia degli scavi di Roma e notizie intorno le collezioni romane di antichità (c. 1000–1530)*, vol. I (Rome: 1902).

Magister S., "Pomponio Leto collezionista di antichità. Note sulla tradizione manoscritta di una raccolta epigrafica nella Roma del tardo Quattrocento", *Xenia Antiqua* 7 (1998) 167–196.

Magister S., "Pomponio Leto collezionista di antichità", in Miglio M. (ed.), *Antiquaria a Roma. Intorno a Pomponio Leto e Paolo II* (Rome: 2003) 51–121.

Medioli Masotti P., "L'Accademia Romana e la congiura del 1468", *Italia medioevale e umanistica* 25 (1982) 89–204.

Medioli Masotti P., "L'Accademia Romana e la congiura del 1468", in *Callimaco Esperiente poeta e politico del '400* (Florence: 1987) 169–179.

Medioli Masotti P., "Codici scritti dagli Accademici romani nel carcere di Castel S. Angelo (1468–1469)", in Avesani R. – Ferrari M. – Foffano T. – Frasso G. – Sottili A.(eds), *Vestigia. Studi in onore di Giuseppe Billanovich* (Rome: 1984) 451–459.

Michele Ferno, *Iulii Pomponii Laeti Elogium historicum*, in Mansi G.D. – Fabricius J.A. (eds.), *Bibliotheca Latina mediae et infimae aetatis*, VI (Florence: 1859).

Miglio M., *Scritture, scrittori e storia II, Città e corte a Roma nel Quattrocento* (Manziana: 1993).

Miglio M., "Homo totus simplex. Mitografie di un personaggio", in Modigliani A. – Osmond P. – Pade M. – Ramminger J. (eds), *Pomponio Leto tra identità locale e cultura internazionale* Proceedings of the International Conference (Teggiano, 3–5 October 2008) (Rome: 2011) 1–16.

Modigliani A. – Osmond P. – Pade M. – Ramminger J. (eds), *Pomponio Leto tra identità locale e cultura internazionale.* Proceedings of the international; conference (Teggiano, 3–5 October 2008) (Rome: 2011) 1–16.

Momigliano A., "Ancient History and the Antiquarian", *Journal of the Warburg and Courtauld Institutes* 3.4 (1950) 285–315.

Ribouillault D., "Hortus academicus: les académies de la Renaissance et le jardin", in Jakob M. (ed.), *Des jardins & des livres* (Geneva: 2018) 3–12.

Samperi R., "La città delle vigne, dei giardini e delle ville (fine XV–XVI secolo)", in Simoncini G. (ed), *Roma. Le trasformazioni urbane nel Cinquecento*, vol. 2 (Florence: 2011) 105–157.

Settis S., *Sopravvivenza dell'antichità*, Quaderni del Dipartimento di scienze dell'antichità, Università di Padova, vol. 5 (Padua: 2001) 105–115.

Settis S., *Futuro del "classico"* (Turin: 2004).

Stok F., "Perotti e l'Accademia Romana", in Jensen R. – Pade M. (eds.), *On Renaissance Academies.* Proceedings of the International Conference "From the Roman Academy to the Danish Academy in Rome" (The Danish Academy in Rome, 11–13 October 2006) (Rome: 2011) 77–90.

Vasoli C., "Le 'Accademie'. L''Accademia Platonica', l''Accademia Romana'", *Storia della Letteratura italiana, III, Il Quattrocento* (Rome: 1996) 78–82.

Viti P., "L'Umanesimo nell'Italia settentrionale e mediana", *Storia della Letteratura italiana, III, Il Quattrocento* (Rome: 1996) 615–620.

Zabughin V., *Giulio Pomponio Leto*, vols. 1–2 (Rome: 1909–1912).

Zabughin V., *Giulio Pomponio Leto. Saggio critico*, vol. 2 (Grottaferrata: 1910–1912).

Zippel G. (ed.), *Le vite di Paolo II di Gaspare da Verona e Michele Canensi* (Castello: 1904–1911).

CHAPTER 4

Horti Colotiani: the Gardens of Angelo Colocci, the Casino del Bufalo and the Birthday of Rome

Giorgio Mangani

Abstract

Angelo Colocci's Roman Horti and Casino was the site of the poetic symposiums he organized as president of the Roman Academy and the place where he kept his important collection of ancient art, including the famous sculpture of Atlas, later sold to the Farnese family. The decorative cycle of the Casino, painted in the 1520s by Polidoro da Caravaggio and Maturino da Firenze, was dedicated to the myth of Perseus and Andromeda, whose homonymous constellations also appear on the celestial globe carried by the figure of Atlas. This essay offers a new interpretation of the political and humanistic significance of this cycle as a celebration of Rome's birthday (April 21), a celebration particularly dear to the members of the Roman Academy in the fifteenth century, and a means of presenting the semi-deterministic astrological theories of Giovanni Pontano (Colocci's master in his Neapolitan period), which were also of great interest to Colocci.

Keywords

Angelo Colocci – Giovanni Pontano – Farnese Atlas – Pomponio Leto – Tommaso Radini Tedeschi

1 Angelo Colocci

A pioneer in the study of Romance languages, Angelo Colocci (1474–1549) is primarily recognised in specialist linguistic circles for his theories on the origin of the Italian vernacular [Fig. 4.1]. Among the greatest collectors of antiquities and books in Rome, Colocci was Pomponio Leto's heir as president of the Roman Academy and devoted himself to Latin and vernacular poetry. He edited contemporary poets, collected epigraphs and worked to shed light on ancient measurement systems. He wished to devote a treatise (*De opifice*

 | DOI:10.1163/9789004517547_005

FIGURE 4.1 Anonymous, *Portrait of Angelo Colocci as Bishop of Nocera Umbra*, 18th century, oil on canvas, 49 × 64 cm, Bologna, University of Bologna, Monti Collection
IMAGE © UNIVERSITY OF BOLOGNA

mundi) to this field of study, which he called *podismus*, and collected enormous volumes of documentation throughout his life including the *Corpus Agrimensorum Romanorum*, a manuscript that had once belonged to Petrarch.[1]

1 The manuscript is one of the most important documents of ancient Roman metrology. It is now known as the *Codex Arcerianus* and passed through many properties after the 1527

Unable to complete the treatise, he committed himself to identifying the measure of the Roman foot – forever known as the "piede colocciano" (Coloccian foot). The Vatican Apostolic Library preserves many of Colocci's drafts and compositions, annotated manuscripts, transcriptions and *excerpta*, as well as numerous works from his famous library.[2] In the seventeenth century, the noble Federico Ubaldini of Urbania, secretary to Cardinal Francesco Barberini, dedicated a biography to Colocci.[3] In 1772, Gianfrancesco Lancellotti, a fellow citizen from the Marche region, followed, publishing a selection of his Italian and Latin compositions.[4] However, significant studies devoted to the Roman collector and theorist only developed in the mid-1970s. In particular, Vittorio Fanelli focused on his linguistic interests and the diversity of his activities as an official of the Roman Curia, including his practices as a collector and his leadership in literary and scientific circles.[5] Fanelli also highlighted Colocci's skill as an investor and financier – tracing the accumulation of his wealth and the flashy building renovations he commissioned during the papacies of Julius II and Leo X. In this period, Colocci claimed ownership over most of the land on the Quirinal Hill adjacent to the main Roman aqueduct. He built the villa where most of his impressive collections would be kept there near the Aqua Virgo.

For many years, studies on Angelo Colocci were primarily negative. The breadth of his interests, spanning language, cosmology, Vitruvian theory, Pythagorean principles, and cosmic harmonies and correspondences, diluted his perceived significance in any one field. Similarly, his obsessive collection of relics and documents related to the history of ancient measures was considered excessively fragmentary. Scholars tend to characterise Colocci as an amateur flirting with a multitude of impossible research projects. In a

---

Roman Sack, when Colocci's library was dispersed, until arriving at its current location, the Wolfenbüttel Library; see Petoletti M., "Un nuovo manoscritto della Biblioteca di Petrarca: il più antico codice degli Agrimensores (Wolfenbüttel, Aug. Fol. 3623)", *Studi Petrarcheschi* 24 (2011) 1–28.

2 Marco Bernardi identified Colocci's remaining books and manuscripts in the Vatican Library; see Bernardi M., "Angelo Colocci, la biblioteca e il milieu napoletano: nuovi interventi, qualche precisazione e un frammento inedito", *Roma nel Rinascimento* (2008) 59–78; Bernardi M., *Lo Zibaldone Colocciano Vat. Lat. 4831. Edizione e commento* (Vatican City: 2008); Bernardi M., "Il Postillato Colocciano delle Prose della volgar lingua: l'Ambrosiano S.R. 226 e il pensiero linguistico di Angelo Colocci", *L'Ellisse* IV (2009) 65–86; Bernardi M., "Angelo Colocci", in Notolese M. – Procaccioli P. – Russo E. (eds.), *Autografi dei letterati italiani. Il Cinquecento*, vol. 2 (Rome – Salerno: 2013) 75–110.

3 Ubaldini F., *Vita di mons. Angelo Colocci*, V. Fanelli (ed.) (Vatican City: 1969).

4 Lancellotti G., *Poesie italiane e latine di monsignor Angelo Colocci* (Jesi: 1772).

5 Fanelli V., *Ricerche su Angelo Colocci e sulla Roma cinquecentesca* (Vatican City: 1979).

meaningful portrait from 1961, the well known Italian literary historian Carlo Dionisotti described him as 'a collezionista […], fino alla mania, di testi letterari d'ogni genere e qualità' ('an obsessive collector of literary texts of all kinds and qualities').[6] This perception began to shift, however, with Ingrid Rowland's essays, which explored the complexity of his thought and highlighted the central role he played in Bramante's Roman cultural projects and urban renewal. The discovery that Colocci and Raphael collaborated on a joint translation of Vitruvius was particularly significant in this regard. Colocci's likely editorial contribution to Raphael's "Letter on Architecture" dedicated to Leo X further introduced novel interpretations of Vitruvian architecture and the concept of "order", which carried cosmological and philosophical meaning.[7]

In 2018, developing on some of Rowland's insights, I tried to clarify in my book *La bellezza del numero. Angelo Colocci e le origini dello stato nazione* (The Beauty of the Number. Angelo Colocci and the Origins of the Nation State) that, like other pioneers of Roman humanism, Colocci's varied interests aligned with his political affiliations.[8] Like most members of the Roman Academy since Leto's time, Colocci supported the pro-Greek Roman party founded by Cardinal Bessarion a generation prior. This affiliation might explain, for example, Colocci's interest in the so-called "Roman courtesan" language. The type of vulgar spoken at the papal court to Colocci and others, "Roman courtesan" could be considered a "super language" capable of shaping the conditions for a unified Italian nation, in contrast to Pietro Bembo's purist linguistic theories. Influenced by his political leanings, Colocci believed that the popular language spoken in ancient Rome, which he considered very close to Greek, was the origin of the Italian vernacular. Developed by Janus Lascaris, a Byzantine scholar and friend of Colocci's, this interpretation was part of a larger diplomatic strategy of rapprochement between Rome, the Christian West and the Greek world, with an ultimate view toward reconquering the ancient Byzantine empire that had fallen to the Ottomans. Colocci's cosmological theories were

6 Dionisotti C., "Per Francesco Colonna", *Italia medioevale e umanistica* 4 (1961) 323–326; Dionisotti C., *Appunti su arti e lettere* (Milan: 1995) 77–80, p. 78 for the quote.

7 Rowland I., "Angelo Colocci e i suoi rapporti con Raffaello", *Respublica Litterarum. Studi Umanistici Piceni* 11 (1991) 217–225; Rowland I, "Raphael, Angelo Colocci, and the Genesis of the Architectural Orders", *The Art Bulletin* 76 (1994) 81–104; Rowland I., *The Culture of the High Renaissance. Ancients and Moderns in Sixteenth-Century Rome* (Cambridge: 1998); Rowland I., "Angelo Colocci alla caccia del 'vestigium Dei'", in Colonna S. (ed.), *Roma nella svolta tra Quattro e Cinquecento*. Atti del Convegno internazionale di studi (Rome: 2004) 87–93.

8 Mangani G., *La bellezza del numero. Angelo Colocci e le origini dello stato nazione* (Ancona: 2018; second enlarged edition, Ancona: 2021).

similarly political, and his views on musical correspondences and harmonies were reflected in his ideas about society and the nation-state. Ultimately, he imagined an integration of the upper and lower classes according to a set of shared values and principles. Colocci's idea of a living, integrated society, like his views on a national language composed of different linguistic contributions, can be traced back to his teacher Giovanni Pontano. In his youth, Colocci studied with Pontano in Naples while in exile with his uncle Francesco. Both Pontano and Francesco had been high officials of the Aragonese state and shared scientific and astrological interests. The young Colocci, who grew up in a refined environment that cultivated the esoteric tradition, encouraged scientific and literary study and emphasised moral and political philosophy, inherited these interests.[9]

Colocci's later involvement with the Pope's political programs, especially in the years of Julius II and Leo X, most likely stemmed from his connections to the Roman Academy. In the years of Pomponio Leto, the Academy played a role in the Curia's communication strategies. For example, they defined the iconographic cycle of the Stanza della Segnatura; the name Tommaso Inghirami, a Vatican librarian and supporter of the Academy, is frequently mentioned in documents on this design.[10] Colocci was familiar with the cosmological and geopolitical arguments that the cycle proposes about philosophy, justice, poetry, and religion; we can also hypothesise Colocci and Inghirami's roles in the project based on their closeness to Raphael. These four themes, originally identified by Manfredo Tafuri and later by Nicholas Temple, seem to have represented the four strategic poles of Bramante's planned urban *renovatio* of Rome as well as the four strategic domains of Julius II's cultural, military and geopolitical programme.[11] The insistence on the geometry of the square and the obvious references to Pythagorean and Vitruvian theories in the cycle further affirm its correspondence with Colocci's views. It should also be noted

9 On Pontano and his academy in Naples see George Brocklehurst's essay in this book.

10 Among the most authoritative essays on this topic, see Pfeiffer H. (S.J.), *Zur Ikonographie von Raffaels Disputa. Egidio da Viterbo und die Christlich-Platonische konzeption der Stanza della Segnatura*, Miscellanea Historiae Pontificiae, vol. 37 (Rome: 1975); Reale G., *Raffaello: La "Scuola di Atene"* (Milan: 1997); Most G.W., *Leggere Raffaello. La Scuola di Atene e il suo pre-testo* (Turin: 2001); Joost-Gaugier C.L., "Ptolemy and Strabo and Their Conversation with Apelles and Protogenes: Cosmography and Painting in Raphael's *School of Athens*", *Renaissance Quarterly* 51 (1998) 761–787; Joost-Gaugier C.L., *Raphael's Stanza della Segnatura. Meaning and Invention* (Cambridge: 2002).

11 Tafuri M., "Roma Instaurata", in Frommel C.L. – Ray S. – Tafuri M. (eds.), *Raffaello Architetto* (Milan: 1984) 59–106; Temple N., *Renovatio urbis. Architecture, Urbanism, and Ceremony in the Rome of Julius II* (London: 2011).

that Colocci financially invested in the papal urban planning programme. To recognise his contribution to the project, Raphael's *School of Athens* finds Inghirami on the left side of the painting in the guise of Epicurus. The artist's self-portrait is placed on the same line on the right side of the fresco. Next to the bent figure in the act of drawing, generally accepted to be Bramante, we find a bearded individual holding a celestial sphere. In my book, I suggest that this figure is Colocci, based on this association with the cosmological studies that made him famous in the Curia.

2 The Political Implications of Colocci's Estate and Collections

Colocci owned many properties in Rome. He had a home in the Parione quarter as well as a *vigna* on the Quirinal hill where he held his collections. The Quirinal was the urban quarter traditionally linked to the baronial Colonna family; their Loggia dei Colonnesi, which housed their antiquarian collections, was established on the Quirinal in the fifteenth century.[12] The family were the primary patrons of the Greek party in Rome, thus taking up residence on the Quirinal in the first half of the sixteenth century likely carried political significance. We can observe this phenomenon among other leaders of the Roman Academy. Platina and Leto, for example, chose to locate their homes on the Quirinal. Cardinal Bessarion was also given a home next to pope Martin V Colonna's residence near the Basilica of Santi Apostoli where he was eventually buried. Under his financial and political patronage, Bessarion's home became the site of the first academic meetings. Following Leto's death, Colocci restored his home and *vigna* to host academic meetings; the physical proximity of those living on the Quirinal was thus mirrored by their social connections.[13] Take, for example, the shared border of Colocci's villa with the Del Bufalo family's estate. Collectors of antiquities and patrons of the Roman Academy, the Del Bufalo became very close to the Della Valle and eventually the Colonna in the seventeenth century. The lawyer and humanist Lelius Della Valle (1400–1475) served as Leto's and the Roman Academy's defence counsel when they were tried for impiety in 1468 by Pope Paul II.

The collection of ancient sculptures decorating Colocci's Horti Colotiani was also exploited politically. They celebrated the great, persuasive power of the literary, antiquarian and rhetorical culture cultivated by the academicians. Metamorphosis figured as the underlying theme of the collection and

12 On this monument and the Colonna family, see also Tiziana Checchi's essay in this book.
13 On Colocci's *vigna* on the Quirinal, see also Alessia Dessì's essay in this book.

the Casino's painted cycle represented poetry as a tool capable of building "monuments" more lasting than marble ones. Members of Colocci's humanist circle advanced the idea that silent, ancient sculptures could no longer be understood and needed the help of modern interpreters. Acting as ventriloquists of sorts, they composed lines that they imagined the sculptures might speak. By using this rhetorical technique, which was widely popular in the papal court, the poets of the Roman Academy were able to transform ancient sculptures into 'talking devices'.[14] This enabled them to diffuse their re-interpretations of Roman history and consolidate the pope's political power by painting his reign as a natural inheritance of the Roman *imperium*, like in the Vatican Belvedere.

3 The Colocci/Del Bufalo Casino

In this essay I will expand on my previous studies and consider the meaning of the decorative cycle of Colocci's Casino on the Pincio, where today's Via del Tritone is located. Surrounded by his well-known Horti Colotiani, the scholar received friends for symposiums and dove into poetic and scientific compositions at this site. The Casino fresco was undertaken *c.*1521–1522 by Polidoro da Caravaggio and Maturino da Firenze and depicted stories of Perseus and Andromeda.[15] The following scenes have been identified and studied: Perseus frees Andromeda, Andromeda's sacrifice, Perseus petrifies Phyneus and his soldiers, the Muses at the source of the Hippocrene with the poets, and Fortune and a Caryatid. They were positioned across from the following scenes: Danae receives Jupiter in the tower as golden rain, the Garden of the Hesperides, Perseus petrifies Atlas, and Mars and Venus [see Figs. 4.5–4.9].[16] Rolf Kultzen's work demonstrates that the Casino was often considered a Del Bufalo property,

14 An example of this is the well-known case of the Belvedere *Sleeping nymph*, then recoded as Cleopatra by Evangelista Maddaleni Capodiferro and Baldassare Castiglione. See Muecke F., "Poetry on Rome from the Ambience of Pomponio Leto: Topography, History, Encomium", *L'Ellisse. Studi storici di letteratura italiana* 2 (2007) 101–126. On this topic, see also Denis Ribouillault's essay in this book, "The Poetic Fortune of the Nymphaeum of Villa Giulia".

15 Silvia Ginzburg proposes a dating of 1521–1522 for the frescoes, which are traditionally dated 1525; see Ginzburg S., "Rinascenza dell'antico e lingua moderna negli affreschi del Casino Del Bufalo: Polidoro e Maturino per Angelo Colocci", in Bock N. – Foletti I. – Tomasi M. (eds.), *Survivals, revivals, rinascenze. Studi in onore di Serena Romano* (Rome: 2017) 407–419.

16 Colucci I. – Masini P. – Miracola P. (eds.), *Dal giardino. Bellezza e scienza al museo. Polidoro da Caravaggio nel Casino del Bufalo. Studi e restauro* (Rome: 2013) offers a detailed history

while Silvia Ginzburg has proposed that, while the Casino initially belonged to the Colocci, it was sold to the neighbouring Del Bufalo prior to Colocci's death.[17]

This sale would have included several pieces from Colocci's collections, which were dispersed at that time. Several studies suggest that the ancient statue of *Atlas* that would come to be known as the *Farnese*, given to the noble family in 1562 by Paolo Del Bufalo and now conserved in the National Museum of Naples, could be found among these pieces [Fig. 4.2]. Although no official documentation indicates that Colocci previously owned the sculpture, several arguments can be made in support of this claim. First, the meanings of the Casino cycles are in perfect harmony with the astronomical and astrological themes found on the celestial globe supported by Atlas, particularly the constellations of Perseus, Andromeda, Cepheus, Pegasus, and Cassiopeia [Fig. 4.3]. Documents identified by Kristen Lippincott show that the sculpture was already known by those close to Colocci in the 1520s, the years when the Casino Colocci environment was being organised.[18] Among Lippincott's earliest sources, we find a drawing by Michelangelo dated *c.*1532–1533, if not earlier, that depicts a man holding a globe in both arms. Its surface is marked with an ecliptic, as can also be seen on the Farnesian sphere. She also cites a drawing produced in 1527–1528 by Baldassarre Peruzzi, a close acquaintance of Colocci and the Roman Vitruvian circle that is now considered the model for Sigismondo de Fanti's *Triompho di Fortuna* frontispiece. De Fanti's book on astrology, a subject of great interest to Colocci, also recalls the *Fortuna* theme represented in the Casino cornerstone. And the noteworthiness of the Atlas

of the Casino Del Bufalo and many pictures of the restored and removed frescos, which are now in the Museo di Roma in Palazzo Braschi.

17 Kultzen R., "Die Malereien Polidoros da Caravaggio im Giardino del Bufalo in Rom", *Mitteilungen des Kunsthistorischen Institutes in Florenz* 9 (1960) 97–120. Kultzen thought a 1531 document of sale in the Roman state archive ('Atti del Notaio Stephano De Amannis', Collegio Notai Capitolini, vol. 81, cc. 139–142) signed by Antonio Del Bufalo and Angelo Colocci which speaks of a *viridarium* should not be considered evidence of the change of property of this Casino, although this is probable because of the damage caused to Colocci's Pincian estates during the 1527 Sack. In addition, all previous Roman antiquarian literature agreed on Colocci's ownership of the building, which was referred to as "Ninfeo Colocci-Del Bufalo" at its first location in the Museo di Roma. Ginzburg S., "Rinascenza dell'antico". Vincenzo Farinella, one of the most authoritative Italian scholars of Raphael and his school, emphasises an even earlier dating of Polidoro's paintings to 1520–1521, the most productive years of Colocci's cultural life. Farinella V., "Perseo, le Muse e la poesia: un ciclo di favole ovidiane nell'orto letterario di Angelo Colocci", in Mangani G. (ed.), *Raffaello e Colocci. Bellezza e scienza nella costruzione del mito della Roma antica*, exh. catalogue, Jesi, Musei Civici di Palazzo Pianetti (Rimini: 2021) 76–93.

18 Lippincott K., "The Early Reception of the Farnese Atlas: An Addendum to Bober and Rubinstein's Renaissance Artists and Antique Sculpture", *Schifanoia* 52–53 (2017) 227–238.

FIGURE 4.2
Farnese Atlas, Roman copy of Hellenistic original, 2nd Century AD. Naples, Museo Nazionale Archeologico, n. 6374
IMAGE © WIKIMEDIA COMMONS

FIGURE 4.3 Reconstruction of the ancient constellations on the Farnese Atlas' globe, from Richard Bentley (ed.), *M. Manilii Astronomicon* (London, H. Woodfall: 1739)
IMAGE © GIORGIO MANGANI

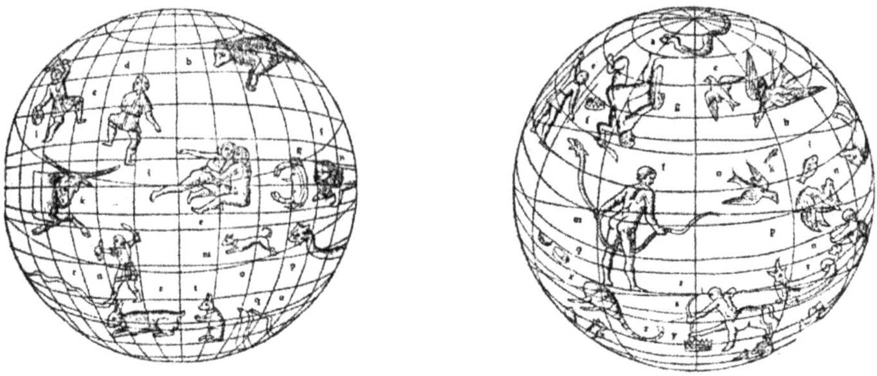

FIGURE 4.4 Reconstruction of the celestial constellations in Book 9 of Vitruvius's *De architectura*, Fra Giovanni Giocondo (ed.) (Venice, Giovanni Tacuino: 1511)
IMAGE © GIORGIO MANGANI

sculpture [Fig. 4.4] becomes even more evident when considering influence on Fra Giocondo's illustration of Vitruvius's treatise published in 1511. A close friend of Colocci with many shared interests, Fra Giocondo's celestial globes and their constellations show marked similarities with the Farnese.[19]

Ulisse Aldrovandi, who visited Rome around 1550, noted the significance of Colocci's assembled works. Describing the ancient sculptures preserved in palaces and gardens, Aldrovandi writes that Colocci's nephews inherited the remains of his collection and records the sculpture of Atlas on display in Bernardino de Fabi's residence instead of in the Del Bufalo villa.[20] His observation contradicts an account by the Dutch writer Stephanus Venandus Pighius at the end of the 1540s, who was at that time the secretary to Cardinal Cervini, himself a friend of both Colocci and Fulvio Orsini.[21] Pighius was particularly interested in the iconography of Atlas, which he tended to interpret, like Colocci and Orsini, as referring to Hercules. He recorded, for example, the

19 Michelangelo's drawing is in the Ashmolean Museum, Oxford (WA. 1846.63*v*). Colocci was in touch with Michelangelo and requested that he make a portrait of one of his lovers; Colocci's lines were published in 1772 in Lancellotti, *Poesie italiane* 203. Peruzzi's drawing is in the Metropolitan Museum of Art, New York (acc. 1992.304). The illustrated edition of Vitruvius's *De architectura*, edited by Fra Giovanni Giocondo, was printed by Giovanni Tacuino's Venetian printing press in 1511, and Colocci owned a copy; see Bernardi, "Angelo Colocci" 85.

20 Aldrovandi U., *Delle statue antiche che per tutta Roma in diversi luoghi e case si veggono* (Rome: 1550).

21 Pighius mentioned the sculpture in his *Hercules prodicius, seu principis iuventutis vita et peregrinatio* (Antwerp: 1587) 360.

existence of a similar sculpture in Ravenna and executed (or commissioned) drawings of the Del Bufalo Atlas/Hercules prior to its sale that recalled original elements that had disappeared during ancient restorations.[22] Henning Wrede accounts for Aldrovandi's misunderstanding regarding the ownership of the sculpture, explaining that it had been moved to a restorer's workshop close to Bernardino Fabi's residence at that time.[23] Wrede also notes the strong correlation between the sculpture and the decorative cycle of Perseus and Andromeda in the Casino fresco.[24] However, he underestimates Colocci's role and fails to consider the fact that the sculpture was already in the garden when it was Colocci's property. Basing his claims on the assumption that the sculpture had always belonged to the Del Bufalo, Wrede argues that Perseus was a sort of hero to the family. Perseus is indeed portrayed on the globe in the act of killing the sea monster with the Medusa near the sign of Taurus, much like the scene painted on the façade of the Casino. Since the bull was the heraldic figure of the Del Bufalo family, Wrede identifies the iconography as a commemorative reminder. According to Kathleen Wren Christian, this explanation is unreliable. It rests on too detailed an analysis of the globe, which would have hardly been visible to observers and therefore ineffective in performing the function attributed to it by Wrede.[25] However, Perseus's location in the zodiacal space belonging to Taurus might evidence the astrological meaning that Colocci ascribed to his system of constellations, which constituted the *paranatellonta* of Taurus – an idea that I will analyse more closely later.

The *Sleeping Nymph* echoed the role played by Atlas as a sculpture in dialogue with the frescoes, admonishing those who entered the sacred Colocci garden with an epigram that evoked a sleepy state of consciousness as metaphor for the poetic invasion of the Horti Colotiani ('Huius nympha loci, sacri custodia fontis / Dormio, dum blandae sentio murmura quae / Parce meum quisquis tangis cava marmora somnum / Rumpere, sive bibas, sive lavere, tace'; 'I, Nymph of this place, guardian of this sacred spring, / I sleep while listening to the soft murmur of the water. / Do not trouble my sleep, you, who touches this sculpted marble! / Whether you drink or bathe, be quiet'). The

22 A drawing of the sculpture without a head or arms is in the *Codex Pighianus* (Ms Lat. 2°, A. 61, f. 226r), Staatsbibliothek Preussischer Kulturbesitz, Berlin.

23 Wrede H., *Der Antikengarten der del Bufalo bei der Fontana Trevi* (Mainz am Rhein: 1983); Wrede H., "Perseus als Heros der del Bufalo", in Maier F.G. (ed.), *Alt-Paphos auf Cypern* (Mainz am Rhein: 1984) 52–54; Wrede H. – Harprath R., *Der Codex Coburgensis: Das erste systematische Archäologiebuch* (Coburg: 1986).

24 Wrede H., *Der Antikengarten*; Wrede H., "Perseus als Heros" 52–54.

25 Christian K.W., *Empire Without End. Antiquities Collections in Renaissance Rome, c. 1350–1527*, (New York – London: 2010) 284.

statue recalled the meeting of the Muses at the Hippocrene spring represented on the walls of the Casino. Both sculptures introduced the meaning of the Casino cycle, dialoguing with the paintings in the same way the poets hosted by Colocci in his convivial meetings orated their lines and made them into 'talking devices'.[26] Aby Warburg, who included the Atlas in his *Mnemosyne* repertory, noted that among the forty-five constellations depicted on the globe, the story of Perseus was the only one represented with all of its characters (Perseus, Andromeda, Cepheus, Pegasus and Cassiopeia).[27] The figure of Atlas is also recalled in a scene on the lateral façade of the building in which Perseus petrifies the giant – though a drawing by Cherubino Alberti, which is a copy of an engraving, remains the only documentation of this façade [see Fig. 4.9].[28] Ovid's *Metamorphoses*, printed by Aldo Manuzio in 1502 and held in Colocci's library, inspired the Casino cycle.[29] Like many of Colocci's other books and collection pieces, *Metamorphoses* passed to Fulvio Orsini, who mediated the Farnese family's purchase of the *Atlas* sculpture – further attesting to the undeniable relationship between the sculpture and the frescoes.

In *Urania*, published in 1505 as part of a series of posthumous editions, Colocci's master Pontano recalled the stories of Perseus and Andromeda. Alongside Pontano's pupil Giovanni Antonio Summonte, Colocci played a decisive role in the publication of these editions, making suggestions and corrections to his master's texts based on manuscript versions he possessed, as was custom at the time. The Casino narratives emblematised Colocci and Pontano's understanding of astrological semi-determinism developed by Pontano. Located close to each other in the boreal hemisphere, the constellations of Perseus and Andromeda, Pegasus, Cepheus and Cassiopeia were, in the Eastern astrological tradition, referred to as the *paranatellonta* of nearby Taurus. *Paranatellonta* were extra-zodiacal constellations that rose together and were considered relevant in determining the character of those born under their signs. Based on their position, they played a supplementary role to other signs located in the zodiacal circle. Among the Perseus and Andromeda, Cepheus, Cassiopeia, and

26 Ubaldini, *Vita di mons. Angelo Colocci* 54. On the tradition of the sleeping nymph, see Denis Ribouillault's essay in this book, "The Poetic Fortune of the Nymphaeum of Villa Giulia".

27 Warburg collected the Farnese Atlas with the detail of the constellations pictured on the celestial globe and representations of the myth of Perseus and Andromeda in table 2 of his *Mnemosyne atlas*.

28 Cherubino Alberti (1553–1615), *Atlas is turned to stone by Perseus*, copy of an engraving in turn a copy from Polidoro da Caravaggio's fresco in the Casino Colocci/Del Bufalo (Florence, Gabinetto dei disegni e delle stampe, Uffizi, Inv. 9297S).

29 Ovid, *Metamorphoses* IV, 663–768, 769–803; V, 294–340.

Pegasus constellations – also linked characters in Ovidian mythology – Perseus shone the brightest and was therefore the most influential. It rose exactly at the confluence of Aries with Taurus on 21 April, the date of the Roman Christmas feast. Long opposed by religious authorities on the basis that it was pro-pagan, the feast was eventually re-introduced by the Academy and legitimised by Pope Sixtus IV in 1483.

The two figures in the corners of the casino façade, *Fortune* and a *Caryatid*, are further evidence of its astrological framework [Figs. 4.5–4.6 and see Fig. 4.10]. These two figures are depicted as the "cornerstones" of the building, corroborating that the imagery was not only intended as a literary evocation, but also as a means of highlighting the relationship of Fortune to astrological determinism, one of Pontano's topics of study. The *Caryatid* referred to Vitruvius, who had explained how this Greek architectural figure served as a reminder of the betrayal of the maiden of Karyes, enslaved for siding with Persian enemies. The *Caryatid* represented yielding to fear and the absence of will and determination, as well as recalling Medusa's "petrification". It stands in opposition to *Fortune* who governed human events. Placed in a strategic position, Fortune acted against the deterministic force that supposedly immobilized the will exercised by the stars in the guise of the *Caryatid*. In the end, the date of 21 April provides a key to unravelling the iconographic meaning of this narrative, synthesising the legacy of Leto and Pontano on the Roman Christmas.

The relationship of astrological determinism to chance and fortune was central to Pontano's reflections during his years in Naples when Colocci had known him. He described his thinking in several works: *De fortuna*, written at the end of the fifteenth century and published posthumously by Pietro Summonte in 1512; *De prudentia*, considered an addition to the former and also published posthumously in 1508; and *Aegidius*, written around 1500, published in 1507, and dedicated to Pontano and Colocci's mutual friend Egidio da Viterbo. Like Colocci, Pontano had tried to synthesise the stoic astrological determinism of the ancient world and the concepts of human action and free will central to Christian theology, seeking their reconciliation with ideas of divine providence. Pontano's elaboration became the most widespread interpretation of this issue in the Renaissance and seems to have indirectly influenced the idea of fortune so central to Machiavelli's thought.[30] Pontano believed that the stars exerted an influence on human character and behaviour in relation to their position at one's birth. This force was strengthened or weakened by various

30 Ginzburg C., *Nondimanco. Machiavelli, Pascal* (Milan: 2018) 67–77.

FIGURE 4.5
Polidoro da Caravaggio and Maturino, *Fortuna* in the Colocci/Del Bufalo Casino in Rome, fresco, *c.*1521–1522
IMAGE © GIORGIO MANGANI

FIGURE 4.6
Polidoro da Caravaggio and Maturino, *Caryatid* in the Colocci/Del Bufalo Casino in Rome, fresco, *c.*1521–1522
IMAGE © GIORGIO MANGANI

factors including geographical location and, thus, not all places were impacted by the stars in the same way. Colocci closely studied this subject and his notes and drafts of manuscript essays show how he combined Pontano's influence with Ptolemy's *Tetrabiblos*.

As the combined product of fortune and will, the "human factor" also played a role in determining behaviour within the realm of astral influences.

The possibility of divine intervention (premonitory messages, for example) was also thought to intervene in this mechanical process, occurring indirectly through the aegis of the stars, which acted as intermediaries between divine providence and free choice. Some humanists believed that men made instinctive choices (the product of the inertial forces produced by the stars), as well as intentional decisions based on the desire to follow fortune with the backing of will and reason. Without the help of reason, one could fall into the realm of passion, which was fully dominated by the stars. By relying on reason and will, however, it was possible to manage external influences and to ride the circumstances generated by chance. This paradigm became the proverb: 'fortune favours the brave'. Based on his observations as a seasoned statesman, Pontano pointed out in the fourth book of *De prudentia* that one must adapt to circumstances and take advantage of them. Fortune is a dynamic and contradictory force, in which surrendering to passion alone can be harmful. Ungoverned by reason, impetuous and dynamic action can alter the order of things and lead to unpredictable results. The impulsive, daring action generally typical of young people can sometimes create favourable conditions, yet this dynamism can also lead to ruin. Considering Pontano's theory of fortune, the overall meaning of the Casino's decorative cycle is rendered more comprehensible.

The astronomical framing of Roman Christmas confers a sort of messianic and cosmological status to the traditional Roman celebration that had been relaunched from Leto's time. It confirmed the parallels between Perseus and Romulus by highlighting their shared illegitimate births – Perseus and his mother were abandoned by their father in a basket like Romulus and Remus – and impetuous natures as two strong-willed human actors opposed to fate. This framing also provided an opportunity to pay homage to and further disseminate Pontano's astrological philosophy. Following the mythological tradition established by Ovid as translated into an astrological language by Pontano, the story of Perseus provided an origin for the Muses, which the academic meetings in the Horti Colotiani promoted. The Muses emerged from the accidental blow of Pegasus's hoof, which had been generated by Medusa's blood. Pegasus's myth was linked to the celebration of Fame, which flew, like the winged horse. It was also the subject represented in the Farnesina by Raphael's school, which was created to celebrate Agostino Chigi, Colocci's collaborator in Roman land speculation and friend and colleague at the Papal Secretariat.[31]

31 Rowland I., *The Culture of the High Renaissance. Ancients and Moderns in Sixteenth-Century Rome* (Cambridge: 1998) 68–108.

FIGURE 4.7 Polidoro da Caravaggio and Maturino, *Andromeda*, Colocci/Del Bufalo Casino in Rome, fresco, c.1521–1522
IMAGE © GIORGIO MANGANI

In the Casino's decorative cycle, Polidoro and Maturino depict Perseus as an impetuous and impulsive young man riding into events with insistent dynamism. The narrative references the combination of chance and determined, youthful impulsiveness: the casual meeting with Andromeda (she in chains, he free) [Fig. 4.7], the fight with the monster, and the depiction of Andromeda's betrothed petrified with his men by Medusa [Fig. 4.8]. The story further seems to exemplify the theme of an eternal struggle between chance, will and necessity as represented by the Fortune and Caryatid's cornerstones. While Athena's gift – the supernatural head of Medusa – evokes divine intervention, it is only activated in response to the hero's strategic and conjunctural decisions – further exemplifying the interplay of reason and the ability to seize opportunity. The theme of the façade was likely replicated and developed on the other side of the building. There, Perseus and Atlas and the garden of the Hesperides seem to have been represented, according to Cherubino Alberti's surviving drawing of the lost frescoes (*Atlas is turned to stone by Perseus*) [Fig. 4.9]. Atlas was considered the father of astronomy/astrology. Transformed into a mountain bearing the world, his figure was often used in architecture as a telamon – the male counterpart of the caryatid.

FIGURE 4.8 Polidoro da Caravaggio and Maturino, *Perseus fighting with Phyneus*, Colocci/Del Bufalo Casino in Rome, fresco, *c.*1521–1522
IMAGE © GIORGIO MANGANI

FIGURE 4.9 Cherubino Alberti (1553–1615), *Atlas is turned into stone by Perseus*, drawing, *c.*1521–1522 Florence, Gallerie degli Uffizi, Gabinetto dei disegni e delle stampe (Inv. 9297S)
IMAGE © UFFIZI

4 The Political Meaning of the Casino's Decoration

The astrological-mythological themes of the Casino also carried political significance and implicitly made critical references to the pontificate of Julius II consistent with the ideas of Colocci and the Roman Academy. Tommaso Radini Tedeschi (1488–1527), a Dominican from Piacenza who arrived in Rome around 1514 and died prematurely during the Sack of 1527, confirmed this political-iconographic function. Despite lack of evidence attesting to his relationship with Colocci, Radini Tedeschi's acquaintances, his role in the Curia and the subjects of his main works make him a natural interlocutor of Colocci's. This is especially true if one considers the restrained papal environment in the years of Pope Leo X, during which Colocci was at the height of his fame and influence in the papal court. Radini Tedeschi was an astrologer who sought to make astrological tradition compatible with Christian religion, an interest shared by Pontano and Ludovico Lazzarelli a generation earlier. In his youth, Lazzarelli was financially supported by Colocci, who may have edited his publication *Bombyx* in 1498.[32] In *Sidereus abyssus*, which was published in Pavia in 1511 and reprinted in Paris by Le Fevre in 1514 and referenced esoteric culture, the mysticism of numbers and Pythagorean and musical philosophy, Radini Tedeschi translated the characters produced by the various asterisms of ancient astrological classics into models adapted to Christian morality. It is difficult to imagine that such a personality was not in contact with Colocci, who shared a similar philosophy and was already a member of the Papal Secretariat.

Radini Tedeschi began his career as a professor of theology at the Studium Urbis and became Master of the Sacred Palace in 1521. A year prior in St. Peter's Basilica, he made a solemn oration in which he celebrated the season of Leo X's pontificate as the age of meekness – the so-called *Pax Leonina*. He had previously treated this subject in the *Hexasphaerium*, published by Colocci's usual publisher Mazzocchi in 1516; in this volume he referred to himself as a *pansophilos*. In his oration at St. Peter's, Radini Tedeschi made abundant use of the Pythagorean numerological mysticism of three and four, mentioning the seven candelabra symbols and the myth of Atlas. He used musical philosophy in *Calipsychia, sive de pulchretudine animae*, another theme central to Colocci's thought.[33]

One source seems to confirm the possible relationship between the two men. Dedicating a portrait to his brother in *De viris illustribus ordinis praedicatorum*, Leandro Alberti, also a Dominican, attributed to Radini Tedeschi the merit of having introduced to Rome the idea that the classical Muses frequently met with Apollo on Mount Parnassus.[34] This idea was incorrect but unanimously accepted in the Curia's milieu. Mount Parnassus was near Corinth and the *Coricia* cave dedicated to Pan and the nymphs was located on its summit; the nymphs had been confused with the Muses since Cyriac of

32 Lazzarelli L., Bombyx *ad Ang. Colotium honestate indolis puerum* (Rome: E. Silber, 1498); Lazzarelli L., *Opere ermetiche,* Moreschini C. – Saci M.P. – Troncarelli F. (eds.) (Pisa – Rome: 2009).

33 Radini Tedeschi T., *Calipsychia, sive de pulchretudine animae* (Milan, Gerardo Pontico: 1511).

34 Alberti L., *De viris illustribus ordinis praedicatorum* (Bologna, Girolamo Benedetti for Giovanni Battista Lapi: 1517).

Ancona's time. The true source of the Muses, the Hippocrene, was located on the nearby Mount Helicon. Radini Tedeschi and Colocci both perpetuated their erroneous location on Parnassus, which then passed to Raphael who painted the Muses' Parnassus meeting place in the Stanza della Segnatura (see Fig. 1.4 in Chap. 1). This shared, incorrect interpretation may suggest that Colocci and Radini Tedeschi had some acquaintance.

Sidereus abyssus reveals much regarding Radini Tedeschi's thought – a figure who deserves more careful consideration for the role he played in the Coloccian environment during Leo X's pontificate. It offers a contextualisation of the moral, astrological and political interpretation of the story of Perseus and Andromeda, identifying Perseus with *Justice*, Andromeda with the Gnome/Synesis of Aristotle's *Nicomachean Ethics* (i.e., Wisdom), and Pegasus with sacrifice to God.[35]

Colocci's Casino was conceived in his prime philosophical years following Leo X's pontificate, which Radini Tedeschi had recognised as a golden age of docility and wisdom after the years of obscurantism and political regret suffered under Hadrian VI. In a veiled way, though clear in the restricted milieu of his friends, the Casino superimposed on the story of Perseus and Andromeda a judgement of Julius II and Leo X grounded in the opinions of the Roman curia [Fig. 4.10]. The former was identified with an excess of belligerent activism (Perseus) and the latter with peace and wisdom (Andromeda). In Rome, this interpretation was shared not only by Radini Tedeschi, who celebrated Pope Leo, but also by the nobleman Marcantonio Altieri. Altieri, who was very close to the Roman Academy and a leading member of the Roman aristocracy, a group that had been marginalised by the papacy since its return to Rome, included complaints in his *Bacchanalia*, describing Julius II's pontificate as an age that had not enjoyed 'pur un hora de quiete' ('even an hour of tranquillity').[36] Though suggested in the Casino, the identification of Perseus with military virtues was not restricted to characterisation of Julius II. Recent studies have shown that this iconography had been employed a few years earlier in the funerary chapel dedicated to Andrea Carafa in the church of San Domenico Maggiore in Naples. The Count of Santa Severina and Royal Chancellor of the Aragonese, Carafa was celebrated in a cycle of astrological figures, now destroyed, inspired by Pontano, who used illustrated works by Aratus, Manilius,

---

35 Radini Tedeschi Thomae, *Sideralis abyssus* (Paris: Thome Kees, pour Hemon Le Fevre, 1514) I, 15–17.

36 Miglio M., "Come introduzione. Marco Antonio Altieri tra curia e municipio", in Cantatore F. – Chiabò M. – Farenga P. – Gargano M. – Morisi A. – Modigliani A. – Piperno F. (eds.), *Metafore di un pontificato. Giulio II (1503–1513)* (Rome: 2010) 1–14; Kolsky S., "Culture and Politics in Renaissance Rome: Marco Antonio Altieri's Roman Weddings", *Renaissance Quarterly* 40.1 (1987) 49–90.

FIGURE 4.10 The façade of the Colocci/Del Bufalo Casino in Rome, engraving by E. Maccari and G. Iannoni, 1876
IMAGE © GIORGIO MANGANI

Firmicus Maternus and Hyginus published in Venice in the last years of the fifteenth century, as well as by Radini Tedeschi, sources present in his library. Among these figures, the noble Carafa was presented in the guise of Perseus, Hercules and Orion as symbols of moral virtues.[37]

Colocci and Pontano's insistence on employing mythological images derived from classicism to represent contemporary political topics reinforces the deep link between Roman and Neapolitan humanists.

5 Hercules Musarum

Colocci's celebration of Pontano, which informed the design of his humanistic garden, is made further evident when we consider the reference to *Atlas Petrified by Perseus in the Garden of the Hesperides* found among the decorations of the Roman Casino [see Fig. 4.9]. Pontano dedicated a work titled *De hortis Hesperidum* to this theme, drawing on the story of Hercules stealing the

37 Naldi R., *Andrea Ferrucci, marmi gentili tra la Toscana e Napoli* (Naples: 2002) 143–152.

golden apples to celebrate the landscape and territory of Campania where he claimed the *mala aurea* (golden apples) of the mythological garden had been taken.[38] Colocci, his Greek friend Janus Lascaris and Pontano were all attached to cultivating the myth of Hercules as friend of the Muses (*Hercules Musarum* or *Musagetes*). This cult was established in ancient Pythagorean circles and introduced to Italy in the second century BCE by thinkers such as the poet Ennius and the consul Fulvius Nobilior.[39] Pontano employed the figure of Hercules to represent himself in his Neapolitan funerary chapel, which he planned; he also defined himself as 'beloved by the Muses' in his epitaph.[40] The relationship of Hercules to the Muses is further highlighted in a passage dedicated to Colocci by his fellow countryman Francesco Panfilo, in which he is called 'friend of the Muses' (*Aonidon philos*). In the *Picenum* (1575), which celebrates glories from "Picenum", the ancient name of the Marche region where Colocci was from, he is described as Hercules fighting with the giant Antaeus.[41] Colocci also used this metaphor (*dilectus Musis*) in an epitaph dedicated to his friend Lascaris.[42]

---

38 Iacono A., "Il De hortis Hesperidum di Giovanni Pontano tra innovazioni umanistiche e tradizione classica", *Spolia. Journal of Medieval Studies* 1 (2015) 186–237.

39 Detienne M., "Héraclès, héros pythagoricien", *Revue de l'histoire des religions* 158.1 (1960) 19–53; De Stefano F., "'Hercules Musarum in Circo Flaminio'. Dalla dedica di Fulvio Nobiliore alla Porticus Philippi", *Archeologia Classica* 65 (2014) 401–431.

40 'Sum enim Iovianus Pontanus, / Quem amaverunt bonae Musae', quoted in Paolo Giovio, *Elogia virorum litteris illustrium* (Basle: Peter Perna, 1577) 61. Giovio's source was Girolamo Borgia's (a previous student of Pontano's in Naples and friend of Colocci's when he moved to Rome) unpublished *De bellis italicis*. See: Roick, M., *Pontano's Virtues. Aristotelian Moral and Political Thought in the Renaissance* (London: 2017) 9, 196. On Pontano's Neapolitan chapel and the role played by him in its project, see De Divitiis B., "PONTANVS FECIT: Inscriptions and Artistic Authorship in the Pontano Chapel", *California Italian Studies* 3.1 (2012) 1–36.

41 'Angelus hac digna fuit Urbe Colotius ortus; / Angelus Aonidum maximus ille philos. / Vertitur in varias formas, totidemque figuras / Carpathiis Proteus, quod Deus egit aquis. / Polypus aequoreis non tot tibi sumit in undis / Quot facies apto tempore cautus agit. / Vinceret Anthaeum ludo, nitidaque palestra / Aurea quique humeris jam tulit astra suis' ('In this worthy city was born Angelo Colocci, / The greatest friend of the Muses; / Proteus, to whom God entrusted the Aegean waters, / Metamorphoses in various shapes and in as many forms. / Cautious Polypus did not dare to bring you between so many waves of the sea, / As many as its shapes, although it was calmest with good weather. / He would even win against Antheus in a contest, in a shiny palestra, / Who is wearing now the golden stars on his shoulders'), quoted in Panfilo Francesco, *Picenum* (Macerata: Sebastianus Martellinus, 1575), quoted in Lancellotti Gianfrancesco, *Poesie italiane e latine di monsignor Angelo Colocci* (Jesi, Pietropaolo Bonelli stampatore pubblico vescovile, e del Sant'Uffizio: 1772) 55.

42 'Dilectus Musis, et rebus natus agendis, / Quem gravida Imperiis protulit alte domus; / Tot freta dum curris, totque invia regna pererras, / Ausonia tandem Lascaris urbe facies. / Sed situs in patria es, Graio licet orbe remotus; / Nulla est virtuti terra aliena tuae' ('Dear

The figure of Hercules was central to Colocci's academic environment for several reasons. Hercules recalled the astronomical-astrological knowledge conveyed from Atlas to Hercules and then from Hercules to humankind. He was also considered the first teacher of Evander, king of the Arcadians and among the first colonisers of *Latium Vetus*. Andrea Fulvio, author of *Antiquitates* and an authoritative exponent of the Colocci cenacle, clarified this topic.[43] A letter of remembrance written by Jacopo Sadoleto and addressed to Colocci recorded that the Roman academicians used to meet at the remains of the Temple of Hercules on the Lungotevere during the years of Colocci's presidency.[44] Furthermore, Hercules was also the emblem of the Roman aristocratic class known as "Bovattieri", who, since the Middle Ages, had prospered because of cattle breeding. This association was forged due to the legend of Hercules fighting against the local deity Cacus, who had stolen his cattle and hidden them in the hollows of the Capitoline Hill.

Colocci, who kept the sculpture of Atlas with the earth on his shoulders in his garden, likely would have favoured an ambiguous interpretation of this image. He may have considered him to be Hercules, who, according to the mythology, had for some time carried the world on his shoulders instead of Atlas in the Garden of the Hesperides. This was a topic to which Colocci and his master Pontano had devoted their attention. Stephanus Vinandius Pighius, the Dutch secretary to Cardinal Cervini, corroborated this interpretation. Notably, Cervini, who later became Pope Marcellus II, was a close friend and the eventual executor of Colocci's will. Fulvio Orsini, who acted as a mediator in the purchase of the sculpture by the Farnese family, also agreed with Pighius's assertion. The Farnese's devotion to the cult of Hercules was the

---

to the Muses, you were born to do great things, / Coming from an ancient imperial family; / While you run among as many narrow straits, as you wander through inaccessible lands, / You will finally become a member of the Ausonian lineage, Lascaris. / Even more, you were adopted by the Homeland, although you're far away from your Greek world;/ Nowhere is your virtue truly alien.'), quoted in Ubaldini, *Vita di mons. Angelo Colocci* 83.

43 'La verità di questa favola è che Hercole hebbe oltre le forze esterne del corpo grandissimo ingegno et particolarmente delle cose naturali si dilettò, havendo notitia dell'astrologia et del corso dei cieli, che da Atlante principal maestro di quei tempi haveva appresa, la quale dopo la morte di quello essercitò' ('The truth of this tale is that Hercules was not only a vigorous hero, but also smart in knowing the natural world, practicing with astrology and astronomy as a student of Atlas, the most expert of that time, topics that he cultivated after his master's death'), quoted in Fulvio Andrea, *L'antichità di Roma* (Venice: Per Girolamo Francini Libraro in Roma all'insegna del Fonte, 1588) 312.

44 The letter was sent by Jacopo Sadoleto when he was in Carpentras after the 1527 Sack; see Sadoleto J., *Epistolarum libri sexdecim* (Lyon: 1550) 244. The passage is quoted in Ribouillault's introductory essay in this book.

primary reason driving this hybrid Atlas/Hercules reading, but Orsini was also following an older interpretative stream extending to Colocci's time. With this Herculean identification, the sculpture was placed in the Farnese Palace, inspiring the moral framework of Annibale Carracci's paintings, as explained by Panofsky and confirmed by recent scholars.[45] John Beldon Scott, analysing the Perseus and Andromeda cycle of the Casino and its influence on the paintings of the Farnese Palace in Rome, was surprised that Orsini described the sculpture as Hercules rather than Atlas and thought this was erroneous.[46] However, Carracci and Orsini were adhering to the interpretative lesson cultivated among Colocci's friends. The polysemous sculpture of Atlas/Hercules was linked to a strain of scientific and cultural thought and prestige promoted by Colocci in Roman curial circles – a prestige so conspicuous that he was placed by Raphael in the *School of Athens* with the world in his hands. Aby Warburg would have appreciated this enduring image, like the act performed by Charlie Chaplin in *The Great Dictator*, of the feat of holding the world with ease.

References

Aldrovandi U., *Delle statue antiche che per tutta Roma in diversi luoghi e case si veggono* (Rome: 1550).

Beldon Scott J., "The meaning of Perseus and Andromeda in the Farnese Gallery and on the Rubens House", *Journal of the Warburg and Courtauld Institutes* 51 (1988) 250–260.

Bernardi M., "Angelo Colocci, la biblioteca e il milieu napoletano: nuovi interventi, qualche precisazione e un frammento inedito", *Roma nel Rinascimento* (2008) 59–78.

Bernardi M., *Lo Zibaldone Colocciano Vat. Lat. 4831. Edizione e commento* (Vatican City: 2008).

45 Panofsky E., *Hercules am Schweidewege und andere antike Bildstoffe in der neueren Kunst* (Leipzig: 1930). The influence of the *Atlas* in Carracci's Farnese Palace iconographic cycle has been underlined by C. Loisel in *L'idea del bello*, exh. cat., Rome, Palazzo delle Esposizioni and ex Teatro dei Dioscuri (Rome: 2000) vol. 2, 234–235. Orsini authored the full content of the cycle, as documented in his letter to Cardinal Odoardo Farnese in August 1595 (Ms Vat. Lat. 9064). See: Beldon Scott J., "The Meaning of Perseus and Andromeda in the Farnese Gallery and on the Rubens House", *Journal of the Warburg and Courtauld Institutes* 51 (1988) 250–260.

46 Ibid. 255.

Bernardi M., "Il Postillato Colocciano delle Prose della volgar lingua: l'Ambrosiano S.R. 226 e il pensiero linguistico di Angelo Colocci", *L'Ellisse. Studi storici di letteratura italiana* 4 (2009) 65–86.

Bernardi M., "Angelo Colocci", in Notolese M. – Procaccioli P. – Russo E. (eds.), *Autografi dei letterati italiani. Il Cinquecento*, vol. 2 (Rome – Salerno: 2013) 75–110.

Colucci I. – Masini P. – Miracola P. (eds.), *Dal giardino al museo. Polidoro da Caravaggio nel Casino del Bufalo. Studi e restauro* (Rome: 2013).

Christian K.W., *Empire Without End. Antiquities Collections in Renaissance Rome, c. 1350–1527* (New York – London: 2010).

De Divitiis B., "PONTANVS FECIT: Inscriptions and Artistic Authorship in the Pontano Chapel", *California Italian Studies* 3.1 (2012) 1–36.

De Stefano F., "'Hercules Musarum in Circo Flaminio'. Dalla dedica di Fulvio Nobiliore alla Porticus Philippi", *Archeologia Classica* 65 (2014) 401–431.

Detienne M., "Héraclès, héros pythagoricien", *Revue de l'histoire des religions* 158.1 (1960) 19–53.

Dionisotti C., "Per Francesco Colonna", *Italia medioevale e umanistica* 4 (1961) 323–326.

Dionisotti C., *Appunti su arti e lettere* (Milan: 1995).

Fanelli V., *Ricerche su Angelo Colocci e sulla Roma cinquecentesca* (Vatican City: 1979).

Fulvio A., *L'antichità di Roma* (Venice: Per Girolamo Francini Libraro in Roma all'insegna del Fonte, 1588).

Ginzburg C., *Nondimanco. Machiavelli, Pascal* (Milan: 2018).

Ginzburg S., "Rinascenza dell'antico e lingua moderna negli affreschi del Casino Del Bufalo: Polidoro e Maturino per Angelo Colocci", in Bock N. – Foletti I. – Tomasi M. (eds.), *Survivals, revivals, rinascenze. Studi in onore di Serena Romano* (Rome: 2017) 407–419.

Kolsky S., "Culture and Politics in Renaissance Rome: Marco Antonio Altieri's Roman Weddings", *Renaissance Quarterly* 40.1 (1987) 49–90.

Kultzen R., "Die Malereien Polidoros da Caravaggio im Giardino del Bufalo in Rom", *Mitteilungen des Kunsthistorischen Institutes in Florenz* 9 (1960) 97–120.

Iacono A., "Il De hortis Hesperidum di Giovanni Pontano tra innovazioni umanistiche e tradizione classica", *Spolia. Journal of Medieval Studies* 1 (2015) 186–237.

Joost-Gaugier C.L., "Ptolemy and Strabo and Their conversation with Apelles and Protogenes: Cosmography and Painting in Raphael's 'School of Athens'", *Renaissance Quarterly* 51 (1998) 761–787.

Joost-Gaugier C.L., *Raphael's Stanza della Segnatura. Meaning and Invention* (Cambridge: 2002).

Lancellotti Gianfrancesco, *Poesie italiane e latine di monsignor Angelo Colocci* (Jesi: 1772).

Lazzarelli L., *Opere ermetiche*, Moreschini C. – Saci M.P. – Troncarelli F. (eds.) (Pisa – Rome: 2009).

L'idea del bello, exh. cat., Rome, Palazzo delle Esposizioni and ex Teatro dei Dioscuri (Rome: 2000).

Lippincott K., "The Early Reception of the Farnese Atlas: An Addendum to Bober and Rubinstein's Renaissance Artists and Antique Sculpture", *Schifanoia* 52–53 (2017) 227–238.

Mangani G., *La bellezza del numero. Angelo Colocci e le origini dello stato nazione* (Ancona: 2021, second edition Ancona: 2021).

Mangani G. (ed.), *Raffello e Colocci. Bellezza e scienza nella costruzione del mito della Roma antica,* exh. cat., Jesi, Palazzo Pianetti Musei Civici (Rimini: 2021).

Miglio M., "Come introduzione. Marco Antonio Altieri tra curia e municipio", in Cantatore F. – Chiabò M. – Farenga P. – Gargano M. – Morisi A. – Modigliani A. – Piperno F. (eds.), *Metafore di un pontificato. Giulio II (1503–1513)* (Rome: 2010) 1–14.

Most G.W., *Leggere Raffaello. La Scuola di Atene e il suo pre-testo* (Turin: 2001).

Muecke F., "Poetry on Rome from the Ambience of Pomponio Leto: Topography, History, Encomium", *L'Ellisse. Studi storici di letteratura italiana* 2 (2007) 101–126.

Pighius Stephanus Vinandius, *Hercules prodicius, seu principis iuventutis vita et peregrinatio* (Antwerp: 1587).

Radini Tedeschi T., *Calipsychia, sive de pulchretudine animae* (Milan, Gerardo Pontico: 1511).

Radini Tedeschi T., *Sideralis abyssus* (Paris: Thome Kees, pour Hemon Le Fevre, 1514).

Naldi R., *Andrea Ferrucci, marmi gentili tra la Toscana e Napoli* (Naples: 2002).

Panofsky E., *Hercules am Schweidewege und andere antike Bildstoffe in der neueren Kunst* (Leipzig: 1930).

Panfilo F., *Picenum* (Macerata: Sebastianus Martellinus, 1575).

Pfeiffer H. (S.J.), *Zur Ikonographie von Raffaels Disputa. Egidio da Viterbo und die Christlich-Platonische konzeption der Stanza della Segnatura,* Miscellanea Historiae Pontificiae, vol. 37 (Rome: 1975).

Reale G., *Raffaello: La "Scuola di Atene"* (Milan: 1997).

Rowland I., "Angelo Colocci e i suoi rapporti con Raffaello", *Respublica Litterarum. Studi Umanistici Piceni* 11 (1991) 217–225.

Rowland I., "Raphael, Angelo Colocci, and the Genesis of the Architectural Orders", *The Art Bulletin* 76 (1994) 81–104.

Rowland I., *The Culture of the High Renaissance. Ancients and Moderns in Sixteenth Century Rome* (Cambridge: 1998).

Rowland I., "Angelo Colocci alla caccia del 'vestigium Dei'", in Colonna S. (ed.), *Roma nella svolta tra Quattro e Cinquecento.* Atti del Convegno internazionale di studi (Rome: 2004) 87–93.

Sadoleto Jacopo, *Epistolarum libri sexdecim* (Lyon: Sebastian Gryphius, 1550).

Tafuri M., "Roma Instaurata", in Frommel C.L. – Ray S. – Tafuri M. (eds.), *Raffaello Architetto* (Milan: 1984) 59–106.

Temple N., *Renovatio urbis. Architecture, Urbanism, and Ceremony in the Rome of Julius II* (London: 2011).
Ubaldini F., *Vita di mons. Angelo Colocci*, Fanelli V. (ed.) (Vatican City: 1969).
Wrede H., *Der Antikengarten der del Bufalo bei der Fontana Trevi* (Mainz am Rhein: 1983).
Wrede H., "Perseus als Heros der del Bufalo", in Maier F.G. (ed.), *Alt-Paphos auf Cypern* (Mainz am Rhein: 1984) 52–54.
Wrede H. – Harprath R., *Der Codex Coburgensis: Das erste systematische Archäologiebuch* (Coburg: 1986).

CHAPTER 5

The Unknown "Pleasure Grounds" of the Cardinals Colonna at the Abbey of Subiaco

Tiziana Checchi

Abstract

This study analyses the relationship of some cardinals of the Colonna family to humanistic culture and investigates the hitherto unknown 'pleasure grounds' that these cardinals, as commendatory abbots, enjoyed at the Abbey of Subiaco between the 15th and 16th centuries. Here, surrounded by the writers of their entourage, who dedicated poetic and celebratory compositions to them, the Cardinals Colonna entertained themselves in joyful gatherings. The analysis of these poems and of old descriptions found among the archival documents, never before considered in the scholarly literature, has made it possible not only to understand the nature of these now vanished places and their uses, but also to rediscover some unknown representations of them in a cycle of frescoes that decorate the so-called "Studiolo di Marcantonio Colonna" in the Rocca Abbaziale (Abbey Fortress) of Subiaco.

Keywords

Colonna family – Cardinals – pleasure grounds – humanistic gatherings – poetry – Abbey of Subiaco – landscape frescoes

In the fifteenth century, alongside the proliferation of antiquarian studies, it became standard for Roman humanists to host literary activities gatherings and academic meetings in gardens, mirroring the customs and practices of antiquity.[1] Basil Bessarion imported the concept of the academy to Rome, understood in the classical Greek tradition as an open meeting among intellectuals marked by an emphasis on dialogue.[2] From the beginning of his time in

1 See the introduction in this book and the essays by George Brocklehurst, Alessia Dessì and Giorgio Mangani.

2 Manzoli D., "La fortuna di Bessarione e lo sviluppo delle accademie a Roma", in Gutkowski A. – Prinzivalli E. (eds.), *Bessarione e la sua Accademia* (Rome: 2012) 81–99. In addition to this

 | DOI:10.1163/9789004517547_006

Rome in the early 1440s, Bessarion welcomed men of letters and friends to the garden next to his house adjoining the Basilica dei Santi Apostoli.[3] Formerly owned by the Colonna, the residence was confiscated from the family after conflicts with the papacy following the death of Pope Martin v Colonna in 1431.[4] Visitors were among the most illustrious cultural figures of the time and included Teodoro Gaza, Pomponio Leto, Lorenzo Valla, Bartolomeo Platina, Francesco Filelfo, Biondo Flavio and many others.[5]

Shortly thereafter, the Roman or Pomponian Academy was established on the adjacent Quirinal Hill and named for its founder, Pomponio Leto, himself a former visitor to the *Domus Bessarionea*.[6] The Monte Cavallo area was among those favoured by humanists and home to Bartolomeo Sacchi, also called Platina. The site lent itself particularly well to the recovery of *antiquitates* and the evocation of ancient landscapes and unusual atmospheres, as it combined a still-rustic nature with strong antiquarian reference. This was achieved in part due to the presence of numerous ruins including the monumental remains of

---

volume, see especially the important studies by Concetta Bianca: Bianca C., "L'Accademia del Bessarione tra Roma e Urbino", in Cerboni Baiardi G. – Chittolini G. – Floriani P. (eds.), *Federico di Montefeltro. Lo stato/Le arti/La cultura* (Rome: 1986) 61–79; *Da Bisanzio a Roma. Studi sul cardinale Bessarione* (Rome: 1999); idem, "Roma e L'Accademia Bessarionea", in Fiaccadori G., with the assistance of Cuna A. – Gatti A. – Ricci S. (eds.), *Bessarione e l'Umanesimo*, exh. cat., Biblioteca Nazionale Marciana (Naples: 1994) 119–127; idem, "Le accademie a Roma nel Quattrocento", in Jensen R. – Pade M. (ed.), *On Renaissance Academies. Proceedings of the International Conference 'From the Roman Academy to the Danish Academy in Rome. Dall'Accademia Romana all'Accademia di Danimarca a Roma', The Danish Academy in Rome, 11–13 October 2006* (Rome: 2010) 47–59.

3 Manzoli, "La fortuna di Bessarione" 83.

4 Finocchi Ghersi L., "Bessarione e la basilica romana dei Santi XII Apostoli", in Fiaccadori, *Bessarione e l'Umanesimo* 129. On the works commissioned by the cardinal in the residence, see Magister S., *Arte e politica: la collezione di antichità del cardinale Giuliano della Rovere nei palazzi ai Santi Apostoli* (Rome: 2002) 404–414; Schelbert G., *Der Palast von SS. Apostoli und die Kardinalsresidenzen des 15. Jahrhunderts in Rom* (Norderstedt: 2007) 134–152; Gatti I., *Il palazzo della Rovere ai SS. Apostoli: uomini, pietre, vicende. Parte I: I Cardinali Bessarione e Riario* (Rome: 2015).

5 Bianca, *Da Bisanzio a Roma* 170–172.

6 See most recently, with preceding bibliography, Dessì A., "Pomponio Leto e la topografia del Quirinale nei protocolli notarili", *Rivista dell'Istituto di Storia dell'Europa Mediterranea* 9/I n.s. (December 2021), Special Issue: Colesanti G.T. – Piñol D. – Sakellariou E. (eds.), *Il Notaio nella società dell'Europa mediterranea (secc. XIV–XIX). The Notary in the Mediterranean European Society (14th–19th centuries)* 287–310, as well as her essay in this book; Rowland I., "L'Accademia Romana fra Pomponio Leto e Angelo Colocci", in Mangani G. (ed.), "Raffaello e Angelo Colocci. Bellezza e Scienza nella costruzione del mito della Roma antica", exh. cat, Musei Civici di Palazo Pianetti (Santarcangelo di Romagna: 2021) 28–39. For the biography of Pomponio Leto see Accame M., *Pomponio Leto. Vita e insegnamento* (Rome: 2008); Accame M., "Pomponio Leto, Giulio", *Dizionario biografico degli italiani* 84 (Rome: 2015) 711–716.

the Baths of Constantine and those of the great temple of the Severan period (late 2nd–early 3rd century), known at the time as the "Frontispicium Neronis".[7] Vineyards, gardens, small houses, defensive structures, religious buildings, and more recent villas and lodges (*Casini*) were found among the ruins.[8]

Fifteenth-century academies took on a new form in the so-called literary *horti* as Rome emerged as a universal centre of literature and art in the early sixteenth century.[9] Like many patrons of the arts and members of the Curia, the Colonna opened their residences and gardens to men of letters, offering opportunities for groups of *sodales* to meet and converse. These convivial meetings customarily included lavish banquets, an abundance of jokes and witticisms, poetry, scholarly discourse on a wide variety of topics, and the composition of verse in praise of the lord and patron.[10] In the "Loggia dei Colonnesi" ("Loggia of the Colonna family") on the slopes of the Quirinal Hill, Cardinal Prospero received his circle of humanist and antiquarian friends, including Poggio Bracciolini, Leon Battista Alberti and Biondo Flavio.[11] The Loggia may also have housed the cardinal's valuable collection of antiquities, which included the celebrated *Belvedere Torso* and the *Three Graces* now in the Piccolomini Library in Siena.[12] Drawn to the charms of antiquity, Prospero organised excavations

7 On the temple see most recently, with preceding bibliography, Capanna M.C., "Ancora sul complesso templare severiano del Quirinale", *Mélanges de l'École française de Rome* 133.1 (2021) 173–208; Sommella P., "Sull'identificazione del tempio dei Giardini Colonna al Quirinale", *Atti dell'Accademia Nazionale dei Lincei. Rendiconti della Classe di Scienze Morali, Storiche e Filologiche* 9.31 (2020) 5–77; Pensabene P., "Il tempio gigantesco del Quirinale", in Picozzi M.G. (ed.), *Palazzo Colonna. Giardini. La storia e le antichità* (Rome: 2018) 11–44; Gallocchio E., "Le strutture romane lungo il pendio del Quirinale", in Picozzi M.G. (ed.), *Palazzo Colonna* 45–72; Picozzi M.G., "Templum Salutis et Serapis", in Picozzi M.G. (ed.), *Palazzo Colonna* 73–100; Spila A., "Palladio e Antonio da Sangallo il Giovane sul grande tempio del Quirinale", *Annali di architettura* 29 (2017) 135–142.

8 On the hill and its urban development over the centuries, see especially Negro A., *Guide rionali di Roma. Rione II – Trevi*, 2.1 (Rome: 1985); Mazzetti di Pietralata C., "Il Colle del Quirinale", in Campitelli A. – Cremona A. (eds.), *Atlante storico delle ville e dei giardini di Roma* (Milan: 2012) 71–81.

9 Gnoli D., "Orti letterari nella Roma di Leone X", *Nuova Antologia* 7.269 (January–February 1930) 3–19, 137–148.

10 Manzoli, *La fortuna di Bessarione* 86.

11 Petrucci F., "Colonna, Prospero", *Dizionario biografico degli italiani* 27 (Rome: 1982) 416–418; Rehberg A., "Pio II e i Colonna: fra 'amicizie' personali e interessi della Chiesa", in Di Paola R. – Antoniutti A. – Gallo M. (eds.), *Enea Silvio Piccolomini. Arte, storia e cultura nell'Europa di Pio II* (Rome: 2006) 434–436, with bibliography.

12 On the "Loggia" and ancient sculptures, see Musso L., "Le antichità Colonna nei secoli XV e XVI", in Carinci F. – Keutner H. – Musso L. – Picozzi M.G. (eds.), *Catalogo della Galleria Colonna in Roma. Sculture* (Rome – Busto Arsizio: 1990) 13–16; Finocchi Ghersi L., "Le residenze dei Colonna ai Santi Apostoli", in Chiabò M. – D'Alessandro G. – Piacentini P. – Ranieri C. (eds.), *Alle origini della nuova Roma. Martino V (1417–1431)* (Rome:

on the slopes of the Quirinal and restored what remained of the temple stairs and precious coloured marble floors, which were likely used to decorate the Loggia.[13] Advised by Leon Battista Alberti, the cardinal also attempted the recovery of the Roman ships in Lake Nemi.[14]

Prospero's nephew Giovanni Colonna, son of Antonio, Prince of Salerno, and Imperiale, daughter of Stefano Colonna, succeeded him in his ecclesiastical career [Fig. 5.1].[15] Born in 1453, Giovanni held the post of apostolic protonotary and in 1480 was elevated to cardinal with the title of Santa Maria in Aquiro, becoming bishop of Rieti in the same year.[16] He seems to have owned fragments of Egyptian granite sculptures whose whereabouts are currently unknown, likely originating from the Quirinal Serapeum.[17] Objects of this type suggest that Cardinal Colonna had an appreciation for "exotic sculptures", a growing interest among collectors of this period.[18] Like Cardinal Prospero before him, Giovanni was also in close contact with writers and humanists, including Marco Antonio Casanova and, above all, Evangelista Maddaleni Capodiferro, for whom he was 'patrono suo praecipuo' ('the principal patron').[19] Called

---

1992) especially 61–62, 67–68; Safarik E.A., *Palazzo Colonna* (Rome: 1999) 48, 60; Magister, *Arte e politica* 403–404; Palma Venetucci B., *Dallo scavo al collezionismo: un viaggio nel passato dal Medioevo all'Ottocento* (Rome: 2007) 35–36; Strunck C., *Berninis unbekanntes Meisterwerk. Die Galleria Colonna in Rom und die Kunstpatronage des römischen Uradels* (Munich: 2007) 59, 62; Schelbert, *Der Palast von SS. Apostoli* 159–166; Picozzi M.G., "Le sculture degli appartamenti. Contributo alla storia delle antichità della famiglia Colonna", in Picozzi M.G. (ed.), *Palazzo Colonna. Sculture antiche e dall'antico* (Rome: 2010) 13–14; Christian K.W., *Empire Without End: Antiquities Collections in Renaissance Rome, c. 1350–1527* (New Haven: 2010); Lucci M., "Le vicende dei Giardini Colonna tra XIV e XVI secolo", in Picozzi, *Palazzo Colonna* 104–111; Spila A., "The Double Stairway of the Temple of Serapis on the Quirinal Hill: Fame and Oblivion in the early modern Period", *Pegasus* 18/19 (2018) 43.

13 Biondo Flavio, "Book 1, Chapter 1," *Roma Instaurata* (Verona: 1481) in Valentini R. – Zucchetti G., *Codice topografico della città di Roma*, IV (Rome: 1953–1954) 283–284.

14 Petrucci, "Colonna, Prospero" 417; Musso, "Le antichità Colonna" 13.

15 Petrucci F., "Colonna, Giovanni", *Dizionario biografico degli italiani* 27 (Rome: 1982) 342–344; Scatizzi P., "I Colonna signori di Genazzano", in Bureca A. (ed.), *Il Castello Colonna a Genazzano. Ricerche e restauri* (Rome: 2000) 57, plate 6, footnote 9.

16 Scatizzi, "I Colonna signori di Genazzano" 57, pl. 6, n. 9.

17 Fusco L. – Corti G., *Lorenzo de' Medici Collector and Antiquarian* (Cambridge: 2006) 289–290, with preceding bibliography; Picozzi, "Le sculture degli appartamenti" 14.

18 Palma Venetucci B., "Antichità esotiche nel collezionismo del XV e XVI secolo", in Palma Venetucci B. (ed.), *Culti orientali tra scavo e collezionismo* (Rome: 2008) 73–88, with bibliography.

19 Ballistreri G., "Casanova, Marco Antonio", *Dizionario biografico degli italiani* 21 (Rome: 1978) 171; Ballistreri G., "Capodiferro, Evangelista Maddaleni, de', detto Fausto", *Dizionario biografico degli italiani* 18 (Rome: 1975) 621–625. See also Vatican Apostolic Library, Vat. Lat. 3351, fol. 1v. On the relations between the poet, Cardinal Giovanni Colonna, and various members of the Roman family, see especially: Castelli G.P., "*Ante diem clade Urbis*

FIGURE 5.1
Cardinal Giovanni Colonna. Engraving, 15.5 × 10.3 cm. From: De Santis Domenico, *Columnensium procerum imagines, et memorias nonnullas* [...] (Rome, Angelo Bernabò: 1675). Private collection
IMAGE © TIZIANA CHECCHI

Fausto, Capodiferro was Leto's student and participated in Roman humanist circles from an early age. He was recognised for the facility with which he composed Latin verse, including dialogic eclogues for performance at banquets and celebrations; we owe to Fausto, among other contributions, *Marcus Antonius*.[20] A dramatic epithalamium, *Marcus Antonius* was composed for the wedding of Marcantonio I Colonna and Lucrezia Franciotti Della Rovere, niece of Julius II, at the Roman residence at Santi Apostoli on 2 January 1508. Fausto also composed two eclogues in honour of the pontiff, the first of which, *Umber*, was performed in Cardinal Colonna's palace during a banquet given for Guidobaldo, Duke of Urbino.[21] As we shall see, this accomplished poet was also responsible for some of the works examined in this study.

interiit. Fausto Evangelista Maddaleni Capodiferro: il suo tempo, la sua famiglia e il Sacco di Roma", in *Dall'Archivio Segreto Vaticano. Miscellanea di testi, saggi e inventari*, 9 (Vatican City: 2016) 147–329 and *passim*.

20 Ballistreri, "Capodiferro" 622. Vatican Apostolic Library, Vat. Lat. 3351, fols. 108v–111r, in Ballistreri, "Capodiferro" 624 and in Castelli, "*Ante diem clade Urbis interiit*" 167–172.

21 Vatican Apostolic Library, Vat. Lat. 3351, fols. 24r–v; Ballistreri, "Capodiferro" 624.

Following the death of Innocent VIII on 25 July 1492, Rodrigo Borgia was supported by Cardinal Giovanni Colonna at the conclave and elected Pope on 11 August with the name of Alexander VI. In thanks, he gifted the rich *commenda* of the Abbey of Subiaco to the cardinal along with numerous other benefices.[22] The fate of the Abbey of Subiaco was closely tied to the alternating fortunes of the Colonna family in the century that followed, often becoming the setting for the frequent struggles that saw the powerful Roman baronial clan oppose the papacy in the broader context of the wars of supremacy between France and Spain in Italy.[23] Intensifying conflicts with Pope Alexander VI forced Cardinal Giovanni to flee Rome for Sicily in 1499; his escape was followed in 1501 by the confiscation of properties, stripping of titles, and exile of his relatives.[24] Alexander VI eventually assigned the *commenda* of Subiaco to his own family in perpetuity.[25] Cardinal Giovanni returned to Rome following the death of the pope in 1503 and Alexander VI's successor, Julius II, restored to him his properties, including the Abbey of Subiaco. Elected in part thanks to the vote of Cardinal Colonna, the new pope initially showed generosity towards the family. After the death of Cardinal Giovanni on 26 September 1508, his nephew Pompeo (1479–1532), son of Vittoria Conti and Girolamo Colonna, the illegitimate son of Antonio, Prince of Salerno, succeeded him as proprietor of the Abbey [Fig. 5.2].[26]

---

22 Archivio Colonna, III BB 6, 22, cit. in Scatizzi, "I Colonna signori di Genazzano" 19 and footnote 68. The Colonna family Archive is currently housed at the Biblioteca Statale del Monumento Nazionale of S. Scolastica at Subiaco. On Giovanni Colonna as commendatory abbot of Subiaco, see Mirzio Cherubino da Treviri, *Chronicon sublacense (1628–1630)*, ed. L. Branciani, vol. 2 (Subiaco: 2014), fols. 153r–v; Jannuccelli G., *Memorie di Subiaco e sua Badia* (Genoa: 1856) 231–232; Mariani L., *Storia di Subiaco e suo distretto Abbaziale*, ed. M. Sciò (Subiaco: 1997) 175–177; Carosi G.P., *Badia di Subiaco* (Subiaco: 1970) 77–78. Scatizzi, "I Colonna signori di Genazzano" 19.

23 Scatizzi, "I Colonna signori di Genazzano" 19–29; Serio A., *Una gloriosa sconfitta. I Colonna tra papato e impero nella prima età moderna* (Rome: 2008) 119–335; Carosi, *Badia di Subiaco* 77–78, 80–83.

24 Scatizzi, "I Colonna signori di Genazzano" 20.

25 Carosi, *Badia di Subiaco* 77–78.

26 Giovio Paolo, *Le vite di Leon Decimo et D'Adriano Sesto Sommi Pontefici, et del Cardinal Pompeo Colonna* [...] *tradotte da M. Lodovico Domenichi* (Venice, Giovanni de' Rossi: 1557), fols. 138r–184v; Consorti A., *Il cardinale Pompeo Colonna* (Rome: 1902); Petrucci F., "Colonna, Pompeo", *Dizionario biografico degli italiani* 27 (Rome: 1982) 407–412. On his period of rule over the Subiaco Abbey, see Mirzio, *Chronicon sublacense*, fols. 153v–165r; Jannuccelli, *Memorie di Subiaco* 232–234; Mariani, *Storia di Subiaco* 177–191; Carosi, *Badia di Subiaco* 78–82.

FIGURE 5.2
Cardinal Pompeo Colonna. Engraving, 15.7 × 10.6 cm. From: De Santis Domenico, *Columnensium procerum imagines, et memorias nonnullas* [...] (Rome, Angelo Bernabò: 1675). Private collection
IMAGE © TIZIANA CHECCHI

After an education focusing on literary studies, Pompeo distinguished himself by his military valour, daring character, and burning desire for glory.[27] Soon, however, he was forced to embrace a career in the Church to preserve the family's representation in the Sacred College of Cardinals and maintain the enormous profits connected to the resulting benefices. As Paolo Giovio recounts in the *Life* of his friend Colonna, shortly after Cardinal Giovanni died of 'una febre autunnale' ('an autumn fever'), Pompeo was awarded the *commenda* of Santa Maria di Grottaferrata and the Bishopric of Rieti, as well as the 'molto ricca' ('very rich') Abbey of Subiaco and 'alcuni piacevoli luoghi a Subiaco' ('some pleasant places at Subiaco').[28] From then on, Pompeo lived in:

27 Consorti, *Il cardinale Pompeo Colonna* 14.
28 Giovio, *Le vite di Leon Decimo* [...] *et del Cardinal Pompeo Colonna*, fol. 147r.

> maggior pompa et più delicatamente, abbracciando le lettere con grandissimo ardore, senza però intermettere punto i pensieri, i quali appartenevano alle cose.[29]

> greater pomp and more delicately, embracing the study of letters with great ardour, without however abandoning his thoughts of more important things.

Although he was now a man of the Church, Colonna retained the ardent and turbulent spirit that had always characterised him, defending his family's interests in the complex political situation of the time and often warring in the ranks of the Hispano-imperial faction against the papacy. When Julius II became seriously ill in 1511, to the extent that his death seemed imminent, Pompeo led a popular movement in opposition to the pontiff with Antimo Savelli, Roberto Orsini, and Giorgio Cesarini.[30] When Julius II unexpectedly regained his health, Pompeo was forced to leave Rome and seek refuge first in Nemi and then in Subiaco, where he began to strengthen the fortress to defend against possible attacks by the papal army [Fig. 5.3].[31] In 1512, a papal sentence declared him a rebel and deprived him of the bishopric and its ecclesiastical benefices. Pompeo, unwilling to submit, barricaded himself in the Cervara stronghold in the territory of the Abbey of Subiaco, where he remained until he obtained the forgiveness of the pontiff, albeit without official posts and prebends.[32] Colonna only regained possession of his properties in 1513 at the behest of Leo X, successor to Julius II.[33] That same year, Pompeo formally handed over the *commenda* of Subiaco to his nephew Scipione Colonna, but retained the rights of usufruct and return and, in practice, also the running of the Abbey.[34] Regarding the period that followed these events, Giovio relates:

> D'alhora in poi Pompeo honorato habitò Roma et la Corte per due anni, di modo che la state andava spesso a Subiaco et Frascati senza lasciar

29 Giovio, *Le vite di Leon Decimo*, fol. 147r.

30 Petrucci, "Colonna, Pompeo" 407; Giovio, *Le vite di Leon Decimo*, fol. 147v. ff.; Petrucci, "Colonna, Pompeo" 407–408; Serio A., "Pompeo Colonna tra papato e 'grandi monarchie', la *pax* romana del 1511 e i comportamenti politici dei baroni romani", in Visceglia M.A. (ed.), *La nobiltà romana in età moderna* (Rome: 2001) 63–87.

31 Giovio, *Le vite di Leon Decimo*, fols. 150r–151r; Mirzio, *Chronicon sublacense*, fol. 154r.

32 Giovio, *Le vite di Leon Decimo*, fol. 151v; Mirzio, *Chronicon sublacense*, fol. 154v.

33 Coppi A., *Memorie Colonnesi* (Roma: 1855) 266; Petrucci, "Colonna, Pompeo" 408.

34 Mirzio, *Chronicon sublacense*, fol. 154v. See also Jannuccelli, *Memorie di Subiaco* 234–238; Mariani, *Storia di Subiaco* 187–191; Carosi, *Badia di Subiaco* 80–82.

FIGURE 5.3 Abbey Fortress of Subiaco
IMAGE © PIERO ANTONUCCI

> punto la domestichezza della gioventù Romana, et l'amicitia de' cittadini grandi, et massimamente di quei Cardinali, i quali erano nobili et d'animo grandi. Percioche egli si dilettava di cacciare et uccellare: di maniera che spesso con mirabile apparato et grandissima spesa faceva conviti publici in mezzo le campagne et appresso le fontane con gran piacevolezza et allegrezza.[35]
>
> From then on, the honoured Pompeo lived in Rome and at the Court for two years, so that in summer he would often go to Subiaco and Frascati without in any way giving up his familiar relations with the youth of Rome and his friendship with important citizens, especially those Cardinals who were noble and of elevated spirit. He enjoyed hunting wild animals and birds and, for this reason, often held marvellously lavish and expensive public banquets in the countryside and near springs, which were extremely pleasant and festive.

35 Giovio, *Le vite di Leon Decimo* [...] *et del Cardinal Pompeo Colonna*, fol. 152v. On hunting as aristocratic entertainment see especially Ribouillault D., *Rome en ses jardins: Paysage et pouvoir au XVIe siècle* (Paris: 2013) 123–126, with bibliography.

In appearance, Pompeo was 'molto eccellente con generosa fronte et con bella presenza di corpo' ('very handsome, with an ample forehead and a fine physique') and showed himself:[36]

> in tutte le cose ch'egli toglieva a fare et trattare, et con splendor civile molto galante, et sontuoso prelato, et sopra tutti gli altri huomini in ogni giuoco et negotio con certa libertà soldatesca allegrissimo.[37]

> a very gallant and elegant prelate in everything that he did, possessed of great civility, and above all other men very cheerful in both play and work, with a certain soldierly nonchalance.

Shortly after his elevation to cardinal in 1517, Pompeo commissioned a painting on canvas from Raphael depicting *Young St. John the Baptist*, which he later gave to his doctor Jacopo Berengario da Carpi in gratitude for having cured him of an illness.[38] Between 1517 and 1524, he commissioned a missal composed of seven beautifully illuminated volumes.[39] This precious work emerged in the context of contemporary circles of bibliophile intellectuals and aligned with antiquarian interests of the time, as evidenced by its vast repertoire of motifs derived from Greco-Roman and Egyptian antiquities. Among them we find floral trails with reproductions of medals, coins and bas-reliefs, 'almost to the extent of constituting a catalogue of the best-known and most frequently represented works in contemporary notebooks of the great archaeological discoveries held in Roman collections of the sixteenth century.'[40]

---

36 Giovio, *Le vite di Leon Decimo* [...] *et del Cardinal Pompeo Colonna*, fols. 152v–153r.

37 Ibidem, fol. 152v.

38 Vasari Giorgio, *Le Vite de' più eccellenti architetti, pittori, et scultori italiani, da Cimabue insino a' tempi nostri. Nell'edizione per i tipi di Lorenzo Torrentino. Firenze 1550*, Bellosi L. – Rossi A. (eds.), vol. 2 (Turin: 2015) 637. The painting then passed into the hands of the Florentine Francesco Benintendi and was displayed in 1589 in the Tribuna degli Uffizi. The work, currently no. 1446 of the Galleria degli Uffizi, is hotly debated by critics, divided between those who consider it entirely by Raphael and those who see in it the intervention of his most important pupils, Giulio Romano and Giovan Francesco Penni. It is thought to date between 1517, the year in which Pompeo Colonna was appointed cardinal, and about 1518.

39 Mochi C., "Il messale di Pompeo Colonna: antichità ed egizianismi a Roma", in Colonna S. (ed.), *Roma nella svolta tra Quattro e Cinquecento* (Rome: 2004) 439–452. At present, six volumes are in Manchester (England) at the John Rylands University Library, while the seventh is in Camarillo, Los Angeles, in the St. John Seminary. On Pompeo Colonna as a patron see Baker Bates P., "A Portrait of Cardinal Pompeo Colonna, Rival and Imitator of the Papal Caesars", *Papers of the British School at Rome* 76 (2008) 183–199, 356.

40 Mochi, "Il messale di Pompeo Colonna" 440.

Although critics have suggested Pompeo's involvement in the so-called Nymphaeum of Genazzano, it is difficult to attribute responsibility for the commission to him.[41] Pompeo never ruled over this village, nor did he inherit it, being the son, as has been mentioned, of Girolamo Colonna, who was the illegitimate child of Antonio (d. 1472), Prince of Salerno and prefect of Rome. In his 1470 will, Antonio named his legitimate sons Cardinal Giovanni (d. 1508), Pietro Antonio (d. 1479), and Prospero (d. 1523) as the heirs to the estate, undivided and indivisible.[42] From this brief genealogy, it is clear that Cardinal Giovanni and his brother Prospero, and later Prospero's son and heir, Vespasiano, exercised dominion over Genazzano, and were probably responsible for the construction of the nymphaeum. In confirmation of this likelihood, it is interesting to note what Giovio said about Pope Clement VII's destruction of Colonna possessions in revenge for the Colonna troops' looting of the Vatican palaces in September 1526. Giovio relates how, during the devastation and plundering of fourteen villages, the papal forces acted with:

> molta crudeltà su Subiaco ch'era proprio le delitie di Pompeo, et anco a Ghinazzano, ma con alquanto più piacevole ira et ciò per usare copertamente rispetto al bellissimo palazzo di Vespasiano, che Prospero aveva edificato.[43]

> great cruelty against Subiaco, which was the pleasure ground of Pompeo, and also against Genazzano, but with a somewhat more tempered wrath, and this in order to act with covert respect towards the beautiful palace of Vespasiano which Prospero had built.

---

41 See especially Frommel C.L., "Bramantes 'Ninfeo' in Genazzano", *Römisches Jahrbuch für Kunstgeschichte* 12 (1969) 159; Bruschi A., *Bramante architetto* (Bari: 1969) 1050–1051; Thoenes C., "Note sul 'Ninfeo' di Genazzano", in *Studi Bramanteschi* (Rome: 1974) 578–579; Borsi S., "Il ninfeo di Genazzano", in Borsi F., with the assistance of Borsi S., *Bramante* (Milan: 1989) 326–329; Barucco P., "L'ordine tuscanico nel Ninfeo e nel Castello. Problemi di datazione e attribuzione", in Bureca (ed.), *Il Castello Colonna a Genazzano* 153–154; Döring M., "La nascita della rovina artificiale nel Rinascimento italiano ovvero il 'Tempio in rovina' di Bramante a Genazzano", in Di Teodoro F.P. (ed.), *Donato Bramante. Ricerche, proposte, riletture* (Urbino: 2001) 382; Frommel C.L., *Architettura alla corte papale nel Rinascimento* (Milan: 2003) 234–236, with further bibliography.

42 Scatizzi, "I Colonna signori di Genazzano" 18.

43 Giovio, *Le vite di Leon Decimo* [...] *et del Cardinal Pompeo Colonna*, fol. 168v. Vespasiano was the son and heir of Prospero Colonna, also a man of arms, and died in 1528. For the biography of these two figures see especially Petrucci F., "Colonna, Prospero", *Dizionario biografico degli italiani* 27 (Rome: 1982) 418–426; Idem, "Colonna, Vespasiano", Petrucci, "Colonna, Prospero" 447–448; Scatizzi, "I Colonna signori di Genazzano" 18–23, 57 plate 6.

Of sophisticated and learned eloquence, Cardinal Pompeo was also a poet and writer.[44] He was known to demonstrate courtesy and generosity towards men of letters and poets such as Vincenzo Pimpinella, who was also his secretary although he was later dismissed after being accused of theft, Francesco Maturanzio, Tamira Macaroccio, Marco Antonio Casanova, Camillo Porzio, "Phedro da Volterra", namely Tommaso Inghirami, also called Fedra or Fedro.[45] In addition to Giovio, who wrote the *Life of Pompeo*,[46] authors such as Giano Anisio, Lucio Domizio Brusoni and Girolamo Balbi dedicated poems to him. Writers including Pietro Gravina, Pietro Corsi da Carpineto, Antonio Tebaldeo, Gerolamo Borgia, and Matteo Bandello also composed verses in his honour.[47]

Among the many poems dedicated to Cardinals Giovanni and Pompeo Colonna by the men of letters in their service or belonging to their entourage are those dedicated to the places where these poets accompanied their protectors and patrons in the territory of the Abbey of Subiaco.[48] Some of Fausto Evangelista Maddaleni Capodiferro's poems, set in Subiaco, celebrate the river Aniene, its valley and the magnificent villa built on its banks by the Emperor Nero in the early years of his reign.[49] The villa, known as the *Simbruina Stagna*, consisted of a series of pavilions and residential buildings arranged along the banks of three artificial lakes created by blocking the river with three dams.

44 Colonna was the author of *Apologiae mulierum* published in Zappacosta G., *Studi e Ricerche sull'Umanesimo italiano* (Bergamo – Milan – Florence – Rome – Messina: 1972) 159–246 and recently by Franco Minonzio (Como: 2015). Also by Pompeo is an epigram published a century after his death by Giulio Cesare Capaccio in *Il Forastiero* (Naples, Giovanni Domenico Roncangliolo: 1634) 446–455.

45 Serio, *Una gloriosa Sconfitta* 322 footnote 6; Ballistreri G., "Casanova, Marco Antonio" 173. Falzone P., "Maturanzio (Mataratius), Francesco", *Dizionario biografico degli italiani* 72 (Rome: 2009) 338–341. Ballistreri, "Casanova, Marco Antonio" 171–174. Giovio, *Le vite di Leon Decimo* […] *et del Cardinal Pompeo Colonna*, fol. 146v. In my opinion, the person mentioned by Giovio should be identified as Tommaso Inghirami, called Fedra (or Fedro), and not as the unknown 'Pietro da Volterra' as claimed by Consorti and Petrucci. For the biography of Tommaso Inghirami, see Benedetti S., "Inghirami, Tommaso, detto Fedra", *Dizionario biografico degli italiani* 62 (Rome: 2004) 383–387. For the identification as 'Pietro da Volterra': Consorti, *Il cardinale Pompeo Colonna* 30; Petrucci, "Colonna, Pompeo" 411.

46 Giovio Paolo, *De uita Leonis decimi pont. max. libri 4. His ordine temporum accesserunt Hadriani sexti pont. max. et Pompeii Columnae cardinalis vitae* […] (Florence, Lorenzo Torrentino: 1548).

47 Petrucci, "Colonna, Pompeo" 411.

48 I thank the archivist Piero Scatizzi for bringing this to my attention and for his valuable advice.

49 Vatican Apostolic Library, Vat. lat. 3419, fols. 11r–v, 12v, 75v; Vatican Apostolic Library, Vat. lat. 3351, fols. 32r, 57v.

This history explains the names *Sublaqueum* or *Villa neroniana Sublaquensis*, given to the villa and later to nearby Subiaco.[50] Capodiferro composed a poem in praise of Cardinal Pompeo 'in arce Cervariae', that is, in the fortress overlooking the village of Cervara.[51] Due to its elevated position on the edge of the Simbruini mountains, the fortress was a fundamental strategic stronghold guarding the road from the upper Aniene valley into the border area between the Kingdom of Naples and the Papal State – an important site for ensuring control and defence of the Abbey's territory. After the completion of the aforementioned restoration project, Pompeo sought refuge from the wrath of Julius II della Rovere in Subiaco in 1512. Fausto Evangelista Maddaleni Capodiferro likely also belonged to the cardinal's entourage and drew inspiration from the site's extraordinary position while composing poems in praise of his patron:

In arce Cervariae
Victor inaccessam[52] Pompeius condidit arcem:
naturae atque artis munere pollet opus.
Arx habet haec muros, hic inexpugnabile pectus
stans[53] hos haeroas ferre Columna solet.[54]

In the fortress of Cervara
Pompeo the victor founded the inaccessible fortress:
the work is strong thanks to nature and art.
This fortress has walls, here the invincible breast,
Colonna, unmoved, supports these heroes.

50 On the villa see, most recently and with further bibliography, Parolini C., "Simbruina Stagna. Una nuova lettura alla luce dell'analisi geomorfologica", *Scienze dell'Antichità* 25.1 (2019) 173–186; Zaccaria M., "Il Sublaqueum. La villa di Nerone a Subiaco", in Valenti M. (ed.), *Residenze imperiali nel Lazio* (Monte Porzio Catone: 2008) 43–52; Di Matteo F., *Villa di Nerone a Subiaco. Il complesso dei 'Simbruina Stagna'* (Rome: 2005); Fiore Cavaliere M.G., "La villa di Nerone a Subiaco e la fondazione del monastero benedettino di S. Clemente", in Mari Z. – Petrara M.T. – Sperandio M. (ed.), *Il Lazio tra antichità e medioevo. Studi in memoria di Jean Coste* (Rome: 1999) 341–367; Fiore Cavaliere M.G., *Sublaqueum – Subiaco. Tra Nerone e S. Benedetto* (Rome: 1994).

51 Vatican Apostolic Library, Vat. Lat. 3351, fols. 123r, 125r. See: Castelli, "*Ante diem clade Urbis interiit*" 199–200.

52 In place of 'Invictam Invictus'.

53 In place of: 'Stans'.

54 Vatican Apostolic Library, Vat. Lat. 3351, fol. 123r.

In arce Cerbereae
Cerberus heic spernit medicatam frugibus offam,
tempnit et herculeas semideumque manus.
Non heic Persephone servatur, at orta Columna
substinuisse polum pondere nixa suo.

In the fortress of Cerberus
Here Cerberus refuses to eat the drugged cake,
scorns the hands of Hercules and the demigod.
Here Persephone is not guarded, but Colonna rises
who, with his great strength, strives to support the heavens.

Ibidem
Cerbere, custodi solido ex adamant [Adamante] Columnam!
Haec quovis frangi dura cruore nequit,
at possunt humiles lacrumae mollire precantum;
hostiles nequeunt ac violare manus.
Haec, haec Palladium est Urbi patriaeque saluti:
haeic, Ichare, Alcyde, deciperisque, Sinon.[55]

In the same place
O Cerberus, guard from the hard diamond [Adamante] Colonna!
It, hard, cannot be broken by any blood,
but the humble tears of those who pray can soften it;
Enemy hands cannot violate it.
This, this is the Palladium for Rome and for the salvation of the fatherland:
here, O Icarus, O Alcides, and you, O Sinon, are deceived.

Subtly alluding to the fortified site, the poet makes a mythological reference to Cerberus and the salvation of the homeland, imagining the impregnable Colonna fortress at Cervara transformed into the place where the Palladium of Rome is now found.[56] Capodiferro also composed verses in honour of his primary patron, Cardinal Giovanni Colonna, one of which reads:

Sublaquei. Ad fontem Reginae[57] Baccho et Naiadi sacrum.
Non heic Moecenas dominus, sed lympha merumque:
Ni mecum coenas fit meus heic dominus.

55 Vatican Apostolic Library, Vat. Lat. 3351, fol. 125r.
56 Castelli, "*Ante diem clade Urbis interiit*" 199–200 footnote 155.
57 Added between the lines.

Sextili potius sint saturnalia mense,
et liceat servis libera verba loqui.[58]

In Subiaco. At the Queen's Fountain, sacred to Bacchus and the Naiad.
Here the master is not Maecenas, but water and pure wine:
and unless you do not dine with me here those will become my master [i.e. water and pure wine].
Let the Saturnalia be held in August,
and let the servants speak freely.

This poem celebrates one of Cardinal Giovanni Colonna's pleasure grounds near Subiaco, a spring where he enjoyed himself in the company of his court at pleasant gatherings enlivened by food, wine, verse and erudite entertainments. The poet notably states that his patron, Cardinal Colonna, is absent, an observation that aligns the work with the period during which Pompeo was forced to flee to Sicily to escape Pope Alexander VI's persecution.[59] A verse by Marco Antonio Casanova, who worked throughout his life in the service of various members of the Colonna family, particularly Cardinal Pompeo, is likely also dedicated to this spring.[60] One of his poems dedicated to the Cardinal reads as follows:

De Pompeiana lympha, olim augusta.
Olim augusta fuit, nunc Pompeiana vocatur,
Assidue magnis lympha dicata viris.
Fons celeber dominis,
Augusto non fuit ullus
Maior, Pompeio nullus erit melior.[61]

Pompeian water, once Augustan.
Once Augustan, now known as Pompeian,
water always dedicated to great men
famous fountain for (his) lords (masters),
there was none greater than Augustus,
none will be better than Pompeo.

58 Vatican Apostolic Library, Vat. Lat. 3351, fol. 57v.

59 As can be deduced from some of his poems, Fausto also spent at least some time in Sicily with Cardinal Giovanni. See Castelli, "*Ante diem clade Urbis interiit*" 173.

60 Giovio, *Le vite di Leon Decimo* [...] *et del Cardinal Pompeo Colonna*, fol. 146v; Ballistreri, "Casanova, Marco Antonio" 171–174. Casanova also composed verses dedicated 'Ad amicos Sublacum incolentes': Vatican Apostolic Library, Vat. Lat. 2833, fol. 193r–v.

61 Vatican Apostolic Library, Vat. Lat. 2833, fol. 192r.

In these verses, as in those by Fausto Evangelista Maddaleni Capodiferro cited above, the place celebrated is a spring that, this time, Casanova refers to as 'Pompeiana lympha', or 'Pompeian waters/spring' in tribute to Pompeo Colonna, which he states were once called 'Augustan'.

In the twelfth chapter of *curator aquarum* Sextus Julius Frontinus' treatise *De Aquae Ductibus Urbis Romae*, composed between 97 and 98 AD, we learn that the name 'Fons Augusta' or 'Augustan Water' was given to a spring that emperor Octavianus Augustus had channelled into the Marcian aqueduct in the last decade of the first century BC.[62] The *Aqua Marcia* had its springs in the upper Aniene valley in the territory of the Abbey of Subiaco near the village of Marano.[63] The 'Augusta' spring celebrated by the Colonna family poets lay slightly upstream, and gave its name to the mountain called Agosta above and the *Castrum* that sprang from there in the following centuries.[64] It also inspired the village's coat of arms, which features a sequence of arches alluding to the aqueduct. In my opinion, this motif is also recognisable in the flag depicted in a fresco from the early fourteenth century that decorates the Cosmatesque cloister of the monastery of Santa Scolastica at Subiaco.[65] These observations support an attribution of the pleasure grounds near Subiaco praised by Fausto Evangelista Maddaleni Capodiferro as the spring consecrated 'to Regina, to Bacchus and to the Naiad', celebrated by Marco Antonio Casanova as 'Pompeian water, once Augustan. Once Augustan, now known as Pompeian'. The venue for the grand banquets of Giovanni Colonna's court is none other than the spring formerly known as 'Augusta'; it formed a charming, little lake commonly referred to as the 'Peschiera' ('Fishpond') beneath the village of Agosta in the territory of the Abbey of Subiaco.[66]

This splendid site had been a place of relaxation for several popes, including among them, Pope Pius II (Enea Silvio Piccolomini, 1405–1464; elected pontiff on 19 August 1458), a well-known humanist who, in mid-September 1461 during the journey from Tivoli to Subiaco, stopped for refreshments near the clear

62 Panimolle G., *Il fiume Aniene e i suoi castelli* (Subiaco: 2016) 90; idem, *Le acque e gli acquedotti di Roma antica* (Subiaco: 2007) 43–46.

63 Panimolle, *Le acque e gli acquedotti* 43–60.

64 Idem, *Il fiume Aniene* 90–99, especially 90.

65 Hitherto identified as the coat of arms of the town of Ponza, now Arcinazzo Romano, for the reasons stated the flag should be interpreted as that of the town of Agosta. Fabio Mari independently reached the same identification of the flag as the coat of arms of Agosta, see Mari F., 'Il paesaggio legittimante. Intorno alle pitture del Chiostro cosmatesco di Santa Scolastica a Subiaco' in press.

66 Vatican Apostolic Library, Vat. Lat. 3351, fol. 57v; Vatican Apostolic Library, Vat. Lat. 2833, fol. 192r.

spring whose waters formed the Peschiera.[67] He describes the beauty of the place in his *Commentaries*:

> ['Beneath the village of Agosta ...]
>
> sgorga dal vivo sasso una vena d'acqua con tanto impeto quanto ne basterebbe per mandare continuamente quattro o più macine di mulino. La fonte si stende fra le roccie (sic) con l'ampiezza di un'aia, ma in profondità non supera mai l'altezza di un uomo, brillante nel fondo per la ghiaia, che in più posti a seconda degli zampilli della sorgente, saltella con dolce mormorio. L'acqua è cristallina e tanto trasparente che mostra nettamente il fondo, gelida e dolce al gusto.[68]
>
> a stream of water flows from the living rock with as much force as would suffice to continually drive four or more millstones. The spring extends among the rocks to the size of a threshing floor, but in depth it never exceeds the height of a man, shining at the bottom because of the gravel, which in several places, depending on the gushes of the spring, leaps up with a sweet murmur. The water is crystalline and so transparent that it clearly reveals the bottom, cold and sweet to the taste.

The *Commentaries* reveal how the pontiff consumed his meal near the fountain with the cardinals and dignitaries of the papal retinue and personally drank from the spring fresh waters. Numerous people rushed from nearby villages to see the pontiff and stopped to eat quietly in the meadows on the other side of the spring. After the meal, the local peasants began to fish in the river Aniene, whose course runs along the road to Subiaco, as a sort of tribute to the pontiff. And upon setting out again, the pope watched them with pleasure; each time they caught a fish, they greeted him with loud cries and, as a sign of devotion, brought the trout to his servants. Because of this, Pius II spent most of his journey 'with the greatest delight'.[69]

In the unpublished manuscript *Descrittione et Historia dell'Abbadia di Subiaco* dated 1615, the painter Rutilio Scotti speaks of the entertainments

---

67 Panimolle G., *Agosta castello dell'Abbazia sublacense* (Rome: 1986) 65. Pellegrini M., "Pio II", in *Enciclopedia dei Papi*, II (Rome: 2000) 663–685; Pellegrini M., "Pio II, papa", *Dizionario biografico degli italiani* 83 (Rome: 2015) 794–803.

68 Pio II (Enea Silvio Piccolomini), *I Commentari*, ed. G. Bernetti, II (Siena: 1972) 273–274.

69 Ibidem 274.

enjoyed by the Colonna in the territory of the Abbey of Subiaco.[70] With reference to Affile, his home village, Scotti wrote of the wealth of its game:

> li Posteri Prencipi, come ne vien da nostri Veterani detto, […] si dilettavano grandemente di cacciare et uccellare, di maniera che spesso con lieti animi, et mirabile apparato facevano recreationi per le fontane domesticamente con i Gentilhuomini convittori, et altri simili loro famigliari. Particolarmente gl'Ill.mi Colonnesi, quali per la gran delettatione l'havevano già fatta caccia riservata sotto grandissime pene […]. E specialmente dicono il Sig. Pompeo Cardinale che era uso nel cacciare […]. Di qui poi cacciando con picciolo viaggio scendendo nel paese dell'Augusta ove si facevano reali Conviti per laamenità e piacevolezza di una vaga peschiera, dove con real apparato solevano essere raccolti da persone di qualità della nostra Abbadia, et quasi sempre nell'uscir dell'estate tosto che […] l'aria cominciava a farsi temperata.[71]

> The Princes, as is recounted by our old men, […] greatly enjoyed hunting animals and birds, so that they often with cheerful spirit and wonderful luxury entertained themselves near the springs, in familiar company with the gentlemen of the Abbey and others close to them. In particular the illustrious Colonnesi so loved hunting that they reserved the area for themselves, on pain of serious punishment. […] And they say that Cardinal Pompeo especially, who was a keen hunter […]. Furthermore, while hunting a short journey would bring them down to the village of Augusta where lavish banquets were held thanks to the beauty and pleasantness of a charming fishpond, where they were greeted with royal

70 Scotti Rutilio, *Descrittione et Historia dell'Abbadia di Subiaco illustrata da Ruttilio Scotti Affilano Pittore e Cosmografo. Consegnata all'Ill.mo et Rev.mo Sig.re il Sig.r Scipione Card.le Burghese. Perpetuo Commendatario. Data in Affile, sotto l'anno del 1615*, unpublished mss. held at the Vatican Apostolic Library, Chig.I.23, where there is also a copy (Vatican Apostolic Library, Chig.I.24). Another transcription of the manuscript was made in 1928 by 'Giggi Rossi' and was kept in the Municipality of Affile. A copy is in the Biblioteca Statale of S. Scolastica at Subiaco. In the quotations given here, reference has been made to the latter. The painter Rutilio Scotti was originally from Affile, a village near Subiaco. Almost unknown to historians, the artist worked in the monastery of Santa Scolastica at Subiaco in the nearby Benedictine monastery of San Donato and the monastery of the Sacro Speco of San Benedetto, based on accounts from the 1640s. Cfr. Checchi T., "L'opera di Vincenzo Manenti nei monasteri benedettini di Subiaco e il *San Francesco* nella chiesa di S. Maria *in arce* a Jenne", in Paniccia C. – Pistilli P.F. (eds.), *Dalla baronia al papato. Jenne al tempo di papa Alessandro IV* (Roma: 2023) 169.

71 Scotti, *Descrittione e Historia dell'Abbadia di Subiaco* 146–147.

> luxury by persons of quality from our Abbey and almost always at the end of summer as soon as [...] the air began to cool down.

As Giovio wrote, Subiaco and its territory was 'the pleasure ground of Pompeo'.[72] To this effect, when Pope Clement VII had the houses and possessions of the Colonna attacked and burned in revenge for the September 1526 looting of the Vatican palaces, he acted with particular cruelty against these places.[73] Describing the same events, Valesio says that during the devastation, the papal troops destroyed 'un luogo di delizie, che egli [Pompeo] s'era fabricato appresso Subiaco' ('a pleasure ground that he [Pompeo] had built for himself near Subiaco').[74]

The Jesuit Domenico Antonio Pierantoni, writing in the early eighteenth century, advanced a better impression of the appearance of the fishpond near Agosta:

> Alle radici della collina di duro sasso, sopra cui sta fondato il castello dell'Agosta, così denominato *ab Aqua Augusta* [...], col commodo di molti fonti di acqua, che vi sgorgano, vi è stata fabricata una Peschiera per vivaio, e conserva delle Trotte; la quale se bene dicesi fatta da gli Abbati Commendatarii Sublacensi della famiglia Colonna, circa l'anno 1570, come alcuni lo arguiscono dalla loro arma, cioè di una colonna coronata intagliata in marmo, posta sopra di una gran Porta, in mezzo ad una longa muraglia, che si distende fino al letto del Fiume, nulladimeno si riconosce più antica; mercé che, più di un secolo prima, si legge descritta da Papa Pio secondo, anno 1461, quando egli, nel viaggio fatto da Tivoli a Subiaco, alla Visita del Sacro Speco di S. Benedetto, invitatovi dal primo Abbate Commendatario Cardinale Giovanni Torrecremata, attesta di haver desinato alle sponde di questa Peschiera.[75]

---

72 Giovio, *Le vite di Leon Decimo* [...] *et del Cardinal Pompeo Colonna*, fol. 168v. The destruction also extended to the 'delitie di Pompeo' ('pleasure grounds of Pompeo') in the territory of Subiaco, as Scotti recalls regarding the village of Affile, reporting what 'ricontono i nostri antichi veterani' ('our aged veterans recount'), that is the village elders. Scotti, *Descrittione e Historia dell'Abbadia di Subiaco* 164.

73 On the events of 1526 and those of the more tragic "Sack" of Rome the following year see Serio, *Una gloriosa Sconfitta* 308–320; Scatizzi, "I Colonna signori di Genazzano" 22; Piergentili P.P. – Venditti G. (eds.), *Scorribande, Lanzichenecchi e soldati ai tempi del sacco di Roma. Papato e Colonna in un inedito epistolario dall'Archivio Della Valle-Del Bufalo (1526–1527)* (Rome: 2009).

74 Archivio Colonna, II A 40: Valesio Francesco, *Della Istoria di Casa Colonna*, mss. (ca. 1735–1740), III, 329.

75 Pierantoni Domenico Antonio, *Aniene illustrato*, ed. G. Giammaria (Anagni: 2003) 220.

> At the feet of the hill of hard rock on which the village of Agosta stands, thus named for the *Acqua Augusta* [...] with the convenience of the many springs that gush out there, a fishpond was built to raise and keep trout. Although it is said to have been constructed by the Commendatory Abbots of Subiaco of the Colonna family, in around the year 1570, as some deduce from their coat of arms, namely a column with a crown carved in marble, placed above a great door in the middle of a long wall that extends right down to the bed of the river, it is nonetheless clearly older; indeed, more than a century earlier, we read a description of it by Pope Pius II, of the year 1461, when, during his journey from Tivoli to Subiaco to visit the Holy Cave of St Benedict on the invitation of the first Commendatory Abbot Cardinal Giovanni Torrecremata, he attests that he dined on the shores of this fishpond.

The writer continues:

> Comunque si sia, questa Peschiera ella si è in forma riquadrata per longhezza di circa cento passi con grosso muro intorno, ben terrapienato al di fuori, e con quattro aperture, o bocche per l'esito dell'acqua, fresca, e limpida.[76]
>
> Whatever the case, this fishpond is square in shape with sides about a hundred paces long with a substantial wall around them, well embanked on the outside and with four apertures or mouths for the water to flow out, fresh and clear.

---

76 Pierantoni, *Aniene illustrato*. At the end of the eighteenth century during the rule of Pope Pius VI Braschi, commendatory abbot of Subiaco from 1773, the fishpond was listed among the assets of the Abbey Mense, which also noted its boundaries. At the time, the area took the form of a 'possessione arativa con albori di Celsi, et altri fruttiferi, et infruttiferi' ('agricultural property with mulberry trees and other fruitbearing and non-fruitbearing trees') with a 'Peschiera di trotte in loco detto Matignano di capacità di Rubbia venti in circa, confinante le pendici, e balse di detto Castello, le mura da capo a detta possessione, il Fiume da una parte, la forma dell'acqua dall'altra da capo a piedi, et altri ecc.' ('fishpond for trout in the place called Matignano with a capacity of about twenty *Rubbia*, next to the slopes and cliffs of the aforementioned village, the walls at the top of this property, the river on one side, the pond on the other from top to bottom, and others, etc'), see Archivio storico dell'Abbazia territoriale di Subiaco, I.B.1.2: *Registro de' beni, diritti, e proventi dell'Abbazia sublacense compilato per ordine di Nostro Signore Papa Pio Sesto da Francesco Tranquilli geometra e perito publico* [...] (1785) 898.

In the mid-nineteenth century, Jannuccelli also mentions the pond, describing it as follows:

> Sgorga sotto la strada romana una copiosa e fresca fonte che raccolta in un artificiale bacino formava una estesa e deliziosa peschiera, vivaio un giorno per le trote a comodo dei cardinali abbati.[77]
>
> next to the Roman road a copious and fresh spring gushes forth; collected in a manmade basin it once formed an extensive and delightful fishpond, formerly a place where trout were reared for the use of the Cardinal Abbots.

He distinguishes it from the other 'grossi capi d'acqua' ('large springs') owned by the monastery of Santa Scolastica at Subiaco and located 'sulla via rotabile verso l'occidente [...] che riuniti e scorrendo per artefatti canali imprimono il moto a due molini a grano' ('along the roadway towards the west [...] which united and flowing through manmade canals drive two mills for grain').[78]

I propose that an unidentified depiction of this pleasure ground can be found in the Abbey Fortress of Subiaco in the so-called "Studiolo of Cardinal Marcantonio Colonna" [Fig. 5.3]. Situated on a rock spur overlooking the town of Subiaco and the upper Aniene valley, the Fortress was built between 1075 and 1077 by Abbot Giovanni VII (also known as Giovanni V) and served as a bulwark of monastic authority over the village and its surrounding territory.[79] An isolated fortification, it featured defensive structures, prisons, watchtowers, three circuits of walls and a residential building with an adjoining church dedicated to St. Thomas. Radical restorations were undertaken in 1476 after

77 Jannuccelli, *Memorie di Subiaco* 432.

78 Ibidem 433.

79 For the history of the Abbey Fortress see: Gori F., "Residenza, nascita e fortificazioni de' Borgia nella Rocca di Subiaco", *Archivio storico, artistico, archeologico e letterario della città e provincia di Roma* 4.3 (1880) 126–140; *La Rocca di Subiaco* (Rome: 1912); Coccanari G., "Lucrezia e la fortezza dei Borgia di Subiaco", *Bollettino di studi storici ed archeologici di Tivoli e regione* 55 (1932) 2017–2022; Coccanari O., "La Rocca sublacense e la nascita di Cesare e Lucrezia Borgia", *Bollettino di studi storici ed archeologici di Tivoli e regione* 60 (1933) 2239–2242; Caronti L. – Cignitti B., *L'abbazia nullius sublacense* (Rome: 1956) 179–182; Caronti L., "La Rocca Abbaziale di Subiaco", *La rassegna del Lazio* 10–12.13 (1966) 35–56; Curcio G. – Garibaldi V., "Subiaco: Rocca Abbaziale", in *Dall'abbandono al riuso*, exh. cat., Palazzo Braschi (Milan: 1981) 166–174; Caronti L. – Orlandi G. – Priori L., *La rocca di Subiaco* (Subiaco: 1989); Minasi M., *I Colonna nella rocca di Subiaco. La decorazione cinquecentesca* (Rome: 1997); De Angelis M. – Orlandi A. – Orlandi M. – Sbraga G. – Renzetti E. (eds.), *La Rocca di Subiaco. Arte, Storia e Architettura* (Subiaco: 2017).

the establishment of the *commenda* in 1456. Cardinal Rodrigo Borgia had the massive square tower, the so-called "Borgia Tower", erected to defend the older medieval structure to which it was connected by a walkway with a trap door. It is likely that Lucrezia Borgia, the daughter of the pontiff and Vannozza Catanei, was born here. When the commendatory abbots of the Colonna family replaced Rodrigo in 1492, the manor became a Ghibelline outpost in the continuous clashes between the family and the papal authority until the Peace of Cave in September 1557.[80] As has been discussed, these frictions included moments of marked tension during the papacy of Julius II and resulted in genuine armed conflicts during the pontificates of Clement VII and Paul IV. The Fortress of Subiaco was attacked and looted on at least three different occasions: 1526, 1528, and again in 1556.[81] These clashes mainly involved Cardinal Pompeo and his nephews Scipione and Francesco Colonna. The latter two would go on to rule the Abbey together with their uncle Pompeo after his brother Scipione's death in battle in 1528. Following Pompeo's death, possibly from poisoning, in Naples where he was Viceroy in 1532, Francesco obtained full control over the fortress.[82] Among other observations, Giovio said that in Naples, Pompeo had had a villa adorned close to the sea in the 'contrada' of Chiaia, 'nobilissima per la vaghezza dei giardini' ('most exquisite for the beauty of its gardens') with a cedar grove and beautiful marble fountains.[83] Taking up the *topoi* of humanistic culture, Pompeo's biographer also describes the cardinal's life and activities in this pleasant place, which he greatly enjoyed. Giovio speaks of Colonna as being:

> peritissimo dell'agricoltura et sempre vestito da Cardinale, innestava arboscelli, et alla presenza del popolo, ragionava con Napolitani, i quali si dilettano molto della piacevolezza de' giardini, delle varie sorte di semenze, et del modo di piantare herbaggi nobili et fiori stranieri. L'altre hore, et massimamente innanzi mangiare, si dava a gli studi, ne' quali si dilettava tanto, che felicemente ancora essercitava lo stile.[84]

80 For an overview, see Scatizzi, "I Colonna signori di Genazzano" 19–28.

81 Carosi, *Badia di Subiaco* 80–83; Minasi, *I Colonna nella rocca di Subiaco* 163.

82 Petrucci, "Colonna, Pompeo" 411. For Francesco Colonna's rule over the Abbey of Subiaco cfr. Mirzio, *Chronicon sublacense*, fols. 164v–169v; for the correct date: ibidem 605 footnote 1, 606 footnote 1; Jannuccelli, *Memorie di Subiaco* 238–241; Mariani, *Storia di Subiaco* 191–197; Carosi, *Badia di Subiaco* 82–83.

83 Giovio, *Le vite di Leon Decimo* [...] *et del Cardinal Pompeo Colonna*, fol. 183r.

84 Ibidem. For the history of the palace and the garden at Chiaia cfr. Archivio Colonna, II A 17, especially fol. 577r ff.; Archivio Colonna, III BB, 44, 36.

> a great expert on agriculture and always dressed as a Cardinal, would graft trees and in the presence of the people, would converse with the Neapolitans, who take great delight in the charms of the gardens, the various sorts of plants and the manner in which noble herbs and foreign flowers are planted. At other times of day, and particularly before eating, he devoted himself to his studies, which delighted him to such an extent that he still possessed a fine style.

Addressing the destruction caused by the papal army, the Colonna commendatory abbots restored and redecorated the Fortress of Subiaco on several occasions. These works were most notably undertaken by Cardinal Pompeo [see Fig. 5.2], Archbishop Francesco, and his successor in 1559, Cardinal Marcantonio Colonna (1523–*c.*1597) [Fig. 5.4].[85] In addition to massive

FIGURE 5.4
Cardinal Marcantonio Colonna, engraving, 15.3 × 10.4 cm. From: De Santis Domenico, *Columnensium procerum imagines, et memorias nonnullas* [...] (Rome, Angelo Bernabò: 1675). Private collection
IMAGE © TIZIANA CHECCHI

85 The son of Marcello and Vittoria Colonna, Marcantonio was probably born in 1523. In 1549 his uncle Francesco Colonna renounced the Abbey of Subiaco in his favour, but he took possession of it only after the death of his predecessor in 1559. For the biography

fortification works, Pompeo's patronage was also likely responsible for the now vanished paintings that Rutilio Scotti attributed to his fellow townsman:[86]

> Chirico di Casa Monaci eccellentissimo pittore da porsi a paragone di quanti n'hebbe quell'età, il quale si ritrovò a dipingere nella sua gioventù nel Palazzo della Rocca di Subiaco, dove si vedono alcune vaghe opere, il quale passò a miglior vita nell'anni ultimi delle guerre di Clemente VII.[87]
>
> Chirico from the Monaci family, a highly skilled painter who was the equal of any of that time, who in his youth found himself painting in the Palace of the Fortress of Subiaco, where some charming works can be seen; he died in the final years of the wars of Clement VII.

Following the cessation of conflicts with Paul IV in 1557, Francesco Colonna intervened in the fortification and ordered other important reconstruction and decoration projects.[88] His successor Marcantonio Colonna continued these undertakings [Fig. 5.4]. Among other projects, the renovations included the construction of a new apartment on the *piano nobile* of the Borgia Tower consisting of four rooms, two of which likely served as reception rooms while the remainder had more intimate functions. These included a room perhaps intended as the abbot's bedroom and a second, smaller space connected to the former by a narrow corridor and also accessed by a spiral staircase within a cylindrical tower. With the completion of these works, the Rocca took the form of a 'fortissimo luogo' ('very well-defended place') overlooking the town with 'comodissime habitationi' ('very comfortable living quarters') and 'belli appartamenti, ornati di sale, loggie (sic) et camere dipinte di vaghissime figure [...]' ('fine apartments, adorned with halls, loggias and rooms painted with very fine figures').[89]

The frescoes that decorate the last space in the apartment – a small, almost square room found in the small tower abutting the north side of the Borgia

---

of Colonna see Petrucci F., "Colonna, Marcantonio", *Dizionario biografico degli italiani* 27 (Rome: 1982) 368–371. For the rule of Cardinal Marcantonio Colonna over the Abbey of Subiaco cfr. Mirzio, *Chronicon sublacense*, fols. 170r–180r; Jannuccelli, *Memorie di Subiaco* 241–245; Mariani, *Storia di Subiaco* 197–207; Carosi, *Badia di Subiaco* 83–85.

86 Giovio, *Le vite di Leon Decimo* [...] *et del Cardinal Pompeo* Colonna, fol. 151r; Mirzio, *Chronicon sublacense* fol. 154r.

87 Scotti, *Descrittione e Historia dell'Abbadia di Subiaco* 140–141.

88 Mirzio, *Chronicon sublacense*, fol. 169v; Scotti, *Descrittione e Historia dell'Abbadia di Subiaco* 161.

89 Scotti, *Descrittione e Historia dell'Abbadia di Subiaco* 158–159.

Tower – are particularly interesting for the purposes of this study. Overlooking the surrounding landscape from three windows, the room was perhaps used as a study at the time.[90] The decorations were probably carried out between 1559 and 1565 on the orders of the Abbot Marcantonio Colonna, who had not yet been made Cardinal, as is shown by the bishop's hat that surmounts the coat of arms at the entrance to the small room. The dating of the frescoes can perhaps be further narrowed down to 1562 – the year of Colonna's departure to the Council of Trent, from which he only returned in 1564.[91] The frieze that runs along the walls presents alternating depictions of cardinals and theological virtues, to which the allegorical figure of *Religio*, decorative panels, the mottos of the Colonna family (*Nec mersa superstat, Flectimur non rupimur undis, Immota manet, Et in adversis lucet*), and landscapes were added. Mara Minasi has interpreted the landscapes and dedicated a study to this cycle [Fig. 5.5].[92] Minasi describes some scenes as generic landscapes (*Landscape with a Chase Scene, Landscape with Ruins and a Bivouac Scene, Snowy Landscape, Landscape with a Fishing Scene, Landscape with a Small Temple*). She identifies two panels respectively as a view of the *Valley of the Roman Forum* and a representation of *Subiaco*. Finally, she interprets the *Landscape with a Maze* as alluding to inner meditation as a path to progressive ennoblement [Figs. 5.6–12].[93] This latter metaphor emphasises the typical function of an early modern study, which was understood as a space for solitude and constructive meditation.[94]

While the allegorical figures like the *imprese* aimed to glorify the patron and his family, I contend that the landscapes celebrate places at the Abbey of Subiaco that had been renowned since antiquity and were particularly significant to the Colonna family. Typical for the time, these sites were represented not with topographical fidelity, but rather recall and juxtapose elements drawn from real life, reinterpreting the landscape and architecture for rhetorical and

90 For this interpretation: Minasi, *I Colonna nella rocca di Subiaco* 184–185.

91 In 1560, Marcantonio was appointed archbishop of Taranto, while two years later he went to Trent to participate in the sessions of the Council. He returned to Rome in 1564 and on 12 March of the following year was appointed cardinal. Cfr. Petrucci, "Colonna, Marcantonio" 369. The cardinal was also responsible for commissioning other remnants of paintings on the second floor of the Borgia Tower that recently emerged during restoration work. Cfr. Orlandi M., "L'arte negli appartamenti Colonna e Macchi", in De Angelis – Orlandi – Orlandi – Sbraga – Renzetti (eds.), *La Rocca di Subiaco* 62.

92 Minasi, *I Colonna nella rocca di Subiaco* 184–192.

93 Sara Segatori also suggests a possible allusion to the seasons: cfr. Segatori S., *Affreschi cinquecenteschi nell'appartamento Colonna della Rocca abbaziale di Subiaco* (undergraduate thesis, Università degli studi di Roma "La Sapienza": 2006) 121–125.

94 Minasi, *I Colonna nella rocca di Subiaco* 192.

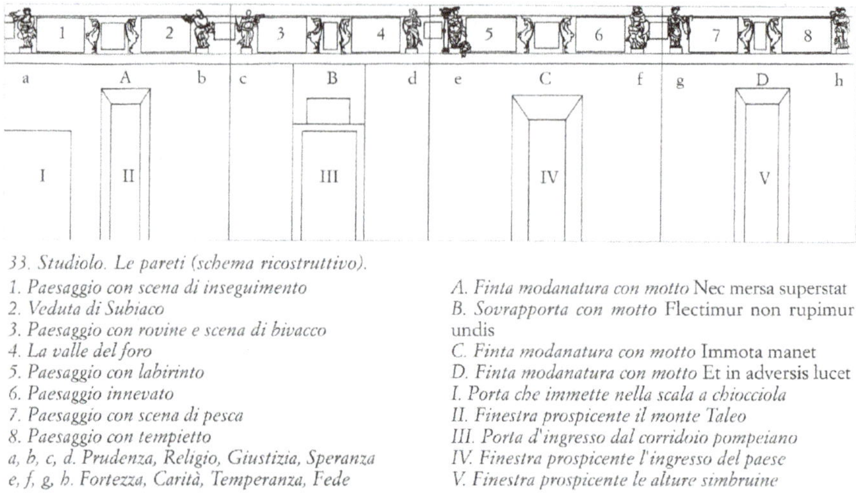

FIGURE 5.5 The so-called Studiolo of Cardinal Marco Antonio Colonna: reconstruction drawing of wall decorations. From: Minasi M., *I Colonna nella Rocca abbaziale di Subiaco. La decorazione cinquecentesca* (Rome: 1997) 185

FIGURE 5.6 *Landscape with a Temple*, probably *Landscape with Nero's Villa Sublaquensis*, *c.*1559–1565 (or 1562). Fresco. Subiaco, Abbey Fortress, Studiolo of Cardinal Marco Antonio Colonna
IMAGE © PIERO ANTONUCCI

FIGURE 5.7 *Landscape with Ruins and a Bivouac Scene*, probably *Landscape with the Ruins of Nero's Villa Sublaquensis*, *c*.1559–1565 (or 1562). Fresco. Subiaco, Abbey Fortress, Studiolo of Cardinal Marco Antonio Colonna
IMAGE © PIERO ANTONUCCI

FIGURE 5.8 *Landscape with a Fishing Scene*, probably *View of Marano*, *c*.1559–1565 (or 1562). Fresco. Subiaco, Abbey Fortress, Studiolo of Cardinal Marco Antonio Colonna
IMAGE © PIERO ANTONUCCI

FIGURE 5.9 *Valley of the Roman Forum*, *c*.1559–1565 (or 1562). Fresco. Subiaco, Abbey Fortress, Studiolo of Cardinal Marco Antonio Colonna
IMAGE © PAOLO SBRAGA

FIGURE 5.10 *Snowy Landscape*, probably *View of Cervara*, *c*.1559–1565 (or 1562). Fresco. Subiaco, Abbey Fortress, Studiolo of Cardinal Marco Antonio Colonna
IMAGE © PIERO ANTONUCCI

FIGURE 5.11 *Landscape with a Chase Scene*, probably *View of Agosta*, c.1559–1565 (or 1562). Fresco. Subiaco, Abbey Fortress, Studiolo of Cardinal Marco Antonio Colonna
IMAGE © PIERO ANTONUCCI

FIGURE 5.12 *View of Subiaco*, c.1559–1565 (or 1562). Fresco. Subiaco, Abbey Fortress, Studiolo of Cardinal Marco Antonio Colonna
IMAGE © PIERO ANTONUCCI

propagandistic purposes.[95] Small figures who introduced elements of daily life into the scene enliven each view. They allude to some of the main tasks and productive activities that characterised the territory of Subiaco at the time: fishing, hunting and agriculture, notably the cultivation of cereals, vines, maize, olives and fruit trees. The frescoes were probably retouched in the eighteenth century and are currently in a very poor state of conservation. Despite this, it is possible to confirm the proximity of these landscapes to Northern European culture, although they are tempered by clear references to the classical world and Roman art, as confirmed by quotations to the Castel Sant'Angelo Farnese decorations.[96] I propose that water is the unifying element across the landscapes represented in the frieze. Visible in each scene, it was likely intended to evoke and celebrate the Aniene river, whose course crosses and distinguishes the entire territory of the Abbey. Jurisdiction over the river was the prerogative of the commendatory abbot.[97]

Landscape with a Temple [Fig. 5.6] may allude to the famous villa that the emperor Nero (37–68; emperor 54–68) had built on the banks of the Aniene in the early years of his reign. This was perhaps before 60, the year in which, according to Tacitus, a bolt of lightning struck the table of the emperor while he was feasting in the villa. As previously mentioned, the course of the river Aniene was dammed on Nero's orders to create three artificial lakes.[98] In an unpublished manuscript from 1627, Giovanni Camillo Contestabile writes that:

> Nerone Imperatore veniva per suo gusto l'estate ad abitar vicino le nominate acque, dove nella riva del Lago, che quivi era aveva edificato un

95 On this topic and for the relationships between the real landscape and landscapes frescoed inside aristocratic residences I refer to the interesting studies by Denis Ribouillault. See especially Ribouillault D., "Landscape *all'antica* and Topographical Anachronism in Roman Fresco Painting of the Sixteenth Century", *Journal of the Warburg and Courtauld Institutes* 71 (2008) 211–237; "Le Salone de la Villa d'Este à Tivoli: un théâtre des jardins et du territoire", *Studiolo: Revue d'histoire de l'art de l'Académie de France à Rome* 3 (2005) 65–94; "Paesaggio dipinto, paesaggio reale: notes sur une fenêtre de la Villa d'Este à Tivoli", in Venturi G. – Ceccarelli F. (eds.), *Delizie in villa. Il giardino rinascimentale e i suoi committenti* (Florence: 2008) 269–287; "Le ville dipinte del cardinale Ippolito d'Este a Tivoli: l'architettura di fronte all'antico, la tradizione ferrarese e un nuovo documento su Belriguardo", in Ceccarelli F. – Folin M. (eds), *Delizie estensi. Architetture di villa nel Rinascimento italiano ed europeo* (Florence: 2009) 341–371; *Rome en ses jardins*; "De la peinture au jardin (en passant par la poésie): la vallée Giulia à Rome, de Michel-Ange à Poussin", in Brunon H. – Ribouillault D. (eds.), *De la peinture au jardin* (Florence: 2016) 43–96.

96 Minasi, *I Colonna nella rocca di Subiaco* 190–192.

97 Archivio storico dell'Abbazia territoriale di Subiaco, 1.B.1.1, fol. 2r and passim.

98 For the bibliography on the villa, see footnote 50.

> palazzo molto superbo, del quale a' nostri tempi si veggano li vestigii delli muri rovinati.[99]
>
> for his own pleasure the emperor Nero came to live in the summer near the aforementioned waters, where he had a very magnificent palace built on the shores of the lake in the area, of which at our time we see the remains of the ruined walls.

At the time he was writing, it was believed that the villa consisted mainly of 'a very magnificent palace' built on the banks of one of the three artificial lakes.[100] This description is consistent with the scene depicted in the fresco, which shows a sumptuous, colonnaded palace adorned with an imposing statue and overlooking a body of water crossed by small boats.[101] In the early decades of the seventeenth century, one could admire fragments of 'effigiato marmo mischi et porfidi, i quali dan segno, che ivi fussino molte delitiose e superbe fabriche' ('sculpted marble of mixed colour and porphyry, which signal that there were many beautiful and magnificent buildings in that place') among the ruins.[102] Large marble panels and beautiful statues could also be found. Contestabile writes that in 1551, a large marble statue depicting *Pallas*, whose limbs were broken but head remained intact, 'con molto artifizio di arte fatta' ('made with great artifice'), was discovered under the ruins.[103] The then commendatory abbot Francesco Colonna donated this sculpture to Pope Julius III and its representation could be admired 'ritrattata al naturale sopra la porta del Camerino nella Rocca di Subiaco' ('depicted lifelike above the door to the small chamber in the Fortress of Subiaco').[104] In the sixteenth century, 'the remains of ruined walls' were all that was left of this sumptuous villa and the other buildings constructed by Nero.[105] These ruins may be referenced in

---

99 Contestabile Giovanni Camillo, *La Valle Sagra ove si trattano i fatti, e miracoli del Glorioso Patriarca San Benedetto operati da lui nel distretto di Subiaco* [...], 1627, unpublished mss. held as a reproduction at the Biblioteca Statale of Santa Scolastica, 271–272.

100 Ibidem 271.

101 This landscape recreates the background of the fresco depicting the *Farewell of Perseus* in the room of the same name in Castel Sant'Angelo. Reproduced in a pedantic way, a preparatory drawing of the same scene, now in the Louvre, served as a major model for the Subiaco fresco. Minasi, *I Colonna nella rocca di Subiaco* 190–192. The same model was used for a scene on the ceiling of the Palazzo della Loggia at Bagnaia, painted for Cardinal Niccolò Ridolfi *c.*1547. I thank Denis Ribouillault for this suggestion.

102 Scotti, *Descrittione e Historia dell'Abbadia di Subiaco* 183.

103 Contestabile Giovanni Camillo, *La Valle Sagra* 272.

104 Ibidem.

105 Ibidem.

the scene immediately opposite the fresco just described, which was previously interpreted as a *Landscape with Ruins and a Bivouac Scene* [Fig. 5.7]. Here, in the background of a lively popular scene, the remains of ancient buildings along the shores of a large lake can be observed.

The high quality and health-giving nature of the waters of the Aniene were famous in the time of the Romans, who used them to supply the city of Rome. To this end, they constructed four great aqueducts: the *Anio Vetus* (272 BCE), the *Marcius* (144 BCE), the *Claudius*, and the *Anio Novus* (started by Caligula in 38 and completed by Claudius).[106] The so-called *Landscape with a Fishing Scene* [Fig. 5.8] perhaps celebrates this other famous *topos* of the Subiaco area. It depicts the town of Marano, which, according to Pliny, was the starting point of the aqueduct Aqua Marcia, famous among the waters of the world for its freshness and purity.[107] In the mid-nineteenth century, one could still see 'antichi muri, acquidotti e vasche simili a quelle che veggonsi nelle vetuste terme di Roma' ('old walls, aqueducts and basins similar to those visible in the ancient baths of Rome') near this place.[108] These and other ruins scattered throughout the territory may be represented in the fresco, which depicts ruins with arches visible beneath the town. After a long journey, the waters of the river Aniene reach Rome; an allusion to this subsequent course is perhaps suggested by the bridge visible at the centre of the background, possibly identifiable as the Lucano Bridge flanked by the mausoleum of the Plautii. The little scene in the foreground shows one of the main activities that took place, and which still takes place near the river: fishing. Scotti wrote that it was common to see naked fishermen along the river with nets and other implements, happily fishing for delicate trout.[109] Like the jurisdiction over the river waters, fishing rights also belonged to the commendatory abbot.[110] The structure visible in the background may be a mill that once existed beneath Marano.

The view of the *Valley of the Roman Forum* towards the Colosseum faces this landscape on the opposite wall [Fig. 5.9].[111] Minasi identified the temple in the central position as that of Antoninus Pius and Faustina, the religious building behind it as the church of Santi Cosma e Damiano, and the three columns with an entablature fragment as the surviving remnants of the Temple

106 Di Matteo, *Villa di Nerone a Subiaco* 21; Panimolle, *Le acque e gli acquedotti* especially 33–110.
107 Di Matteo, *Villa di Nerone a Subiaco* 21.
108 Jannuccelli, *Memorie di Subiaco* 459.
109 Scotti, *Descrittione e Historia dell'Abbadia di Subiaco* 161–162.
110 Archivio storico dell'Abbazia territoriale di Subiaco, 1.B.1.1, fols. 2r, 105r e passim.
111 Minasi, *I Colonna nella rocca di Subiaco* 198 footnote 78.

of Castor. She interpreted the door in the background as a representation of the Arch of Titus, its medieval additions no longer visible today. Finally, towering in the background are the Flavian Amphitheatre and the line of arches of the Claudian aqueduct which, as mentioned, was fed by the waters of the Aniene.[112] In my opinion, this fresco may allude to the final destination of the river: namely Rome and the Tiber, whose waters are seen flowing in the foreground enlivened by bathing nymphs spied on by a satyr.[113] Among other features, the village of Marano depicted in the opposite scene was the last portion of the Abbey's territory along the road that led to Rome. The springs that fed the Aqua Claudia aqueduct were the exact area between Arsoli and Marano; as previously mentioned, the arches of this aqueduct are alluded to in the fresco of the *Valley of the Roman Forum*.[114]

Notably, the lunettes on the walls of the main reception room of the Abbey Fortress feature Nero's magnificent villa and the three artificial lakes on which it was built. The aqueducts fed by the Aniene river can also be observed in these lunettes, which are attributed to the patronage of the commendatory abbot Francesco Colonna and dated to the years 1557–1559. Minasi interpreted two of the three views, in this case executed in a classicising style that immerses the ancient ruins 'in a light and almost unreal landscape', as a *Landscape with the Villa of Nero at Subiaco* and a *Landscape with the So-Called "Trophies of Marius"*.[115] The latter refers to a monumental fountain located in the north corner of today's Piazza Vittorio in Rome, believed in the Renaissance to be fed by the *Aqua Marcia* represented here at its monumental point of arrival.[116] Denis Ribouillault has offered a convincing interpretation of the other landscape visible in the lunettes of the hall as a depiction of the grandiose Villa of Nero at Subiaco.[117] Minasi formerly interpreted this image as a *Landscape with the Belvedere Courtyard*, given its derivation from the fresco now detached and kept in the Hall of the Festoons in Castel Sant'Angelo.[118]

112 Panimolle, *Le acque e gli acquedotti* 61–82.

113 In Minasi's opinion, this scene derives, albeit not faithfully, from one of the stories of Perseus in the room of the same name in Castel Sant'Angelo; Minasi, *I Colonna nella rocca di Subiaco* 190.

114 Panimolle, *Le acque e gli acquedotti* 61–65.

115 Minasi, *I Colonna nella rocca di Subiaco* 176–179.

116 Ibidem.

117 Ribouillault, "Landscape *all'antica*" 233–234; Ribouillault, *Rome en ses jardins* 59–60, 116–118.

118 Minasi, *I Colonna nella rocca di Subiaco* 176.

Returning to the frescoes of the so-called "Studiolo of Marcantonio", the *Snowy Landscape* [Fig. 5.10] can, in my opinion, be identified as another key site for the Colonna family and its territory near the Abbey of Subiaco: the village of Cervara. The deer – in Italian *cervo* – in the foreground are notable, as they are a very common species in the region and pun on the name of the town. The deer is also found on the town's coat of arms, as well as on the flag that appears in the fresco decorating the Cosmatesque cloister of the Santa Scolastica monastery at Subiaco, which dates to the early decades of the fourteenth century. Known for its 'fairly harsh climate', the village is shown perched on a mountain with the outline of an ecclesiastical complex beneath it.[119] This is perhaps identifiable as the church of Santa Maria Maddalena, near which the old hospital was located.[120] Compared to its actual location, the village is represented much lower in order to reveal the larger *leitmotif* of the complete cycle: the course of the river Aniene, seen flowing at the foot of the mountain on which the village stands. This area, specifically the Barco district, marked the start of another of the four ancient aqueducts: the *Anio Novus* (52 BC), which was subsequently extended directly below the three lakes of the Neronian villa at the time of the emperor Trajan.[121] The opening visible in the bottom left of the fresco may allude to this structure.

Hunting, another popular activity in the area, is referred to in the scene in the foreground of the fresco. Here, a hunter is depicted in the act of stalking deer. Behind is a small figure frightened by a wolf and a wild boar, other animals commonly found in the Abbey's territory. As I have already noted, hunting was one of the favourite entertainments of the Cardinals Colonna in the area and another sphere over which they enjoyed jurisdictional power as commendatory abbots. Another panel until now identified as *Landscape with a Chase Scene* [Fig. 5.11] faces this panel. In my opinion it depicts another site closely linked to the Colonna family's activities in the Subiaco area: the pleasure ground that they owned below the village of Agosta. The profile of the village can be recognised at the top of this scene, with the bell tower of the parish church towering above and the pleasure ground where the Colonna enjoyed banquets and merry gatherings below. To repeat the above-mentioned words of Pierantoni,

> Alle radici della collina di duro sasso, sopra cui sta fondato il castello dell'Agosta [...] vi è stata fabricata una Peschiera [...] in forma riquadrata

---

119 Jannuccelli, *Memorie di Subiaco* 429.

120 Ibidem 430.

121 Panimolle, *Le acque e gli acquedotti* 83–101.

> per longhezza di circa cento passi con grosso muro intorno, ben terrapienato al di fuori, e con quattro aperture, o bocche per l'esito dell'acqua.[122]

> at the feet of the hill of hard rock on which the castle of Agosta stands […] a fishpond was built […] square in shape with sides about a hundred paces long, with a substantial wall around them, well embanked on the outside and with four apertures or mouths for the water to flow out.

A long 'wall that extends right down to the bed of the river' surrounded the area with a gateway at its centre surmounted by the family coat of arms, namely a column with a crown carved in marble.[123] In my opinion, this door can be identified with the one that is now in the village of Agosta, where it was transported and reassembled in later centuries. To the right of the fresco, a procession on horseback approaches the fishpond, led by a figure whose head is covered by a broad-brimmed hat resembling a cardinal's galero. This scene closely recalls Scotti's account of Cardinal Pompeo who, after hunting, used to travel the short distance down to the village of Agosta to hold lavish banquets in the company of eminent figures from his own court and the Abbey of Subiaco, owing to the beauty and pleasantness of its charming fishpond.[124]

The *View of* Subiaco, the centre of the power for the commendatory abbot in the Abbey's territory, is placed next to this scene [Fig. 5.12]. The Fortress dominates the view, pictured in its updated form after the restorations carried out by Archbishop Francesco Colonna a few years prior between 1557 and 1559. The *Landscape with a Maze* that faces this scene was also probably meant to allude to a real place. At the moment, however, no mention or documents have emerged of any villas owned by the Colonna family that correspond to the building in the background, nor evidence of the presence of a labyrinth or formal garden in the Sublacensis area. Certainly, this image was also laden with symbolic meanings, probably intended to underline the value of the study located inside the Abbey Fortress, which, in the eyes of its patron Cardinal Marcantonio Colonna, was a space of solitude and meditation.[125] In the depiction of the Fortress in the *View of Subiaco* [Fig. 5.12], a medieval core

---

122 Pierantoni, *Aniene illustrato* 220.

123 Ibidem.

124 Scotti, *Descrittione e Historia dell'Abbadia di Subiaco* 147.

125 Minasi, *I Colonna nella rocca di Subiaco* 192. In the context of allegorical scenes, the motif of the labyrinth with a deer is found, for instance, in a drawing by Baptiste Pellerin. See Fig. 8.5, Chap. 8 in this book. On the labyrinth in Renaissance art, see Brunon H., "Le paradigme labyrinthique dans l'histoire des jardins: exemples italiens aux XVe et XVIe siècles, ou du 'cosmogramme' au 'mésocosme'", in Brunon H. (ed.), *Le jardin comme labyrinthe du*

with walls punctuated by defensive towers rises on the left. The residential portion of the structure that would have hosted the old apartments of the abbot and the church of San Tommaso is perhaps higher up. Further to the right looms the hulking Borgia Tower and lower down in the scene are the bridge of San Francesco and the church of Sant'Andrea Apostolo, formerly dedicated to Sant'Abbondio and rebuilt by Francesco and later Marcantonio Colonna.[126] On the banks of the river Aniene, a garden is divided into flowerbeds and surrounded by walls, at the centre of which is an open gateway.

Cardinal Ascanio Colonna succeeded his uncle Marcantonio to become Commendatory Abbot in 1592.[127] This site was described as follows in 1597:

> Il S.r Abbate have il Giardino con le sue mura intorno et una possessione sotto dett[a] il boschetto dove ci sono alcuni albucci confine il fiume [...] quale giardino et possessione detto S.r Abbate tiene in suo potere, et se ne dispone a suo arbitrio.[128]
>
> The Abbot has a garden with walls surrounding it and a property beneath known as the grove where there are some poplars along the river [...] the Abbot owns this garden and property and makes use of them as he wishes.

An unpublished document referring to works carried out in the garden by Cardinal Scipione Caffarelli Borghese reveals the presence of a fishpond.[129] Borghese was the nephew of Pope Paul v and became commendatory abbot of the Abbey after the death of Cardinal Ascanio Colonna in 1608, which marked the end of the Colonna rule over the territory of Subiaco. As with the fishpond at Agosta, nothing is left of this other unknown pleasure ground. The verses of poets, archival documents and fine frescoes that decorate the so-called "Studiolo of Marcantonio Colonna" in the Fortress of Subiaco are all that remains to remember the beauty and the ancient renown of these sites.

---

monde: permanence et métamorphoses d'un imaginaire de la Renaissance à nos jours (Paris: 2008) 69–131, 303–304.

126 Contestabile, *La Valle Sagra* 286–287; Jannuccelli, *Memorie di Subiaco* 242–243.

127 Petrucci F., "Colonna, Ascanio", *Dizionario biografico degli italiani* 27 (Rome: 1982) 275–278. For his period of rule of the Abbey of Subiaco: Mirzio, *Chronicon sublacense*, fols. 180v–181v; Jannuccelli, *Memorie di Subiaco* 247–250; Mariani, *Storia di Subiaco* 207–211; Carosi, *Badia di Subiaco* 85.

128 Archivio storico dell'Abbazia territoriale di Subiaco, 1.B.1.1, fol. 4v.

129 Archivio storico dell'Abbazia territoriale di Subiaco, 1.A.42.a.4.

References

Accame M., *Pomponio Leto. Vita e insegnamento* (Rome: 2008).

Accame M., "Pomponio Leto, Giulio", *Dizionario biografico degli italiani* 84 (Rome: 2015) 711–716.

Baker Bates P., "A Portrait of Cardinal Pompeo Colonna, Rival and Imitator of the Papal Caesars", *Papers of the British School at Rome* 76 (2008) 183–199, 356.

Ballistreri G., "Capodiferro, Evangelista Maddaleni, de', detto Fausto", *Dizionario biografico degli italiani* 18 (Rome: 1975) 621–625.

Ballistreri G., "Casanova, Marco Antonio", *Dizionario biografico degli italiani* 21 (Rome: 1978) 171–174.

Barucco P., "L'ordine tuscanico nel Ninfeo e nel Castello. Problemi di datazione e attribuzione", in Bureca A. (ed.), *Il Castello Colonna a Genazzano. Ricerche e restauri* (Rome: 2000) 141–159.

Benedetti S., "Inghirami, Tommaso, detto Fedra", *Dizionario biografico degli italiani* 62 (Rome: 2004) 383–387.

Bianca C., "L'Accademia del Bessarione tra Roma e Urbino", in Cerboni Baiardi G. – Chittolini G. – Floriani P. (eds.), *Federico di Montefeltro. Lo Stato/ Le Arti/ La Cultura* (Rome: 1986) 61–79.

Bianca C., *Da Bisanzio a Roma. Studi sul cardinale Bessarione* (Rome: 1999).

Bianca C., "Roma e L'Accademia Bessarionea", in Fiaccadori G., with the assistance of Cuna A. – Gatti A. – Ricci S. (eds.), *Bessarione e l'Umanesimo*, exh. cat., Biblioteca Nazionale Marciana (Naples: 1994) 119–127.

Bianca C., "Le accademie a Roma nel Quattrocento", in Pade M. (ed.), *On Renaissance Academies. Proceedings of the International Conference 'From the Roman Academy to the Danish Academy in Rome. Dall'Accademia Romana all'Accademia di Danimarca a Roma', The Danish Academy in Rome, 11–13 October 2006* (Rome: 2010) 47–59.

Biondo Flavio, "Book 1, Chapter 100," *Roma Instaurata* (Verona: 1481), in Valentini R. – Zucchetti G., *Codice topografico della città di Roma*, IV (Rome: 1953–1954) 283–284.

Borsi S., "Il ninfeo di Genazzano", in Borsi F., with the assistance of Borsi S., *Bramante* (Milan: 1989) 326–329.

Brunon H., "Le paradigme labyrinthique dans l'histoire des jardins: exemples italiens aux XVe et XVIe siècles, ou du 'cosmogramme' au 'mésocosme'", in Brunon H. (ed.), *Le jardin comme labyrinthe du monde: permanence et métamorphoses d'un imaginaire de la Renaissance à nos jours* (Paris: 2008) 69–131, 303–304.

Bruschi A., *Bramante architetto* (Bari: 1969).

Capaccio Giulio Cesare, *Il Forastiero* (Naples, Giovanni Domenico Roncangliolo: 1634).

Capanna M.C., "Ancora sul complesso templare severiano del Quirinale", *Mélanges de l'École française de Rome* 133.1 (2021) 173–208.

Caronti L., "La Rocca Abbaziale di Subiaco", *La rassegna del Lazio* 10–12.13 (1966) 35–56.

Caronti L. – Cignitti B., *L'abbazia nullius sublacense* (Rome: 1956).

Caronti L. – Orlandi G. – Priori L., *La rocca di Subiaco* (Subiaco: 1989).

Carosi G.P., *Badia di Subiaco* (Subiaco: 1970).

Castelli G.P., "Ante diem clade Urbis interiit. Fausto Evangelista Maddaleni Capodiferro: il suo tempo, la sua famiglia e il Sacco di Roma", in *Dall'Archivio Segreto Vaticano. Miscellanea di testi, saggi e inventari* 9 (Vatican City: 2016) 147–329.

Checchi T., "L'opera di Vincenzo Manenti nei monasteri benedettini di Subiaco e il *San Francesco* nella chiesa di S. Maria *in arce* a Jenne", in Paniccia C. – Pistilli P.F. (eds.), *Dalla baronia al papato. Jenne al tempo di papa Alessandro IV* (Roma: 2023) 153–182.

Coccanari G., "Lucrezia e la fortezza dei Borgia di Subiaco", *Bollettino di studi storici ed archeologici di Tivoli e regione* 55 (1932) 2017–2022.

Coccanari O., "La Rocca sublacense e la nascita di Cesare e Lucrezia Borgia", *Bollettino di studi storici ed archeologici di Tivoli e regione* 60 (1933) 2239–2242.

Consorti A., *Il cardinale Pompeo Colonna* (Rome: 1902).

Contestabile Giovanni Camillo, *La Valle Sagra ove si trattano i fatti, e miracoli del Glorioso Patriarca San Benedetto operati da lui nel distretto di Subiaco* [...], 1627, unpublished mss. held as a reproduction at the Biblioteca statale of S. Scolastica at Subiaco.

Christian K.W., *Empire Without End: Antiquities Collections in Renaissance Rome, c. 1350–1527* (New Haven – London: 2010).

Curcio G. – Garibaldi V., "Subiaco: Rocca Abbaziale", in *Dall'abbandono al riuso*, exh. cat., Rome, Palazzo Braschi (Milan: 1981) 166–174.

De Angelis M. – Orlandi A. – Orlandi M. – Sbraga G. – Renzetti E. (eds.), *La Rocca di Subiaco. Arte, Storia e Architettura* (Subiaco: 2017).

Dessì A., "Pomponio Leto e la topografia del Quirinale nei protocolli notarili", *Rivista dell'Istituto di Storia dell'Europa Mediterranea* 9/I n.s. (December 2021), Special Issue: Colesanti G.T. – Piñol D. – Sakellariou E. (eds.), *Il Notaio nella società dell'Europa mediterranea (secc. XIV–XIX). The Notary in the Mediterranean European Society (14th–19th centuries)* 287–310.

Di Matteo F., *Villa di Nerone a Subiaco. Il complesso dei 'Simbruina Stagna'* (Rome: 2005).

Döring M., "La nascita della rovina artificiale nel Rinascimento italiano ovvero il 'Tempio in rovina' di Bramante a Genazzano", in Di Teodoro F.P. (ed.), *Donato Bramante. Ricerche, proposte, riletture* (Urbino: 2001) 343–406.

Falzone P., "Maturanzio (Mataratius), Francesco", *Dizionario biografico degli italiani* 72 (Rome: 2009) 338–341.

Finocchi Ghersi L., "Le residenze dei Colonna ai Santi Apostoli", in Chiabò M. – D'Alessandro G. – Piacentini P. – Ranieri C., *Alle origini della nuova Roma. Martino V (1417–1431)* (Rome: 1992) 61–75.

Finocchi Ghersi L., "Bessarione e la basilica romana dei Santi XII Apostoli", in Fiaccadori G., with the assistance of Cuna A. – Gatti A. – Ricci S. (eds.), *Bessarione e l'Umanesimo*, exh. cat., Biblioteca Nazionale Marciana (Naples: 1994) 129–136.

Fiore Cavaliere M.G., *Sublaqueum – Subiaco. Tra Nerone e S. Benedetto* (Rome: 1994).

Fiore Cavaliere M.G., "La villa di Nerone a Subiaco e la fondazione del monastero benedettino di S. Clemente", in Mari Z. – Petrara M.T. – Sperandio M. (ed.), *Il Lazio tra antichità e medioevo. Studi in memoria di Jean Coste* (Rome: 1999) 341–367.

Frommel C.L., "Bramantes 'Ninfeo' in Genazzano", *Römisches Jahrbuch für Kunstgeschichte* 12 (1969) 137–160.

Frommel C.L., *Architettura alla corte papale nel Rinascimento* (Milan: 2003).

Fusco L. – Corti G., *Lorenzo de' Medici Collector and Antiquarian* (Cambridge: 2006).

Gallocchio E., "Le strutture romane lungo il pendio del Quirinale", in Picozzi M.G. (ed.), *Palazzo Colonna. Giardini. La storia e le antichità* (Rome: 2018) 45–72.

Gatti I., *Il palazzo della Rovere ai SS. Apostoli: uomini, pietre, vicende.* Parte I: *I Cardinali Bessarione e Riario* (Rome: 2015).

Giovio Paolo, *De uita Leonis decimi pont. max. libri 4. His ordine temporum accesserunt Hadriani sexti pont. max. et Pompeii Columnae cardinalis vitae* [...] (Florence, Lorenzo Torrentino: 1548).

Giovio Paolo, *Le vite di Leon Decimo et D'Adriano Sesto Sommi Pontefici, et del Cardinal Pompeo Colonna* [...] *tradotte da M. Lodovico Domenichi* (Venice, Giovanni de' Rossi: 1557).

Gnoli D., "Orti letterari nella Roma di Leone X", *Nuova Antologia* 7.269 (January–February 1930) 3–19, 137–148.

Gori F., "Residenza, nascita e fortificazioni de' Borgia nella Rocca di Subiaco", *Archivio storico, artistico, archeologico e letterario della città e provincia di Roma* 4.3 (1880) 126–140.

Gutkowski A. – Prinzivalli E. (eds.), *Bessarione e la sua Accademia* (Rome: 2012).

Jannuccelli G., *Memorie di Subiaco e sua Badia* (Genoa: 1856).

Lucci M., "Le vicende dei Giardini Colonna tra XIV e XVI secolo", in Picozzi M.G. (ed.), *Palazzo Colonna. Giardini. La storia e le antichità* (Rome: 2018) 101–118.

Magister S., *Arte e politica: la collezione di antichità del cardinale Giuliano della Rovere nei palazzi ai Santi Apostoli* (Rome: 2002).

Manzoli D., "La fortuna di Bessarione e lo sviluppo delle accademie a Roma", in Gutkowski A. – Prinzivalli E. (eds.), *Bessarione e la sua Accademia* (Rome: 2012) 81–99.

Mari F., 'Il paesaggio legittimante. Intorno alle pitture del Chiostro cosmatesco di Santa Scolastica a Subiaco', in press.

Mariani L., *Storia di Subiaco e suo distretto Abbaziale*, ed. M. Sciò (Subiaco: 1997).

Mazzetti di Pietralata C., "Il Colle del Quirinale", in Campitelli A. – Cremona A. (eds.), *Atlante storico delle ville e dei giardini di Roma* (Milan: 2012) 71–81.

Minasi M., *I Colonna nella rocca di Subiaco. La decorazione cinquecentesca* (Rome: 1997).

Mirzio Cherubino da Treviri, *Chronicon sublacense (1628–1630)*, ed. L. Branciani, 2 vols. (Subiaco: 2014).

Mochi C., "Il messale di Pompeo Colonna: antichità ed egizianismi a Roma", in Colonna S. (ed.), *Roma nella svolta tra Quattro e Cinquecento* (Rome: 2004) 439–452.

Musso L., "Le antichità Colonna nei secoli XV e XVI", in Carinci F. – Keutner H. – Musso L. – Picozzi M.G. (eds.), *Catalogo della Galleria Colonna in Roma. Sculture* (Rome – Busto Arsizio: 1990) 13–18.

Negro A., *Guide rionali di Roma. Rione II – Trevi*, II, 1 (Rome: 1985).

Palma Venetucci B., *Dallo scavo al collezionismo: un viaggio nel passato dal Medioevo all'Ottocento* (Rome: 2007).

Palma Venetucci B., "Antichità esotiche nel collezionismo del XV e XVI secolo", in Palma Venetucci B. (ed.), *Culti orientali tra scavo e collezionismo* (Rome: 2008) 73–88.

Panimolle G., *Agosta castello dell'Abbazia sublacense* (Rome: 1986).

Panimolle G., *Le acque e gli acquedotti di Roma antica* (Subiaco: 2007).

Panimolle G., *Il fiume Aniene e i suoi castelli* (Subiaco: 2016).

Parolini C., "Simbruina Stagna. Una nuova lettura alla luce dell'analisi geomorfologica", *Scienze dell'Antichità* 25.1 (2019) 173–186.

Pellegrini M., "Pio II", in *Enciclopedia dei Papi*, II (Rome: 2000) 663–685.

Pellegrini M., "Pio II, papa", *Dizionario biografico degli italiani* 83 (Rome: 2015) 794–803.

Pensabene P., "Il tempio gigantesco del Quirinale", in Picozzi M.G. (ed.), *Palazzo Colonna. Giardini. La storia e le antichità* (Rome: 2018) 11–44.

Petrucci F., "Colonna, Ascanio", *Dizionario biografico degli italiani* 27 (Rome: 1982) 275–278.

Petrucci F., "Colonna, Giovanni", *Dizionario biografico degli italiani* 27 (Rome: 1982) 342–344.

Petrucci F., "Colonna, Marcantonio", *Dizionario biografico degli italiani* 27 (Rome: 1982) 368–371.

Petrucci F., "Colonna, Pompeo", *Dizionario biografico degli italiani* 27 (Rome: 1982) 407–412.

Petrucci F., "Colonna, Prospero", *Dizionario biografico degli italiani* 27 (Rome: 1982) 416–418.

Petrucci F., "Colonna, Prospero", *Dizionario biografico degli italiani* 27 (Rome: 1982) 418–426.

Petrucci F., "Colonna, Vespasiano", *Dizionario biografico degli italiani* 27 (Rome: 1982) 447–448.

Picozzi M.G., "Le sculture degli appartamenti. Contributo alla storia delle antichità della famiglia Colonna", in Picozzi M.G. (ed.), *Palazzo Colonna. Sculture antiche e dall'antico* (Rome: 2010) 9–84.

Picozzi M.G., "Templum Salutis et Serapis", in Picozzi M.G. (ed.), *Palazzo Colonna. Giardini. La storia e le antichità* (Rome: 2018) 73–100.

Pierantoni Domenico Antonio, *Aniene illustrato*, ed. G. Giammaria (Anagni: 2003).

Piergentili P.P. – Venditti G. (eds.), *Scorribande, Lanzichenecchi e soldati ai tempi del sacco di Roma. Papato e Colonna in un inedito epistolario dall'Archivio Della Valle-Del Bufalo (1526–1527)* (Rome: 2009).

Pio II (Enea Silvio Piccolomini), *I Commentari*, ed. G. Bernetti, 5 vols. (Siena: 1972).

Rehberg A., "Pio II e i Colonna: fra 'amicizie' personali e interessi della Chiesa", in Di Paola R. – Antoniutti A. – Gallo M. (eds.), *Enea Silvio Piccolomini. Arte, storia e cultura nell'Europa di Pio II* (Rome: 2006) 434–436.

Ribouillault D., "Le Salone de la Villa d'Este à Tivoli: un théâtre des jardins et du territoire", *Studiolo: Revue d'histoire de l'art de l'Académie de France à Rome* 3 (2005) 65–94.

Ribouillault D., "Landscape *all'antica* and Topographical Anachronism in Roman Fresco Painting of the Sixteenth Century", *Journal of the Warburg and Courtauld Institutes* 71 (2008) 211–237.

Ribouillault D., "Paesaggio dipinto, paesaggio reale: notes sur une fenêtre de la Villa d'Este à Tivoli", in Venturi G. – Ceccarelli F. (eds.), *Delizie in villa. Il giardino rinascimentale e i suoi committenti* (Florence: 2008) 269–287.

Ribouillault D., "Le ville dipinte del cardinale Ippolito d'Este a Tivoli: l'architettura di fronte all'antico, la tradizione ferrarese e un nuovo documento su Belriguardo", in Ceccarelli F. – Folin M. (eds), *Delizie estensi. Architetture di villa nel Rinascimento italiano ed europeo* (Florence: 2009) 341–371.

Ribouillault D., *Rome en ses jardins: Paysage et pouvoir au XVI*[e] *siècle* (Paris: CTHS-INHA, 2013).

Ribouillault D., "De la peinture au jardin (en passant par la poésie): la vallée Giulia à Rome, de Michel-Ange à Poussin", in Brunon H. – Ribouillault D. (eds.), *De la peinture au jardin* (Florence: 2016) 43–96.

Ribouillault D., "Hortus Academicus: les académies de la Renaissance et le jardin", in Jakob M. (ed.), *Des jardins & des livres* (Geneva: 2018) 23–34.

La Rocca di Subiaco (Rome: 1912).

Rowland I.D., "L'Accademia Romana fra Pomponio Leto e Angelo Colocci", in Mangani G. (ed.), *Raffaello e Angelo Colocci. Bellezza e Scienza nella costruzione del mito della Roma antica*, exh. cat., Jesi, Musei Civici di Palazzo Pianetti (Santarcangelo di Romagna: 2021) 28–39.

Safarik E.A., *Palazzo Colonna* (Rome: 1999).

Scatizzi P., "I Colonna signori di Genazzano", in Bureca A. (ed.), *Il Castello Colonna a Genazzano. Ricerche e restauri* (Rome: 2000) 13–70.

Schelbert G., *Der Palast von SS. Apostoli und die Kardinalsresidenzen des 15. Jahrhunderts in Rom* (Norderstedt: 2007).

Scotti Rutilio, *Descrittione et Historia dell'Abbadia di Subiaco illustrata da Ruttilio Scotti Affilano Pittore e Cosmografo. Consegnata all'Ill.mo et R.mo Sig.re il Sig.r Scipione Card.le Burghese. Perpetuo Commendatario. Data in Affile, sotto l'anno del 1615*, Vatican Apostolic Library, Chig.I.23, where there is also a copy (Vatican Apostolic Library, Chig.I.24). Another transcription of the manuscript was made in 1928 by 'Giggi Rossi' and was kept in the Municipality of Affile. A copy is in the Biblioteca Statale of S. Scolastica at Subiaco.

Segatori S., *Affreschi cinquecenteschi nell'appartamento Colonna della Rocca abbaziale di Subiaco* (Tesi di laurea, Università degli studi di Roma "La Sapienza": 2006).

Serio A., "Pompeo Colonna tra papato e 'grandi monarchie', la *pax* romana del 1511 e i comportamenti politici dei baroni romani", in Visceglia M.A. (ed.), *La nobiltà romana in età moderna* (Rome: 2001) 63–87.

Serio A., *Una gloriosa Sconfitta. I Colonna tra papato e impero nella prima età moderna* (Roma: 2008).

Sommella P., "Sull'identificazione del tempio dei Giardini Colonna al Quirinale", *Atti dell'Accademia Nazionale dei Lincei. Rendiconti della Classe di Scienze Morali, Storiche e Filologiche* 9.31 (2020) 5–77.

Spila A., "Palladio e Antonio da Sangallo il Giovane sul grande tempio del Quirinale", *Annali di architettura* 29 (2017) 135–142.

Spila A., "The Double Stairway of the Temple of Serapis on the Quirinal Hill: Fame and Oblivion in the early modern Period", *Pegasus* 18/19 (2018) 37–71.

Strunck C., *Berninis unbekanntes Meisterwerk. Die Galleria Colonna in Rom und die Kunstpatronage des römischen Uradels* (Munich: 2007).

Thoenes C., "Note sul 'Ninfeo' di Genazzano", in *Studi Bramanteschi* (Rome: 1974) 575–583.

Vasari Giorgio, *Le Vite de' più eccellenti architetti, pittori, et scultori italiani, da Cimabue insino a' tempi nostri. Nell'edizione per i tipi di Lorenzo Torrentino. Firenze 1550*, ed. Bellosi L. – Rossi A., 2 vols. (Turin: 2015).

Zaccaria M., "Il Sublaqueum. La villa di Nerone a Subiaco", in Valenti M. (ed.), *Residenze imperiali nel Lazio* (Monte Porzio Catone: 2008) 43–52.

Zappacosta G., *Studi e Ricerche sull'Umanesimo italiano* (Bergamo – Milan – Florence – Rome – Messina: 1972).

CHAPTER 6

The Poetic Fortune of the Nymphaeum of Villa Giulia (with New Poems by Giovan Francesco Commendone)

Denis Ribouillault

Abstract

This article brings together for the first time a collection of poems written to celebrate the famous Nymphaeum of the Villa Giulia, the villa of Pope Julius III in Rome. Some of the verses were written by Giovanni Francesco Commendone, and the Pope ordered them to be engraved in the Nymphaeum. Other poems were written by Ippolito and Lelio Capilupi, Joachim Du Bellay, Fausto Sabeo, Miguel da Silva, Lorenzo Gambara, Francesco Franchini, and Andreas Masius. An examination of these poems reveals their continuity with a well-known tradition of poetic composition about statues of sleeping nymphs, established decades earlier and found in the gardens of Angelo Colocci, Johann Goritz, and the Vatican. The article thus highlights the fundamental use of Renaissance villa gardens for poetic composition and the inseparability between the material and spatial dimension of the Renaissance garden and the literary imagination it helped to create.

Keywords

Poetry – Sleeping Nymph – Pope Julius III – Garden – Renaissance Rome

1 Introduction: Villa Poetry in the Renaissance

In 1552, the second year of Julius III's pontificate, a medal featuring the pope's new villa was issued under the name "*Fons virginis Villae Giuliae*" (Fountain of the Virgin of the Villa Giulia), referencing the ancient aqueduct, the Aqua Virgo or Acqua Vergine, that supplies the fountains of its nymphaeum [Fig. 6.1]. The medal commemorates the final model of the project, which had begun

 | DOI:10.1163/9789004517547_007

FIGURE 6.1
Drawing of the Villa Giulia medal, after Letarouilly
IMAGE © DENIS RIBOUILLAULT

shortly after the accession of the pope in 1550.[1] In addition to what it reveals about the genesis of the villa, the medal testifies to the total subordination of its architecture to the spring, and to the significance of its thematic and decorative program to Julius III's pontificate.[2] Indeed, it was relatively rare for a private building like a suburban villa to be the object of such public commemoration – especially considering that the villa was transferred to the pope's brother, Baldovino del Monte, as early as 1553. The other medals produced during Julius' pontificate correspond to the events and monuments (basilica, Holy Door) of the 1550 Jubilee, to the larger papal political orientations (*anglia resurgens, securitas, hilaritas pontificia, annona pontificia*, etc.), and to devotions to the Virgin and Christ (for example, *Virgo tua gloria partus*).[3] How can we then explain the centrality of the fountain in the villa and its importance

1 Coolidge J., "The Villa Giulia: A Study of Central Italian Architecture in the Mid-Sixteenth Century", *The Art Bulletin* 25 (1943) 185–187 and later scholars date the medal to 1553. Richard Tuttle advances the dating to 1552. Tuttle R., *Jacopo Barozzi da Vignola* (Milan: 2002) 167. A second version of the medal, accompanied by the mention "ann. IIII", is dated 1554. The medal was struck by Gian Federico Bonzagna after a design by Bartolomeo Ammannati.

2 See especially Davis C., "Villa Giulia e la 'Fontana della Vergine'", *Psicon* 8.9 (1976) 131–141. Concerning the Villa Giulia, an essential reference is the thesis of Alessandro Nova, who holds that the nymphaeum was completed in the Spring of 1553: Nova A., *The Artistic Patronage of Pope Julius III (1550–1555): Profane Imagery and Buildings for the De Monte Family in Rome* (New York – London: 1988).

3 *Choix historique des médailles des papes* (Paris: 1839) 10–12.

FIGURE 6.2
Girolamo Sicciolante (school of), *Pope Julius III (Giovanni Maria Ciocchi del Monte)*, c.1550. Oil on canvas, 106 × 86 cm. Amsterdam, Rijksmuseum, inv. SK-A-3413
IMAGE © RIJKSMUSEUM

in the pope's political and cultural agenda? What artistic and cultural models informed its genesis and what role did the nymphaeum play in the life of the court?

Limited documentation exists to answer these questions. A rare account in Giovanni Ferro's *Teatro d'Imprese* (1623) speaks of the importance of the medal by describing 'the purity of the site, which is not tainted by vices' – an interpretation that contradicts Julius III's reputation in the historiography, that of a pope embroiled in scandals [Fig. 6.2].[4] The nymphaeum underwent such grave alterations following the death of Julius III that it is extremely difficult

4 Ferro Giovanni, *Teatro d'imprese* (Venice, G. Sarzina: 1623) 188: 'La Villa Giulia ch'era un palazzo suburbano vicino a Roma, era di Giulio terzo col *fons virgo*, per mostrare la purità del luogo non macchiato da vitij' ('The Villa Giulia, a suburban palace near Rome, was built by Julius III with the Fountain of the Virgin to show the purity of the place, unspoiled by vices'). According to most specialists, Rabelais stages the coarse appetites of Julius III, personified by the character of Bishop Homenaz, in *Quart Livre*. See Marichal R., "Le dernier séjour de Rabelais à Rome", *Comptes rendus des séances de l'Académie des Inscriptions et Belles-Lettres* 4 (1980), 694–696. Julius III's love for Innocenzo del Monte, a poor beggar whom he made

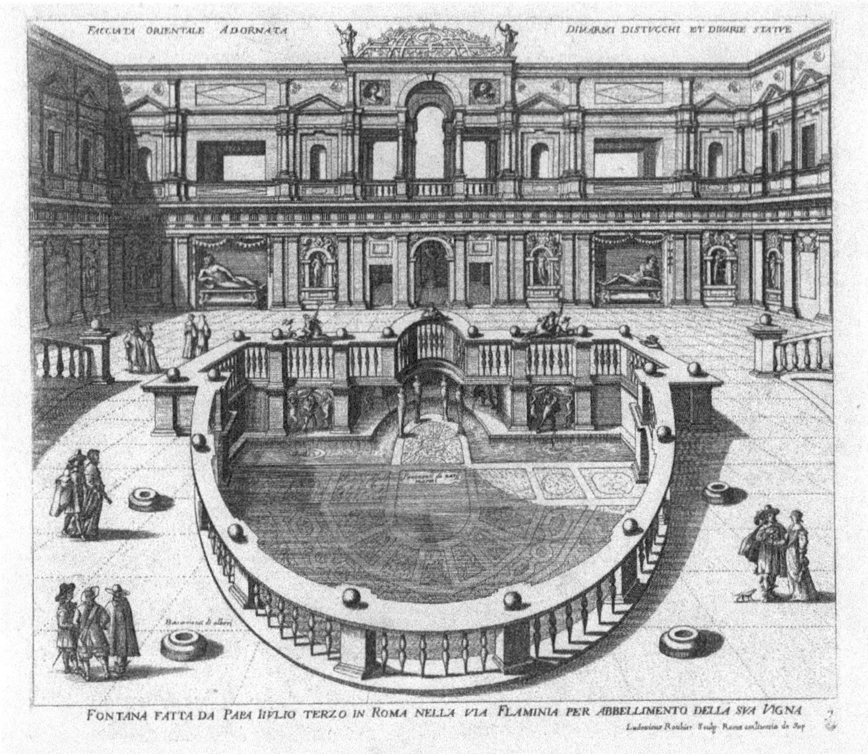

FIGURE 6.3 Louis Rouhier (engraver), view of the Villa Giulia Nymphaeum, *c.*1653–1691. Etching, 25.1 × 28.2 cm. From: *Le fontane ne' Palazzi e ne' giardini di Roma* [...] *parte terza* (Rome, Giovanni Giacomo de' Rossi: *c.*1653–1691). Amsterdam, Rijksmuseum, inv. BI-1893-A39-58
IMAGE © RIJKSMUSEUM

to advance a reliable reconstitution of its original state. Through remaining descriptions and a few engravings, we know that paintings and numerous sculptures ornamented the walls and loggias that surrounded the nymphaeum, notably a sculpture of a sleeping nymph or Venus. Four large plane trees, the placements of which were carefully determined, protected it from the ardour of the sun [Fig. 6.3]. One of the two loggias that opened onto the nymphaeum, which sheltered a statue of Venus, evoked the classical temple close to where a young virgin discovered the ancient water source, according to Frontinus' writings on classical aqueducts (end of the 1st century AD). Indeed, this story is reproduced in a bas-relief, attributed to Bartolommeo Ammannati, that still decorates the ceiling of this small room. In a subtle mise en abyme, the

cardinal in an enormous scandal, did little to improve his reputation. Historians have often retained these inglorious facts from his pontificate.

bas-relief itself recalls the painting of the same story that adorned the classical temple, according to Frontinus.[5]

Beyond the limited visual and documentary sources that have been handed down to us, a more significant quantity of poems dedicated to the nymphaeum shed new light on our understanding of this famous space and its original function. These poems have been little studied, and this article undertakes to bring them together for the first time.[6] Recent historiography on the literature and poetry produced in the context of villas and their gardens amply justifies our interest in the proposed topic.[7] "Villa poetry", as it may be called, was a rich poetic genre during the Renaissance, deeply influenced by the ancient models of Horace, Martial, Pliny, and Statius and pursued by great humanists and poets such as Angelo Poliziano, Jacopo Sannazaro, Giovanni Pontano, and Paolo Giovio. We can distinguish two major literary forms within the genre: the first derives from the classical models of *ekphrasis* and are committed to describing a villa and its garden, and the second attempts to resuscitate the art of epigrammatic poetry. Notable examples in the former category include descriptions of Agostino Chigi's villa (the Farnesina) by Egidio Gallo (*De viridario Augustini Chigii Vera libellus*, 1511) and by Blosio Palladio (*Suburbanum Chisii*, 1512)[8] or the later description of the Villa d'Este at Tivoli by the great French poet Marc-Antoine Muret, composed in 1575 in the tradition of Statius' *Silvae* and Poliziano.[9] Others include the encomiastic descriptions of the Villa Medici at Careggi by Alessandro Braccesi (1493), the Villa Montalto in Rome by Aurelio Orsi (1588) and, in the seventeenth century, Villa Aldobrandini

5 Davis, *Villa Giulia*. See also Ribouillault D., "La villa Giulia et l'âge d'or augustéen", in Morel P. (ed.), *Le miroir et l'espace du prince dans l'art italien de la Renaissance* (Paris: 2012) 339–388 (with bibliography) and Ribouillault D., "De la peinture au jardin (en passant par la poésie): la vallée Giulia à Rome, de Michel-Ange à Poussin", in Brunon H. – Ribouillault D. (eds.), *De la peinture au jardin* (Florence: 2016) 43–92.

6 In 1955, Jacob Hess announced the publication of a study on the poetic fortune of the nymphaeum but, to my knowledge, it was never published. See Hess J., "Amaduzzi und Jenkins in Villa Giulia", *English Miscellany* 6 (1955) 175–204 republished in his *Kunstgeschichtliche Studien zu Renaissance und Barock*, vol. 1 (Rome: 1967) 316. Ibidem, vol. 2, 221–226.

7 See Brunon H., "De la littérature au jardin", in Venturi G. – Ceccarelli F. (eds.), *Delizie in villa: il giardino rinascimentale e i suoi committenti* (Florence: 2008) 5–31.

8 On these poems and many others, see Caruso C., "Poesia umanistica di villa", in Crivelli T. – Caruso C. (eds.), *Feconde venner le carte. Studi in onore di Ottavio Besomi*, vol. 1 (Bellinzone: 1997) 272–294. Blosio Palladio also testifies to his passion for gardens in a series of letters, most of which remain unpublished, conserved at the Corsiniana Library in Rome. The digitised, handwritten letters are available online.

9 Tucker G.H., "La description de la villa d'Este à Tivoli chez M.-A. Muret (*Poematia*, 1575), dans la mouvance des *Silves* de Stace et de Politien", *Camenae* 22 (2018) accessible online. On the poetic descriptions of villas, see also Manzoli D., "Ville e palazzi di Roma nelle descrizioni latine", *Studi romani* 56 (2008) 109–166.

at Frascati by Giovanni Batttista Ciampoli.[10] The ekphrastic *all'antica* genre was so prestigious that descriptions of lost villas were also produced, such as the description of Lucullus' villa at Tusculum by Lelio Manfredi, composed in 1515 for Isabella d'Este, based on a similar work by Varro about his villa in Cassino.[11] Later, Giovanni Battista Campeggi would call his villa near Bologna "Il Tuscolano" in reference to Cicero's ancient villa. Between 1567 and 1582, numerous *letterati* gathered there to write poems about villas, celebrating their owners as new Ciceros.[12] Selected quotes sometimes figured among the decoration of villas, like in the case of the loggia of the Villa Lante on the Janiculum (1533) or the Villa Mondragone in Frascati (1616–1618). At the Villa Lante, the architect Giulio Romano had a passage by the Roman poet Martial, concerning Julius Martialis' classical villa, engraved in the elegant loggia opening on a breathtaking view of Rome: *Hinc septem dominos uidere montis/et totam licet aestimare Romam* [from here we can see the seven queen hills and take in the entire city of Rome] (*Ep.* iv. 64, 11–12), thus closely linking the poetic memory of the panorama of ancient Rome with the modern landscape. Architects also participated in this literary craze by writing descriptions in Italian. Most were inspired, sometimes to the point of plagiarism, by the letters of Pliny the Younger describing his famous villas, rediscovered in 1419 by Guarino da Verona at the court of Ferrara. The best-known example is Raphael's letter concerning the Villa Madama,[13] which is at once a Plinian pastiche, a guided walk, and a claim to authorship over the villa's design. In the same vein, Bartolomeo Ammannati wrote a description of the Villa Giulia in 1555, which is today among the documents on the building most cited by historians.[14]

---

10 Braccesi Alessandro, "Ad eundem descriptio horti Laurentii Medicis", in *Alexandri Bracii Carmina*, ed. A. Perosa (Florence: 1943) 75–77; Orsi Aurelio, *Perettina, sive Sixti V, horti exquilini* (Rome, J. Martinellum: 1588); Caruso, *Feconde venner le carte* 279–282; Ciampoli Giovanni Battista, *Rime di Monsignor Giovanni Ciampoli* [...] (Rome, Carbelletti: 1648) 54–72.

11 Mattei F., "L'invenzione dell'antico alla corte dei Gonzaga. Lelio Manfredi e la descrizione della villa di Lucullo a Tusculum", *Mitteilungen des Kunsthistorischen Institutes in Florenz* 60.1 (2018)107–126.

12 Aksamija N., "Architecture and Poetry in the Making of a Christian Cicero: Giovanni Battista Campeggi's Tuscolano and the Literary Culture of the Villa in Counter-Reformation Bologna", *I Tatti Studies* 13 (2010–2011) 127–199.

13 Di Sanzio R., "Letter Describing Villa Madama: Transcription, Translation and Commentary", in Dewez G., *Villa Madama: A Memoir Relating to Raphael's Project* (London: 1993) 21–31.

14 Ammannati Bartolomeo, letter dated 2 May 1555 to Marco Benavides in Padua (Pesaro, Bibl. Oliveriana) MS 374, II, fols. 91–96, cited in Falk T., "Studien zur Topographie und Geschichte der Villa Giulia in Rom", *Römisches Jahrbuch für Kunstgeschichte* 13 (1971) 171–173.

As with the topographical views of villas and gardens that accompanied the development of this literary genre, it is important to question the lack of synchronism between construction sites and their descriptions. Among the many painted or engraved *vedute* of villas and gardens, many depict projects that were only partially realised. These images nevertheless played an important role in the development of the buildings, ensuring that they were completed according to the initial design sanctioned by the image.[15] In a recent book on the Villa Madama in Rome, Yvonne Elet has shown that poetic and literary descriptions may have played a similarly active role in the development of the architectural project. In addition to analyzing Raphael's letter, Elet specifically focuses on a long poem in verse written by the humanist Francesco Sperulo in 1518, when the construction work on the villa had barely begun. Titled "*Villa Medicea versibus fabricate*" – literally, "The Villa Medici constructed in verse" – the poem played an important role in elaborating what Elet calls the "conceptual structures" or "meta structures" of the architectural project, which refer both to its architecture and decoration and to the archaeological, poetic, philosophical, and ideological values they embody.[16] As the poem's title makes clear, poetry is one of the key materials of the villa's conceptual structure.

Epigrammatic poetry forms a second category of literary production inseparable from Renaissance villa and garden culture. Originally an inscription in prose or verse attached to a tomb, statue, or monument, the epigram came to designate a short poem of an ingenious, witty (*argutia*), and scholarly character building on cultivated ambiguity and shared cultural references.[17] This type of poetic composition was practised at garden banquets modelled after the tradition of philosophical symposia.[18] It drew wide interest partially due to the poet's capacity for improvisation and the seemingly spontaneous and informal character of their compositions, marked by *sprezzatura*.[19] A letter from

15 On this question, see Ribouillault D., "Toward an Archaeology of the Gaze. The Perception and Function of Garden Views in Italian Renaissance Villas", in Beneš M. – Lee M.G. (eds.), *Clio in the Italian Garden: Twenty-First Century Studies in Historical Methods and Theoretical Perspectives* (Washington D.C.: 2011) 205–208; Ribouillault D., *Rome en ses jardins. Paysage et pouvoir au XVIe siècle* (Paris: 2013) 300–315.

16 Elet Y., *Architectural Invention in Renaissance Rome: Artists, Humanists, and the Planning of Raphael's Villa Madama* (Cambridge: 2017) with a bibliography on the poetic dimension of Renaissance villas.

17 On the genre, see Carey C. – Kanellou M. – Petrovic I. (eds.), *Greek Epigram from the Hellenistic to the Early Byzantine Era* (Oxford: 2019), and De Beer S. – Enenkel K.A.E. – Rijser D. (eds.), *The Neo-Latin Epigram: A Learned and Witty Genre* (Leuven: 2009), notably the introduction by Karl A.E. Enenkel.

18 On the importance of this tradition to Renaissance gardens, see George Brocklehurst's essay in this book.

19 Caruso, *Feconde venner le carte* (Poesia umanistica di villa) 274–275.

Cardinal Jacopo Sadoleto from 1529 addressed to the humanist Angelo Colocci captures the spirit of these poetic jousts:

> Oh, how often do those colloquiums, those banquets which we used to hold, come back to mind, when, either in your gardens [orti suburbani], or in mine on the Quirinal, or in the Circus Maximus, or on the banks of the Tiber at the Temple of Hercules, or elsewhere, we would gather together among learned men, respectable no less for virtue than in reputation: where, after the informal banquets, seasoned less with gluttony than with spirit, we would recite poems and declaim orations, to the great satisfaction of all because they revealed a greatness of spirit [*altezza d'ingegno*], tinged nevertheless with joy and grace.[20]

Epigrammatic poems must also be understood as an integral part of the villas and gardens that inspired them. Kathleen Wren Christian's studies on the collections of ancient sculpture in Roman gardens have emphasised how epigrams seemed to make the statuary speak (*prosopopoeia*), as if the poems often affixed on or near the statues activated the voices, spirits, and vital breaths (*pneuma*) that were often attributed to these works.[21] Tangible vestiges of these ancient practices persist in Rome today, even if the beliefs attached to them have evolved. On the base of the ancient statue known as Pasquino located near Piazza Navona, pieces of paper featuring poems or pamphlets that satirise or critique power and authority (*pasquinades*) continue to surround and give voice to the figure, as they have for centuries.[22] The sculpture known as Marforio, another "marble mouth" now restored and exhibited in the Capitoline Museums, operated according to the same principle [Fig. 6.4]. In an engraving made by Nicolas Beatrizet around 1550, a young man is depicted sitting on a fragment of ruin, either drawing the statue or composing a poem – perhaps the one that can be read on the wall behind the statue. In any case, the image shows the direct relationship between the statue, the ancient and modern inscriptions scattered around the city, and the artistic and poetic inspiration and agency they provided to the public.

---

20 Sadoleto Jacopo, Letter to Angelo Colocci (Rome: Biblioteca Apostolica Vaticana) Vat. lat. 4102, cc. 16v–17r. See Gnoli D., "Orti letterari nella Roma di Leone x", *Nuova antologia* (1930) 15.

21 Christian K.W., "Poetry and 'Spirited' Ancient Sculpture in Renaissance Rome", in Kenda B. (ed.), *Aeolian Winds and the Spirit in Renaissance Architecture* (London: 2006) 103–124.

22 See Dickinson G., *Du Bellay in Rome* (Leiden: 1960) 155–164, and more recently Spagnolo M., *Pasquino in piazza. Una statua a Roma tra arte e vituperio* (Rome: 2019). A certain number of texts are translated into French in Lafon J.B.M., *Pasquino et Marforio. Les bouches de marbre de Rome* (Paris: 1877).

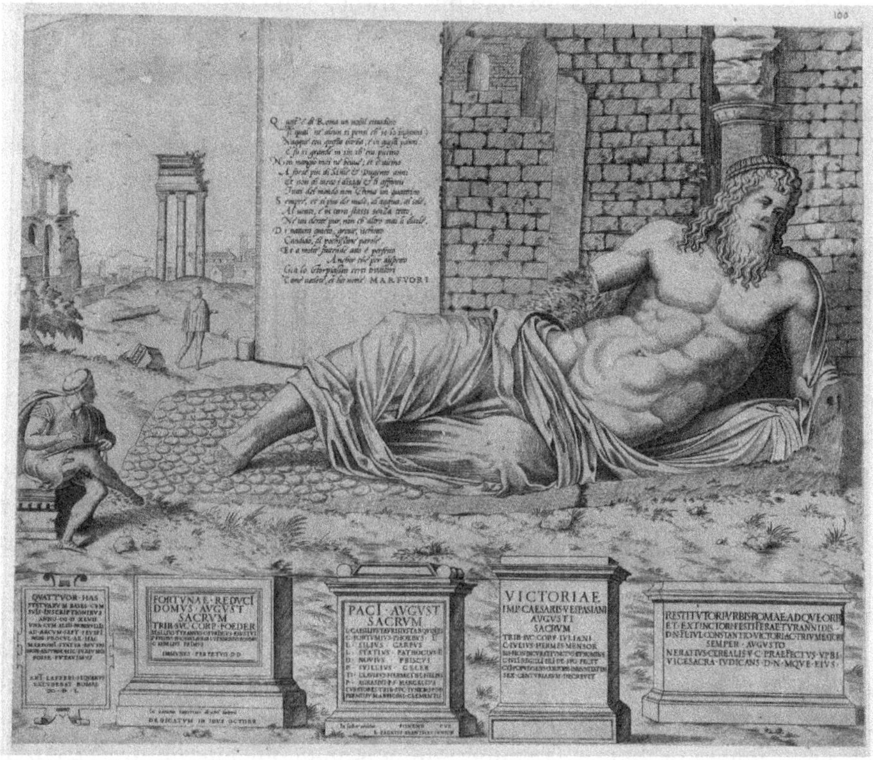

FIGURE 6.4 Nicolas Beatrizet (attributed to), *Marforius*. Engraving, 37.2 × 43.5 cm. From: *Speculum Romanae Magnificentiae* (n.p., Antonio Lafreri: 1550). New York, The Metropolitan Museum of Art, inv. 41.72 (2.100)
IMAGE © MET

According to specialists, the practice of attaching poems to statues is thought to derive from ancestral votive practices. In ancient Greece and Rome, prayers to the gods were left on lead tablets that were sometimes nailed to statues or the walls of sanctuaries. In the *Ars poetica* (75–7), Horace indicates that the model of epigrams used for poetic epitaphs derives from these prayer rituals, which persisted in a Christian context, as testified by the ex-votos that continue to be attached to the tombs of saints in churches.[23] The Platonic Academy provided another important model emulated by Renaissance poets assembled in gardens. Diogenes Laertius (3, 43–45) recalls that after Plato's death, his disciples buried him in the garden of the Academy and composed epigrams in his honour, which they attached to his tomb. Some of these verses

23 Wellington Gahtan M., "Appended Epitaphs", in Steiner-Weber A. – Enenkel K.A.E. (eds.), *Acta Conventus Neo-Latini Monasteriensis. Proceedings of the Fifteenth International Congress of Neo-Latin Studies (Münster, 2012)* (Leiden: 2015) 214–226.

made their way to Renaissance humanists through the *Greek Anthology*, a collection of epigrams written in memory of famous poets and philosophers. It was published in Florence in 1494 in the Byzantine Maximus Planudes version and quickly became an object of great interest.

Having considered these literary precedents, let us now turn our attention more specifically to the poems about the nymphaeum of the Villa Giulia.

2 The Poems of Giovan Francesco Commendone (1524–1584)

A little-explored episode in the history of the Villa Giulia is that of the poetic production of Giovan Francesco Commendone.[24] Born in Venice on 17 March 1524 to a family of doctors with humanist interests, he studied philosophy at the University of Padua, Plato in particular, before turning to law. Commendone arrived in Rome in time for the 1550 Jubilee and presented by Venetian political figures, began to frequent the papal court as well as an important circle of scholars, which included Giulio Poggiani, Guglielmo Sirleto and Paolo Manuzio. He developed a close friendship with Annibale Caro (1507–1566), who participated in elaborating the programme of Julius III's nymphaeum and is one of the key figures in the present study; Caro later dedicated an extremely laudatory sonnet to Commendone. Placed in the service of the Secretary of State, the young Commendone quickly rose in the curial hierarchy. Sent on various diplomatic missions to England and Portugal, he was appointed protonotary and bishop of Zante and Cefalonia in 1555. In 1564, he was made cardinal by Pope Pius IV.[25]

The most precious information about Commendone's Roman youth is found in the biography written by his secretary, Antonio Maria Graziani, who later became Bishop of Amelia. Published in Latin in 1669, it was translated into French two years later.[26] In the biography, we learn of Commendone's

24 To my knowledge, only Nova, *The Artistic Patronage* 105–107 and Tschäpe R., *Die Villa Giulia: Rekonstruktion des Bildestandes und ein Versuch, die Strukturen des ikonographischen Bestandes zu fassen* (Aachen: 1995) 207–210, have briefly examined some of Commendone's poems.

25 See Caccamo D., "Commendone, Giovanni Francesco", *Dizionario biografico degli italiani* 27 (1982). Annibale Caro's sonnet is published in Caro Annibale, *Opere*, ed. U.A. Amico (Florence: 1864) 436. On Annibale Caro's contribution to the nymphaeum programme, see Ribouillault, *La villa Giulia*, 339–340.

26 Gratianus Antoine, *De vita Joannis Francisci Commendoni Cardinalis libri iv* (Paris: 1669) book 1, chapter 5; Gratiani Antoine Maria, *La vie du cardinal Jean François Commendon* (Paris, Chez Sébastien Mabre-Carmoisy: 1671). Following quotes: my translation.

extreme talent for poetic composition. According to Graziani, at the age of 14, "he allowed himself to be charmed by the sweetness of poetry, and applied himself so happily to reading and writing verse, that one was astonished to see that in his great youth, not only did he surpass those of his age, but he could even be compared to the most excellent poets of his century".[27] In chapter VIII, Graziani explains how the young man fell into the pope's good graces:

> Pope Julius had a beautiful pleasure home built with great care and expense outside the Flaminian Gate. He had the pure waters of an ancient spring conducted through long canals. A marble Nymph, of ancient workmanship, who as if dozing, and leaning over an urn, collected them, and distributed them in the gardens. Julius [sic.] wanted various little poems to be made in honour of this statue, to be engraved according to the disposition and spaces of the marbles. All the fine minds, not only of Rome, but also of the whole of Italy, worked on this subject. Commendone had an extreme passion for verse since childhood. It is true that it had diminished a little: but feeling animated by the desire to please his master, and by this commendable emulation, which excited all the learned men of Italy, he composed some epigrams, which were highly suited to the subject, and which filled the empty spaces of the marbles very appropriately. The Pope having read them, as he had a strong predilection for these kinds of works, not only approved them, but ordered them to be engraved, and preferred them to all others.[28]

Graziani does not transcribe any of these poems, however, explaining that the admiring pope's promotion of Commendone had the paradoxical consequence 'that he renounced poetry entirely', destroying many works in the process, 'burning them with more pleasure than he previously had in writing them'.[29] These poems have nevertheless survived his acts of self-censorship and are preserved in several manuscripts now held in various Italian archives.[30]

27 Ibidem 13.

28 Ibidem 36–37.

29 Ibidem 38. Even if Graziani does not mention it explicitly, such an attitude might believably correspond to the unexpected death of Julius III in March 1555 and the abrupt shift in the moral and religious climate at the start of the Counter Reformation under Marcellus II and then Paul IV.

30 *Carmina reperta in Bibliotheca Cardinalis Commendoni quae ex legato R.P.D. Antonii Cauchi (Cocco) Commendoni eius Nepotis pervenerunt ad Stephanum Benassium. Et creduntur composita ab eodem Cardinale* (Rome, Archivio Storico della Pontificia Università Gregoriana) Ms. APUG 1219, I, 1–54. A mention of Commendone's poems can be found in a note in a

In the only study devoted to Commendone's poetic oeuvre, the eminent Latinist P. Emilio Springhetti examined the copy preserved in the archives of the Gregorian Pontifical University of Rome, which contains fourteen epigrams of different genres or metres, enumerated in the manuscript from 15 to 28.[31] A very interesting dissertation by the Jesuit poet Girolamo Lagomarsini, dated 6 June 1768, is placed at the beginning of the codex, which attests to the authenticity of the poems and discusses various copies.[32] The first epigram concerning the nymphaeum was published in 1783 in the fourth volume of the *Anecdota litteraria ex mss. codicibus eruta*, edited by the scholar and abbot Giovanni Cristoforo Amaduzzi who also authored one of the first studies on the Villa Giulia, though it was unfortunately never completed. One of his chapters intended to deal with 'Poesie varie in celebrazione di questa villa, especialmente del ninfeo annesso al palazzo principale'.[33] Amaduzzi indicates that he copied the epigram 'from the manuscript codex of the archives of Cornelius Mussius, bishop of Bitunto [Bari], formerly in the library of the Sylvester Friars of Sentium or of Sassoferrato, and now of Cardinal Rector Stefano Borgia' – therefore a manuscript in the Vatican Library.[34]

---

brief article from 1844. The author indicates that the verses 'si trovano in un codice vaticano unitamente a quattro libri di lettere di Antonio Graziano vescovo di Amerino' ('are found in a Vatican codex together with four books of letters by Antonio Graziano, Bishop of Amerino'). See "Degli orti di papa Giulio III", *Il Saggiatore Giornale Romano* 3 (1844) 91–92. Unfortunately, I was not able to find this manuscript at the Vatican Library, which seems to have been the source consulted by both Amaduzzi and Angelo Mai (see *infra*).

31 Springhetti A., "Ioannes Franciscus Commendone (1524–1584). Legatus Pontificius et Cardinalis, Poëta Latinus", *Archivum Historiae Pontificiae* 8 (1970) 231–232. The Gregorian archives conserve Springhetti's own transcription of Ms. apug 1219. See Fondo Springhetti, file 11, chap. 5.

32 Lagomarsini Girolamo, *Dimostrazione dell'autenticità dei carmi contenuti in APUG 1219 attribuiti al Card. Giovan Francesco Commendone*, vol. 2 (Rome, Archivio Storico della Pontificia Università Gregoriana) Ms. apug 1219 1–24 (1768). On Lagomarsini, see Franco Arato's notice in *Dizionario degli italiani*, 63 (2004).

33 Amaduzzi Giovanni Cristoforo, *Storia della villa di Giulio III fuori della Porta del Popolo nella via Flaminia opera lasciata imperfetta dall'abate Gio. Cristofano Amaduzzi co' monumenti e notizie sulla Medesima* (Savignano sul Rubicone, Biblioteca della Rubiconia Accademia dei Filopatridi) Ms. Am. 49, cc. 1r–249r (ca. 1777–1782). On Amaduzzi and the Villa Giulia, see, besides Hess, "Amaduzzi und Jenkins in Villa Giulia", Benocci C., "Amaduzzi e Villa Giulia: un erudito settecentesco interpreta il paradiso di papa Giulio III", in Delbianco P. (ed.), *Atti della nona Giornata amaduzziana* (Cesena: 2012) 57–109. Amaduzzi learned of Graziani's biography on Commendone and, on this basis, wrongly attributed the *Lex Hortorum* of the nymphaeum to Commendone. He also mentions the epigrams on the nymphaeum by Fausto Sabeo. This attribution is further discussed below.

34 Amaduzzi Cristofano Giovanni, *Anecdota litteraria ex mss. codicibus eruta*, vol. 4 (Rome, Apud Antonium Fulgonium: 1783) 429–430, 445–446.

Adonium Francisci Commendonii
Ad fontem celeberrimum Iulii iii. P.M.

Spectator nitidi loci,
Et fontis, geminas vides
Sacro numine virgines.
Saxo labitur altera,
Lymphis accubat altera.
Isti iniurius oscula
Figes, ut magis aestues,
Illi innoxius, ut calor
Urens, ut sitis arida
Pellatur, bibe, prolue,
Hauri, nudaque pectora
Caste respiciens, abi.

An Adonic[35] *poem by Francesco Commendone*
For the very famous fountain of Julius III

Pontifex Maximus / You, who look at this splendid place / And this fountain, see these twin girls / Virgins [created] by the grace of God. / One flows on the stone, / The other is stretched out near the fountain. / To this one [the second] you will give rather illicit kisses, / In order to burn more with desire; / To that one [the first] you will give rather innocent ones, / To soothe the burning heat and your dry thirst / Drink, bathe, / Quench yourself, and, after casting a pure glance upon / Their uncovered breasts, go away!

Other poems in the series were only published in 1842, appearing in the *Spicilegium Romanum* of Cardinal Angelo Mai, a famous philologist and tireless decipherer of manuscripts. They are presumably from the same codex belonging to Graziani and preserved in the Vatican Library.[36] Following are

35 A type of verse consisting of dactyls and spondyls.

36 Mai A., *Spicilegium Romanum*, vol. 8 (Romae, Typis collegii urbani: 1842) 479–487. Mai also mentions five brief epigrams by Commendone published in 1554 in the extensive publication about fish by Ippolito Salviani, personal physician to Julius III: Salviani Ippolito, *Aquatilium animalium historiae, liber primus: cum eorumdem formis, aere excusis* (Romae, Apud eundem Hippolytum Saluianum: 1554), fol. 5v. On this topic, see also Lagomarsini, 1768, 3–5.

the poems in the order and numbering chosen by Mai, with their translation below:

I

Nympha loci niveosque sinus pectusque pudicum
Tecta iacet, crines fusa, supina caput.
Virginei flos oris hiat, patulisque labellis
Halat inoffenso blanda sopore quies.
Dextra iniecta humero tereti labentia tergo
Colligit in parvos lintea pressa sinus.
Laeva comes lateri pronae superincubat urnae,
Qua nitidis virgo desilit unda vadis.
Et fovet algentes latices levis halitus almi
Pectoris; haud alio mitior aura loco est.
Ede diu molles animas, genialia fontis
Flumina ! Sic felix irriget ora sopor !
Sic et nympharum princeps celebrere choreis !
Sic tibi perpetua defluat amnis aqua !
Rexque paterque hominum totum qui temperat orbem
Saepe suam dicat, semper Iulus amet !

The Nymph of this place lies, her chest, white as snow, is covered, covered also her modest throat, / [She] lies, hair spread, head back. / The flower of her virginal mouth opens, and through her open lips, / Her sweet repose exhales a peaceful slumber. / Her right hand rests on a vase leaning sideways, / From where the virgin water descends in its crystalline waves. / The light breath of her nourishing chest warms the cold waters; / No breeze is sweeter anywhere else. / Let the world see [, Prince,][37] the joyous waves of this spring, / Let be seen her long-sleeping soul! That a deep, happy sleep may thus refresh the face, / And so, Prince, that you may be celebrated by the dances of the Nymphs! So that the eternal water of this river may flow unceasingly for you; / That Julius, king and father of men, who watches over the whole world, / Often designates it [as] his own, and that he loves her forever.

II

Tibure ducta, patris sub Iuli montibus aureis,
Foeta undis virgo delicias pario.

37 This obviously refers to Pope Julius III.

From Tibur,[38] [gushing] down from the golden hills of Father Julius, / I, the virgin abundant in water, give birth to delights.

III

Casta verecundis oculis pete virginis hospes
Limina, virgineas et pete purus aquas.
Labitur una vadis virgo, cubat altera virgo,
Sacra vides geminis marmora virginibus.
Labilis in blandos dum murmurat unda susurros,
Sub Iuli dormit munere nympha loci.
Et somnum et violare sacrae vada virginis, hospes,
Parce, bibe, et domino laeta precatus abi.[39]

You who pass here, seeking with your modest gaze the chaste abode of this virgin / And direct yourself unstained towards her virgin waters. / One of these virgins flows at the bottom of the basin, the other rests at its edge. / You see here a monument in marble consecrated to two twins. / While the spring that flows murmurs softly / The nymph of this place sleeps under the work of Julius. / Guard yourself, visitor, from disturbing the sleep and the waters of this sacred virgin, / Drink and, after having joyfully saluted the master of the place, go away.

IV

Quisquis ab aurorae populis et luminis ortu,
Quisquis ab occasu pelagoque, ruente quadriga,
Sive aestu nimio, sive ustis frigore ab oris,
Cernis inexpletis oculis haeresque tuendo
Eximia forma recubantis virginis ora,
Compositosque in somnum artus, propriumque quieti
Obrepens per membra decus, placidisque labellis
Molliter halantes tranquilli flaminis auras;
Hoc volo te monitum paucis, ignarus Iuli
Neforte in legem commisso, principis iram
Offendas, graviter teneros violare sopores,

38 This is the former name of the small city of Tivoli.

39 Note that Angelo Mai here has combined two separate epigrams into a single poem. See Fondo Springhetti 11, caput v, 84–85. The two poems are numbered 17(3) (Casta verecundis [...]) and 18(4) (Labilis [...]).

Virgineosque manu latices temerare profana
Ausis, Oreadum cum sit castissima princeps.
Virgo est quam spectas efflantem pectore somnum,
Virginis ad blandae crepitantia murmura lymphae.
Scilicet haec olli pater ocia fecit Iulus.

[That you come] from the people of the aurora, there where the light is born, / Or from the regions of the sunset, [whether you arrive] by sea or [by land] in swift carriages / [Whether you were born] in the lands of too great heat or by the cold, / You discern with insatiable eyes and you fixedly look / The face of remarkable beauty of this virgin reclined / [And] her body asleep, / [And] this beauty proper to sleep that slides stealthily through her limbs; / [And you distinguish] the soft murmur of the quiet breath that passes through her peaceful lips / I want to warn you in few words, since you do not know Julius: / Do not arouse the rage of the prince because of a crime against his law! / Dare not disturb their peaceful sleep / And profane these virginal waters with your impious hand, / Since their princess is the most chaste of the Oreads.[40] / She is that virgin whom you watch peacefully breathe in her sleep, / To the sound of the rustling murmurs of her soft virginal water. / Evidently, it is Father Julius who assures her all this reposefulness.[41]

V
Sum fons virgineus, fluente pura,
Aqua virginea virginis puellae,
Nitori sacer, et sacer pudori.
Immundi procul este et impudici.

I am a virginal fountain, a Pure source, / A virgin water of a Virgin girl, / Sacred thanks to my beauty, Sacred thanks to my modesty. / [You], abject and indecent, keep yourself far from me!

VI
Virgo puella, virgo aqua est.
Aut purus hospes adi, aut abi.

40 The Oreads are nymphs of the mountains and caves, sisters of the satyrs.

41 Here, the term translates the Latin *otium*, a central concept in ancient and Renaissance villa culture.

This young girl is a virgin, this water is a virgin.
/ Come with a pure soul stranger or be gone!

Springhetti, for his part, was content to publish only one additional poem in his 1970 article, supplementing the seven already published by Amaduzzi and Mai:[42]

Este procul impurae manus,
Et impudica lumina.
Est enim aqua Virgo, quae fluit;
Virgo est puella, quae cubat.

Hold yourself far, impure hands, / And you, indecent light / For this water that flows is the Aqua Virgo: / This Virgin is this girl who sleeps.

In the original French version of this article published in 2020, I only included the eight poems already published by Amaduzzi, Mai and Springhetti. With the reopening of the Gregorian archives after the pandemic of Covid-19, it was finally possible to consult Ms Apug 1219, which contains the remaining five poems. We publish them here with the numbers Springhetti gave them on the manuscript itself and his own transcription of it, together with their translation by Alin Mocanu.[43]

5 (19)
Virginis hem placidum spectator voce soporem,
Ore labella, manu pectora ne violes.
At puro ore, manu pura, sive ureris aestu,
Sive siti, tacitus perge lavare, bibe.

Oh, you, the one who contemplates the Virgo, with your voice, her serene sleep / Do not disturb! Do not touch her lips with your mouth, nor her breasts with your hand. / However, with unsullied mouth and pure hands, either ardent because of the heat, / Or dying of thirst, silently, go on and bathe, and drink.

---

42 Springhetti, "Ioannes Franciscus Commendone" 232.

43 Rome, Archivio Storico della Pontificia Università Gregoriana, Fondo Springhetti, file 11, chap. 5, 85–87, 89: (5) 19, (7) 21, 8 (22), 12 (26) and (13) 27. I am very grateful to Alin Mocanu, not only for his translations, but also for his help in the transcription of Commendone's manuscript.

7 (21)
Qui nitidos circum spirantia marmora fontes
Delicias Circi ridentia divitis horti,
Aequoraque, et clivos opulentis prodiga ramis
Poma suos numero, ut forma superantia frondes,
Tanta loci genio, et congestis undique donis,
Tanta peregrinis opibus, tantoque nitore,
Eximiique operis cultu spectacula rerum
Praeteris, obtutuque nimis defixus in uno
Oris Virginei florem, Pariique papillas
Marmoris, et niveas nudato e pectore flammas,
Fragrantesque auras, geminique latentis ocelli,
Inque leves somnos sopito e lumine lumen
Mollius incendens hospes miraris, et ardes,
Fer levipes gressum, tacitus bibe, cautior hauri,
Ne mage sicca sitis, ne torridus aestus adurat,
Praemoneo caveas, nec te malus auferat error,
Aemula si soli strepitu commota recludat
Lumina, flagrabunt fontes, atque ipsa Theatri
Marmora, flamma erit omne nemus, tecum omnia flammae.

You, who goes past the breathing marbles surrounding the shining fountains, / Past the marbles smiling at Circe of this garden rich in delights / And in waters, past its hills [adorned] with fruit trees abundant by their number / In marvelous branches, given the exceeding beauty of their leaves, / Past so many wonders of this place *genius*', its piled-up ubiquitous enchantments, / Its exotic wealth and its splendor, / Past the display of refinement of this extraordinary work of art, / Fixing obsessively your gaze at the same time / On the flower of her virginal mouth, on her breasts / Made of Paros marble, on her snowy flames emanated from her naked chest, / On her scented breath, on the brightness of her mysterious eyes, / During her delicate sleep in a dim light, / Passionate and sensitive visitor, you admire her and burn with desire. / Approach gently, drink quietly, carefully take a bath, / Do not let the dried thirst nor the torrid heat consume you. / I warn you to be careful: do not let a bad mistake deprive you of all these. / If disturbed only by a sound on the ground she opens her jealous eyes, / The fountains will be on fire, as well as the marble of the Theatre. / The entire wood will be ablaze and the flames will burn you, with everything else around.

8 (22)
Heus hospes genius loci
Te paucis volo, ne inscius
Erres. Haec chori Oreadum est
Formosissima Virginum
Et castissima. Virgo aqua est,
Cuius fonte lavit sacras,
Et mensis superûm manus
Assuetas hominum Pater
Idem maximus, optimus,
Idem Iulius. Haec satis.
Tu nunc ore, oculis, manu
Pius, purus adi, aut abi.

Hey, guardian spirit of this place, / I shall only tell you this, so that, on account of your ignorance, / You do not make any error: this [Nymph], belonging to the Oreads' chorus, is / The most beautiful amongst the virgins / And the most chaste. This fountain is Virgo. / In her waters, the Father of all humans, / The mighty, the best Julius / Purifies his sacred hands / Accustomed to the divine [ceremonial] tables. She is abundant. / Now, you [who passes by], with pure mouth, eyes and hands, / Approach her piously, or else, go away!

12 (26)
Qui ruperit placidum soporem Virginis,
Aut polluerit aquae fluenta Virginis,
Somno gravatum vigilia angat pertinax,
Et sicca anhelum semper exurat sitis,
Nec spectet umquam lumina, quibus optimus
Pater suum serenat orbem Iulius.

The one who disturbed Virgo's placid sleep, / Or soiled her fluid waters, / Let an obstinate insomnia torment him when he feels the weight of sleep; / Let a desiccated thirst always dry him up when he gasps for air; / Let him never, ever see the daylight again, with which the best / Father, Julius, brightens his world.

13 (27)
Soluta curis omnibus
Dormi puella molliter.
Iulus orbi consulens
Insomnis excubat Pater.

Exempt of all concern, / Sleep, young girl, peacefully. / Julius takes care of the world [and] / Relentlessly he keeps watch, [our] Father.

Let us now examine these poems briefly. They are articulated around three main themes. First, they highlight the linguistic confusion of the term Virgo, signifying both the virgin nymph and the name of the spring. The poems insist on the coexistent presence of two virgins: one, the marble statue, personifying the other, the spring water. In other words, their poetic force comes from the repeated use of paronomasia, the use of words with similar consonances but different meanings in the same sentence to create a deliberate indeterminacy between signifier and signified. A second, classic theme in these poems involves the intrusion of the spectator, notably referenced in *Spectator*, the first poem published by Amaduzzi, which reactivates the imagery of the voyeuristic and

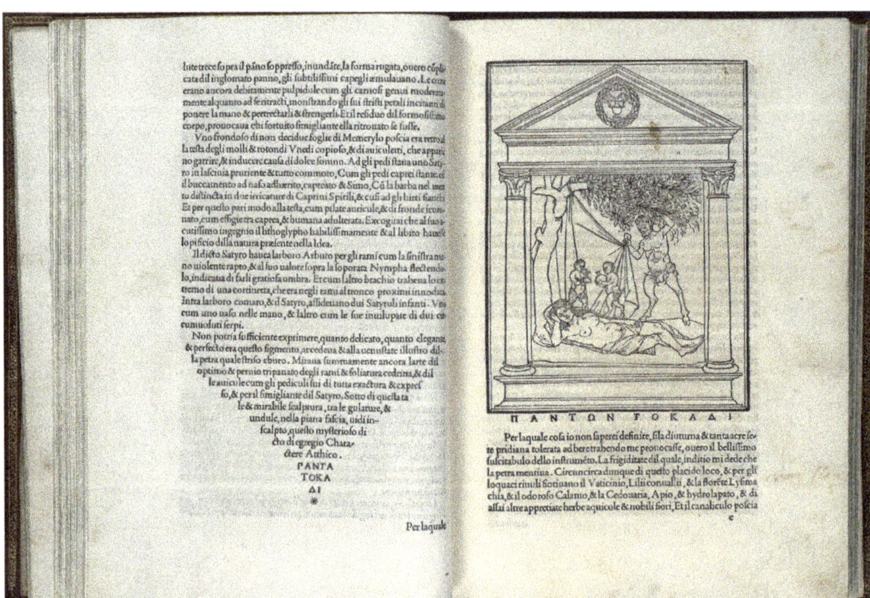

FIGURE 6.5 Benedetto Bordone (attributed to), *Nymph and Satyr*. Engraving, 29.5 cm × 22.4 cm. From: Colonna Francesco, *Hypnerotomachia Poliphili* (Venice, Aldo Manuzio: 1499). New York, The Metropolitan Museum of Art , inv. 23.73.1
IMAGE © MET

impure gaze that transgresses the virginal space of the nymphaeum. Following the tradition of Francesco Colonna's *Hypnerotomachia Poliphili*, published in Venice in 1499 in Commendone's homeland, the motif of the satyr lusting after a sleeping nymph introduces an erotic charge to the poems while activating a discourse on artistic reception [Fig. 6.5].[44] Finally, the spectator's desire as a figure thirsting for fresh water is tempered by the quasi constant presence of the pope at the conclusion of the poems. He appears benevolent but severe, the true guardian of the nymph's virginal virtues and of the sacred dimension of the spring. Thus, the visitor is presented with a guided choice between *virtus* and *voluptas*, which references the humanist ideal of the *Vir virtutis*, namely the human ability to influence the whims of fortune through virtuous action. Unsurprisingly, this ideal figures as the key element of Pope Julius III's personal *impresa*, which pictures Prudence (or Virtue) subduing Fortune, and is an omnipresent motif in the villa's decor [Fig. 6.6].[45] In addition to the

FIGURE 6.6
Federico Brandani, after Prospero Fontana, *Virtue Subduing Fortune*, 1553. Stucco, Rome, Villa Giulia
IMAGE © WIKIMEDIA COMMONS

44 On the gaze of satyrs, see François Lavocat's fundamental study, *La syrinx au bûcher. Pan et les satyres à la Renaissance et à l'âge baroque* (Geneva: 2005) in particular 284 *et sq.*

45 Skinner Q., *The Foundations of Modern Political Thought: Volume 1, The Renaissance* (Cambridge: 1978) 94, 101 (The powers of the *Vir virtutis*). On the presence of *Fortuna* at Villa Giulia, see Ribouillault, *La villa Giulia* 360, 364–365, 372.

FIGURE 6.7
Virtue Subduing Fortune and Nymph and Satyr, entrance to the Nymphaeum. Rome, Villa Giulia
IMAGE © DENIS RIBOUILLAULT

ceiling of the main dining room, the allegory is also found at the entrance to the Nymphaeum, together with the motif of the satyr and the sleeping nymph shown immediately below, a motif directly inspired by the famous engraving in the *Hypnerotomachia Poliphili* [Figs. 6.5 and 6.7]. The theme of the choice between *virtus* and *voluptas* was taken up almost immediately in the Villa d'Este in Tivoli, which was built beginning in 1550 and stages the famous episode of Hercules at the crossroads through the distribution of the statues in the garden and in the painted decorations of the villa.[46]

46 Ribouillault D., "Le Salone de la Villa d'Este à Tivoli: un théâtre des jardins et du territoire", *Studiolo* 3 (2005) 65–94. On this theme, see the famous study by Panofsky E., *Hercules am Scheidewege* (Leipzig: 1930).

3 The Sources of Inspiration for the Nymphaeum: the *huius nympha loci* Tradition

Commendone's poems belong to a poetic tradition already well established in Rome a generation earlier involving the composition of epigrams celebrating – and often making speak – statues of sleeping nymphs associated with springs: the so-called *huius nympha loci* motif. This is also a famous iconographic topos, whose complex ramifications both south and north of the Alps, particularly in the work of Albrecht Dürer and Lucas Cranach, has inspired countless studies [Fig. 6.8].[47] The origin of this poetic tradition can be traced back to a pseudo-classical epigram dated to *c.*1459–1464 and attributed to the humanist Giannantonio Campano, a member of Pomponio Leto's Roman Academy. Betraying the influence of classical poetic models, notably Martial (*Nympha sacri regina lacis* [...], *Ep.* 9, 58) and Ovid (*Fast.* 3, 18–19), Campano's epigram gave a voice to a statue from the Roman collection of Cardinal Giacomo Ammannati Piccolomini, then Bishop of Pavia:

47 On the subject of the *huius nympha loci*, the most frequently referenced studies are: Kurz O., "Huius Nympha Loci: A Pseudo-Classical Inscription and a Drawing by Dürer", *Journal of the Warburg and Courtauld Institutes* 16 (1953) 171–177; Meiss M., "Sleep in Venice: Ancient Myths and Renaissance Proclivities", *Proceedings of the American Philosophical Society* 110.5 (1966) 348–382; Liebmann M., "On the Iconography of the Nymph of the Fountain by Lucas Cranach the Elder", *Journal of the Warburg and Courtauld Institutes* 31 (1968) 434–437; Macdougall E.B., "The Sleeping Nymph: Origins of a Humanist Fountain Type", *The Art Bulletin* 57.3 (1975) 357–365; Bober P.P., "The *Coryciana* and the Nymph *Corycia*", *Journal of the Warburg and Courtauld Institutes* 40 (1977) 223–239; and Coffin D., *Gardens and Gardening in Papal Rome* (Princeton: 1991) 32–39. See also more recent publications on the subject with extensive bibliographies: Ruvoldt M., *The Italian Renaissance Imagery of Inspiration: Metaphors of Sex, Sleep, and Dreams* (Cambridge: 2004) 65–89, 90–121; Pataki Z.A., '*Nympha ad amoenum fontem dormiens*' (*cil vi/5, 3*e*)*: Ekphrasis oder Herrscherallegorese? Studien zu einem Nymphenbrunnen sowie zur Antikenrezeption und zur politischen Ikonographie am Hof des ungarischen Königs Matthias Corvinus* (Stuttgart: 2005); Christian K.W., *Empire Without End. Antiquities Collections in Renaissance Rome c. 1350–1527* (New Haven – London: 2010) 134–142, 178–182; Leroux V., "L'érotisme de la belle endormie", *Seizième siècle* 7 (2011) 15–35; Espluga X., "La primera tradición textual de *Carmina Latina Epigraphica vi* 3*e (1)", in Martínez C.F. – Belén A.L. – Gómez Pallerés J. – Del Hoyo Calleja J. (eds.), *Ex officina. Literatura epigráfica en verso* (Sevilla: 2013) 121–155; Baert B., "The Sleeping Nymph Revisited: Ekphrasis, *Genius Loci* and Silence", in Enenkel K.A.E. – Traninger A. (eds.), *The Figure of the Nymph in Early Modern Culture* (Leiden: 2018) 149–176; Agnoletto S., "Giocare a fare I Classici. L'epigramma 'Huius Nympha Loci', l'invenzione dell'Antico e l'Arianna/Cleopatra dei Musei Vaticani", *Engramma. La tradizione classica nella memoria occidentale* 163 (2019) accessible online.

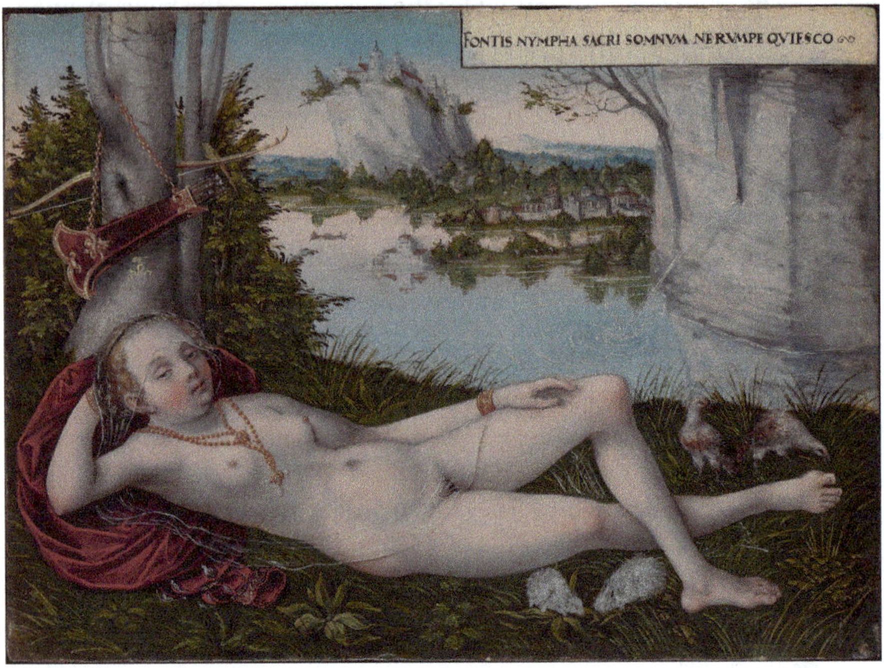

FIGURE 6.8 Lucas Cranach, *Nymph at the Source*, c.1545–1550. Oil on wood, 15.2 × 20.3 cm. New York, The Metropolitan Museum of Art, inv. 1975.1.136
IMAGE © MET

> Huius nympha loci, sacri custodia fontis
> Dormio, dum blandae sentio murmur aquae.
> Parce meum, quisquis tangis cava marmora, somnum
> Rumpere. Sive bibas sive lavere tace.

> I, Nymph of this place, guardian of this sacred spring, / I sleep while listening to the soft murmur of the water. / Do not trouble my sleep, you, who touches this sculpted marble! / Whether you drink or bathe, be quiet.[48]

At the end of the fifteenth century, one could admire similar statues in the gardens of the Cardinal of San Clemente, Domenico della Rovere, at the Palazzo dei Penitenzieri, at the Palazzo della Cancelleria, home of Cardinal Rodrigo Borgia (the future Alexander VI), and perhaps also in the house of

48 On Campano's epigrams, see De Beer S., "The *Pointierung* of Giannantonio Campano's Epigrams: Theory and Practice", in De Beer S. – Enenkel K.A.E. – Rijser D. (eds.), *The Neo-Latin Epigram* (Leuven: 2009) 137–163.

Agostino Maffei. But the most famous and best-studied sculpture gardens are those at the Villa Belvedere in the Vatican installed under the pontificate of Julius II and in the gardens of Angelo Colocci and Johann Goritz. The choices observed at the Villa Giulia cannot be fully understood without considering these precedents.

In 1512, Julius II acquired a large statue of a reclining woman from Agostino Maffei's collection of antiques, which was placed in the so-called Cortile delle statue in the Villa Belvedere and set in a fountain specially designed in the form of a rustic grotto, alongside the Laocoön and the Apollo Belvedere. This statue of a Sleeping Ariadne inspired numerous poems that clearly extend, sometimes to the point of paraphrasing, the tradition inaugurated by Campano. The poetic mise-en-scène that characterises these poems of the spectator admiring the nymph but ordered not to disturb her sleep also finds a visual parallel in a now famous drawing by Francesco de Holanda (*Libro*, fol. 8v) dated *c*.1538–1539, who was also interested in epigraphy and poetic composition [Fig. 6.9].[49] As this drawing shows, the new nymph was identified as Queen Cleopatra by the snake bracelet she wore on her left arm, which made her easy to fit into a discourse promoted by the panegyrists of Julius II, who celebrated the Pope as a new Julius Caesar.[50]

The statue was again carefully staged during the pontificate of Julius III, who followed in the footsteps of his namesake (when the statue was first installed in the Vatican, the young Giovan Maria Ciocchi Del Monte, the future Julius III, was Julius II's *cameriere*). In 1550, Julius III had the statue of Ariadne/Cleopatra moved from the *Cortile delle statue* to a room now called the *Sala di Cleopatra* at the end of the Belvedere corridor. The fountain maker Curzio Maccarone was commissioned to design a new rustic niche for the statue. Daniele da Volterra was assigned to paint the vault with scenes focusing on the Christian theme of

49 On Francisco de Holanda at the Belvedere, see Deswarte-Rosa S., "Francisco de Holanda et le Cortile di Belvedere", in Andreae B. – Pietrangeli C. – Winner M. (eds.), *Il Cortile delle statue. Der Statuenhof des Belvedere im Vatikan* (Mayence: Rome, 1998) 402; on his interest in epigraphy, see ibid. "Sous la dictée de la *Sibylle*. Épigraphie et Poésie. Un exemplaire des *Epigrammata Antiquae Urbis* annoté par André de Resende et Francisco de Holanda", in González Germain G. (ed.), *Peregrinationes ad inscriptiones colligendas. Estudios sobre epigrafía de tradición manuscrita* (Barcelona: 2016) 73–134.

50 See especially Barkan L., "The Beholder's Tale: Ancient Sculpture, Renaissance Narratives", *Representations* 44 (1993) 133–166. For a more recent study, featuring an ample bibliography, see Wolf C.M., *Die Schlafende Ariadne im Vatikan: ein hellenistischer Statuentypus und seine Rezeption* (Hamburg: 2002) and Miletto L., "Una celebrazione poetica del Cortile delle Statue e della 'Cleopatra' in Vaticano: Aurelio Serena da Montopoli", *Prospettiva* 165–166 (2017) 3–19 about a Latin poem by the humanist Aurelio Serena dated 8 March 1512, which would thus constitute the first known poetic celebration of the Sleeping Ariadne.

FIGURE 6.9 Francisco de Holanda, *The Fountain of Cleopatra/Sleeping Nymph*, drawing, 1538–1539. From: De Holanda Francisco, *Antigualhas*, fol. 8v. El Escorial, Library of San Lorenzo
IMAGE © ESCORIAL

water and salvation, thus re-orienting the humanist theme of the *huius nympha loci* to follow Counter-Reformist priorities.[51] Julius II's *Cortile delle statue* thus emerges as a crucial model for the nymphaeum of Julius III's villa. In addition to numerous antique statues, we find in both the central presence of the

51 On Julius III's program for the Vatican Belvedere, see Ribouillault D., "Jeux de mots et d'images: les frises peintes de l'appartement de Jules III au Vatican (vers 1550–1553)", in Fenech Kroke A. – Lemoine A. (eds.), *Frises peintes. Les décors des villas et palais au Cinquecento* (Paris – Rome: 2016) 112–113.

river gods (the Tiber and the Nile in the Vatican; the Tiber and the Arno at Villa Giulia [Fig. 6.3]) and that of the sleeping nymph.

In 1513, the humanist and Apostolic Secretary Angelo Colocci purchased a property near the area of the Trevi Fountain and San'Andrea delle Fratte, where the ruins of the Claudian aqueduct that brought the water of the Aqua Virgo to Rome were located.[52] In this garden, which was used for the meetings of Pomponio Leto's Roman Academy, numerous statues and inscriptions were displayed, one of which reproduced the epigram *huius nympha loci*.[53] Visiting in 1559, the Frenchman Jean-Jacques Boissard copied the inscription and later associated it with a statue of a sleeping nymph in an engraving published in his *Romanae urbis topographia et antiquitates*, though scholars no longer unanimously accept the authenticity of his rendering [Fig. 6.10].[54] In any case, the nymph sculpture became the subject and voice of many poems and stimulated research into the ancient history of the spring and its associated cult of nymphs.[55] In one of his poems *Ad Nympham Salonida*, Colocci was the first to identify the correct origin of the spring in Salone, a village in the Roman countryside.[56] Colocci possessed Frontinus' volume on ancient aqueducts in his library, whose *editio princeps* (1484–1492) had been edited by Pomponio Leto and Johannes Sulpicius and integrated into their edition of Vitruvius' *De architectura*.[57] The humanist discussions surrounding the restoration of the

52 The arch of the Claudian aqueduct figures in a drawing by Fra Giocondo (Uffizi, 1541A), who also published Frontinus' text in his 1513 edition of Vitruvius: *Vitruvius iterum et Frontinus a Iocundo revisi repurgatique* [...] (Florence, F. Giunta: 1513). On Colocci's garden, see also Giorgio Mangani's essay in this book.

53 On Colocci, in addition to the bibliography *supra*, see the entry in *Dizionario biografico degli italiani* 27 (1982). On his interest in epigrams, see Wellington Gahtan, "Angelo Colocci as a Collector of Epigrams" 277–288.

54 Agnoletto, "Giocare a fare i classici".

55 On this subject, see Davis, *Villa Giulia* and Fehrenbach F., "Nymph and Corpuscle. Transformations of the Aqua Virgo", in Böhme H. – Rapp R. – Rösler W. (eds.), *Übersetzung und Transformation*, Transformationen der Antike 1 (Berlin: 2007) 463. They cite the works of Pomponio Leto, Bartolomeo Marliano, Lucio Fauno, Lucio Mauro, Pirro Ligorio, Bernardo Gamucci, Fioravante Martinelli, and Faminio Nardini. The most advanced studies on the aqueduct were undertaken by Agostino Steuco, who published them in 1547 in his *De revocanda in Urbem Aqua Vergine*, in *Opera omnia quae extant omni, e veteribus bibliothecis* [...] (Paris: Michaëlem Sonnium, 1578) fol. 293v–296r. The link between the *Aqua Virgo* and a Roman nymph cult is the topic of a later work by Chiflet Jean, *Aqua Virgo fons Romae celeberrimus, et prisca religione sacer* (Antwerp, Balthasar Moretus: 1652; Officinia Plantiniana: 1662) (which mentions Campano's epigram, 17).

56 Ubaldini Federico, *Vita Angeli Colotii episcopi Nucerini* (Rome: 1673), 32–40, which reports numerous poems on the *Aqua Virgo*.

57 A copy of Frontinus is mentioned in the inventory of Colocci's library. See Lattès S., "Recherches sur la bibliothèque d'Angelo Colocci", *Mélanges d'archéologie et d'histoire* 48 (1933) 316.

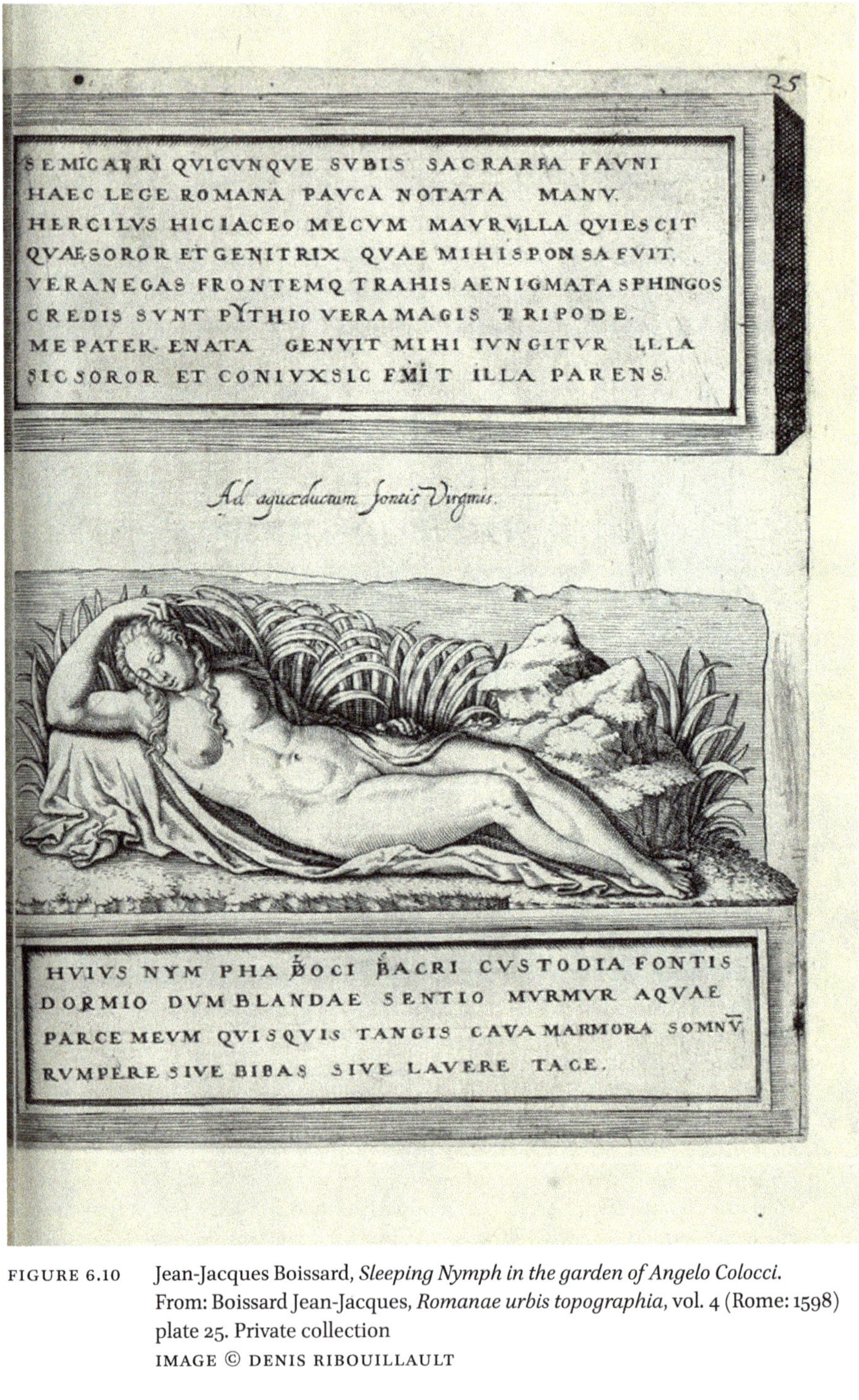

FIGURE 6.10 Jean-Jacques Boissard, *Sleeping Nymph in the garden of Angelo Colocci.* From: Boissard Jean-Jacques, *Romanae urbis topographia*, vol. 4 (Rome: 1598) plate 25. Private collection
IMAGE © DENIS RIBOUILLAULT

Aqua Virgo and the epigraphic and poetic staging of these by the members of the Roman Academy in the Horti Colotiani allow us to better understand the architectural and thematic choices of the Villa Giulia. In this respect, the initial idea of organising the villa around the spring probably came from Julius III's uncle, Antonio Maria Ciocchi del Monte, who was made a cardinal by his friend Julius II in March 1511.[58] Indeed, in 1519 he acquired the land *fuori Porta Flaminia* where the spring emerges.[59] A distinguished patron of the arts – he notably played a role in the construction of the tomb of Julius II by Michelangelo[60] – he was also an appreciated humanist in Roman academic circles. Angelo Colocci, for example, gave him Virgil's famous Codex Mediceus, which he himself had inherited from Pomponio Leto before it passed into the hands of Julius III.[61]

A third Roman garden provided the humanist and poetic *sodalitates* with a suitable meeting place: that of the Luxembourg cleric Johann Goritz, who was also attached to the Curia during the pontificate of Julius II. Located near the Basilica of Maxentius and Constantine, it contained a grotto with a statue of a sleeping nymph that can be associated with an abridged version of Campano's epigram (*Nymphae, Loci/ Bibe. Lava / Tace*). The poetic celebrations of the nymph statue took on a special significance on the day of St. Anne, Goritz' patron saint. He had an altar erected for this saint in the church of Sant'Agostino, which was associated with a statue of the Virgin and Child by Andrea Sansovino and Raphael's famous fresco of the Prophet Isaiah. The statue was displayed on a pillar near this altar, marking the site of Goritz' future tomb and upon which humanists attached their poems like votive offerings. After gathering in the Church, the poets assembled in Goritz's garden where they placed other epigrams: some attached to the cedar trees, others to the walls of the garden, and to the wells and many statues found in the space. These poems were the subject of a 1524 publication by Blosio Palladio, titled *Coryciana* – a double allusion to Goritz's Latinised name (derived from Virgil's *Georgics* 4, 127) and to Corycia, the nymph of Mount Parnassus mentioned by

58 Vasari notes 'E per Antonio cardinale di Monte cominciò una grande fabbrica alla sua vigna fuor di Roma in sull'Acqua Vergine' ('And for Antonio Cardinal di Monte he began a large building on his vineyard outside Rome on the Acqua Vergine'): Vasari Giorgio, *Le Vite de' più eccellenti pittori, scultori ed architettori*, ed. G. Milanesi, vol. 7 (Florence: 1878–1885; 1981) 497.

59 Nova, *The Artistic Patronage*, 58.

60 Agosti B. – Beltramini M., "Antonio del Monte e la maniera moderna", in Serra A. (ed.), *Humanitas. Studi per Patrizia Serafin* (Rome: 2015) 433–451.

61 (Florence, Biblioteca Medicea Laurenziana), cod. Mediceus laurentianus latinus 39 1; De Nohlac P., *La bibliothèque de Fulvio Orsini* (Paris: 1887) 272–273.

Pausanias (10, 6, 3; 10, 32, 2).[62] They form a precedent and an indispensable model for the inscriptions in the Villa Giulia commissioned by Julius III from the young Commendone.

4 Cultivating Ambiguity: the Multiple Identities of the Nymph

The epigrams invented for Colocci and Goritz frequently merge the imagery of the nymph, the statue, the spring, as well as Venus and the Virgin Mary. This should come as no surprise if we consider, for example, that the epigrams in the *Greek Anthology* are characterised by the repeated use of figures of speech that play on the sonorities, positions, and relationships between various words (paronomasia, alliteration, chiasmus, asyndeton, homeoteleuton, polysyndeton, hyperbaton) and their meaning (metonymy, metaphor, simile, periphrasis or synecdoche).[63] In the case of the Villa Giulia poems, their intertwining references to multiple and various virgins admirably testify to the religious undercurrents informing practices of Renaissance poetic invention about nymphs or, as Phyllis Pray Bober expressed, 'the ease with which poets slip from monotheism to polytheism'.[64] In short, the poems point to the space of the garden not so much as a sacred landscape in the Christian sense of the term, but as a sacral landscape where "the humanist temptation of poetic paganism" comes to light, as evoked by Hervé Brunon.[65] This polysemy is also present in the Villa Giulia. But even more than alluding to the Virgin Mary, the nymphaeum refers to the Virgin as a constellation – the zodiacal Virgin. In fact, the pope was born on 10 September, 1487 under the sign of Virgo, a fact that doubtlessly informed the decoration of the small loggia with an astrological cycle that opens directly

62 Palladio Blosio, *Coryciana*, ed. J. IJsewijn (Rome: 1997). See Blosio Palladio's letter of dedication: '[…] Ubi alius ad arbores citrias, alius ad hortenses parietes, alius ad puteos aut signa, quae illic plurima sunt et speciosa, omnia antiqui operis et gloriae plena, hoc illac temere et varie carmina affigunt, tuas statuas, tuam pietatem liberalitatemque eius diei tam in deos quam in homines tantam uno ore concelebrant?' ('Where [else] does someone affix here and there, randomly, to the citruses [m.n., also lemon trees], to the walls of the garden, to the fountains or to the sculptures, which in this place are numerous and appealing, all of them glorious and abundant in ancient works, various poems that celebrate with the same voice your statues, your piety and the generosity of this day towards gods, as much as towards humans?'). Translation by Alin Mocanu.

63 *Les Épigrammes de Maxime Planude*, ed. – trans. I. Taxidis (Berlin: 2017) 17.

64 Bober, "The *Coryciana* and the nymph Corycia" 228, with further examples.

65 Brunon H., "Du jardin comme paysage sacral en Italie à la Renaissance", in Ribouillault D. – Weemans M. (eds.), *Sacred Landscape. Landscape as Exegesis in Early Modern Europe* (Florence: 2011) 283–316.

onto the nymphaeum where, according to Ammannati, the four elements and the four seasons were also represented. The cosmological dimension of the nymphaeum thus took on a very personal dimension.[66] Virgo also refers here to Virgil's Virgo-Astraea-Fortuna, interpreted in a Christian context which heralds a new golden age (*Ultima Cumaei venit iam carminis aetas* [...] *iam redit et Virgo*; *Ecl.* 4, 6) and figured as an essential theme of the pontificate of Julius III. Thus, in 1550, on the occasion of a spectacle given in Bologna in honour of his election, an artificial mount – an allusion to Pope Del Monte and associated with the dwellings of nymphs – was erected with the motto '*Aurea pax toto surget sub montibus orbe*', a phrase referring to the famous fourth eclogue.[67] In these same years, the papal physician and poet Girolamo Fracastoro composed a long pastoral poem in honour of Julius III based on the Virgilian model (*Maxime divini pastor gregis, inclite Iuli*; *Carm.* 12, 1). Narrated through the voice of the god Pan, it is punctuated by a refrain announcing a new golden age (*Dicite Io, satyrisci, et Io geminate, coloni:/ aurea felici nascuntur saecula mundo*) and frequently plays with references to the mountains so dear to Pope Del Monte's panegyrists.[68]

5 The Nymphs of Hieronymus Cock and Paolo Graziani

Despite their highly suggestive visual character, Commendone's poems do not make it easy to determine the appearance of the nymph statue and its original staging in the lower part of the nymphaeum or *fonte bassa*. One of Commendone's poems, *Nympha loci niveosque* [...], seems to describe a typology corresponding to the nymphs sculpted on the fountains of ancient Rome,

66 On this loggia, Julius III, and astrology at the Villa Giulia, see Ribouillault, *La villa Giulia*, 360. Ammannati's description of the nymphaeum can be found in Falk, *Studien*, 171–173.

67 Gioccarello Anselmo, *Le triomphali Feste e giostre fatte nell'inclitta Città di Bologna con le publicazione d'un altra giostra che se havrà da fare di Santa Croce di maggio prossimo per la creation di M.S. Papa Giulio iii in Bologna* (n.p.: 1550); Ribouillault, *De la peinture au jardin*, 75–77.

68 Fracastoro Girolamo, *Girolamo Fracastoro. Latin Poetry*, trans. J. Gardner (Cambridge: 2013) 246–255 and Houghton L.B.T.H., *Virgil's Fourth Eclogue in the Italian Renaissance* (Cambridge: 2019) 148–150. He writes p. 171: 'Neither the supreme pontiffs of the Renaissance nor their publicists seem to have felt the need to choose between devotion to *Virgo Maria* and the cultivation of *Virgo Astraea* – and indeed the conflation of these two figures elsewhere from at least the time of Constantine suggests that they were not alone'.

like those now in the Vatican Museums or in the Archaeological Museum of Tarragona.[69]

An engraving of the nymphaeum by Hieronymus Cock from *c.*1558 and Paolo Graziani's interpretation of it from 1582 show the nymph in completely different poses, though in both cases she inhabits a curtained alcove and is surrounded by putti. In Cock's engraving, the nymph raises her left arm and faces a second figure, perhaps a satyr in keeping with the famous engraving of the *Hypnerotomachia Poliphili* [Fig. 6.5 and 6.11]. The image is largely fanciful, as I have shown elsewhere, and should not be relied upon as evidence of the actual statue.[70] The later engraving published by Graziani, active in Rome and therefore perhaps more accurate, shows the nymph surrounded by cupids and leaning on an urn, a detail mentioned several times in the poems on the Villa Giulia nymphaeum [Fig. 6.12]. Alessandro Nova suggests that the statuary group in this engraving is inspired by the painting of Venus and Cupid in Philostratus' *Imagines* (i–6), a plausible hypothesis since this text served as the basis for several other paintings in the villa. In any case, the pose is very different from that of the nymphs in the fountains of ancient Rome, of the Ariadne at the Vatican, as well as most other known reclining female nudes. Above all, this nymph does not sleep, whereas the poems almost all allude to her sleeping. On the other hand, she resembles Correggio's *Danaë* and the reclining Venuses of Titian and Primaticcio. Even more, her pose and elongated physiognomy clearly recall the so-called Fontainebleau Nymph by Rosso Fiorentino, copied in the beautiful engraving by Pierre Milan and René Boyvin dated 1545–1554 [Fig. 6.13]. If we compare the two compositions [Fig. 6.12 and 6.13], it is easy to establish that the two nymphs' poses and their elongated limbs are identical, whereas the Fontainebleau dog is replaced by a winged Cupid in the Roman engraving, perhaps inspired by Correggio. To assess these similarities, we may recall that at the beginning of 1540 and the end of 1545, Primaticcio – one of the Italian masterminds of Fontainebleau – travelled to Rome from the court of Francis I with the goal of stocking up on antiquities and executing bronze copies of the most famous sculptures in the Belvedere Court. He made a copy of the Sleeping Ariadne that was then placed in Fontainebleau in the Gallery of Francis I before being relocated to the Queen's Gardens.[71] A young Giacomo Barozzi da Vignola assisted with this work, who would design the nymphaeum

69 Inv. MV_982_0_0 et inv. MNAT 12 275. See the photos in Kurz, *Huius Nympha Loci*, pl. 23c; MacDougall, *The Sleeping Nymph*, [Fig. 6.3] 360.

70 Ribouillault, *De la peinture au jardin*, 68–71.

71 Pressouyre S., "Les fontes de Primatice à Fontainebleau", *Bulletin Monumental* 127.3 (1969) 223–239.

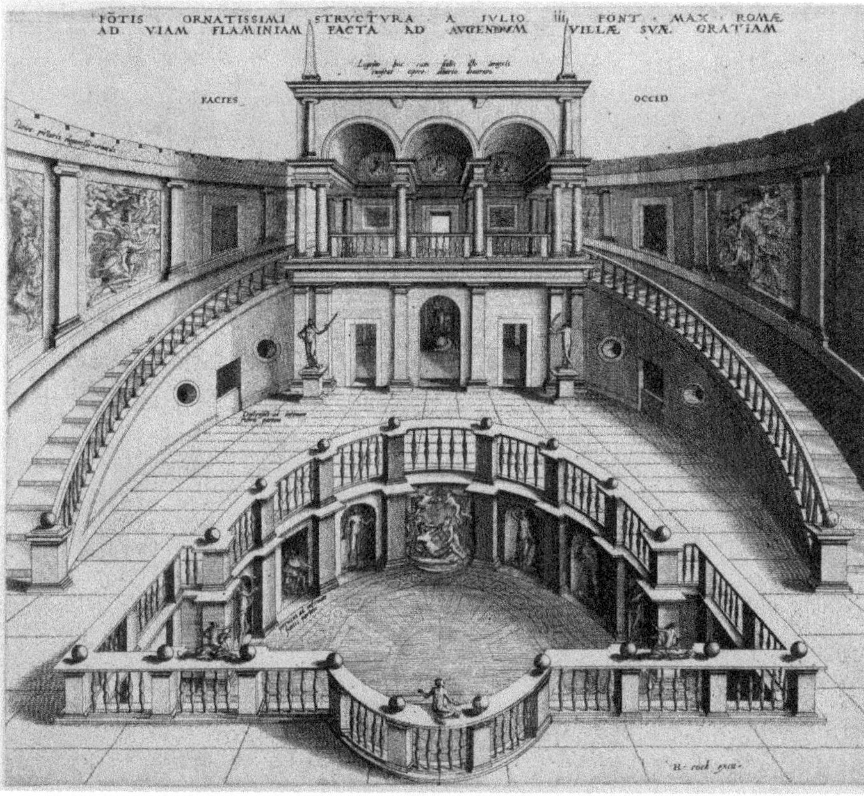

FIGURE 6.11 Hieronymus Cock, after Bartolomeo Ammannati, engraved by Joannes and Lucas van Doetecum, *Nymphaeum of the Villa Giulia; view of the west side*, c.1558. Etching and chisel, 26.1 × 28.9 cm. Brussels, Royal Library of Belgium, Prints and Drawings Room, inv. s.v. 88763

of the Villa Giulia with Giorgio Vasari and Bartolomeo Ammannati a few years later. Moreover, the inscription under the engraving by Milan and Boyvin indicates that the Fontainebleau nymph was a painted model after a drawing conceived for a sculpture. It is believed that Primaticcio, to whom this project is attributed, intended to produce a copper figure to adorn the door of the castle, according to the wishes of the king. This project would have aligned with the time he spent working with Vignola on casting antique models brought back from Rome.[72] Was Vignola inspired by the prestigious royal residence to

72 'Ô Phidias, ô Apelle, a-t-on pu concevoir de votre temps quelque chose de plus beau que cette sculpture dont vous voyez ici l'image, que François Ier […] laissa inachevée à sa maison sous la forme d'une statue de Diane se reposant de la chasse et répandant l'urne de

FIGURE 6.12 Paolo Graziani, after Hieronymus Cock, *Nymphaeum of the Villa Giulia*. From *Speculum Romanae Magnificentiae*, 1582. Engraving, 30.1 × 40.6 cm. London, British Museum, inv. 1950,0211.67
IMAGE © TRUSTEES OF THE BRITISH MUSEUM

FIGURE 6.13 René Boyvin and Pierre Milan, *The Fontainebleau Nymph*, c.1545–1554. Engraving, 31.8 × 52.7 cm. New York, The Metropolitan Museum of Art, inv. 32.105
IMAGE © MET

design the nymphaeum of the new pope, recalling the legend of Fontainebleau to stage the Aqua Virgo?[73] Despite the seductive nature of such a hypothesis, we must remain cautious and suggest that the creator of the engraving was inspired by pre-existing engravings in his depiction of the nymphaeum, which had then been stripped of most of its statues. Indeed, following Julius III's death, his exceptional collection was largely dispersed and the fate of the enigmatic sculpture, separated from its famous guardian, remains unknown.[74]

6 The Author of the *Lex hortorum*: the Portuguese Humanist Aquile Estaço (Achilles Statius, 1524–1581)?

No inscriptions seem to have been added to the decoration of the nymphaeum except for the two long texts engraved on marble slabs or *Lex hortorum* that still welcome visitors at the foot of the two large semicircular entrance ramps. The first consists of a charming Latin description of the villa that praises the garden and its sculptures, water features, fishponds and aviaries. Visitors are entreated to enjoy these wonders provided they do not cause damage. As in the poems by Commendone, the author of the *Lex* personifies the spring by referring to it as *virgo* and asks the visitor not to disturb her by using the verb *temerare*, which means to defile, to fade, or to violate. We here find a warning to the spectators typical of the *huius nympha loci* genre, but also a more precise idea of the function of the spring from which visitors were evidently allowed to drink:

> Aquam hanc, quod est virgo, ne temeranto;
> sitimque fistulis non flumine, poculis non osculo aut volis extinguunto

la fontaine Belle eau?' ('Oh Phidias, oh Apelles, could anything more beautiful have been conceived in your time than the sculpture whose image you see here, which Francis I [...] left unfinished in his house in the form of a statue of Diana resting from the hunt and pouring the urn of the Belle Eau fountain ?'). See Jestaz B., "Benvenuto Cellini et la Cour de France: (1540–1545)", *Bibliothèque de l'École des Chartes* 161.1 (2003) 101–103.

73 On the influence of Vignola's stay in France on his subsequent architectural projects, notably the Villa Giulia, see Frommel S., "Fontainebleau", in Tuttle R. (ed.), *Jacopo Barozzi da Vignola* 129–138; idem, "Vignola in Francia: un episodio incerto del suo itinerario artistico", in Affanni A.M. – Portoghesi P. (eds.), *Studi su Jacopo Barozzi da Vignola* (Rome: 2011) 215–236.

74 During the pontificate of Pius IV, most of the sculptures were moved to the lower courtyard of the Belvedere and to Pius' new Casino. See Falk, *Studien*, 132–134.

> That they do not disturb this water, because she is virgin; / that they quench their thirst in the fountains, not in the river, with glasses, not with their mouth or the palm of their hand.[75]

The second engraved text derives more from legal and testamentary administration than from ekphrastic poetry, commemorating the transfer of the property to the pope's brother Baldovino del Monte in November 1553.[76] When Lagomarsini visited the ruined villa in 1768 and Mai in 1842 [Fig. 6.14], they found no trace of Commendone's poems; it might be assumed that the project evoked by Antonio Maria Graziani had been abandoned following the death of the pope.[77]

The idea of featuring engraved verses in a sculpture garden was not new. According to Francesco Albertini, the garden of Cardinal Carafa on the Quirinal was decorated with "numerous epigrams" from Virgil, Pliny, Hesiod, Cato, Varro, and Columella. In the 1520s, poetic inscriptions and ancient and modern statues were artfully arranged in the garden of Cardinal Andrea della Valle. These two gardens each had a *Lex hortorum*, which anticipates that of the Villa Giulia.[78] The latter garden also possessed a very large number of ancient inscriptions; Amaduzzi indicates that the garden was surrounded by a high wall adorned with more than one hundred inscriptions and bas-reliefs.[79] Rodolfo Lanciani notes the frequency of the phrase '*in hortis Iulii iii pont. max*' in volume VI of the CIL (*Corpus inscriptionum Latinarum*) dedicated to the city of Rome.[80]

Many visitors copied the *Lex hortorum* of Villa Giulia as part of their travel diaries.[81] As far as I know, the earliest mentions of the text, apart from those

75 The Latin text is quoted from the transcription of ibidem 170.

76 Ibidem. The notarized act is published in Tesoroni D., *Il Palazzo Firenze e l'eredità di Balduino del Monte* (Rome: 1889) 89–92.

77 Mai, *Spicilegium romanum* 480. Only Hubert Robert succeeded in sublimating the feeling of sadness when faced with the lamentable state of the villa [Fig. 6.14].

78 Coffin D., "The 'Lex Hortorum' and Access to Gardens of Latium during the Renaissance", *The Journal of Garden History* 2.3 (1982) 202–206.

79 Amaduzzi, *Storia della villa di Giulio III*, c. 19r.

80 Lanciani R., "Il Codice barberiniano XXX, 89 contenente frammenti di una descrizione di Roma del secolo XVI", *Archivio della Società romana di Storia patria* 6 (1883) 232.

81 Some examples of travellers who copied the inscription directly or copied it from older works include Gabriele Simeoni (visit in 1557; published in *Les illustres observations antiques du seigneur Gabriel Symeon florentin. En son dernier voyage d'Italie l'an 1557* (Lyon, Jean de Tournes: 1558) 54–55); Jean-Jacques Boissard (visit in 1559; published in *Romanae urbis topographia et antiquitates* (Frankfurt, Johann Feyrabend: 1597–1602)); Nicolas Audebert (visit around 1574–1578); Giovanni Galeazzo Rossi (mentioned in a letter from 1576. See Aksamija, *Architecture and Poetry* 148); anonymous author of the *Codice barberiniano XXX 89* (visit c.1580); Jean Sarrazin (text reported in 1582); Caspar Janthesius (published in *Itinerarium* (Rhete: 1637)), Martin Zeiller (published in *Topographia Italiae*

FIGURE 6.14 Hubert Robert, *The Nymphaeum of the Villa di Papa Giulio, Rome*, *c.*1761. Red chalk drawing, 39.8 × 53 cm. New York, The Metropolitan Museum of Art, inv. 59.23.71
IMAGE © MET

by Gabriele Simeoni and Jean-Jacques Boissard who visited the villa respectively around 1557 and 1559, are found in two manuscripts preserved in the Vallicelliana Library in Rome. They are interesting as they relate a text that diverges slightly from the one that can be read in the nymphaeum, and which was then widely published. The first manuscript, cod. Vall. G. 51 (*c.*333r–*c.*334r) is a miscellany about the city of Rome.[82] The second manuscript, cod. Vall. B. 106, is better known since it contains a collection of texts by the famous Portuguese humanist Aquiles Estaço (Achilles Statius), whose library and writings formed the basis for the establishment of the Vallicelliana Library.[83] Jozef

(Frankfurtt: 1688)), etc. The text is also reported in Paul-Marie Letarouilly (1840), Angelo Mai (1842), Tilman Falk (1971), and David Coffin (1982).

82 *Scritture spettanti alla città di Roma. Tomo ii. Quale contiene le notizie concernenti la città medesima, suo governo, stato economico, entrate, provedimenti per le inondazioni del Tevere, tasse, composizioni, spese del Palazzo Pontificio et altro* (Biblioteca Vallicelliana) G. 51 (1740). The poem is presented as "Testo [?] Vigna di Papa Giulio".

83 *Achillis Statii Lusitani Orationes Epistolae et Opuscula omnia quae in foliis sparsa habebantur* (Biblioteca Vallicelliana) B. 106. On this manuscript, see Tella Bartoli M., "A Proposito di Aquile Estaço e dei Carmina del codice Vallicelliano B. 106", *Annali dell'Istituto*

IJsewijn refers to the poem on Villa Giulia titled *Lex hortensis* (fol. 178 = 187v; instead fol. 179v = 188v) as an invention of Estaço's "in archaic Latin", apparently unaware that it corresponds, apart from a few details to that engraved in the nymphaeum.[84] It could be argued that Estaço did not compose the poem but copied the inscription as part of his extensive studies of ancient epigraphy, which led him to visit many gardens in Rome. However, this hypothesis does not explain the differences, however minor, between the two versions of the text. The poem of Ms. B. 106 does not contain any abbreviations, and several words and conjugations vary. Additionally, Estaço's epigraphic collection, based on ancient and not modern inscriptions, is contained in another manuscript, cod. Vall. B. 104, which varies greatly in its conception and its genesis from cod. Vall. B. 106. The latter contains no epigraphic transcriptions, essentially featuring the *carmina sacra* and *carmina profana* composed by the poet.[85] Although further research is needed, it is thus not impossible that the elegant *Lex hortorum/Lex hortensis* of the Villa Giulia was composed by Aquiles Estaço, one of the best Latinists of his time, and not by Commendone, as Ammaduzzi has suggested. Estaço's official establishment in Rome is generally dated to 1559, but he could have travelled there while Julius III was on the papal throne around 1555, according to IJsewijn, or even sent the poem from Paris, the Netherlands, or Padua, where he sojourned during his years of *peregrinatio academica* before settling permanently in the Eternal City, where he had a brilliant career in the Roman Curia.[86] Poetry played an important role

Universitario Orientale. Sezione Romana 17 (1975) 293–362; IJsewijn J., "Achilles Statius, a Portuguese Latin Poet in Late 16th-Century Rome", in *Humanismo português na época dos descobrimentos: congresso internacional Coimbra 9 a 12 de outubrou de 1991* (Coimbra: 1993) 109–123 (the most important contribution to date, which criticizes the preceding publication); Miralles Maldonado J.C., "Un poema inédito de Aquiles Estaço", *Euphrosyne. Revista de filologia clássica*, 29 (2001) 389–408; Iannuzzi I., "Tra Portogallo e Roma: note per un profilo di Achille Stazio", in Andretta E. – Valeri E. – Visceglia M.A. – Volpini P. (eds.), *Tramiti: figure e strumenti della mediazione culturale nella prima età moderna* (Rome: 2015) 180.

84 IJsewijn, *Achilles Statius* 110–111, 122–123.

85 The page with the *Lex Hortensis* (fol. 179v = 188v and not fol. 178 = 187v as IJsewijn indicates) is inserted between two indices at the end of two collections of poems entitled *Carmina sacra et profana, de Deo, de Sanctis, et de aliis rebus variis* (fol. 102r–170v) and *Paraphrases variae nonnulorum Psalmorum* (144r–170v). The writing seems to be the authors' own. On the work of Estaço concerning epigraphy, see Almagro A.G., "A Portuguese Contribution to 16th century Roman Antiquarianism: the case of Aquiles Estaço (1524–1581)", in Berbara M. – Enenkel K.A.E. (eds.), *Portuguese Humanism and the Republic of Letters* (Leiden: 2012) 353–376; Idem, "Iscrizioni romane di tradizione manoscritta: il codice epigrafico di Aquiles Estaço", *Euphrosyne. Revista de filologia clássica* 44 (2016) 307–320.

86 IJsewijn, *Achilles Statius* 111 notes with caution: 'Whether this may be interpreted as a sign that Statius has seen Julius III is open to debate'. In the same codex, he also quotes

in his work, and his closest friends in Rome were poets. The German Paulus Melissus called him "the pride of the poets", though most of his compositions have yet to be published. His interest for the villa Giulia is also documented in a series of engravings of ancient herms displayed in the pope's garden, which he published in 1569.[87]

7 Ippolito and Lelio Capilupi (1511–1580; 1497–1560): a Virgilian Villa

The Mantuan-born poet Ippolito Capilupi, secretary to the Duke of Mantua, and his brother Lelio Capilupi, nicknamed the "second Virgil of Mantua" were among the courtiers who played important political and diplomatic roles at the court of Julius III.[88] George Hugo Tucker recently and keenly analysed the poetic description of Julius III's Rome in Cento X of Lelio Capilupi's *Centones ex Virgilio*, published in Rome in 1555 and dedicated to Joachim Du Bellay (*Ad Iulium iii. Pont. Max. Cento x. Romae Prosopopeia*). The poem is composed entirely of fragments from Virgil taken mainly from the *Aeneid*, thereby reinscribing a prestigious and pseudo-genealogical parallel between Pope Julius III and Iulus-Ascanius, son of Aeneas and mythical founder of Rome.[89] The inaugural cento, composed in the spirit of Statius's *Silvae* and Poliziano and dedicated specifically to the Villa Giulia (*Ad Villam Iuliam Iulii iii. Pont. Max*), and

---

(f. 218v) from a letter dated 29 July 1553 that mentions Aquiles' father's desire to see him depart for Italy. On this period in his career, also see Ramalho A.D.C., "Notas sobre a formação de Aquiles Estaço", *Biblos* 54 (1978) 239–252 (again in *Estudos sobre o século xvi* (Paris: 1980) 293–310).

87 Maurach G. – Echinger-Maurach C. (eds.), *Herme oder Terme?: Texte und Bilder zur Klärung der Begriffe: aus: Achilles Statius, Inlustrium viror[um] ut extant in urbe expressi vultus, Romae 1569/Formis Antonij Lafrerj*, Heidelberg 2021 (Fontes 89), accessible online ; Ribouillault D., "De la 'Vigna Poggio' à la villa Balestra: 'Una delle più pittoresche e interessanti vedute di Roma'", in Fagiolo M. – Mazza A. (eds.), *I Monti Parioli e il "Nuovo Campo Marzio della cultura internazionale"* (Rome: 2022) 110–123.

88 Concerning their respective careers, I refer to the bibliographies contained in *Dizionario biografico degli italiani* 18 (1975) (Gaspare de Caro, Ippolito C.; Claudio Mutini, Lelio C.).

89 Capilupi Lelio, *Laelii Capilupi Patritii Mantuani Centones ex Virgilio*, ed. A. Possevino, (Rome, Valerius Doricus: 1555). See Tucker G.H., "La Poétique du centon, une poétique de la Silve ?", in Galand P. – Laigneau S. (eds.), *La Silve: histoire d'une écriture libérée en Europe de l'Antiquité au XVIIIe siècle*, Latinitates 5 (Turnhout: 2013) 525–564. The poem is mentioned by Coffin D.,*The Villa in the Life of Renaissance Rome* (Princeton: 1979) 163. Donatella Manzoli recently published a translation into Italian, "La villa e le forbici. Villa Giulia in un centone di Lelio Capilupi", *Studi romani*, 60.1–4 (2012) 57–85. A summary centred on the Villa Giulia can be found in Ribouillault, *De la peinture au jardin* 60–63. On Capilupi, see also Tucker G.H., "A Roman Dialogue with Virgil and Homer: Capilupi, the *Cento* and Rome", in Caruso C. – Laird A. (eds.), *Italy and the Classical Tradition. Language, Thought and Poetry 1300–1600* (London: 2009) 204–238.

preceded by three distichs by Ippolito which dedicate the poem to the pope galvanises, as in Fracastoro, the image of the papal *vigna* as a Virgilian landscape:

> Virgilius cecinit divini ruris honores,
> Et sacri fontis murmur IVLE tui.
> Nulla fugit tam dulce sonans pede lympha per herbas.
> Non usquam floret tam bene cultus ager.
> Haec igitur merito celebravit Musa Maronis.
> Non alio poterant dulcius ore cani.
>
> Virgil himself praised the beauty of this divine countryside / And from the murmur of your sacred fountain, O Julius. / No other spring streams in the grass while singing on such a / soft rhythm; / No other domain cultivated with so much care is covered with such beautiful colours / The Muse of Virgil celebrated them thus with good reason / And no other poet could have sung them more delicately.[90]

The Aqua Virgo, celebrated here as a sacred spring, obviously occupies a central place in the description of the nymphaeum (v. 50–58), whose majestic plane trees are especially praised. I provide here the original Latin passage with the explanatory notes and translations by Tucker, which offer an insight into the extreme virtuosity of this poem. Throughout, Capilupi has indicated the sources of the Virgilian extracts he stitched together. The note in the left-hand margin clearly indicates that the spring in question is that of the pope's nymphaeum:

[…] *prospectat euntem*[91] *AE.* VII.
G. IIII. Illa quidem,[92] *largoque humectat flumine vultum.*[93] *AE.* I.

---

90 Tucker, *La Poétique* 539. The translation has been slightly revised by Alin Mocanu.

91 '[To tell the truth, this one] follows her/him with her/his eyes'. Virgil, *Aeneid*, VII 813: [*illam omnis* […] *iuven- tus/*] *turbaque miratur matrum et* prospectat euntem, [*/attonitis inhians animis*] (the Rutilian youth and matrons look with wonder at the beautiful warrior Camille, following her with their eyes).

92 'To tell the truth, this one [follows him/her with her/his eyes]'. Idem, *Georgics*, IV 457: illa quidem, *dum te fugeret per flumina præceps*, [*/immanem ante pedes hydrum moritura puella/*[…] *non vidit in herba*] (fleeing, Eurydice does not see the hidden snake that will bite and kill her).

93 'And bathes his face in a great flood of tears'. Virgil, *Aeneid*, I 465: *multa gemens*, largoque umectat flumine voltum (moved, having seen in Carthage a depiction of the fall of Troy, Aeneas bursts into tears).

Marmoream imagi[-] *nem virginiis quae fonti prae est, alloquitur*[94]	AE. I.	O quam te memorem Virgo[95] *que maxima Sacro*	*AE*. VII.
		Fonte sonans:[96] caelo ve quibus te laudibus aequem?[97]	AE. XI.
	AE. I.	*Nympharum domus hic*,[98] Vivoque sedilia saxo.[99]	AE. I.
	G. II.	*Inter enim labuntur aquae*,[100] mirabile visu.[101]	AE. XII.
	AE. VIII.	*Nymphae laurentes*,[102] praestanti corpore Nymphae,[103]	AE. I.
	G. IIII.	*Caesariem effusae nitidam per candida colla*:[104]	
	AE. VII.	Corpora sub ramis deponunt arboris altae:[105,106]	

94 'He addresses the marble statue of the Virgin in front of the water of the spring.'

95 'O what shall we call you, Virgin?'. Ibidem 327: O quam te memorem virgo ? *namque haud tibi voltus* [*/mortalis*] (Aeneas, dazzled by the divine beauty of his mother Venus disguised as a huntress, speaks to her).

96 'Who, the greatest, making the air resound with the waters of your sacred spring'. Ibidem, VII 83–84: [*lucosque sub alta/*] *consulit Albunea, nemorum* quæ maxima sacro/ fonte sona*t sævamque exhalat opaca mephitim* (King Latinus consults the oracle of Faunus near the forest of Albune and near its sacred spring).

97 'Or by what praise to make you equal to the sky?'. Ibidem, XI 125: *vir Troiane*, quibus cælo te laudibus æquem ? (Drances, Turnus' enemy, addresses flattering words to Aeneas).

98 'The home of the nymphs here'. Virgil, *Aeneid*, I, 168: [*intus aquæ dulces vivoque sedilia saxo,/*] Nympharum domus. Hic *fessas non vincula navis* [*/ulla tenent*] (description of the natural harbour where the Trojan fleet took refuge on the coast of Carthage).

99 'And seats [carved] in living stone'. Ibidem, I 167: *intus aquæ dulces* vivoque sedilia saxo, [/ Nympharum domus] (same description).

100 'Because in the middle flow the waters [of the Nymphaea]'. Idem, *Georgics*, II 349: inter enim labentur aquæ, *tenuisque subibit* [*/ halitus*] (advice on planting and irrigation of the vine).

101 'Marvelous thing to see'. Virgil, *Aeneid*, XII 252: [*cunctæque volucres /*] *convertunt clamore fugam* (mirabile visu) [*/ætheraque obscurant pinnis, hostemque per auras/*[...] *premunt*] (the Rutulians see the omen of a flock of birds).

102 'Nymphs of Lazio' [subject, in the nominative, in Capilupi]. Ibidem, VIII 71: Nymphæ, laurentes *Nymphæ, genus amnibus unde est* [*/Tuque, o Thybri*] (On the banks of the Tiber Aeneas prays to the Nymphs of this river [in the vocative]).

103 'Nymphs with extraordinarily beautiful bodies'. Ibidem, II 71: *sunt mihi bis septem* præstanti corpore Nymphæ (the statements of Juno who promises to give as a wife to Aeolus, one of the fourteen most beautiful nymphs known to her).

104 '[Nymphs] with hair cascading down a snow-white neck'. Idem, *Georgics*, IV 337 (description of the hair of the nymphs of Cyrene, mother of the shepherd Aristaeus).

105 '[Nymphs] recline in the shade of the branches of a great tree'. Idem, *Aeneid*, VII 108 (Aeneas, his captains, and his son Iulus-Ascanius are resting, lying at the foot of a large tree).

106 Capilupi Lelio, "Cento ex Virgilio In Villam Iuliam Iulii III. Pont. Max.", in idem, *Centones ex Virgilio* (1555), fol. A[i]^{vo}–ii^{ro}, v. 50–58.

Ippolito also composed two sonnets about the Aqua Virgo. Published in 1590 in the *Capiluporum carmina*, they may date back to the period when the Capilupi brothers were frequent visitors to the Villa Giulia or, more likely, to when Pius IV decided to restore the aqueduct in June 1561.[107]

8 Joachim Du Bellay (1522–60): a Dionysian Villa

Joachim Du Bellay arrived in Rome in 1553 as secretary and relative of Cardinal Jean Du Bellay, who was also an eminent Latinist and patron of one of the most beautiful *all'antica* gardens in Rome, the Horti Bellaiani.[108] Du Bellay became close friends with Lelio Capilupi and other members of the Accademia dei Vignaiuoli like Annibale Caro, to whom he dedicated several laudatory poems.[109] The Villa Giulia was undoubtedly a privileged meeting place for these scholars united by a common love for rural pleasures, which

---

107 Capilupi Alphonso – Capilupi Camilio – Capilupi Giulio – Capilupi Ippolito – Capilupi Lelio, *Capiluporum* [Hippolyti, Laelii, Camilli, Alphonsi, Julii] *carmina*, eds. Julius Capilupus and Josephus Castalio, (Rome, ex typ. haeredum J. Lilioti: 1590) 64–65. In a subsequent edition, the second poem is attributed to Lelio (Bottari Giovanni Gaetano, *Carmina illustrium poetarum italorum* (Florence, apud J.C. Tartinium and S. Franchium: 1719) 199–200). The poems are mentioned by Nova, *The Artistic Patronage* and reproduced in Blänsdorf J., "Aspekte eines poetischen Themas in der neulateinischen Dichtung Italiens und Frankreichs", in Blänsdorf J. – Janik D. – Schäfer E. (eds.), *Bandusia. Quelle und Brunnen in der lateinischen, italienischen, französischen und deutschen Dichtung der Renaissance* (Stuttgart: 1993) 68. On the restoration of the Aqua Virgo aqueduct, see Karmon D., "Restoring the Ancient Water Supply System in Renaissance Rome: The Popes, the Civic Administration, and the Acqua Vergine", *The Waters of Rome* 3 (2005); Rinne K.W., "Between Precedent and Experiment: Restoring the Acqua Vergine in Rome (1560–1570)", in Roberts L.L. – Schaffer S.– Dear P. (eds.), *The Mindful Hand: Inquiry and Invention from the Late Renaissance to Early Industrialisation* (Amsterdam: 2007) 95–116 and Vagenheim G.,"Pirro Ligorio et 'l'histoire secrète' de la restauration de l'Acqua Vergine sous le pontificat de Pie IV (1559–1565)", in Calvelli L. (ed.), *La falsificazione epigrafica. Questioni di metodo e casi di studio*, Antichistica 25 | Storia ed epigrafia 8 (2019) 263–286.

108 On the gardens of Cardinal Du Bellay and the cult of antiquity, see Bardati F., "Jean du Bellay et les *Horti Bellaiani*. Bâtir à l'antique, sur l'antique et pour l'antique", in Lurin E. – Burlot D. (eds.), *L'artiste et l'antiquaire. L'étude de l'antique et son imaginaire à l'époque moderne* (Paris: 2017) 171–189. On the poetic vision of Julius III's Rome, see Tucker G.H., "*Roma instaurata* en dialogue avec *Roma Prisca*. La représentation néo-latine de Rome sous Jules III (1553–1555) chez Janus Vitalis, Joachim Du Bellay et Lelio Capilupi (de l'ekphrasis à la prosopopée)", *Camenae* 2 (2007).

109 Cooper R., *Litterae in tempore belli. Études sur les relations littéraires italo-françaises pendant les guerres d'Italie* (Geneva: 1997) 243. On the poems of Du Bellay addressed to Caro, see Chamard Henry, *Joachim Du Bellay, 1522–1560* (Geneva: 1978) 342–343.

the decorative programme of Villa Giulia – centred on water, but also wine – highlighted more than any Renaissance villa. It is worth mentioning that Caro adapted Longus' famous Greek pastoral novel *Daphnis and Chloé* to Italian in 1538, which Charles Davis has suggested probably influenced the villa's decorations, especially the nymphaeum, and in which frequent allusions to the cult of Bacchus supplement the central narrative of the nymph and the satyr.[110] This work was known to both Capilupi and Du Bellay[111] and it is alongside this link that we point to one of Du Bellay's early epigrams concerning the nymphaeum fountain, published in 1558 in Paris in his *Poematum libri quatuor*

> *In fontem Iulii III. P.M.*
> Cur has mampineis iugis Iulus
> Fontis Virginei Deas sacrarit,
> Hospes scire cupis? Ferunt Lyæum
> Ignibus patriis adhuc rubentem
> Nymphas culmina montium colentes
> Hoc pellucidulo abluisse fonte.
>
> *On the Fons Virginis [of the Villa Giulia] of the sovereign pontiff Julius III /*
>
> Why did Julius immortalize the goddesses / Of the *Fons Virginis* on these hills covered with grapevines, / You would like to know, visitor? Because of Lyaeus [Bacchus-Dionysus], it is said, / Who, still red from the paternal fires, / Was washed by the Nymphs inhabiting the summits of these mountains / In this crystalline spring.[112]

110 Davis C., "Per l'attività romana del Vasari nel 1553: incisioni degli affreschi di Villa Altoviti e la Fontanalia di Villa Giulia", *Mitteilungen des Kunsthistorisches Institutes in Florenz* 23 (1979) 215; Lavocat, *La syrinx au bûcher*; Ribouillault, *De la peinture au jardin* 74–75.

111 According to Nolhac, *La Bibliothèque* 164, 342, 394, Fulvio Orsini, who wrote a postface to Capilupi's *Centones* dedicated to Du Bellay, possessed a Greek manuscript by Longus in his library as well as Caro's translation. See Tucker G.H., "Érotisme, parodie, et l'art du centon dans le *gallus* (1543; *centones ex virgilio*, 1555) de Lelio Capilupi", in Sacré D. – Papy J. (eds.), *Syntagmatia. Essays on Neo-Latin Literature in Honour of Monique Mund-Dopchie and Gilbert Tournoy* (Leuven: 2009) 330.

112 Du Bellay Joachim, *Epigrammata* [42], *Poematum libri quatuor* [...] (Paris, Fédéric Morel: 1558), fol. 26v. The translation given in Tucker, *La description de la Villa d'Este* 22 has been slightly revised here by Alin Mocanu.

The poet here refers to the large number of vineyards making up the immense park of Villa Giulia. The verses also evoke the decoration of the large semi-circular loggia vault, designed as a magnificent pergola, and several over-door paintings of bacchic scenes. The theme of wine and water is especially advanced in the ground floor room to the north of the entrance vestibule, decorated between 1553 and 1555 by Prospero Fontana and his workshop and entirely dedicated to Bacchus and the nymphs [Fig. 6.15]. Preparatory drawings by Giorgio Vasari for the paintings in the nymphaeum similarly allude to Bacchanalian references, here organised according to the Virgilian themes of *Cerealia*, *Baccanalia* and *Fontanalia*.[113] The poet Marc-Antoine Muret, a friend of Du Bellay's, also elaborated a Bacchanalian theme when he composed a praise of the fountains at the Villa d'Este in Tivoli some twenty years later.[114]

It also fell to Du Bellay, the author of *La Deffence et Illustration de la Langue Francoyse* (1549), to translate the prestige and *argutia* attached to the poetic tradition of the sleeping nymph to his mother tongue, as the poets of the Pléiade had done a few years earlier in their poems on the fountain of the Château d'Arcueil (1549).[115] In addition to Sonnet XII of *Le songe* (1558), which features nymphs and satyrs gathered by a spring, Du Bellay produced two other sonnets in French about the *Fons Virginis*, one written prior to and the other following the Pope's death at the Villa Giulia on 23 March 1555. They were first published in Paris in 1569 in the *Œuvres françaises* edited by Fédéric Morel, evoking the range of poetic *topoi* of the *nympha loci*:[116]

La Nymphe Dormante
À la fontaine de Pape Iules III
Bien fut iadis la chasteté craintive,
Seule n'osant par les bois s'égarer,
Où sur les eaux, de peur d'y demeurer
De quelque Dieu peu chastement captive.
Des Dieux cornus la grand'troppe lassive

113 Morel P., "L'eau et le vin à la villa Giulia", in Cazzato V. – Roberto S. – Bevilacqua M. (eds.), *La festa delle arti. Scritti in onore di Marcello Fagiolo per cinquant'anni di studi*, vol. 2 (Rome: 2014) 956–961.

114 Tucker, *La description de la Villa d'Este*. Note that Commendone also dedicated one of his poems to Muret. See Lagomarsini, 17.

115 Blänsdorf, *Aspekte eines poetischen Themas* 59–64: Dorat Jean, *Ad fontem Arculii*; De Ronsard Pierre, *Les Bacchanales, ou le folastrissime voyage d'Hercueil pres Paris, dédié à la joyeuse trouppe de ses compagnons*; De Baïf Jean Antoine, *La ninfe Bièvre*.

116 *Divers poemes de I. Du Bellay*, in *Les Œuvres francoises de Ioachim Du Bellay* [...] (Paris, Fédéric Morel: 1569), fol. 4r–v.

FIGURE 6.15 Prospero Fontana and workshop, *Scenes with Bacchus and Nymphs*, 1553–1555. Villa Giulia, Room of Fortune
IMAGE © WIKIMEDIA COMMONS

Ne permettoit les Nymphes s'asseurer,
Feust au repos, feust pour desalterer
Du long travail la chaleur excessive.
Donques pourquoy est mon dormir si long,
Ce qu'autre Nymphe en seurté ne feit onc ?
Cesse, passant, de t'en donner merveille.
Iules qui peut les Dieux mesmes fascher,
A commandé qu'au pied de ce rocher
Et seule, & nue, & chaste ie sommeille.

The Sleeping Nymph
At the fountain of Pope Julius III
Chastity was once afraid, not daring to walk alone in the woods or on the water, for fear of being captured by an unchaste god. The great lascivious horn of the satyrs would not allow the nymphs to rest in safety and quench their thirst, caused by the excessive heat. So why is my sleep so long when no other nymph could rest safely? Stop wondering, passer-by. Julius, who can make even the gods angry, has ordered that I could sleep alone, naked and chaste, at the foot of this rock.

Elle mesme apres la mort du Pape.
Ce n'estoit pas le sommeil, qui fermoit
Si longuement ma paupiere serree:
Donques pourquoy suis-ie tant demeuree
Tenant fermé l'œil qui point ne dormoit ?
Iadis mon eau, qui craintive souloit
Des yeux mortels se tenir separee,
Pour estre plus des hommes asseuree
Dessoubs ces monts secretement couloit.
Depuis voyant que l'honneur de mon onde,
Iules, par toy estoit publique au monde,
Mes yeux honteux n'ont ozé voir le iour.
Mais puis qu'aux tiens la lumiere est faillie,
Pour n'estre plus de vergongne assaillie,
Ie m'en retourne à mon premier sejour.

The same after the death of the Pope
It was not sleep that kept my eyelid closed for so long. Why then have I so long kept closed the eye that never sleeps? Once, my water, which I used to keep away from mortal eyes for fear of men, flowed secretly beneath

these mountains. Since I have seen the honour of my water made known to the world by you, Julius, my ashamed eyes have not dared to see the day. But since the light has left yours, so that shame may no longer assails my eyes, I return to my first sojourn.

9 Fausto Sabeo (1475–1559): the Triumph of the Epigram

An important series of poems on Julius III's nymph was compiled by one of the most eminent librarians of the Vatican Library, Fausto Sabeo. Called to Rome by Leo X, he was fully enmeshed in the Roman humanist milieu that consecrated the triumph of poetry in the garden in the period preceding the Sack. He addressed poems to Angelo Colocci and Blosio Palladio and was charged, alongside Cardinal Antonio del Monte, with surveying the damage suffered by the library after the Sack, completing the re-cataloguing of its holdings in 1533. Sabeo was particularly attached to France and held high praise for Cardinal Jean Du Bellay. His epigrams were compiled in a magnificent collection dedicated to King Henry II of France, *Epigrammatum Fausti Sabaei Brixiani Custodis Bibliothecae Vaticanae Libri Quinque*, published in Rome in 1556 and now preserved in the Bibliothèque Nationale de France (Lat. 17908). Among his countless epigrams inspired by Catullus, Martial, and the *Greek Anthology* – some of which, titled "e Graeco", are Latin translations of Greek epigrams from the *Anthology* – many concern works of art and, notably, the sculptures that adorned the gardens of Rome: the famous ones in the Belvedere courtyard, but also those in Blosio Palladio's garden on Monte Mario, Cardinal Cervini's on Mount Amiata, and those at the Villa Giulia.[117] These poems were published in a careful German edition with a commentary on the statues.[118] Of interest to the present study, Sabeo dedicated a series of four epigrams to the Fountain of the Virgin, all titled "*In fontem Virginis In Hortis Iulii Tertii Pont. Max*":

117 See his biographical profile by Stefano Benedetti in *Dizionario biografico degli italiani* 89 (2017). On his poems dedicated to art, see Occhipinti C., "Fausto Sabeo e le arti figurative. Tra Vasari e Michelangelo", in Agosti B. – Ginzburg S.– Nova A. (eds.), *Giorgio Vasari e il cantiere delle Vite del 1550* (Venice: 2013) 47–59.

118 Maurach G. – Echiger-Maurach C. – Töns U. (eds.), *Studien zu Faustus Sabaeus, « Epigrammatum Fausti Sabaei Brixiani Custodis Bibliothecae Vaticanae Libri Quinque », Rom 1556, mit Auszügen aus dem Text*, Fontes 31 (Heidelberg: 2009) 75–111.

[37] (p. 136)
Quae pedibus liquidis vitreas percurritis undas
Atque susurratis carmine dulcisono,
Nympha loci, o Nymphae, hic dormit, vos parcite somnum
Rumpere, dum blandae luditis et canitis.

You, who run across the crystalline waves with your limpid feet, / And who murmur a melodious poem, / O Naiads, do not trouble the slumber of the Nymph of this place who sleeps here, / When you play and sing tenderly.

[38] (p. 144)
Huc veniens taceas, hic Cypria lassula dormit,
Alludunt blando murmure Nereides.
Sola licet, secura cubat, non numine tantum,
Sed vigili cura principis et studio.

If you pass by this place, be quiet, for Cypris sleeps here weary, / Jest the Nereids with a soft murmur. / Though alone, she rests in total safety, not so much thanks to her divine nature, / But especially by the care and zeal of this attentive prince.

[39] (p. 148)
Ne me vexato, sum lassa puella quiescens,
Iulius errantem repperit hicque locat,
Ut sim Virginei fontis custodia virgo;
Montia Nympha vocor, quae Charis una fui.

Do not trouble me! I am just a tired young girl resting. / Julius found me wandering and brought me to this place, / So that I could become myself, a virgin, the guardian of the Fountain of the Virgin; / We call me the nymph Montia, in other times one of the Charities.

[41] (p. 151)
Hic sese lavit nitidis Venus aurea in undis,
Quum vicit divas lotaque et una duas.

It is here that the radiant Venus washed herself in the scintillating water / Before she vanquished, single-handedly, the two other goddesses.

Beyond the usual puns on the Pope's name (Montia Nympha), the poet develops a series of subtle and erudite allusions. In the first poem, *Nympha loci* refers to the famous epigram by Campano. In the second, Cypria refers to Cythera, the island where Venus was born. In the third epigram, Sabeo equates the papal nymph with Charis, one of the Charities in Greek mythology. Associated with the Graces by the Romans, poetic references to the Charities often involved goddesses personifying life, beauty, and grace. Meaning "source of joy" in Greek, the term may also refer to certain decorative elements of the Villa Giulia, namely the female allegories of the virtues depicted on the ceiling of the Room of Fortune (including Charity), who evoke the Christian alter egos of the pagan nymphs [Fig. 6.16].[119] The female allegories of Religion and Charity are also found in one of the reliefs that adorned the entrance to the loggia of the Nymphaeum, complementing those of Virtue and Fortune mentioned above [Fig. 6.7].[120] Finally, the fourth poem emphasises the association between the nymph and Venus, alluding to the conclusion of the Judgement of Paris when Venus receives the golden apple (*Venus aurea*). This single episode was the catalyst of the Trojan War, the wanderings of Aeneas, and the origins of Rome – a key theme of Julius III's pontificate and, by extension, his villa.[121]

Sabeo's work marks the triumph of the epigram. Rarely had the genre been the subject of such an enormous output – more than three thousand verses! Considered a means of distinguishing oneself and demonstrating one's ingenuity through infinite variations on the same theme, epigrams were mainly practised by young poets who, like Commendone, later abandoned the exercise, though it enabled them to achieve an elevated social position. Indeed, the epigram was above all a courtly exercise, a means for poets to attract attention, and an effective political instrument in service of the patron for whom it was produced or intended. For Sabeo, an octogenarian at the time of the pope's death, the composition of epigrams was less a means to seize social opportunities than an ongoing lifetime pursuit. Sabeo's publication also contributed to the codification of the genre, which had been endorsed in detailed and scholarly treatises published by Francesco Robortello (*In librum Aristotelis de arte poetica explicationes*, 1548) and Julius Scaliger (*Julii Caesaris Scaliger Novorum Epigrammatum, Liber unicus*, 1533; *Poetices libri septem*, 1561).[122] According to

119 Krappe A.H., "Les Charités", *Revue des Études Grecques* (1932)155–162. See also the linguistic analysis of the term, which illuminates the links between Charis, Venus, and virtue: De Callatay D., "Prologue. Gratuité et grâce", *Revue du MAUSS* 35.1 (2010) 53–61.

120 Nova, *The Artistic Patronage* 88, 163, 165–166.

121 Ribouillault, *La villa Giulia*.

122 Töns, *Die Epigramme des Faustus Sabeus* 97–108.

FIGURE 6.16 Federico Brandani, after Prospero Fontana, *Charity*, 1553. Stucco. Villa Giulia, Room of Fortune
IMAGE © WIKIMEDIA COMMONS

Scaliger, Sabeo's epigrams on the nymph of the *Aqua Virgo* evoked that which is *mollis* (soft), *tener* (tender), *languidus* (delicate), and *suavis* (charming) by depicting beautiful figures with graceful movements at rest alongside the imagery of water and its sensual qualities and silent and shady places that invite relaxation – in short, bringing together the *topoi* of the *stylus humilis* of pastoral poetry.[123]

123 Ibid.

10 Miguel Da Silva (1480–1556): the Loving Sculptor

Eduardo del Pino González has recently published another poem about the fountain of Julius III composed by Miguel da Silva, a Portuguese neo-Latin poet, bishop of Viseu, and later cardinal, and arguably one of the greatest humanists of his time. The poem is in a manuscript in the El Escorial library with a later copy preserved in the National Library of Florence.[124] Drawing on Sylvie Deswarte-Rosa's seminal research on Rome in da Silva's time, let us first offer a brief overview of da Silva's background before examining the poem.[125]

As ambassador to the King of Portugal between 1515 and 1525, da Silva established himself as a key figure in Leo X's and Clement VII's Rome. Against the king's advice, he returned to Rome in 1540 where, despite the confiscation of his property and the royal disgrace to which he was subjected, he was appointed cardinal by Paul III the following year. He died in Rome in 1556 with a reputation as a great humanist and man of letters. Da Silva embodied the archetype of the ideal courtier. In fact, Baldassare Castiglione's famous *Libro del Cortegiano* was dedicated to him in 1527. Trained in Siena and steeped in Florentine Neoplatonism, he was an intimate of Leo X and a close friend of Pope Clement VII. His circle included Blosio Palladio, Lattanzio and Claudio Tolomei, Jacopo Sadoleto, and Giovanni Rucellai, whose famous Florentine garden (Orti Oricellari) was the site for the attempted revival of Marsilio Ficino's Academy. His fame was immense; his compatriot André de Resende wrote in his 1531 *Erasmi Encomium* that he was 'extremely well known in the Latin world and in the city of Remus [Rome], equally learned in both languages, embracing in his single breast philosophy and high poetry, and of equal value in both', and in his 1534 *Oratio pro Rostris* referred to 'Miguel da Silva, nurtured by the two Muses, burning with enthusiasm for the whole of antiquity, who filled the whole of Italy with the noise of his erudition'.

Da Silva's scholarly activities were closely linked to the world of the garden, namely those tended to by Angelo Colocci, who was interested in Portuguese literature and with whom he shared many interests, especially regarding Greek studies and epigraphy. He is included in Colocci's list of poets in Ms. Vat. Lat.

124 (Biblioteca del Monasterio de El Escorial), ms. &– IV–22, fol. 198r.; (Florence, Biblioteca Nazionale Centrale), ms. Magl. VII, 377, f. 2v (the copy of the poem, dated 1747, bears the title "Ad statuam puellae dormientis ad Aquam Verginem"). See Del Pino González E. "El poema *Ad statuam puellae* de Cardenal Viseu y la 'Vergine della fontana' de Villa Giulia en Roma", *Ágora. Estudos Clássicos em Debate* 18 (2016) 157–191.

125 Deswarte-Rosa S., *La Rome de D. Miguel da Silva (1515–1525)* (Lisbon: 1988) 177–307 ; published in Italian as *Il "perfetto cortegiano" D. Miguel da Silva* (Rome: 1989). All subsequent quotes concerning Miguel da Silva are extracted from this book.

3450 alongside Pietro Bembo, Jacopo Sadoleto, Egidio da Viterbo, Fabio Vigili, Claudio Tolomei, Janus Lascaris, Ludovico Ariosto, Giangiorgio Trissino, and Girolamo Britonio, among others. Following the Sack of Rome, Da Silva frequented the literary group in Blosio Palladio's garden in the Villa delle Peschiere, built around 1532 near the Vatican, and where attempts were made to recreate the cultivated atmosphere of the pre-1527 period. Marcantonio Flaminio notably evokes these poetic gatherings in his *Carminum Libri VIII* (1727).

The Cardinal of Viseu owned a garden immediately adjacent to Agostino Chigi's famous villa (the Farnesina) praised by his friend Blosio Palladio. The palace façade, close to the small church of San Jacopo in Settignano, overlooked the Via della Lungara, while the gardens bordered the Tiber next to the stables of villa Chigi. The property is visible on Leonardo Bufalini's 1551 plan of Rome, marked "*vinea. de. r. c. lisevm*", i.e., "vigna of the cardinal of Viseu" [Fig. 6.17]. This garden is not well known, but its reputation in the sixteenth century seems to have been well established if we are to believe the testimony of the Silesian traveller Seyfried Rybisch who visited Rome in September 1554. Besides the Villa Giulia, for which he provides a Latin description, Rybisch writes of the da Silva's gardens:

> There is on the banks of the Tiber the very beautiful palace of the Spanish cardinal [sic] of Viseu: his gardens possess a number of very pleasant delights and have been adorned at great expense and following an admirable arrangement of a multitude of varied trees, of groves of orange trees, of a profusion of beautiful fountains; they are embellished with birds enclosed in wire cages.[126]

126 *Les voyages de Seyfried Rybisch, étudiant silésien: Itinéraire (1548–1554), Quatrième partie*, ed. J. Hiernard (Pessac: 2017) 26. In Rome in September 1554, Rybisch provided one of the oldest surviving descriptions of the Villa Giulia 66–68, translated into French by Jean Hiernard, and here into English by Mariah O'Brien: 'Leaving through the Gate of the People on the Via Flaminia, a thousand paces from the city, we find the Villa Giulia, named after Pope Julius III and vulgarly called the Vinea del Papa. Julius III had purchased several vineyards and meadows in this area and established vast gardens full of incredible delights, very pleasing forests of orange and cedar trees, groves of plane trees, meadows enhanced by the softness of delicate fountains, and very fine monuments; he placed beautiful summer pavilions in various locations, highly suitable for the repose of people weary from long promenades. The villa measures a few thousand steps in circumference. In all, I would like to think that there is nothing in all the world of delight and voluptuousness that is not contained in this place. He also built a very sumptuous palace, the portico of which has received ancient statues, the very elegant one of Cicero

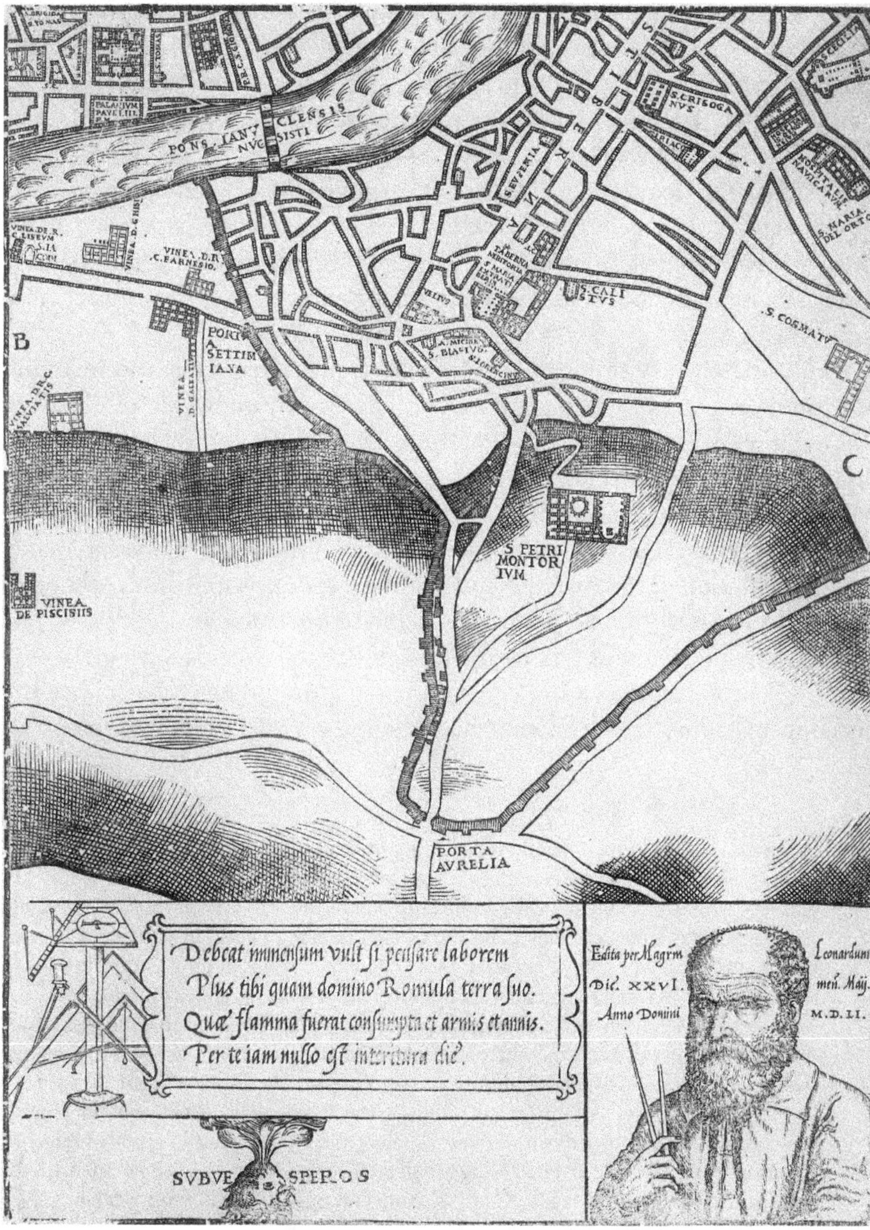

FIGURE 6.17 Leonardo Bufalini, *Plan of Rome*. 1551. Detail featuring the garden of Cardinal Viseu, Miguel da Silva, on the banks of the Tiber (top left corner)
IMAGE © UNIVERSITÄTSBIBLIOTHEK HEIDELBERG

An *avviso* by Angelo degli Oldradi on the terrible Roman floods of 1557 confirms that the cardinal's garden was famous in Rome: 'Ha ruinato anchora il Giardino si famoso di Viseo, posto su la riva del Tevere, la bellezza del quale, a chi non l'a [sic] veduto, come havete voi sarebbe difficil cosa a credere' ('It has also ruined the famous Garden of the Cardinal of Viseu, located on the bank of the Tiber, the beauty of which, to those who have not seen it, as you have, would be difficult to believe').[127] The cardinal was obviously passionate about antiquities and seems to have been the first, in 1525, to have commissioned casts of ancient sculptures to send to Portugal – more than twenty years before King Francis I.[128]

Unfortunately, few of his poems have been preserved. They can be found in various anthologies, such as those compiled by Giovanni Matteo Toscano (*Carmina illustrium poetarum Italorum*, Paris, 1577) and by Tommaso Buonaventura and Giovanni Gaetano Bottari (Florence, 1722). Deswarte-Rosa published the poem *De Agua Argentea* on the aqueduct of Evora addressed to King John III and mentions da Silva's epigram on the city of Rome produced for the exhibition of the *Fasti Capitolini* at the Capitol, as well as his other epigrams addressed to Titian, Camilo Vitelli, and Blosio Palladio.

The poem about Julius III's nymph is significant. While da Silva makes the statue speak according to the conventions of prosopopoeia, he frames the question of its multiple and changing identities as the central focus of his

making a speech, those of Scipio the African, of Brutus and others, very numerous. But what to say of the beautiful fountain that can be seen in the square, the price, beauty, art, and elegance of which cannot be sufficiently appreciated by any of those who behold it? The villa has a portico, alleys, and summer recesses supported by very precious marble columns and whose floor is entirely covered with marbles of different qualities with admirable art and natural colours; it contains an immense quantity of antique statues arranged a bit everywhere for decorative purposes; one sees marble children, admirable for their antiquity and artistry, who pour water from their urns into very beautiful marble basins where one discovers a quantity of varied fish. Among the ancient statues, one can see the representation, astonishing in its age and taste, of a very delicate nymph sleeping nonchalantly on the edge of a murmuring stream, and a very beautiful statue of Mercury, others of Cleopatra, Venus, and Modesty with veiled hands (for this is how the Ancients represented her). But why insist? If I were to recount here all that is worthy of being seen, neither my pen nor my time would suffice, that is certain'.

127 Degli Oldradi Angelo, *Aviso della pace tra la Santita di N.S. Papa Paolo IIII. e la Maestà del Re Philippo: con la narratione del diluuio che è stato in Roma, con le gran ruine de ponti, chiese, palazzi, vigne* [...] (Rome: n.p., 1557).

128 Mendonçà R., "Classical Sculpture Reproductions and Portuguese Gardens", *Gardens & Landscapes of Portugal* 2 (2014) 60–69.

poem, evoking the poetic contributions on the statue that we have gathered in this study, undoubtedly only partially.

> *In statuam puellae ad Aquam Virginem Michaelis Sylvii Cardinalis*
>
> Quae nuper de me multi cecinere caveto,
> Hospes, ne male te carmina decipiant.
> Non dea, non fontis numen, nec Nympha Dianae,
> Sed decet id quod sum dicere simpliciter.
> Sum statua ut cernis specie formata puellae,
> Quam placida oppressit, dum fluit unda, quies.
> Et qui me fecit similem fortassis amavit,
> At surdam precibus, quod sopor ipse notat.
> Iulius hic posuit, nomen quoque Virginis addit
> Conveniens Fonti temporibusque suis.
>
> On the statue of the young girl of the Acqua Vergine by Cardinal Miguel Sylvio
> Attention, visitor, do not let yourself be abused by / The poems that many have recently sung about me. / I am neither a goddess, nor the divinity of this fountain, nor a nymph of Diana. / I should however simply say what I am. / I am a statue, as you can see, chiselled on the beauty of a young girl, / Plunged into a peaceful sleep, as the wave flows [around me]. / He who created me maybe loved a person who resembles me, / But who was deaf to his supplications, hence my own impassivity. / Julius placed me here and gave me the name of Virgin, / Name that is suited simultaneously to the Fountain and to its own era.

Da Silva clearly places himself in competition with the other poets. In an implicit criticism, he contrasts their praise with his naked and humble truth, their poetic imagination with his simple vision of reality: the nymph is only a statue, made in the likeness of a young girl. While he takes an opposite approach to the emphatic language of his contemporaries, using a more dignified *stylus humilis*, he offers in the second part of the poem a new, remarkable, and theoretically significant interpretation: that of the sculptor in love but spurned. Indeed, when reading his poem, it is difficult not to think of the myth of Pygmalion (Ovid, *Metamorphoses* 10, 6) which provides an inverted narrative to da Silva's; Pygmalion falls in love with a statue of a woman and marries her when she comes to life thanks to Venus, while the sculptor of the nymph at

Villa Giulia falls in love with a young woman who rejects him and inspires him to create the statue.[129] Da Silva's anecdote should also be compared to the first monumental female nude in Greek art, the Aphrodite of Knidos by the sculptor Praxiteles, who had used his mistress as a model.[130] The copy preserved in the Vatican was, notably, the subject of an epigram by Fausto Sabeo that drew from the *Anthology of Planudes* (16, 159).[131] In short, Miguel da Silva, familiar with these sources, mastered the techniques of the epigram to open a profound reflection on the relationship between art and nature, illusion and love. The end of his poem, which typically refers to the dedicatee, evokes the positivism with which it began, simply stating that the nymph's name derives from the name of the spring water, Aqua Virgo. Finally, in the last sentence, the reflexive possessive adjective *suis* relates back to Julius and not to the fountain. We may consider this choice as alluding to the pope's astrological sign (Virgo) which, as we have observed, was celebrated in one of the loggias of the nymphaeum and, more generally, to the course of his "just" reign as Virgo Astraea.[132]

11 Lorenzo Gambara (*c.*1495–1585) and Francesco Franchini (1500–59): Poets of the Farnese

Among the intimates of the Farnese court, a family that played a prominent role in Julius III's Rome and influenced his papal election, we find – in addition to Annibale Caro and the librarian Fulvio Orsini – Lorenzo Gambara, a poet and humanist from Brescia.[133] Omnipresent in primary sources, he seems to have been associated with almost every poet in the present study. Probably influenced by Caro, he provided a very liberal translation of Longus' *Daphnis and Chloé* in four books published in 1569. Art historians know him primarily for the poems he addressed to his master, Cardinal Alessandro Farnese, regarding his palace at Caprarola. The *Arcis Caprarolae descriptio* first appeared in the 1569 edition of his *Poemata* and later in a complete work published in 1581.

129 Stoichita V.I., *L'effet Pygmalion. Pour une anthropologie historique des simulacres* (Geneva: 2008).

130 According to the ancient sources, the two myths were related: "Both Clement [of Alexandria] and Arnobius link the story of Pygmalion to the legends woven around the Venus of Knidos, sculpted by Praxiteles. She is said to have attracted young men to the point of 'having commerce with marble'" (my translation); Stoichita, *L'effet Pygmalion*, 21.

131 Maurach – Echiger-Maurach – Töns (eds.), *Studien* 22–23.

132 I would like to thank George Hugo Tucker for pointing this out to me.

133 See the still incomplete profile given by Angela Asor Rosa in *Dizionario biografico degli italiani* 52 (1999).

The description is part of a bucolic narrative that follows the protagonists, Ladon and Aceste, from the Farnese gardens in Rome – the Horti farnesiani on the Palatine and the Farnesina's, which the cardinal had just acquired – to those of Caprarola. According to Carlo Caruso, these characters, disguised in pastoral garb, were certainly inspired by historical figures who visited the villa, notably the scholar Guglielmo Sirleto, who was associated with several poets mentioned in this study and to whom Commendone dedicated a poem.[134] Gambara's poems dedicated to Julius III, published in Rome in 1553 and in Basel in 1555 alongside *Poematum libri VIII* by his friend Basilio Zanchi, are not so well known. Taking the form of pastoral and maritime eclogues and considered among the most significant of the genre, they evoke some of the great decorative and poetic themes of the Villa Giulia.[135]

A long poem, "*In Fontem Iulii III*" entirely focuses on poetic and mythological sublimation of the landscape of the Villa Giulia and its nymphaeum. It features the Oceanid Leuke, mother of the Nereid Clymene, who wanders over hill and dale in search of her missing daughter. The narrative, invented by the poet for the occasion, is obviously inspired by Virgil's story of Ceres and Proserpine in which Leuke is also mentioned, transformed into a poplar tree after her death in the underworld (Serv. *Ad Virg., Eclog.* 4.250 and 7.61). Gambara's narrative choice was judicious, given the many poplars on the grounds of the Villa Giulia. The poem suggests that Clymene, the disappeared nymph, becomes the new guardian of the papal spring described as a marvellous grotto. Like a sibyl, she sings of the advent of a new golden age – an unmistakable reference to the Julian pontificate. The nymph tells her distraught mother, who has by chance been resting near the spring, how she crossed the Tiber where its course widens to escape the amorous assaults of the God Jupiter disguised as a satyr-like shepherd – a rather precise allusion to the ford built on the river to enable the pope to access the villa from the Vatican and the zone of Prati. Having crossed the ford, the chaste nymph is saved by Alcimedon – a reference to none other than Pope Julius III.[136] Numerous allusions run throughout this poem, such

134 Gambara Lorenzo, *Laurentii Gambarae Brixiani Caprarola* (Romae, apud Franciscum Zanettum: 1581). See Caruso, *Feconde venner le carte* (Poesia umanistica di villa) 282–288. Commendone's poem dedicated to Sirleto is published in Mai, *Spicilegium* 485–486.

135 Carrara E., *La poesia pastorale* (Milan: 1909) 396.

136 Zanchi Basilio, *Basilij Zanchij Bergomatis Poematum libri VIII*; *Laurentii Gambarae Brixiani Poematum libri III* (Roma: 1553). I quote from the Basel edition (Basel, Ioannem Oporinum: 1555) 416–420 ("*In Fontem Iulii III*") 420–422; the following poem, titled "*Iulio III. Pontefici Maximi Sacrum*" continues the same descriptive vein of the Villa Giulia as a kingdom of nymphs. On the ford over the Tiber and the area of the *Prati*, see Ribouillault, "Jeux de mots et d'images" 112.

as visual evocations of the aqueduct of the Aqua Virgo, the decor of the nymphaeum, and the bucolic landscape of the villa and its park. The central theme of the fleeing virgin also directly echoes the bas-relief at the entrance to the nymphaeum described by Ammannati as: 'un'Hercole assiso in atto di fiume, ed una femmina appresso in habito di vergine che fuggie. Che dinota l'acqua della fontana secreta [...]' ('A seated Hercules representing the river and a fleeing girl dressed as a virgin representing the water of the secret spring').[137] An English translation of the poem is provided in the appendix.

Another poet, Francesco Franchini, who was at the service of Ottavio Farnese, also left an elegant epigram on the papal fountain. A native of Cosenza and a soldier in the imperial retinue of Charles V during the terrible 1541 Algiers campaign, he moved in the imperial circles with which Ottavio was connected and travelled extensively throughout Europe.[138] In Rome, he associated with Annibale Caro, who praised but sometimes criticised his poetry, as well as Lorenzo Gambara and Basilio Zanchi, alongside whom he was received by the two Du Bellays in the Horti Bellaiani.[139] His poetic oeuvre is known thanks to his *Poemata*, published by Giovanni Onori in Rome in 1554 and again in Basel in 1558, a year before his death.[140] An elegant composition on the *Fons Virgo* was among the satirical and amorous epigrams that earned the work a place in the *Index librorum prohibitorum* in 1559:

IN FONTEM IVLII TERTII PONT. MAX.
Garrula quae manat gelidis e montibus unda,
Quaeque fluit blando per cava saxa sono:

137 Davis, *Villa Giulia* 135–137. Pliny is the source, *Natural History* 31.25. On Alcimedon and his links to the story of Hercules, see Lipka M., *Language in Vergil's Eclogues* (Berlin: 2001) 184.

138 Croce B., "Un poeta latino poco noto. Francesco Franchini", *Quaderni della Critica* 16 (1950) 39–55; Campana D., "Francesco Franchini, 1500–1559", *Galleria dell'Accademia Cosentina*.

139 See Annibale Caro's letter to Franchini in *Lettere scelte dalle familiari del commendatore Annibal Caro* vol. 2 (Verona: 1820) 102. The visit of the group of "Farnesian" poets to the Horti Bellaiani is mentioned in Du Bellay Jean, *Poemata*, trans. G. Demerson – R. Cooper (Paris: 2007) 33.

140 Franchini Francisci, *Francisci Franchini cosentini poetae elegantissimi poemata* [...] (Basel, Petrus Perna typographus [...]: 1558) 215. In his *Poemata*, Franchini notably dedicates poems to Ippolito Salviani, Guglielmo Sirleto, Basilio Zanchi, Gabriele Faerno, Anton Francesco Rainerio, Giulio Poggiano, Janus Vitalis, Pietro Aretino, Cardinal Antonio del Monte, Lorenzo Gambara, Pierio Valeriano, and Annibale Caro. Among the greats, we find the Farnese and their close associates, including Julius III, to whom several poems are devoted (46–49, 52, 193). One of them celebrates his alliance with the powerful family of Paul III, to whom he had just succeeded. We should also mention a series of six epigrams on the statue of a Sleeping Venus in the garden of Cardinal Rodolfo Pio da Carpi (144–146; see Brunon, *De la littérature au Jardin* 19).

Ad sese niveas invitat pura columbas,
Cygne tuos cantus et Philomela tuos:
Formosae Charites, dulcesque accedite Nymphae:
Ad sua vos virgo convocat antra soror.

THE FOUNTAIN OF POPE JULIUS III
The choppy water, flowing from glacial summits, / and whose exquisite babble runs on the hollow rocks, / invites in its purity the snowy doves, / and your songs, Swan, and yours, Nightingale: / Beautiful Graces, come, and you, the tender Nymphs: / The Virgin, your sister, invites you into her lairs.

12 Conclusion

In his portrait of the pontificate of Julius III, Ludwig von Pastor praised the pope for his service to literature, art, and science, namely by appointing the learned cardinal Marcello Cervini (the future Marcellus II) as curator of the Vatican, searching for new Greek manuscripts, and reforming the university. His cultivation of poetry at the Villa Giulia and decision to centre the villa's iconography on the nymph of its spring were directly inspired by his love of letters, nurtured in the humanist milieus of Leo X's and Julius II's Rome. In light of the present study, it is understandable that Alfred von Reumont, following an erroneous reading of an ancient inscription, considered the Villa Giulia as the site of a real academy.[141] If we are to speak of an academy in the villa, it is only in the definition given to it at the time: as an informal meeting place for scholars (*sodales*), united by a common interest in ancient epigraphy and Latin and Greek poetry and who chose various gardens of Rome to serve as the settings for their discussions.[142] These scholars knew each other and belonged to circles of intellectuals who crossed European borders and typically served the Curia, most often as secretaries to eminent and learned cardinals.

The poetic fortune of the Villa Giulia nymphaeum offers rich lessons for the garden historian and the art historian. It demonstrates how intertextuality and intermediality, such as the relationships between poetry, painting, and the garden, as well as scholarship based on the study of antiquity formed the basis

141 von Pastor L., *Histoire des papes depuis la fin du Moyen-Âge*, trans. A. Poizat, vol. 13 (Paris: 1931) 302, 307 (note 1); von Reumont A., *Geschichte der Stadt Rom*, vol. 3 (Berlin: 1870) 705.

142 On the use of gardens by the first academies in Rome, see the introduction and Alessia Dessì's essay in this book.

of villa culture during the Renaissance. To fully understand this culture, the performance and the process of constructing meaning – full of humour and even sarcasm, play and surprises – must be foregrounded above isolated examinations of specific works of art. Spirit, grace, and casualness (*sprezzatura*), the obligatory features of Castiglione's ideal courtier to which our poets adhered, call for relating things and ideas in a witty and erudite manner and for distinguishing oneself intellectually and politically.[143] The nymphaeum might therefore be considered a stage to demonstrate one's spirit (*esprit* in French or *ingegno* in Italian) and achieve success at the papal court. As Commendone's biographer Graziani wrote, Renaissance Rome was a city 'where one can aspire to all honours, when one has wit [*esprit*] and merit'.[144] In addition to Da Silva, this desire for social advancement can be observed in the output of several of our protagonists. Cardinal Antonio Del Monte wrote a short treatise published in 1516 but written as early as 1511, the *Idea del prelato*, on the art of succeeding at the court of Rome for his nephew, the future Julius III. Commendone penned his *Discorso sopra la corte di Roma* in 1554, at the same time he was composing his verses for the nymphaeum.[145] The set of recommendations found in these writings frames the poetic production gathered in the present study within an eminently strategic and political context, without diminishing its artistic relevance, its grace, or its spirit.

13 Addenda to the First Edition

As the initial research for this article was completed in 2019, it was not possible to include another fascinating example of the poetic cult dedicated to the Villa Giulia nymphaeum. In 2020, Maxime Malleux published two poems written in Hebrew by the Orientalist from Louvain Andreas Masius (1514–73), who stayed in Rome between 1551 and 1553. The poems, 'in praise of a statue on the fountain in the garden of Pope Julius III', are composed in the tradition of medieval Andalusian Hebrew poetry, offering 'a unique mixture of the Jewish and Christian poetical traditions'. This is another important and

143 Ricci M.T., "La grâce et la *sprezzatura* chez Baldassar Castiglione", *Bibliothèque d'Humanisme et Renaissance* 65.2 (2003) 233–248.

144 Gratiani, *La vie du cardinal Jean François Commendon* 24.

145 Hurtubise P., *Tous les chemins mènent à Rome. Arts de vivre et de réussir à la cour pontificale au XVIe siècle* (Ottawa: 2009), chapter 5 in particular; Commendone Giovanni Francesco, *Discorso sopra la corte di Roma*, ed. C. Mozzarelli (Rome: 1996); Caccamo, *Giovanni Francesco Commendone*.

fascinating addition to a collection that is likely to grow as research on the subject continues.[146]

Unless otherwise stated, all translations from Latin in this article are by Alin Mocanu, reviewed and revised by George Hugo Tucker and Ginette Vagenheim. I am very grateful to the three of them. Any errors and misinterpretations remain entirely my own.

Appendix

Basilij Zanchij Bergomatis Poematum libri VIII; Laurentii Gambarae Brixiani Poematum libri III (Basel: 1555) 416–420.

Leuke of Laurentum[147] wandered in this green landscape
Which the Tiber separates from the fields of the Tuscan country;
Here, where each of the seven hills, which the swift Albula[148] surrounds with its whirlpools,
Rises to the stars,
She came to the source of the virgin fountain
And, dying, she lay down on a bed of greenery.
Then, weeping for Clymene,[149] once her child
Very cherished, Leuke declaimed these lines:
My daughter, what misfortune has taken you from me, miserable parent that I am?
What foreign land holds you?
My child, during two times ten days, I have sought you in vain across the great forests,

146 Maleux M., "On Hebrew Nymphs and Aqueducts", *Lias* 47.1 (2020) 67–87. The poems were first signaled in Striedl H., "Hebräische Lobgedichte des Andreas Masius auf die Aqua Virgo im Nymphaeum der Villa Giulia in Rom. (Staatsbibliothek Preussischer Kulturbesitz. Ms. or. fol. 13, Bl. 2–10)", ed. Franke H. et al., *Folia rara. Wolfgang Voigt LXV. diem natali celebranti / ab amicis et catalogorum codicum orientalium conscribendorum collegis dedicate* (Stuttgart: 1976), 126–133.

147 Laurentum, the ancient capital of Lazio, is evoked many times at the Villa Giulia. It was there that the Trojans and Aeneas landed and were received by King Latinus, who offered Aeneas his daughter Lavinia in marriage. The villa's sculpture collection included statues of Lavinia and Latinus. See Ribouillault, *La villa Giulia* 361 et idem, *De la peinture au jardin* 58–60.

148 Ancient name of the river Tiber.

149 Name of one of the Nereids.

Without my eyes seeing you.
What mountains have I not filled pitifully with my agonising pleas?
Oh, miserable [that I am], no one has come to my aid!
Even the Naiads who dwell in the nearest meadows
Have heard the high forests resound with my pleas.
You are witnesses, you, verdant forests, and you, neighbouring pastures,
And you, wild beasts, who dwell in the high mountain tops!
Return her to her mother, you Nymphs, return her,
If by chance she is among you!
You, great Pales,[150] and you, divine Faunus,[151]
For whom I have burned incense in my long prayer,
I beg you, show me the way: in what forests has my fugitive daughter found refuge?
In what place does she hide?
Then I will erect altars to you, I will make libations for you as required by the sacred [rites].
And a she-lamb will dye your burning fires with her blood
Clymene recognised the bitter cries of her mother
And concealed her wave-coloured face in the foaming waves.
Indeed, dwelling in a cave, where she hid under the water,
She decorated her house with multicoloured flowers,
Adorned her breast with blue gems
And her hair with green moss.
Then she addressed great praises to the Lord, recalled his glorious deeds
And described the remarkable work of the fountain.
She foretold with her sacred mouth the centuries to come under the orb
Of Hesperia[152] [and] in our time[153]
So that the shepherd might bring the sacrificial victims through the meadows without worry,
And let not the ewe lamb fear to return home alone
And let her not fear the jaws of the wild wolf in the morning when she travels through the forest,
Nor when she shelters under the roof of leaves.
That the farmer no longer bothers to uproot the green forests,

150 Goddess of shepherds and pastures.

151 A god often associated with Pan, he was the deity of the fertility of the herds and fields.

152 The West.

153 All the verses that follow recall the topos of the *aetas aurea* (golden age), so dear to Julius III.

To cleave the fertile earth with iron.
Mars shall wage war on the banks of the Tanais[154]
Or [otherwise] live alone in the forests of Ismarus.[155]
The shepherd will prepare his poem in the verdant shade
So that pious he glorifies the gods with great honour.
Clymene made these divinations in the entrance of the Marmorean lair
Like a prophet who calculates the age of all things.
In another place, her companion nymphs,[156] Actaea,[157] Thoe[158]
And Arethusa[159] with her tender cheeks tinted with purple,
Are heading for the high leafy dwellings of Pallas,
To the caves and under the vaulted roofs of the hidden places.
Cydippe,[160] known for her beauty, now creates a painting at the water's edge,
White lilies mixed with sanguine roses.
Then she combines the colours to weave garlands for her master.
Happy with his return, she returns to see the waters.
She arranges the shining shells in a harmonious order
And scatters in the atrium the corals of the Orient.
Then Clymene encloses the waters under the high arch
And muffles their hoarse murmurs in the vast cellar.
Then Leuke astonished that the wave no longer flows
And that the spring has stopped its course.
She is stunned, and in horror, represses her grave complaints.
A voice resounds [suddenly] from the hoarse pumice stones:
Mother, seek not further Clymene (your daughter),
And do not follow, worried, other long paths.
I am not the prey of the shepherds in the high forests
And no misfortune has befallen me.
I live more happily than others who love the flowering meadows all around
Or the valleys of the high mountains.

154 River that separates Europe from Asia.

155 Mountain in Thrace. In ancient times, Thrace was a wild and barbaric land where war was one of the only activities.

156 It seems that the poet has followed Hyginus' *Fables* here, for only this author lists all the Nereids present in the poem.

157 Nereid of the shores and beaches.

158 Nereid, daughter of Nereus and the Oceanid Doris.

159 One of the Nereids.

160 Nereid, daughter of Nereus and the Oceanid Doris or an Oceanid.

[One morning], when Phoebus awoke, surrounded by a glowing light,
And the Aurora brought the day from the orb,
I came out of the forests and went to the dwellings and
Joyful gatherings of my sisters,
Towards the solitary countryside of Lycidas[161]
I had just taken my usual path under the morning light,
When, frightened by the great Jupiter, I narrowly escaped his assaults.
He had taken on the appearance of the shepherd Aunus,
And often lured the agrarian goddesses with deftness.
As he pursued me across the dewy pastures of Mopsus,
I thought to leave the known road.
Fierce, he shouted, Echo projected [his cry] in the valleys
And the whole wood answered with errant voices.
O virgin, pride of the Naiads, receive these gifts from your lover!
Oh, that I might perish, if I am enamoured of another!
You will always be for me like a light noticed in the dark forests
Or like the clear day in the middle of the night.
I am a shepherd of the race of Faunus and the Sybarite Nymph,
Of clear song in the Alpine pastures.
Soon afterwards, thanks to the arrival of Alcimedon of Tuscany at the river's edge,
She [Clymene] set off quickly to where the waters of the imposing Tiber widen.
She put her feet in the river (what a wonder!)
And crossed the breadth of the stream, swimming with confidence.
After having saved me, Alcimedon placed me in this cave
And anointed me goddess of this sacred spring.
Then he made me guardian of this dear fountain
And gave me to inhabit these banks and this vaulted cave.
I am known among the Nymphs [who dwell] under the water
Which the great Alcimedon directs over a wide expanse.
You, my dear mother, think of the joys that your daughter offers to others
And do not let yourself be distressed by pain because of me.
After she had pronounced [these words] and poured the rippling water from the vase,
The spring [began] to flow with a more joyful murmur than usual.

161 These proper names – Lycidas, Mopsus, Alcimedon – are encountered in Theocritus's and Virgil's pastoral poetry.

References

Agnoletto S., "Giocare a fare i Classici. L'epigramma 'Huius Nympha Loci', l'invenzione dell'Antico e l'Arianna/Cleopatra dei Musei Vaticani", *Engramma. La tradizione classica nella memoria occidentale* 163 (2019), accessible online.

Agosti B. – Beltramini M., "Antonio del Monte e la maniera moderna", in Serra A. (ed.), *Humanitas. Studi per Patrizia Serafin* (Rome: 2015) 433–451.

Aksamija N., "Architecture and Poetry in the Making of a Christian Cicero: Giovanni Battista Campeggi's Tuscolano and the Literary Culture of the Villa in Counter-Reformation Bologna", *I Tatti Studies* 13 (2010–2011) 127–199.

Almagro A.G., "A Portuguese Contribution to 16th century Roman Antiquarianism: the case of Aquiles Estaço (1524–1581)", in Berbara M. – Enenkel K.A.E. (eds.), *Portuguese Humanism and the Republic of Letters* (Leiden: 2012) 353–376.

Almagro A.G., "Iscrizioni romane di tradizione manoscritta: il codice epigrafico di Aquiles Estaço", *Euphrosyne. Revista de filologia clássica* 44 (2016) 307–320.

Amaduzzi Giovanni Cristoforo, *Storia della villa di Giulio III fuori della Porta del Popolo nella via Flaminia opera lasciata imperfetta dall'abate Gio. Cristofano Amaduzzi co' monumenti e notizie sulla Medesima* (Savignano sul Rubicone, Biblioteca della Rubiconia Accademia dei Filopatridi, ca. 1777–1782).

Amaduzzi Giovanni Cristoforo, *Anecdota litteraria ex mss. codicibus eruta*, 4 vols. (Rome, Apud Antonium Fulgonium: 1783).

Anonymous, *Codice barberiniano XXX 89*, ed. R.A. Lanciani (1883).

Arato F., "Lagomarsini, Girolamo", *Dizionario biografico degli italiani* 63 (2004).

Asor Rosa A., "Gambara, Lorenzo", *Dizionario biografico degli italiani* 52 (1999).

Bardati F., "Jean du Bellay et les *Horti Bellaiani*. Bâtir à l'antique, sur l'antique et pour l'antique", in Lurin E.– Burlot D. (eds.), *L'artiste et l'antiquaire. L'étude de l'antique et son imaginaire à l'époque moderne* (Paris: 2017) 171–189.

Barkan L., "The Beholder's Tale: Ancient Sculpture, Renaissance Narratives", *Representations* 44 (1993) 133–166.

Benocci C., "Amaduzzi e Villa Giulia: un erudito settecentesco interpreta il paradiso di papa Giulio III", in Delbianco P. (ed.), *Atti della nona Giornata amaduzziana* (Cesena: 2012) 57–109.

Benedetti S., "Sabeo, Fausto", *Dizionario biografico degli italiani* 89 (2017).

Bober P.P., "The *Coryciana* and the Nymph *Corycia*", *Journal of the Warburg and Courtauld Institutes* 40 (1977) 223–239.

Boissard Jean-Jacques, *Romanae urbis topographia et antiquitates* (Frankfurt, Johann Feyrabend: 1597–1602).

Bottari Giovanni Gaetano, *Carmina illustrium poetarum italorum* (Florence, apud J.C. Tartinium and S. Franchium: 1719).

Braccesi Alessandro, "Ad eundem descriptio horti Laurentii Medicis", in *Alexandri Bracii Carmina*, ed. A. Perosa (Florence: 1943).

Brunon H., "Du jardin comme paysage sacral en Italie à la Renaissance", in Ribouillault D. – Weemans M. (eds.), *Sacred Landscape. Landscape as Exegesis in Early Modern Europe* (Florence: 2011) 283–316.

Brunon H., "De la littérature au jardin", in Venturi G. – Ceccarelli F. (eds.), *Delizie in villa: il giardino rinascimentale e i suoi committenti* (Florence: 2008) 5–31.

Blänsdorf J., "Aspekte eines poetischen Themas in der neulateinischen Dichtung Italiens und Frankreichs", in Blänsdorf J. – Janik D. – Schäfer E. (eds.), *Bandusia. Quelle und Brunnen in der lateinischen, italienischen, französischen und deutschen Dichtung der Renaissance* (Stuttgart: 1993) 1–72.

Caccamo D., "Commendone, Giovanni Francesco", *Dizionario biografico degli italiani* 27 (1982).

Capilupi Alphonso – Capilupi Camilio – Capilupi Giulio – Capilupi Ippolito – Capilupi Lelio, *Capiluporum* [Hippolyti, Laelii, Camilli, Alphonsi, Julii] *carmina*, eds. Julius Capilupus and Josephus Castalio (Rome, ex typ. haeredum J. Lilioti: 1590).

Capilupi Lelio, *Laelii Capilupi Patritii Mantuani Centones ex Virgilio*, ed. A. Possevino, (Rome, Valerius Doricus: 1555).

Caro Annibale, *Lettere scelte dalle familiari del commendatore Annibal Caro*, 2 vols. (Verona: 1820).

Caro Annibale, *Opere*, ed. U.A. Amico (Florence: 1864).

Carrara E., *La poesia pastorale* (Milan: 1909).

Carey C. – Kanellou M. – Petrovic I. (eds.), *Greek Epigram from the Hellenistic to the Early Byzantine Era* (Oxford: 2019).

Carmina reperta in Bibliotheca Cardinalis Commendoni quae ex legato R.P.D. Antonii Cauchi (Cocco) Commendoni eius Nepotis pervenerunt ad Stephanum Benassium. Et creduntur composita ab eodem Cardinale (Rome, Archivio Storico della Pontificia Università Gregoriana).

Caruso C., "Poesia umanistica di villa", in Crivelli T. – Caruso C. (eds.), *Feconde venner le carte. Studi in onore di Ottavio Besomi*, 2 vols. (Bellinzone: 1997) 272–294.

Chamard Henry, *Joachim Du Bellay, 1522–1560* (Geneva: 1978).

Chiflet Jean, *Aqua Virgo fons Romae celeberrimus, et prisca religione sacer* (Antwerp, Balthasar Moretus: 1652; Officinia Plantiniana: 1662).

Choix historique des médailles des papes (Paris: 1839).

Ciampoli Giovanni Battista, *Rime di Monsignor Giovanni Ciampoli* [...] (Rome, Carbelletti: 1648).

Coffin D., *Gardens and Gardening in Papal Rome* (Princeton: 1991).

Coffin D., "The 'Lex Hortorum' and Access to Gardens of Latium during the Renaissance", *The Journal of Garden History* 2.3 (1982) 201–232.

Coffin D., *The Villa in the Life of Renaissance Rome* (Princeton: 1979).

"Colocci, Angelo", *Dizionario biografico degli italiani* 27 (1982), accessible online.

Commendone Giovanni Francesco, *Discorso sopra la corte di Roma*, ed. C. Mozzarelli (Rome: 1996).

Coolidge J., "The Villa Giulia: A Study of Central Italian Architecture in the Mid-Sixteenth Century", *The Art Bulletin* 25 (1943) 177–225.

Cooper R., *Litterae in tempore belli. Études sur les relations littéraires italo-françaises pendant les guerres d'Italie* (Geneva: 1997).

Christian K.W., "Poetry and 'Spirited' Ancient Sculpture in Renaissance Rome", in Kenda B. (ed.), *Aeolian Winds and the Spirit in Renaissance Architecture* (London: 2006).

Christian K.W., *Empire Without End. Antiquities Collections in Renaissance Rome c. 1350–1527* (New Haven – London: 2010).

Croce B., "Un poeta latino poco noto. Francesco Franchini", *Quaderni della Critica* 16 (1950) 39–55.

Davis C., "Villa Giulia e la 'Fontana della Vergine'", *Psicon* 8.9 (1976) 131–141.

Davis C., "Per l'attività romana del Vasari nel 1553: incisioni degli affreschi di Villa Altoviti e la Fontanalia di Villa Giulia", *Mitteilungen des Kunsthistorisches Institutes in Florenz* 23 (1979) 197–224.

De Beer S., "The *Pointierung* of Giannantonio Campano's Epigrams: Theory and Practice", in De Beer S. – Enenkel K.A.E. – Rijser D. (eds.), *The Neo-Latin Epigram. A Learned and Witty Genre* (Leuven: 2009) 137–163.

De Beer S. – Enenkel K.A.E. – Rijser D. (eds.), *The Neo-Latin Epigram: A Learned and Witty Genre* (Leuven: 2009).

De Callataÿ D., "Prologue. Gratuité et grâce", *Revue du MAUSS* 35.1 (2010) 53–61.

De Caro G., "Capilupi, Ippolito", *Dizionario Biografico degli Italiani* 18 (1975).

"Degli orti di papa Giulio III", *Il Saggiatore Giornale Romano* 3 (1844) 91–92.

Del Pino González E., "El poema *Ad statuam puellae* de Cardenal Viseu y la 'Vergine della fontana' de Villa Giulia en Roma", *Ágora. Estudos Clássicos em Debate* 18 (2016) 157–191.

De Nolhac P., *La bibliothèque de Fulvio Orsini* (Paris: 1887).

Dewez G., *Villa Madama: A Memoir Relating to Raphael's Project* (London: 1993).

Deswarte-Rosa S., "Francisco de Holanda et le Cortile di Belvedere", in Andreae B. – Pietrangeli C. – Winner M. (eds.), *Il Cortile delle statue. Der Statuenhof des Belvedere im Vatikan* (Mayence: Rome, 1998) 389–410.

Deswarte-Rosa S., *La Rome de D. Miguel da Silva (1515–1525)* (Lisbon: 1988).

Deswarte-Rosa S., *"Il perfetto cortegiano" D. Miguel da Silva* (Rome: 1989).

Deswarte-Rosa S., "Sous la dictée de la *Sibylle*. Épigraphie et Poésie. Un exemplaire des *Epigrammata Antiquae Urbis* annoté par André de Resende et Francisco de

Holanda", in González Germain G. (ed.), *Peregrinationes ad inscriptiones colligendas. Estudios sobre epigrafía de tradición manuscrita* (Barcelona: 2016) 73–134.

Dickinson G., *Du Bellay in Rome* (Leiden: 1960).

Divers poemes de I. Du Bellay, in *Les Œuvres francoises de Ioachim Du Bellay* [...] (Paris, Fédéric Morel: 1569).

Du Bellay Joachim, *Epigrammata* [42], *Poematum libri quatuor* [...] (Paris, Fédéric Morel: 1558).

Du Bellay Jean, *Poemata*, trans. G. Demerson – R. Cooper (Paris: 2007).

Elet Y., *Architectural Invention in Renaissance Rome: Artists, Humanists, and the Planning of Raphael's Villa Madama* (Cambridge: 2017).

Espluga X., "La primera tradición textual de *Carmina Latina Epigraphica vi* 3*e (1)", in Martínez C.F. – Belén A.L. – Gómez Pallerés J. – Del Hoyo Calleja J. (eds.), *Ex officina. Literatura epigráfica en verso* (Sevilla: 2013) 121–155.

Falk T., "Studien zur Topographie und Geschichte der Villa Giulia in Rom", *Römisches Jahrbuch für Kunstgeschichte* 13 (1971) 103–178.

Fehrenbach F., "Nymph and Corpuscle. Transformations of the Aqua Virgo", in Böhme H. – Rapp R. – Rösler W. (eds.), *Übersetzung und Transformation*, Transformationen der Antike 1 (Berlin: 2007) 455–474.

Ferro Giovanni, *Teatro d'imprese* (Venice, G. Sarzina: 1623).

Fracastoro Girolamo, *Girolamo Fracastoro. Latin Poetry*, trans. J. Gardner (Cambridge: 2013).

Franchini Francisci, *Francisci Franchini cosentini poetae elegantissimi poemata* [...] (Basel, Petrus Perna typographus [...]: 1558).

Frommel S., "Fontainebleau", in Tuttle R. (ed.), *Jacopo Barozzi da Vignola* (Milan: 2002) 129–138.

Frommel S., "Vignola in Francia: un episodio incerto del suo itinerario artistico", in Affanni A.M. – Portoghesi P. (eds.), *Studi su Jacopo Barozzi da Vignola* (Rome: 2011) 215–236.

Gambara Lorenzo, *Laurentii Gambarae Brixiani Caprarola* (Rome, apud Franciscum Zanettum: 1581).

Gioccarello Anselmo, *Le triomphali Feste e giostre fatte nell'inclitta Città di Bologna con le publicazione d'un altra giostra che se havrà da fare di Santa Croce di maggio prossimo per la creation di M.S. Papa Giulio iii in Bologna* (n.p.: 1550).

Gnoli D., "Orti letterari nella Roma di Leone X", *Nuova antologia* (1930) 3–19.

Gratiani Antoine Maria, *La vie du cardinal Jean François Commendon* (Paris, Chez Sébastien Mabre-Carmoisy: 1671).

Gratianus Antoine, *De vita Joannis Francisci Commendoni Cardinalis libri iv* (Paris: 1669).

Hess J., "Amaduzzi und Jenkins in Villa Giulia", *English Miscellany* 6 (1955) 175–204.

Hess J., *Kunstgeschichtliche Studien zu Renaissance und Barock*, 2 vols (Rome: 1967).

Houghton L.B.T.H., *Virgil's Fourth Eclogue in the Italian Renaissance* (Cambridge: 2019).

Hurtubise P., *Tous les chemins mènent à Rome. Arts de vivre et de réussir à la cour pontificale au XVIe siècle* (Ottawa: 2009).

Iannuzzi I., "Tra Portogallo e Roma: note per un profilo di Achille Stazio", in Andretta E. – Valeri E. – Visceglia M.A. – Volpini P. (eds.), *Tramiti: figure e strumenti della mediazione culturale nella prima età moderna* (Rome: 2015) 167–195.

IJsewijn J., "Achilles Statius, a Portuguese Latin Poet in Late 16th Century Rome", in *Humanismo português na época dos descobrimentos: congresso internacional Coimbra 9 a 12 de outubrou de 1991* (Coimbra: 1993) 109–123.

Jacob M. (ed.), *Des jardins & des livres*, exh. cat., Fondation Martin Bodmer (Geneva: 2018).

Janthesius Caspar, *Itinerarium* (Rhete: 1637).

Jestaz B., "Benvenuto Cellini et la Cour de France: (1540–1545)", *Bibliothèque de l'École des Chartes* 161.1 (2003) 71–132.

Karmon D., "Restoring the Ancient Water Supply System in Renaissance Rome: The Popes, the Civic Administration, and the Acqua Vergine", *The Waters of Rome* 3 (2005), accessible online.

Krappe A.H., "Les Charités", *Revue des Études Grecques* (1932) 155–162.

Kurz O., "Huius Nympha Loci: A Pseudo-Classical Inscription and a Drawing by Dürer", *Journal of the Warburg and Courtauld Institutes* 16 (1953) 171–177.

Lafon J.B.M., *Pasquino et Marforio. Les bouches de marbre de Rome* (Paris: 1877).

Lagomarsini Girolamo, *Dimostrazione dell'autenticità dei carmi contenuti in* APUG *1219 attribuiti al Card. Giovan Francesco Commendone* (Rome, Archivio Storico della Pontificia Università Gregoriana, 1768).

Lanciani R., "Il Codice barberiniano XXX, 89 contenente frammenti di una descrizione di Roma del secolo XVI", *Archivio della Società romana di Storia patria* 6 (1883).

Lattès S., "Recherches sur la bibliothèque d'Angelo Colocci", *Mélanges d'archéologie et d'histoire* 48 (1933) 8–344.

Lavocat F., *La syrinx au bûcher. Pan et les satyres à la Renaissance et à l'âge baroque* (Geneva: 2005).

Leroux V., "L'érotisme de la belle endormie", *Seizième* siècle 7 (2011) 15–35.

Les Épigrammes de Maxime Planude, ed. – trans. I. Taxidis (Berlin: 2017).

Les voyages de Seyfried Rybisch, étudiant silésien: Itinéraire (1548–1554), Quatrième partie, ed. J. Hiernard (Pessac: 2017).

Letarouilly P., *Édifices de Rome moderne* [...] (Paris: 1840–1857).

Liebmann M., "On the Iconography of the Nymph of the Fountain by Lucas Cranach the Elder", *Journal of the Warburg and Courtauld Institutes* 31 (1968) 434–437.

Lipka M., *Language in Vergil's Eclogues* (Berlin: 2001).

Macdougall E.B., "The Sleeping Nymph: Origins of a Humanist Fountain Type", *The Art Bulletin* 57.3 (1975) 357–365.
Mai A., *Spicilegium Romanum*, 10 vols. (Rome, Typis collegii urbani: 1842).
Maleux M., "On Hebrew Nymphs and Aqueducts", *Lias* 47.1 (2020) 67–87.
Manzoli D., "La villa e le forbici. Villa Giulia in un centone di Lelio Capilupi", *Studi romani* 60.1–4 (2012) 57–85.
Manzoli D., "Ville e palazzi di Roma nelle descrizioni latine", *Studi romani* 56 (2008) 109–166.
Marichal R., "Le dernier séjour de Rabelais à Rome", *Comptes rendus des séances de l'Académie des Inscriptions et Belles-Lettres* 4 (1980) 686–697.
Mattei F., "L'invenzione dell'antico alla corte dei Gonzaga. Lelio Manfredi e la descrizione della villa di Lucullo a Tusculum", *Mitteilungen des Kunsthistorischen Institutes in Florenz* 60.1 (2018)107–126.
Maurach G. – Echiger-Maurach C. – Töns U. (eds.), *Studien zu Faustus Sabaeus, "Epigrammatum Fausti Sabaei Brixiani Custodis Bibliothecae Vaticanae Libri Quinque", Rom 1556, mit Auszügen aus dem Text*, Fontes 31 (Heidelberg: 2009).
Maurach G. – Echinger-Maurach C. (eds.), *Herme oder Terme? Texte und Bilder zur Klärung der Begriffe: aus: Achilles Statius, Inlustrium viror[um] ut extant in urbe expressi vultus, Romae 1569/Formis Antonij Lafrerj*, Fonte 89 (Heidelberg: 2021).
Meiss M., "Sleep in Venice: Ancient Myths and Renaissance Proclivities", *Proceedings of the American Philosophical Society* 110.5 (1966) 348–382.
Mendonçà R., "Classical Sculpture Reproductions and Portuguese Gardens", *Gardens & Landscapes of Portugal* 2 (2014) 60–69.
Miralles Maldonado J.C., "Un poema inédito de Aquiles Estaço", *Euphrosyne. Revista de filologia clássica*, 29 (2001) 389–408.
Morel P., "L'eau et le vin à la villa Giulia", in Cazzato V. – Roberto S. – Bevilacqua M. (eds.), *La festa delle arti. Scritti in onore di Marcello Fagiolo per cinquant'anni di studi*, 2 vols. (Rome: 2014) 956–961.
Mutini C., "Capilupi, Lelio", *Dizionario biografico degli italiani* 18 (1975).
Nova A., *The Artistic Patronage of Pope Julius III (1550–1555): Profane Imagery and Buildings for the De Monte Family in Rome* (New York – London: 1988).
Occhipinti C., "Fausto Sabeo e le arti figurative. Tra Vasari e Michelangelo", in Agosti B. – Ginzburg S. – Nova A. (eds.), *Giorgio Vasari e il cantiere delle Vite del 1550* (Venice: 2013) 47–59.
Oldradi Angelo degli, *Aviso della pace tra la Santita di N.S. Papa Paolo IIII. e la Maestà del Re Philippo: con la narratione del diluuio che è stato in Roma, con le gran ruine de ponti, chiese, palazzi, vigne* [...] (Rome: n.p., 1557).
Orsi Aurelio, *Perettina, sive Sixti V, horti exquilini* (Rome, J. Martinellum: 1588).
Palladio Blosio, *Coryciana*, ed. J. IJsewijn (Rome: 1997).

Panofsky E., *Hercules am Scheidewege* (Leipzig: 1930).

Pataki Z.A., *'Nympha ad amoenum fontem dormiens'* (*cil vi/5, 3*e*): *Ekphrasis oder Herrscherallegorese ? Studien zu einem Nymphenbrunnen sowie zur Antikenrezeption und zur politischen Ikonographie am Hof des ungarischen Königs Matthias Corvinus* (Stuttgart: 2005).

Pliny the Elder, *Natural History*.

Pressouyre S., "Les fontes de Primatice à Fontainebleau", *Bulletin Monumental* 127.3 (1969) 223–239.

Ramalho A.D.C., *Estudos sobre o século xvi* (Paris: 1980).

Ramalho A.D.C., "Notas sobre a formação de Aquiles Estaço", *Biblos* 54 (1978) 239–252.

Ribouillault D., "De la peinture au jardin (en passant par la poésie): la vallée Giulia à Rome, de Michel-Ange à Poussin", in Brunon H. – Ribouillault D. (eds.), *De la peinture au jardin* (Florence: 2016) 43–92.

Ribouillault D., "Jeux de mots et d'images: les frises peintes de l'appartement de Jules III au Vatican (vers 1550–1553)", in Fenech Kroke A. – Lemoine A. (eds.), *Frises peintes. Les décors des villas et palais au Cinquecento* (Paris – Rome: 2016) 102–129.

Ribouillault D., "La villa Giulia et l'âge d'or augustéen", in Morel P. (ed.), *Le miroir et l'espace du prince dans l'art italien de la Renaissance* (Paris: 2012) 339–388.

Ribouillault D., "Toward an Archaeology of the Gaze. The Perception and Function of Garden Views in Italian Renaissance Villas", in Beneš M. – Lee M.G. (eds.), *Clio in the Italian Garden: Twenty-First Century Studies in Historical Methods and Theoretical Perspectives* (Washington D.C.: 2011) 203–232.

Ribouillault D., *Rome en ses jardins. Paysage et pouvoir au XVIe siècle* (Paris: 2013).

Ribouillault D., "Le Salone de la Villa d'Este à Tivoli: un théâtre des jardins et du territoire", *Studiolo* 3 (2005) 65–94.

Ribouillault D., "De la 'Vigna Poggio' à la villa Balestra: 'Una delle più pittoresche e interessanti vedute di Roma'", in Fagiolo M. – Mazza A. (eds.), *I Monti Parioli e il "Nuovo Campo Marzio della cultura internazionale"* (Rome: 2022) 110–123.

Ricci M.T., "La grâce et la *sprezzatura* chez Baldassar Castiglione", *Bibliothèque d'Humanisme et Renaissance* 65.2 (2003) 233–248.

Rinne K.W., "Between Precedent and Experiment: Restoring the Acqua Vergine in Rome (1560–1570)", in Roberts L.L. – Schaffer S. – Dear P. (eds.), *The Mindful Hand: Inquiry and Invention from the Late Renaissance to Early Industrialisation* (Amsterdam: 2007) 95–116.

Salviani Ippolito, *Aquatilium animalium historiae, liber primus: cum eorumdem formis, aere excusis* (Rome, Apud eundem Hippolytum Saluianum: 1554).

Scritture spettanti alla città di Roma. Tomo ii. Quale contiene le notizie concernenti la città medesima, suo governo, stato economico, entrate, provedimenti per le inondazioni del Tevere, tasse, composizioni, spese del Palazzo Pontificio et altro, Rome, Biblioteca Vallicelliana, cod. Vall. G. 51 (1740).

Simeoni Gabriele, *Les illustres observations antiques du seigneur Gabriel Symeon florentin. En son dernier voyage d'Italie l'an 1557* (Lyon, Jean de Tournes: 1558).

Springhetti A., "Ioannes Franciscus Commendone (1524–1584). Legatus Pontificius et Cardinalis, Poëta Latinus", *Archivum Historiae Pontificiae* 8 (1970) 215–237.

Stoichita V.I., *L'effet Pygmalion. Pour une anthropologie historique des simulacres* (Geneva: 2008).

Steuco Agostino, *De revocanda in Urbem Aqua Vergine*, in *Opera omnia quae extant omni, e veteribus bibliothecis* [...] (Paris: Michaëlem Sonnium, 1578).

Striedl H., "Hebräische Lobgedichte des Andreas Masius auf die Aqua Virgo im Nymphaeum der Villa Giulia in Rom. (Staatsbibliothek Preussischer Kulturbesitz. Ms. or. fol. 13, Bl. 2–10)", ed. Franke H. et al., *Folia rara. Wolfgang Voigt* LXV. *diem natali celebranti / ab amicis et catalogorum codicum orientalium conscribendorum collegis dedicate* (Stuttgart: 1976), 126–133.

Ruvoldt M., *The Italian Renaissance Imagery of Inspiration: Metaphors of Sex, Sleep, and Dreams* (Cambridge: 2004).

Skinner Q., *The Foundations of Modern Political Thought: Volume 1, The Renaissance* (Cambridge: 1978).

Spagnolo M., *Pasquino in piazza. Una statua a Roma tra arte e vituperio* (Rome: 2019).

Tella Bartoli M., "A Proposito di Aquile Estaço e dei Carmina del codice Vallicelliano B. 106", *Annali dell'Istituto Universitario Orientale. Sezione Romana* 17 (1975) 293–362.

Tesoroni D., *Il Palazzo Firenze e l'eredità di Balduino del Monte* (Rome: 1889).

Tschäpe R., *Die Villa Giulia: Rekonstruktion des Bildestandes und ein Versuch, die Strukturen des ikonographischen Bestandes zu fassen* (Aachen: 1995).

Tucker G.H., "Érotisme, parodie, et l'art du centon dans le *gallus* (1543; *centones ex virgilio*, 1555) de Lelio Capilupi", in Sacré D. – Papy J. (eds.), *Syntagmatia. Essays on Neo-Latin Literature in Honour of Monique Mund-Dopchie and Gilbert Tournoy* (Leuven: 2009) 329–344.

Tucker G.H., "La Poétique du centon, une poétique de la Silve ?", in Galand P. – Laigneau S. (eds.), *La Silve: histoire d'une écriture libérée en Europe de l'Antiquité au* XVIIIe *siècle*, Latinitates 5 (Turnhout: 2013) 525–564.

Tucker G.H., "La description de la villa d'Este à Tivoli chez M.-A. Muret (*Poematia*, 1575), dans la mouvance des *Silves* de Stace et de Politien", *Camenae* 22 (2018) accessible online.

Tucker G.H., "A Roman Dialogue with Virgil and Homer: Capilupi, the *Cento* and Rome", in Caruso C. – Laird A. (eds.), *Italy and the Classical Tradition. Language, Thought and Poetry 1300–1600* (London: 2009) 204–238.

Tucker G.H., "*Roma instaurata* en dialogue avec *Roma Prisca*. La représentation néolatine de Rome sous Jules III (1553–1555) chez Janus Vitalis, Joachim Du Bellay et Lelio Capilupi (de l'ekphrasis à la prosopopée)", *Camenae* 2 (2007) accessible online.

Tuttle R. (ed.), *Jacopo Barozzi da Vignola* (Milan: 2002).

Ubaldini Federico, *Vita Angeli Colotii episcopi Nucerini* (Rome, Typis Michaelis Herculis: 1673).

Vagenheim G., "Pirro Ligorio et 'l'histoire secrète' de la restauration de l'Acqua Vergine sous le pontificat de Pie IV (1559–1565)", in Calvelli L. (ed.), *La falsificazione epigrafica. Questioni di metodo e casi di studio*, Antichistica 25 | Storia ed epigrafia 8 (2019) 263–286.

Vasari Giorgio, *Le Vite de' più eccellenti pittori, scultori ed architettori*, ed. G. Milanesi, 9 vols. (Florence: 1878–1885; 1981).

Virgil, *Aeneid*.

Virgil, *Georgics*.

Vitruvius, *Vitruvius iterum et Frontinus a Iocundo revisi repurgatique* [...] (Florence, F. Giunta: 1513).

von Pastor L., *Histoire des papes depuis la fin du Moyen-Âge*, trans. A. Poizat, 21 vols. (Paris: 1931), vol. 13.

von Reumont A., *Geschichte der Stadt Rom*, 4 vols. (Berlin: 1870).

Wellington Gahtan M., "Appended Epitaphs", in Steiner-Weber A. – Enenkel K.A.E. (eds.), *Acta Conventus Neo-Latini Monasteriensis. Proceedings of the Fifteenth International Congress of Neo-Latin Studies (Münster, 2012)* (Leiden: 2015).

Zanchi Basilio, *Basilij Zanchij Bergomatis Poematum libri VIII*; *Laurentii Gambarae Brixiani Poematum libri III* (Basel, Ioannem Oporinum: 1555).

Zeiller Martin, *Topographia Italiae* (Frankfurt: 1688).

CHAPTER 7

The "Academic" Plane Tree in Early Modern Gardens

Denis Ribouillault

Abstract

The iconological method pioneered by Erwin Panofsky and the intellectual circle indebted to Aby Warburg was first applied to the history of gardens in 1960 in David Coffin's famous monograph on the Villa d'Este at Tivoli. However, this approach, which focuses largely on architecture and sculpture, tends to marginalize the study of plants, flowers, and trees, which are often treated separately from the literary and poetic culture that provides a framework for understanding the meaning and function of the Renaissance pleasure garden; as if the natural sciences and literary culture were independent of each other. This essay attempts to bridge this gap by examining the presence and cultural significance of the plane tree (Platanus Orientalis *L.*) in early modern gardens, where it was considered a true Platonic emblem and 'living antique'.

Keywords

Plane tree – Platanus Orientalis L.– Plato – Gardens – Academy – Natural Philosophy

> Delle frondi del Platano coronarono il Genio gli Antichi, e fù detto Arbor Geniale per questo: & hebbe il nome di Platano per la larghezza della foglia, come Platone, per la larghezza degli homeri.
>
> From the leaves of the plane tree the ancients crowned the Genius, and it was therefore called the Genius Tree: & it was called Platanus because of the breadth of its leaves, like Plato because of the breadth of his writing.
>
> PONA FRANCESCO, *Il paradiso de' fiori overo Lo Archetipo de' Giardini* (Verona, Angelo Tamo: 1622) 15.

 | DOI:10.1163/9789004517547_008

'A plant is rarely just a plant. It is also the vegetal incarnation of society's cultural, political and social values'.[1] This phrase, written by a scholar of the gardens of ancient Rome, also applies to the early modern garden. Yet, paradoxically, plants and trees occupy a somehow marginal place in most literature on early modern gardens. There seems to be a division of sorts between the history of gardens and the history of gardening or horticulture, which logically should not exist.[2] The study of the properties of plants, their names, their history, their origin or their use often takes second place to considerations relating to the architecture or iconography of gardens, as articulated through statuary. This may be explained by the fact that most garden historians have a background in architectural history and, to a lesser extent, in art history and literature. Few among them are biologists or professional gardeners. Lise-Margot Dumargne thus perceives in the current disciplinary field of garden and landscape history a duality between the human sciences and the natural sciences.[3] An interdisciplinary and collaborative approach to the garden – or, to borrow Hervé Brunon and Monique Mosser's expression, a more "holistic" approach – is therefore highly desirable, for it may impact not only the preservation of plants, but also the restoration and understanding of historic gardens and landscapes.[4]

In this paper, the plane tree is located at the intersection of the natural and the cultural, being both a living being and the product of a rich culture. My intention is to show that as an "element" in the garden, like a fountain or a statue, the tree can also be subjected to art historical analysis: a "plant iconology" that does not simply examine the symbolism of plants or their mythology or consider plants and trees as mere 'background' or *parerga* but also as lives.[5] In other terms, art history, and even more so garden history, should be

---

1 Macaulay-Lewis E., "Imported Exotica: Approaches for the Study of the Ancient Plant Trade", in Macaulay-Lewis E. – Gleason K.L. (eds.), *The Gardens of the Ancient Mediterranean: Cultural Exchange Through Horticultural Design, Technology, and Plants* (Rome: 2010) 16–26.

2 Of course, several important books have sought to redress this imbalance by focusing on the planting of gardens, for instance Lazzaro C., *The Italian Renaissance Garden* (New Haven – London: 1990) especially chapter 2, "The Planting Reconstructed" or Laird M., *A Natural History of English Gardening, 1650–1800*, (New Haven – London: 2015).

3 Dumargne L.M., "Jardiner à l'épreuve de la dualité sciences humaines / sciences naturelles," *Intermediality* 35 (2020).

4 Brunon H.– Mosser M., "L'enclos comme parcelle et totalité du monde: pour une approche holistique de l'art des jardins", *Ligeia, dossiers sur l'art* 73.76 (2007) 59–75.

5 Michel Weemans and I have stressed the need to reconsider the relationship between landscape elements and figures in early modern landscape representations, the former traditionally interpreted as subordinate to the latter. Ribouillault D. – Weemans M., "Paysage sacré, Livre de la Nature, Exégèse: pour une reconception du paysage dans l'Europe de la première

concerned with such 'living works of art'.[6] The use of live animals and plants in contemporary art installations is now a common occurrence and is receiving increasing attention from scholars. This is but one manifestation of the desire to rethink the long-established hierarchical relationship and separation between man and nature – a paradigm shift driven by the current ecological crisis.[7]

What kind of relationship did the visitor to the garden have with the plane tree? Can it be described as a 'respectful friendship', as Hervé Brunon puts it?[8] More prosaically, the intention of this article is to provide some bibliographical material and begin a survey of the place and significance of plane trees in early modern gardens without claiming to be exhaustive. Since antiquity, the plane tree (*Platanus orientalis L.*) had prestigious associations and, as we will see, was primarily associated with power and eastern luxury on the one hand, and sacredness, philosophical discourse, and poetic inspiration on the other. Hence, its presence in Renaissance gardens contributed to the ideals of aristocratic leisure and academic activities that took place there. To understand how a tree became a literary *topos* and then a Renaissance garden *topos*, one must first return to ancient Greek literature.

1 The Plane Tree in Ancient Sources

Specific mentions of *Platanus orientalis L.* appear in a wide range of ancient Greek sources, from epic and mythological literature to medical science, history, and philosophy. In Homer's *Iliad* (II, 307), the Greeks sacrificed at Aulis,

modernité", in Ribouillault D. – Weemans M. (eds.), *Sacred Landscape. Landscape as Exegesis in Early Modern Europe* (Florence: 2011) ix–xxxi.

6 Ferrari G.R., "The Meaninglessness of Gardens", *The Journal of Aesthetics and Art Criticism* 68.1 (2010) 34: 'when gardening is art, its elements are lives – the lives of plants'.

7 For a contemporary perspective, see, amongst a rich bibliography, Bruce A. – Chandès H. – Couton P.E. – Gaudefroy I. (eds.), *Nous Les Arbres*, exh. cat., Fondation Cartier pour l'art contemporain (Paris: 2019). For a specific example, the work of French artist Céleste Boursier-Mougenot, see Brunon H., "Comme par enchantement", in Lavigne E. (ed.), *Céleste Boursier-Mougenot: rêvolutions*, exh. cat., La Biennale di Venezia 2015, Pavillon français de la 56e exposition internationale d'arts visuels (Paris: 2015) 6–27, and Ramade B., "Ex-Situ", in Bianchini S. – Quinz E. (eds.), *Behavioural Objects 1, A Case Study: Céleste Boursier-Mougenot* (London – Berlin: 2016) 161–178. For a stimulating approach to art history that includes the analysis of the living, see Zhong Mengual E., *Apprendre à voir: le point de vue du vivant* (Paris: 2021).

8 Brunon H., "Amitiés respectueuses. Pour une archéologie de la relation jardinière", *Jardins* 6 (2015) 19–38.

FIGURE 7.1 Ernest Board, *Hippocrates lecturing to his students under the plane tree on the island of Cos*, *c.*1929 (?), oil on canvas, present location unknown
IMAGE © THE WELLCOME COLLECTION

on the way to Troy, 'on the holy altars round the spring [...] beneath a lovely plane-tree, whence bright water ran'.[9] Galen and Dioscorides mention the therapeutic qualities of the plane tree in their medical works and, according to legend, Hippocrates, the father of medicine, taught his students under a large plane tree on the island of Cos [Fig. 7.1]. The therapeutic qualities of the tree were still appreciated at the beginning of the nineteenth century, as the French physician Pierre Joseph Buc'hoz reported after a trip to Iran: 'The Oriental plane tree [...] contributes not a little to the purity of the air and to the healthiness of this country. It is regarded as a real safeguard against the plague and all kinds of airborne corruption. It is assured that Isfahan has not felt the contagion since its streets, gardens, and surroundings were filled with these kinds of trees'.[10] Theophrastus combines the scientific and the mythological when he mentions a plane tree growing near a spring in Gortyna, Crete, under whose branches Zeus is said to have seduced Europa.[11] He specifies that this specimen was always green, whereas the others lost their foliage during the year.

9 Forster E., "Trees and Plants in Homer", *The Classical Review* 50.3 (1936) 99.
10 Buc'hoz P.J., *Dissertations sur le cèdre du Liban, le platane et le cytise* (Paris, Madame Buc'hoz: 1806) 24.
11 Theophrastus, *Historia Plantarum* 1.9.5.

Today, an evergreen plane is placed near the current archaeological site where the story is told to tourists. The Greek historian Herodotus reports a similar sacred fascination with the tree in the Persian Empire. Darius received a gift of a 'golden plane tree' from the Lydian Pythius (VII, 27) and Xerxes, crossing Asia to invade Greece, saw a plane tree by the roadside near the city of Kallatebos so magnificent that he gave it gold jewels and appointed a custodian to look after it (VII, 31).[12] Andrea Sacchi dedicated a beautiful drawing to this subject, which served for an emblematic thesis print of the Roman Seminary, dated 1632.[13] More famously, this episode opens Georg Friedrich Haendel's 1738 opera *Serse* (from the libretto by the Venetian Nicolò Minato) with the recitativo *Fronti tenere e belle / del mio platano amato*, followed by the famous aria *Ombra mai fu*.[14] In Theocritus (18.48), Helen of Troy is presented as a tree goddess, a shady plane on whose bark girls will carve the words 'Adore me; I am the tree of Helen'. These first examples testify to an almost amorous relationship with the tree, with a clear religious dimension.

On the island of Cos, a statue of Theocritus's master, the poet and scholar Philitas of Cos, was placed under a plane tree, in connection with a gymnasium and sanctuary of the Muses. This configuration corresponded well with the two famous Athenian gymnasiums known as the Academy and the Lyceum: the philosophical schools directed respectively by Plato and Aristotle, which were organised as religious societies based on the cult of the Muses.[15] There, the plane trees planted around the gymnasiums offered shade and a natural environment to rest from physical activity and converse.[16] As we

---

12 Stubbings F.H., "Xerxes and the Plane-Tree", *Greece & Rome* 15.44 (1946) 63–67.

13 In the engraving, the phrase "CRESCENT ILLAE CRESCETIS HONORE", taken from Virgil's tenth Eclogue, has been carved on the plane tree. On this print and its meaning, see Rice L., "Pomis sua nomina servant: The Emblematic Thesis Prints of the Roman Seminary", *Journal of the Warburg and Courtauld Institutes* 70 (2007) 195–199. On the Virgilian passage, see Lee R.W., *Names on Trees: Ariosto into Art* (Princeton: 1977) 9–11.

14 Two earlier versions existed, by Francesco Cavalli (1655) and Giovanni Bononcini (1694).

15 On Plato's Academy, see especially Baltes M., "Plato's School, the Academy", *Hermathena* 155 (1993) 5–26.

16 Hardie A., "Philitas and the Plane Tree," *Zeitschrift für Papyrologie und Epigraphik* 119 (1997) 21–36, with further references and bibliography. Links between the Muses, a grove of plane trees, and a library appear in a Greek epigram engraved on a herma from Herculaneum (*CIG*, 6186) found in Rome in 1878. See Curtius E.R., *European Literature and the Latin Middle Ages* (Princeton – Oxford: 2013) 306. More references on the plane tree and trees in culture in general are, amongst a vast literature, Li H.L., *The Origin and Cultivation of Shade and Ornamental Trees* (Philadelphia: 1963) 56–84; Pontoppidan A., *Le platane* (Arles: 1996); Demandt A., *Über allen Wipfeln. Der Baum in der Kulturgeschichte* (Cologne: 2002); Corbin A., *La douceur de l'ombre. L'arbre, source d'émotions, de l'Antiquité à nos jours* (Paris: 2014).

know, the beginning of Plato's *Phaedrus* (229a–b, 230b) brings all these elements together: Socrates and Phaedrus choose a 'charming resting place' outside the city to discuss the merits of rhetoric in the shade of a very extensive and high plane tree, near a stream of fresh water sacred to the nymphs and a grotto of the Muses. This setting is not, however, a simple mise-en-abyme of the Athenian Academy. The plane tree in the prologue of *Phaedrus* puns on the author's proper name, both words being derived from the Greek *platys*, which means "broad." The fortune of this text is immense and there is no doubt that the pun was understood early, as a remark of Timon of Phlius seems to show unequivocally.[17]

The importance of the plane tree in the culture of Persian gardens is largely attested to, and it also garnered a deep fascination among the Romans who decided to bring it from the East to Rome to adorn the great gardens of the aristocracy.[18] In a crucial passage on trees in the *Naturalis Historia*, Pliny the Elder comments with some disdain that the plane tree was initially imported into Italy solely for its shade (*H.N.* 12.6) and not for its fruit, emphasising its status as a luxury item strongly associated with royal or imperial power. Indeed, plane trees, like other exotic plants, were part of the spoils brought back to Rome by victorious generals. 'It is a remarkable fact [Pliny writes,] that ever since the time of Pompey the Great even trees have figured among the captives in our triumphal processions [...]'. Pompey the Great, who paraded living trees during his triumphs, had plane trees planted in his famous portico garden in Rome.[19] The plane tree is one of many plants whose history is linked to

17 Zaslvsy R., "A Hitherto unremarked Pun in the *Phaedrus*", *Apeiron* 15 (1981) 115–116. On the interpretation of this text, see, among a large bibliography, Motte A., "Le pré sacré de Pan et des Nymphes dans le *Phèdre* de Platon", *L'antiquité classique* 32.2 (1963) 460–476; Marder M., "Plato's Plane Tree", in Marder M. (ed.), *The Philosopher's Plant* (New York: 2012) 1–18.

18 A'lam H., "*Čenār*, the Oriental plane (tree), *Platanus orientalis* L. (fam. Platanaceae)", *Encyclopædia Iranica* 5.2 (1990) 129–133; Potts D.T., "Arboriculture in Ancient Iran: Walnut (*Juglans regia*), Plane (*Platanus orientalis*) and the '*Raddle dictum*'", *dabjïr* 6 (2018) 101–109.

19 On this passage and this topic, see especially Kuttner A.L., "Culture and History at Pompey's Museum", *Transactions of the American Philological Association* 129 (1999) 343–373; Macaulay-Lewis E., "The Fruits of Victory; Generals, Plants, and Power in the Roman World", in Bragg E. – Hau L.I. – Macaulay-Lewis E. (eds.), *Beyond the Battlefields: New Perspectives on Warfare and Society in the Graeco-Roman World* (Newcastle: 2008) 205–225 and Marzano A., "Roman Gardens, Military Conquests, and Elite Self-Representation", in Coleman K. (ed.), *Le jardin dans l'Antiquité*, Entretiens sur l'Antiquité classique de la Fondation Hardt 60 (Geneva: 2014) 195–244.

colonialism.[20] In the Roman world, the tree was thus soon connected with the extravagance and luxury associated with the East. The emperor Caligula is said to have transformed a plane tree of enormous size into a dining room accommodating fifteen guests in his so-called "nest". In Lycia, the Roman governor of the province, Gaius Licinius Mucianus, held a banquet for nineteen people in a hollow plane tree, saying that 'he took much more delight to lie within the [...] cabin, than in a stately chamber built of marble'. Pliny's text had a great influence, and Islamic and Italian gardens of the early modern era tried to imitate these legendary trees. In his treatise on trees (*c.*1590), Gianvettorio Soderini mentions a hollow chestnut tree in a place called Silano, into which Pope Paul III Farnese 'could enter [...] with his parasol and his mule, without hitting himself on any side'.[21] Religious versions of such tree architecture still exist today, such as the church of Panagia Plataniotissa in Greece. In the Medici villa of Castello, Niccolò Tribolo arranged and decorated a large oak tree in this way (*c.*1538). Equipped with water jets, it had a square platform accessible by stairs with seats and a marble table with a fountain in the middle. Thirty years later, a similar version was built in the Villa Medici of Pratolino [Fig. 7.2]. It is quite possible that the designers of those Medici gardens, in addition to the well-known chapter by Pliny, were also inspired by a passage from Francesco Colonna's famous *Hypnerotomachia Poliphili* (1499), whose impact on the real gardens of the Renaissance is well known. Francesco Colonna describes an arboreal architecture made up of artfully woven branches of citron, orange, and lemon trees with their fragrant flowers: a palace of greenery with an entrance door and stairs.[22] A contemporary miniature painting from Mughal

20 There now exists an entire research field on "colonial botany" or "botanical imperialism".

21 Soderini Giovanvettorio, "Il trattato degli arbori", ed. A. Bacchi Della Lega, *Opere*, vol. 2 (Bologna: 1903) 569: '[...] sì come papa Pagolo [sic.] III Farnese in un pedale roso d'un Castagno in quel di Silano, con l'ombrello e con la mula si potè raggirare, senza urtare punto da alcuna banda, similmente con comodità in quel platano Licinio Muziano tre volte banchettò con diciotto compagni agiatamente, ponendovisi a tavola nel vuoto di quel pedale, accomodandosi di tutto quello che gli occorse, per riposarsi e sedersi quietamente, con l'abbondanza delle sue foglie e rami, i quali tanto oltre per la campagna si distendevano, che si poteva sotto essi largamente passeggiare e fare esercizio [Just as Pope Paul III Farnese was able to enter the hollow trunk of a chestnut tree in Silano with his parasol and his mule, without hitting himself on any side, so Licinio Muziano three times feasted with eighteen companions in the comfort of this hollow plane tree outfitted with a table and everything needed to rest and sit quietly within its abundant leaves and branches that extended so far outside that it was possible to walk and exercise under their cover]'. The treatise was composed between ca. 1588 and 1596.

22 Colonna Francesco, *Hypnerotomachia. The Strife of Loue in a Dreame*, trans. R. Dallington (London, Simon Waterson: 1592) fol. 45v: 'Vnto which enclosure when wee came, I founde it equilaterall, with three fences like a streight wall, as high as the Cyprus Trees vpon either

FIGURE 7.2 Stefano Della Bella, *The Tree House at Pratolino*, ca. 1653. Engraving, 25 × 36.8 cm. New York, The Metropolitan Museum of Art
IMAGE © MET

India, dated 1588, shows two men eating and drinking whilst discussing poetry on an elaborate wooden platform built atop a large plane tree, reminding us of the profound impact of ancient Roman sources and the Roman world in general on Islamic gardens [Fig. 7.3]. Several similar structures are found in

---

sides of the waye, that wee had passed along in: which was altogither of Cytrons, Orenges and Lymonds, bushing with their leaues one within an other, and artifitially knitte and twisted togither, and the thicknes mee thought of sixe foote: with a Gate in the middest of the same Trees, so wel composed as is either possible to bee thought or done. And aboue in conuenient places were made windowes, by meanes whereof, the bowghes in those places were to be seene bare, but for their greene leaues which yeelded a most sweet and pleasant verdure. Betwixt the curious twistings of the braunches and their greene leaues, the white flowers did aboundantly shewe themselues a singular Ornament, breathing foorth a most delectable and sweete odour. And to please the eye, the faire fruite was in no place wanting, where it should yeelde content. And afterwardes I might perceiue, that in the interstitious thicknes, the bowghes (not without a wonderful woorke) were so artificially twisted and growne togither, that you might assend vp by them, and not bee seene in them, nor yet the way where you went vp'. The original passage is in Colonna Francesco, *Hypnerotomachia Poliphili* (Venice, Aldus Manutius: 1499) fol. 88.

FIGURE 7.3
Basawan (attributed to), *Anvari Entertains on a Tree-top Platform, Folio 109b from a manuscript of the Divan of Anwari, copied for Emperor Akbar* (*r.* 1556–1605), 1588. Ink, opaque watercolour, and gold on paper, 13.8 × 7.5 cm. Cambridge, Harvard Art Museums, no. 1960.117.109
IMAGE © HARVARD MUSEUMS

Mughal paintings, but it is unclear whether they correspond to actual tree pavilions designed in Mughal gardens.[23]

23 On this miniature, see Schimmel A., *Anvari's Divan: A Pocket Book for Akbar* (New York: 1983) 77–80. Another closely related image is a folio representing *Eskandar (Alexander the great) in a tree pavilion*, ca. 1610–1620, at the Aga Khan Museum in Toronto. This scene may derive from an earlier iconography known from several Indian miniature paintings of the second half of the sixteenth century representing Mughal emperor Humayun (d. 1556) in a tree pavilion with a younger companion (often identified as Humayun's heir, the future emperor Akbar). See Melikian Chirvani A.S., *Le Chant du monde: L'art de l'Iran safavide, 1501–1736* (Paris: 2007) 436–437; ibidem (ed.), *Architecture in Islamic Arts.*

In ancient Rome, it was the academic plane tree that enjoyed the greatest fortune both in literature and in actual gardens. The first famous plane tree described by Pliny is one in Plato's Academy, whose roots 'extend far beyond its branches'. In Cicero's *De Oratore* (1.7.28), a plane tree in the gardens of Crassus' villa in Tusculum triggers the memory of Plato's plane tree:

> Crassus, why don't we imitate Socrates, as he appears in Plato's *Phaedrus*? For this plane tree of yours has brought this notion to my mind. It is broad, with spreading branches, and is not less suited to covering this place in shade than that tree whose shadow Socrates sought – a tree that seems to me to have grown not so much because of the stream that is described as because of Plato's words. And what that man, with his very well-hardened feet, did – casting himself down on the grass and thus speaking things which the philosophers say were spoken divinely – surely can be conceded to *my* feet with greater justice.[24]

This is a good example of the mnemonic function of a garden topos, and Cicero's well-known definition in *De finibus* (5.1) sums it up well: 'To be in a place where famous men lived or worked moves us more than reading of them or their deeds and for this reason place is the basis of the art of memory'. The plane tree, the vegetal embodiment of Plato himself and irrevocably linked to the Athenian academy and the *Phaedrus*, functions in the same way. In the Roman estate, the plane-tree-as-Plato-as-Academy becomes the past-but-living witness to the philosophical discussions that take place under its shade. The tree also appears in the poetry of Statius (*Silvae* 2.3), Martial (*Ep.* 9.61), in the agricultural writings of Cato (*De agr.* 133.2), and in Vitruvius (5.11.4), who compares plane trees to the *ambulationes* of the Romans, which offer shade and coolness in summer. The idea that walking in the shade of plane trees, inherited from the Athenian Academy, is beneficial to both body and soul would become central in Renaissance and Baroque gardens and is mentioned, for example, by the physician Girolamo Mercuriale in the early 17th century.[25]

Treasures of the Aga Khan Museum, exh. cat., Aga Khan Trust for Culture (Geneva: 2012) 250–251.

24 For an interpretation of this passage, see Stull W., "Deus ille noster: Platonic Precedent and the Construction of the Interlocutors in Cicero's 'De oratore'", *Transactions of the American Philological Association* 141.2 (2011) 250.

25 On Statius' and Martial's poems, see Newlands C.E., "Martial, Epigrams 9.61 and Statius, Siluae 2.3: Two Branches from the Same Tree?", *Scholia* 20 (2011) 92–109 and Coleman K., "Melior's Plane Tree. An Introduction to the Ancient Garden", in Coleman (ed.), *Le jardin dans l'Antiquité* 1–26. On Cato on the plane tree (and many other Greek and Latin

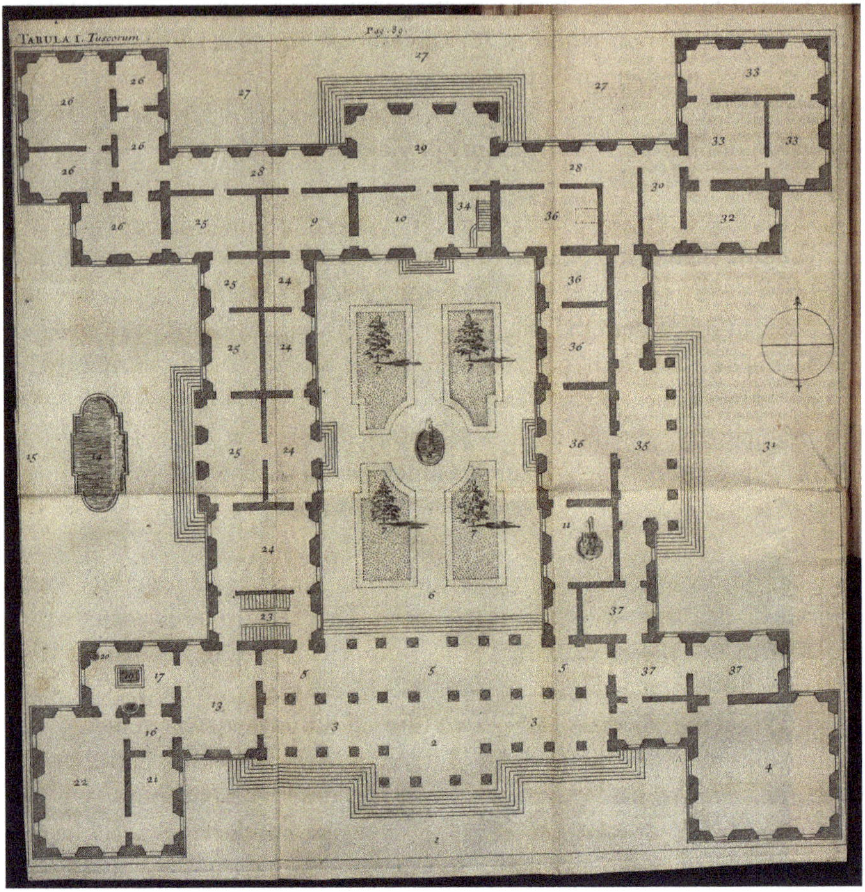

FIGURE 7.4 Jean-François Félibien des Avaux, *Les plans et les descriptions de deux des plus belles maisons de campagne de Pline le consul* [...] (Paris, Florentin & Pierre Delaulne: 1699) 89
IMAGE © IDAI / ARACHNE

A mosaic from Pompeii shows to what degree the *topos* had also taken root in visual culture. The image displays a group of philosophers with objects intimately linked to Plato's academy: the sundial associated with the astronomer Eudoxus of Cnidus, the plane tree for Plato, a portico with vases, and

sources), see Schievenin R., "Sotto il platano", *Incontri di filologia classica* 12 (2012–2013) 239–256. On Vitruvius and plane trees in aristocratic Roman gardens, see Grimal P., *Les Jardins romains* (Paris: 1969) 171–187. On Girolamo Mercuriale's mention of plane trees in Plato's academy in his *De arte gymnastica libri sex* (Venice, Giunta: 1601) 37–38, see Gage F., "Exercise for Mind and Body: Giulio Mancini, Collecting, and the Beholding of Landscape Painting in the Seventeenth Century", *Renaissance Quarterly* 61.4 (2008) 1183.

the circular bench or *schola* [see Fig. 1.1 of the introduction].[26] When Pliny the Younger (*Ep.* 5.6) had a small courtyard in his Tuscan villa 'shaded by four plane trees' with, among them, 'a marble fountain, from which the water plays upon and lightly sprinkles the roots of the plane trees and the grass plot beneath them', he was obviously reproposing the *topos* of the Platonic Academy mise en abyme in Plato's *Phaedrus* [Fig. 7.4]. In the Renaissance, this was considered obvious, as is again made clear in Pirro Ligorio's entry on academies in his great *Enciclopedia del mondo antico*: 'Tra gli altri Atrij, uno vi era piantato di Platani come era nell'Accademia Platonica' ('Among the other entrance halls, one was planted with plane trees, as in the Platonic Academy').[27]

2 The Plane Tree in Renaissance Sources

Unlike classical studies where the plane tree motif has been analysed almost *ad nauseam*, studies devoted to *Platanus Orientalis* L. in the early modern period are rare. As far as I know, the transplantation of the prestigious memories of the Platonic academy on Italian soil first appear in Pietro Bembo's *De Aetna* published at the Aldine Press in Venice in 1496. In this dialogue between the narrator and his son, plane trees are used as metaphors for the ancient Platonic and Aristotelian academies, and the question of their cultivation on Italian soil takes a new importance. They recall that after their introduction in Italy, first by way of Sicily, plane trees abounded in many cities – Bernardo remembers seeing a magnificent specimen near Rome by the lake of Diana of Ariccia – though the carelessness of men left them to wither. This sums up the project of the humanist academies of the Renaissance, namely, to revive the knowledge of the Ancients which had been gradually neglected and forgotten. As such, the *De Aetna* has been considered by scholars as a model of the humanist transmission of knowledge.[28] Marsilio Ficino's analysis of the setting of the *Phaedrus* in his complex commentary written *c.*1466–1468 is surprisingly dry in comparison, stating the obvious: 'Among the ingenious aspects of the dialogue are the following: the description of the spot stands

26 Rashed M., "La mosaïque des philosophes de Naples: une représentation de l'académie platonicienne et son commanditaire", in Noirot C. – Ordine N. (eds.), *Omnia in uno. Hommage à Alain-Philippe Segonds* (Paris: 2013) 36–37.

27 Ligorio Pirro, *Enciclopedia del mondo antico* (Turin, Archivio di Stato) Taur. 20, fol. 42v.

28 Bembo Pietro, *De Aetna* (Rome, Aldus Manutius: 1496). See Williams G., *Pietro Bembo on Etna: The Ascent of a Venetian Humanist* (New York: 2017).

allegorically for the Academy; the plane tree for Plato'.[29] However, the Platonic pastoral appears frequently in the work of Florentine humanists thereafter, as in Cristoforo Landino's *Disputationes Camaldulenses*, composed between 1472 and 1474, where Leon Battista Alberti, Marsilio Ficino, Donato Acciaiuoli, Lorenzo de' Medici, and other humanists retire to a place in the forest where, 'in a flowery meadow, a clear spring was shaded by the wide-spread branches of a mighty beech'. 'My brave friends', said Battista, 'the tree here and the water trickling from the spring with a sweet murmur call to your mind the plane tree and the stream of Socrates'.[30] Even known critiques of the pagan revival such as Girolamo Savonarola appear affected by the mystical Platonism of Ficino, although Christian exegesis of the plane tree and its shade already existed, particularly in that of Augustine.[31] The plane tree on the outskirts of Florence, the setting for his *Dialogus de veritate prophetica* (1497), is yet another avatar of Plato's *locus academicus* in the *Phaedrus*.[32] In 1513, Gianfrancesco Pico della Mirandola, describing the Belvedere Garden of Julius II at the Vatican in Rome, uses the same image of Plato's sacred wood in his *De Venere et Cupidine expellendis carmen*, this time again with a strong critical stance towards Neoplatonism and paganism.[33]

In the sixteenth century, interest in the plane tree took an encyclopedic form. The earliest account on the tree I have found, relying heavily on Pliny's *Natural History* both in form and content, is a chapter of the monumental *Lectiones antiquae* by the Ferrarese professor of Greek and Latin Caelius Rhodiginus (in Italian Lodovico Ricchieri), published in Venice in 1516. The frontispiece of the 1517 Basel edition shows portraits of poets, philosophers, and historians of

29 'Inter haec artificiosissima loci descriptio allegorice signat Academiam, platanus Platonem'. See Ficino Marsilio, *Commentaries on Plato, Phaedrus and Ion*, ed. M.J.B. Allen, vol. 1 (Cambridge MA – London: 2008) 38–41.

30 Landino Cristoforo, *Disputationes Camaldulenses*, ed. P. Lohe (Florence: 1980). The passage is quoted in Biermann H., "War Leon Battista Alberti je in Urbino?", *Zeitschrift für Kunstgeschichte* 65.4 (2002) 501.

31 On Augustine and Christian exegesis of the plane tree, see Schievenin, "Sotto il platano" 251–253. Also, see chapter 49 of Lemnius Levinus, *An Herbal for the Bible* (n.p: 1579).

32 Savonarola Girolamo, *Dialogus de veritate prophetica* (Venice, Bartolomeo de' libri: 1507) fol. 3v. See Wind E., *Pagan Mysteries in the Renaissance* (New Haven: 1958) 154. On the use of the *locus amoenus* in Renaissance dialogues, see Kushner E., "Le rôle structurel du 'locus amoenus' dans les dialogues de la Renaissance", *Cahiers de l'AIEF* 34 (1982) 39–57. The *topos* of the Platonic plane tree also appear in Moravus Augustinus, *Dialogus in defensionem poetices* (Venice, Paul Fridenperger: 1493).

33 Pico della Mirandola Gianfrancesco, *De Venere et Cupidine expellendis carmen* (Rome, Iacobus Mazochius: 1513). For an English translation and a commentary, see Piana M., *Fallax Antiquitas: Gianfrancesco Pico della Mirandola's Critique of Antiquity* (Ph.D. diss., McGill University: 2017) 217–284, 335–339.

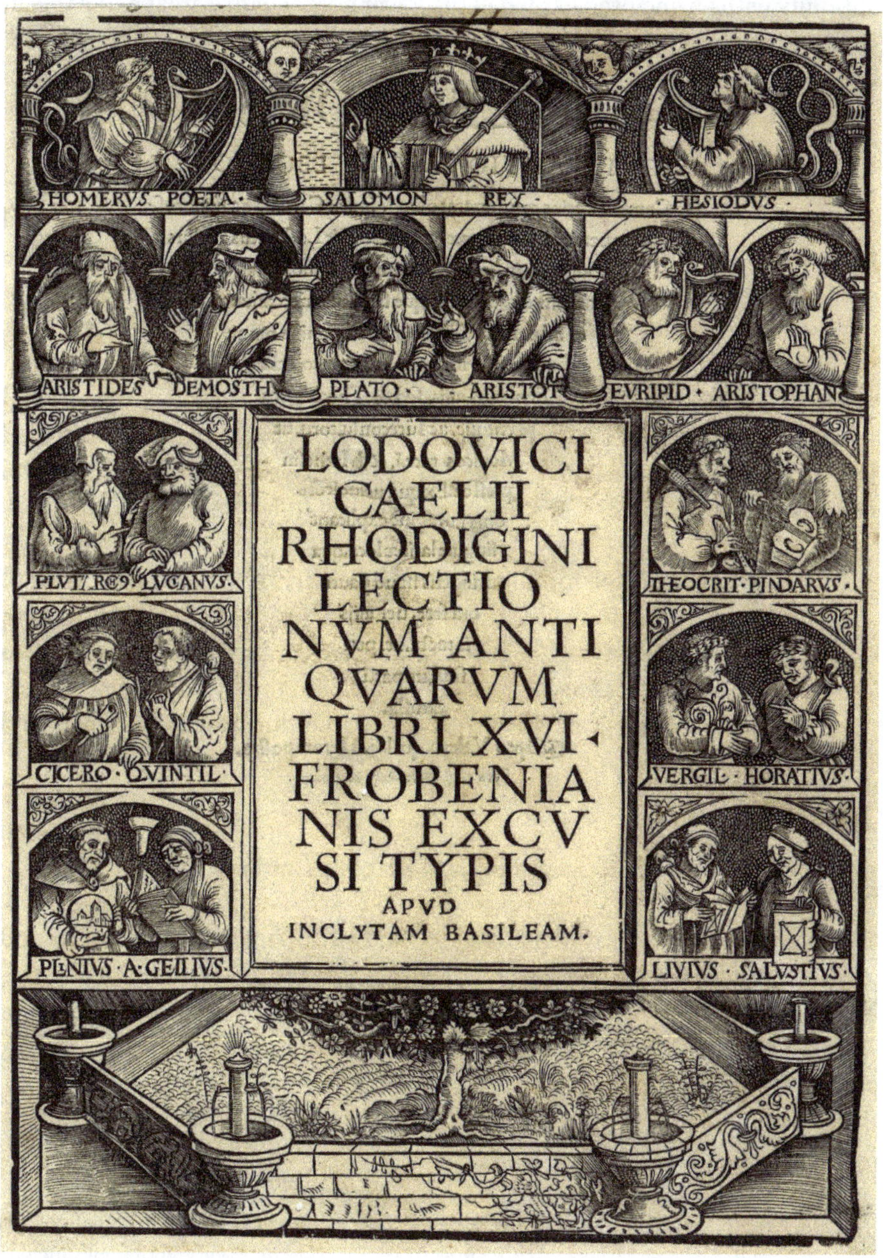

FIGURE 7.5 Lodovico Ricchieri, *Lectionem Antiquarum Libri XVI* (Basel, J. Froben: 1517) frontispiece. London, British Museum, no. 1895,0122.1077

antiquity with an enclosed garden in the lower border with a great plane tree in the middle [Fig. 7.5].[34] Although Pliny still plays a vital role in it, a much more ambitious treatise on the plane tree, the *Liber de Platano*, was composed in 1537 by the Portuguese humanist João Rodrigues de Sá de Meneses and shows how entangled the history of European humanism was with the *Platanus orientalis*. Two editions of the manuscript, dedicated to Luìs Teixeira, a former pupil of Angelo Poliziano, have recently been published in Portuguese and German.[35] The pretext for this fascinating work is the demonstration of the existence of the tree in Portugal, a roundabout way of legitimising and claiming the development of the study of Greek and Latin in Portuguese territory, as the title of the first chapter indicates: 'The plane tree plucked from the darkness and brought to the light. Where it is shown that it existed among us and other things about it'.[36] Taking part in this dialogue is another great Portuguese humanist, Don Miguel da Silva, bishop of Viseu and a soon-to-be cardinal, who confirms that plane trees were still in existence in Italy:

> I saw the plane tree in Italy, more precisely in Florence in the house of Giovanni Rucellai, an illustrious citizen, a very rich man and a great scholar. Finding myself in Tuscany, he showed me the plane tree in his garden, a very pleasant place, and he declared that when he saw it grow pale and dry, he sprinkled it with wine, according to Pliny's recipe, and immediately the tree regained its strength, usual vigor and brilliance.[37]

3 Plane Trees in Renaissance and Baroque Gardens

Let us now turn to real trees planted in gardens, not only for their beauty and the shade they created, but also for their cultural prestige and their status as true "living antiques". The Orti Oricellari of Florence were famous for their

34 Rhodigini Ludovico Caelii, *Lectionum antiquarum* (Frankfurt – Leipzig, Danielis Fieveti: 1566) 1375–1377. See Pignatti F., "Ricchieri, Lodovico Maria", *Dizionario biografico degli italiani* 87 (2016).

35 De Meneses João Rodrigues de Sà, *Die Platane. Liber de platano (1537)*, eds. R. Friedlein – A. Lozar (Leiden: 2008); Tarrío A.M.S., *Paisagem e erudiçao no humanismo português. João Rodrigues de Sá de Meneses – De Platano (1527–1537)* (Lisbon: 2009). Both books include bibliographies on plane trees in the Renaissance.

36 Ibidem 232: 'Eruta a latebris et in lucem producta Platanus. Ostensumque eam apud nostrates hodie reperiri, deque ea inibi nonnula'.

37 Ibidem 24–31, 237.

collection of ancient sculptures. Because they seem to have been designed as a wood or *selva* (a word that has deep literary associations, recalling Petrarch,[38] Bembo (*Asolani* 1.5) and, of course, Horace's well-known expression 'Atque inter silvas Academi quaerere verum ('And seek truth in the groves of the Academe')' (*Ep.* 2.45)), the gardens were an appropriate choice for a place notoriously linked to Ficino's Platonic academy. The Rucellai plane tree is celebrated in a poem about the garden written circa 1504 by Petrus Crinitus. In the spirit of Horace and of the *Silvae* of his master Poliziano, it is a true catalogue of trees, appreciated like the ancient statues displayed in the garden through the prism of ancient poetry.[39] According to Machiavelli's famous Socratic dialogue *The Art of War* (1521), located in the Orti Oricellari, guests were gathered 'in the most secret and shaded part of the garden, where some of us would sit down on the grass, which in that place is cool and fresh, some of us on chairs placed in the shade under the huge trees'.[40] He also confirms that the numerous rare plants in the garden were intended to evoke the species mentioned in classical literature.

Another villa where the reverence for *Platanus orientalis* L. is linked with humanism is the villa of Paolo Giovio on Lake Como, 'entirely dedicated to the Genius and the Muses'. According to the *Descriptio Larii Lacus* of 1559 (xv), Paolo Giovio had a wood of plane trees planted in his garden on the spot where Pliny (*Ep.* 1.3) had planted one (*opacissima Plinii platanus*) [Fig. 7.6].[41]

In Rome, the famous nymphaeum of the villa Giulia was outfitted with special basements for four plane trees, in emulation of Pliny's Tuscan Villa [See Fig. 6.3 in chap. 6] – an idea that had in fact already been materialised several decades earlier in the apartment of Isabella d'Este in the Ducal Palace of Mantua, which was organised around a great, enclosed garden with four

38 On the strong association between Petrarch and the term *selva*, see Brunon H., "*Locus secretus*: topique et topophilie", in Luciani D. – Mosser M. (eds.), *Petrarca e i suoi luoghi. Spazi reali e paesaggi poetici alle origini del moderno senso della natura* (Treviso: 2009) 41–55.

39 On the Rucellai garden, see Comanducci R.N., "Gli Orti Oricellari", *Interpres* 15 (1995–1996) 302–355 and the references given in the introduction. On Crinito's poem, see especially Galand-Hallyn P., "Aspects du discours humaniste sur la villa au XVIe siècle", in Galand-Hallyn P. – Lévy C. (eds.), *La villa et l'univers familial de l'Antiquité à la Renaissance* (Paris: 2008) 133–136, with Latin text and French translation.

40 The English translation is taken from Gatti H., *Ideas of Liberty in Early Modern Europe* (Princeton: 2015).

41 On Giovio's villa, see Lauterbach I., "Gardens of the Milanese 'villeggiatura'", in Dixon Hunt J. (ed.), *The Italian Garden: Art, Design and Culture* (Cambridge: 1996) 132–136.

FIGURE 7.6 Detail of Paolo Giovio, *Larii lacus vulgo comensis descriptio*. Paris, BNF, Département Cartes et plans, GE DD-2987 (5137)
IMAGE © BNF/ GALLICA

plane trees: the so-called *Cortile dei quattro platani*.[42] Guillaume Philandrier, who was admitted to the Accademia della Virtù, mentions several plane trees in Rome in his commentary on Vitruvius' *De architectura*. Referring to Plato's academy and other ancient sources, he mentions the Vigna Salviati for its numerous plane trees, as well as a large and beautiful plane tree on the Pincio hill (*colle hortulorum*).[43] The Vigna Salviati was situated on the site of the former garden of Johann Goritz, between Trajan's column and the Campidoglio; a garden famous for its academic meetings, where one of the first examples of

42 Coffin D., *The Villa in the Life of Renaissance Rome* (Princeton: 1979) 163; Brown C.M., *Isabella d'Este in the Ducal Palace in Mantua* (Rome: 2005) 131–133.

43 Philandrier Guillaume, *M. Vitruvii Pollionis de Architectura Libri decem* (Lyon, Jean de Tournes: 1552) 212. The first edition dates from 1544. On Philandrier, see the many studies by Frédérique Lermele.

FIGURE 7.7 Frances Benjamin Jonhston, *Loggia delle Muse, Villa Lante a Bagnaia, summer 1925*. Hand-coloured photograph, 8.2 × 10.2 cm. Washington D.C., Library of Congress
IMAGE © LIBRARY OF CONGRESS

the iconographic type of the so-called "sleeping nymph" was to be found; it was to become a *topos* of the sixteenth-century garden.[44]

Another Plinian reference appears at the Villa Lante at Bagnaia, whose plane trees are the object of special praise by Cardinal Gianfrancesco Gambara's personal physician, Castore Durante, in his *Herbario nuovo* published in Rome in 1585.[45] Documents show that plane trees were initially planted in the formal garden in relation to the two *Casini* called *Logge delle Muse* located at the highest point of the garden. We show here an interesting hand-coloured photograph from 1925, reproduced as lantern slide by the American photographer Frances Benjamin Johnston (1864–1952) [Fig. 7.7]. Rows of plane trees were also arranged in one of the intermediary terraces built along the main axis,

44 For an overview and a bibliography on this topic, see Ribouillault's essay in this book, chap. 6.

45 Durante Castore, *Herbario Nuovo* (Rome, Bartolomeo Bonfadino: 1585) 366: 'In asia sono molto maggiori. Bellissimi sene vedono hoggi nel amenissimo Barco di Bagnaia, famosissimo luogho dell'Illustriss. & Magnanimo S. Cardinal Gambara, mio Signore'.

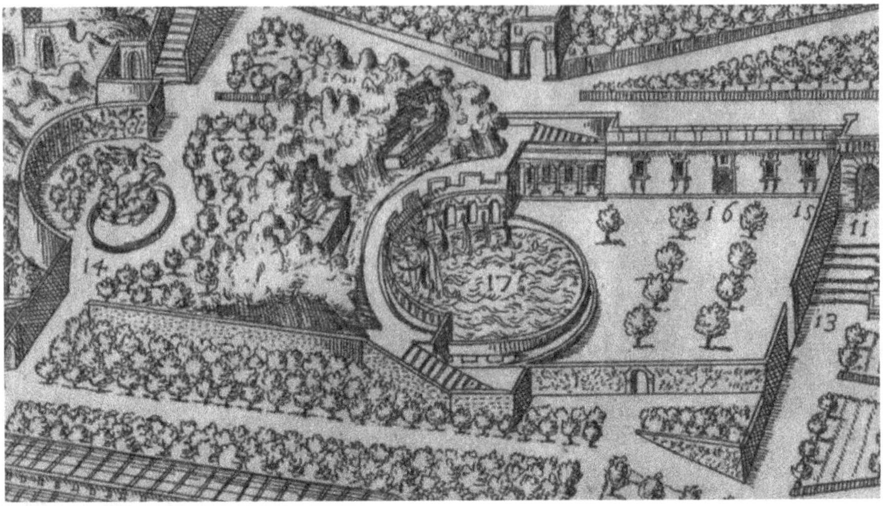

FIGURE 7.8 Giovanni Ambrogio Brambilla (attributed to), detail of the Tivoli Fountain. Etching, 36.7 × 52 cm, made in 1581 after an etching by Étienne Dupérac, *Speculum Romanae Magnificentiae: Tivoli Palace and Gardens*, 1573, 49 × 57.2 cm. From: *Speculum Romanae Magnificentiae* (Rome, Claudio Duchetti: 1581). New York, The Metropolitan Museum of Art, no. 41.72(3.65)

flanking a long table filled with running water, the so-called "Fountain of the Table", this time an evocation of Pliny the Younger's description of floating dinner plates at his villa in Tusci (*Ep.* 5.6).[46]

Also, particularly elaborate is the Tivoli Fountain at the Villa d'Este in Tivoli. There, ten plane trees were planted, corresponding to the ten statues of nymphs or muses placed in the niches of the semi-oval exedra of the large fountain – a Platonic motif further reinforced by the presence of a Parnassus above the fountain [Fig. 7.8]. Furthermore, its oval shape was certainly inspired by the ruins of the so-called Academy of Hadrian's Villa if we refer to their interpretation by Pirro Ligorio, the architect of the garden, who was one of the first to study them. For contemporary poets, the whole garden was celebrated as an ideal academy, as suggested by the French Latinist Marc-Antoine Muret or Ercole Cato who qualified Villa d'Este as 'Un'accademia, un cenacolo, un teatro

46 Alessi A., "La Palazzina Gambara: gli architetti e i pittori", in Frommel S. (ed.), *Villa Lante a Bagnaia* (Milan: 2005) 120–121 (note 13); Lazzaro, *The Italian Renaissance Garden* 254, 257–258, 262–265. In his description of the garden of 1578, Fabio Arditio also mentions plane trees in the garden of the Villa Farnese at Caprarola. A number of old plane trees can still be seen in the garden today.

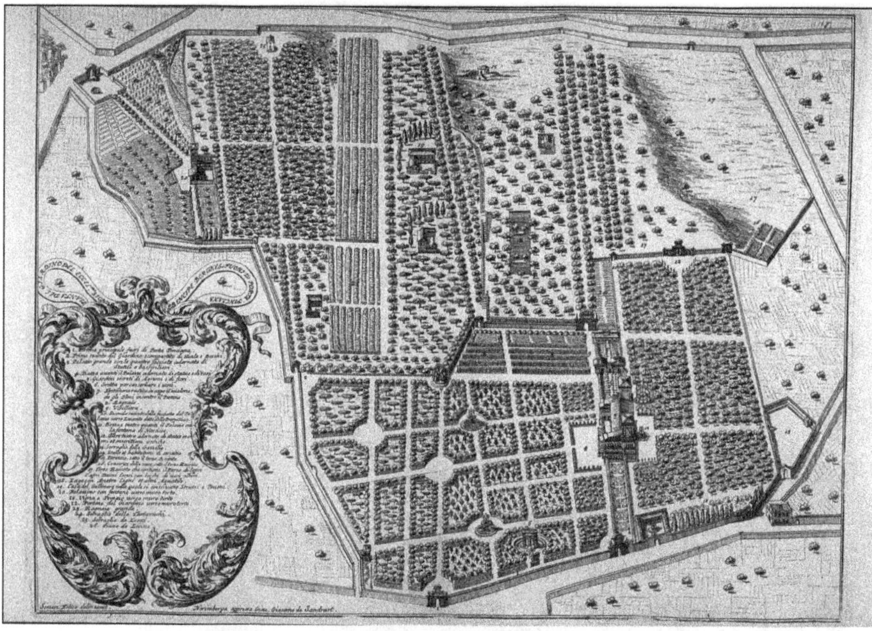

FIGURE 7.9 Giovanni Battista Falda, *Villa Borghese*. From: *Li Giardini di Roma* (Rome, Giovanni Giacomo de' Rossi: *c.*1680). Private collection. The *valle dei platani* is indicated n. 18
IMAGE © DENIS RIBOUILLAULT

del mondo pieno d'uomini singolari atti al governo di ogni repubblica' ('An academy, a cenacle, a theatre of the world full of remarkable men fit for the government of any republic').[47]

Plane trees continued to play a key role in Baroque gardens, especially in Rome. Perhaps less immediately linked with the idea of the academy, the exceptional arrangement of plane trees now called Valletta dei Platani in the great park of the Villa Borghese is worth mentioning. It was designed ca. 1620 for cardinal Scipione Borghese by his gardener Domenico Savino da Montepulciano around a quadrangular artificial lake (ca. 150 × 25 m), where swans and ducks had been introduced. Two small oval islands in the middle of the lake, each with two large plane trees, were used as a shelter for the water birds as can be seen in Giovan Battista Falda's view of the garden (no. 18) [Fig. 7.9]. Today,

47 On the Fountain of Tivoli, Ligorio and ancient academies, see the important chapter in Ranaldi A., *Pirro Ligorio e l'interpretazione delle ville antiche* (Rome: 2001) 98–122. See also the neoplatonic interpretation of the fountain in Desnoyers G., *La Villa d'Este à Tivoli ou Le songe d'Hippolyte. Un rêve d'immortalité héliaque* (Paris: 2002) 218–220.

FIGURE 7.10 A plane tree planted *c.*1620 in the park of Villa Borghese. Photograph. Rome
IMAGE © DENIS RIBOUILLAULT

at least eleven of the forty-eight trees originally planted for the cardinal are still alive. These great, green patriarchs, four hundred years old and said to be the oldest in Rome, have a unique ecological, cultural, social, and historical value [Fig. 7.10].[48]

48 Montelatici Domenico, *Villa Borghese fuori Porta Pinciana* (Rome, G.F. Guagni: 1700) 102–103; Campitelli A., *Villa Borghese. Storia e gestione* (Rome: 2005) 85. On trees at the Villa Borghese, see Beneš M., "The Social Significance of Transforming the Landscape at

In a chapter on plane trees of his *Ristretto delle piante*, published in 1697, Francesco Persio Sacconi discusses another interesting arrangement of sixteen plane trees planted around a fountain at the Villa Mondragone, property of the same cardinal Borghese at Frascati, which the author compares again with Mucianus's famous Lycian plane tree. Sacconi points out that the trees, planted ninety years before, had almost swallowed the stone benches that had been placed at their foot.[49] Frascati has several other interesting examples of old plane trees in the gardens of famous villas. At Villa Falconieri, there is an exceptional oak tree (*Quercus pubescens Willd.*), inscribed on the list of the monumental trees of Latium (2017). The students of the Latin academy that is now based there, the *Accademia vivarium novum*, are used to taking lessons in the open air under its shade, thus continuing a centuries-old tradition. A plane tree (*Platanus orientalis* L.) in the garden of Villa Aldobrandini is registered on the same list.[50] In the garden of Villa Sora, there was also a *Fontana dei Platani*,

the Villa Borghese, 1606–1630: Territory, Trees and Agriculture in the Design of the First Roman Baroque Park", in Petruccioli A. (ed.), *Gardens in the Time of the Great Muslim Empires: Theory and Design* (Leiden: 1997) 1–31.

49 Persio Sacconi Francesco, *Ristretto delle piante con suoi nomi antichi, e moderni, della terra, aria, e sito, ch'amano, consacrato alla S.C.R.M. Leopoldo I Imperatore* (Vienna, Andrea Heyinger: 1697) 78–80: 'Plinio lib. 12 cosi dice: furono i Platani prima celebrati quelli nella Loggia dell'Accademia d'Athene, dove crebbero in altezza più di trenta sei braccia; hora è un bellissimo Platano in Licia sù la strada sopra un fonte il quale à guisa d'un habitatione fà una spelonca d'ottanta un piede coprendosi, con molti rami i quali paiono Alberi, occupa I campi con lunghissima ombra, e accioche paia in tutto spelonca hà dentro à se un cerchio di sasso, che fa grotto, ed'abbraccia molti sassi carichi di musco; (à me non è di meraviglia sapendo, che gli stà à canto): alla Villa di Monte Dragone nel Parco del Eccell. Principe Borghese sono sedici Platani attorno ad'una fontana di grossezza l'abbracciar di due homini, al piede di ciascheduno vi stà un scabello di pietra da essi abbracciati, ch'apena si vedono, sono circa novant'anni, che furono piantati) [...]' ('In Book 12, Pliny says that plane trees were first celebrated in the loggia of the Academy of Athens, where they reached a height of more than thirty-six fathoms; there is a beautiful plane tree in Lycia, on a road above a spring, which forms, like a dwelling, a grotto eighty feet high, with many branches as large as trees; it casts a very wide shadow over the surrounding fields, and to make it appear in every way like a grotto, there are rocks arranged in a circle inside it, forming a grotto, and it is surrounded by many rocks covered with moss; (this does not surprise me, knowing that the tree is very near); at the Villa of Monte Dragone, in the park of his excellency the Prince Borghese, there are sixteen plane trees around a fountain, [and] it takes two men holding hands to surround it; at the foot of each of them there is a stone bench that you can hardly see [because it has been swallowed up by the tree over time]; they were planted about ninety years ago [...]').

50 See European Commission, "Elenco alberi monumentali della Regione Lazio", *The official portal for European data*, https://data.europa.eu/data/datasets/elenco-alberi-monumentali-della-regione-lazio?locale=it [accessed 13.01.2023].

FIGURE 7.11 Detail of the *Fontana dei Platani* in Giovanni Battista Falda, *The Orti Farnesiani*. From: *Li Giardini di Roma* (Rome, Giovanni Giacomo de' Rossi: *c*.1680)
IMAGE © DENIS RIBOUILLAULT

a beautiful fountain in the centre of a circle of large plane trees, decorated with eight statues of grotesque creatures pouring water into the basin.[51]

One garden in Rome shows much more obvious links to the memories of the ancient academies. Under Ranuccio II Farnese, another *Fontana dei platani* in the famous Orti Farnesiani on the Palatine was appropriately transformed into a theatre for the *Accademia degli Arcadi* between 1693 and 1699. The original fountain, with its octalobed basin surrounded by ten plane trees, is visible in the well-known engraving by Giovan Battista Falda (n. 8), above the *Fontana degli specchi* (n. 10) [Fig. 7.11]. Not only did this layout coincide well with the idea of the ancient academy, as shown by the aforementioned example of the oval fountain of the Villa d'Este in Tivoli, but the new Bosco Parrasio was also located in the place where Evander, king of the Arcadians, had chosen to establish his own residence in archaic times. The new theatre, in the form

51 On the garden, see Guerrieri Borsoi M.B., *Villa Sora at Frascati. Una villa romana fra '600 e '700* (Rome: 2000) 99–102. The fountain and the trees are visible in old postcards and in a painting of 1891 by the American impressionist painter Theodore Robinson.

FIGURE 7.12 The Arcadian Theatre in the Farnese Garden on the Palatine. From: Piovene Pietro, *I Cesari in metallo mezzano e piccolo raccolti nel Museo Farnese*, vol. 10 (Parma, Nella Stamperia di S.A.S.: 1727) 31
IMAGE © KATRINA GRANT

of a rotunda with two rows of seats, was shaded by the plane trees and solemnly inaugurated on 7 June 1693. So great was public excitement that 'many, not being able to see the theatre and hear the shepherds' songs in any other form, had occupied even the highest branches of the neighboring trees'. Here, the big plane trees were used as true theatre balconies![52] The best available image is an engraving by Stefano Zucchi showing the Fontana dei Platani with the theatre structure behind, published in Pietro Piovene's *I Cesari* [Fig. 7.12].[53] A drawing of the structure by Pier Leone Ghezzi, himself a member of the academy, is preserved at the Biblioteca Angelica.[54]

52 Crescimbeni Giovanni Mario, *L'Arcadia* (Rome, Antonio de' Rossi: 1711) 18; Boni G., "L'Arcadia sul Palatino", *Bollettino d'Arte* 8 (1914) 373; Dixon S.M., *Between the Real and the Ideal. The Accademia Degli Arcadi and Its Garden in Eighteenth-Century Rome* (Newark: 2006) 56–57.

53 Piovene Pietro, *I Cesari in metallo mezzano e piccolo raccolti nel Museo Farnese*, vol. 10 (Parma, Nella Stamperia di S.A.S.: 1727) 31.

54 Pier Leon Ghezzi, View of the Bosco Parrasio at the Orti Farnesiani, Biblioteca Angelica, Ms. 2136, fol. 164; published in Grant K., "The Bosco Parrasio as a Site of Pleasure and of Sadness", *Histoire culturelle de l'Europe* 3 (2019) 4, accessible online.

4 Plane Trees in Botanical Gardens and Natural Philosophy

Today, we are used to seeing plane trees everywhere in our parks and modern cities, but we must remember that during the Renaissance, plane trees were considered exceptional and exotic. They were therefore of particular interest to naturalists. Chapters on plane trees can be found in the famous works of Johannes de Cuba and Pietro Andrea Mattioli. In this respect, Mattioli is particularly indebted to the Dutch physician Willem Quackelbeen (1527–1561) who sent him information and pictures of plane trees from Constantinople.[55] Like Bembo or Sá de Meneses, Mattioli wanted to know if these large oriental trees were still visible in Italy and recalled those seen in Naples and in the botanical garden of Padua. A plane tree dating from around 1680 can still be seen there, but visitors in the nineteenth century have described others that may date from the sixteenth century.[56] The French naturalist Pierre Belon mentions an exceptional plane tree in the same city but places it rather 'in the garden of Cardinal Bembo'.[57] He also points out that the gardens of Rome were particularly valuable in this respect, and that one garden was named "Vigne du Platan" because all the other plane trees in Rome came from one of its trees.[58] In 1546,

55 Mattioli Pietro Andrea, *I Discorsi* [...] (Venice, Vincenzo Valgrisi: 1568) 148–149. Opsomer J.E., "Un botaniste trop peu connu Willem Quackelbeen (1527–1561)", *Bulletin de la Société Royale de Botanique de Belgique* 93 (1961) 1–2.

56 Valéry A.C., *Voyages en Italie*, vol. 7 (Paris: 1831) 7: 'Le jardin botanique de Padoue, fondé par le sénat de Venise en 1545, est le plus ancien qu'il y ait en Europe. Son emplacement n'a pas changé; un vieux platane oriental, au tronc noueux, aux courts rameaux, mais encore verdoyans, date de sa création. Je ne pouvais le contempler sans une sorte de respect; je trouvais quelque chose de docte à ce contemporain de tant de professeurs illustres, dont les statues de pierre sont à quelques pas, qu'il avait reçus sous son ombre, et il me semblait comme une espèce de doyen parmi les arbres savans des jardins botaniques' ('The Botanical Garden of Padua, founded in 1545 by the Venetian Senate, is the oldest in Europe. Its location has not changed; an old oriental plane tree, with a gnarled trunk and short but still green branches, dates from its creation. I could not look at it without a kind of respect; I found something learned in this contemporary of so many illustrious professors, whose stone statues are a few steps away, whom it had received under its shade, and it seemed to me like a kind of dean among the learned trees of the botanical gardens').

57 Belon Pierre, *Les remonstrances sur le default du labour & culture des plantes, & de la cognoissance d'icelles* (Paris, Gilles Corrozet: 1558), 77v. Bembo's plane tree came from Crete according to Dupuigrenet-Desroussiles F., "Regards et savoirs: images du jardin botanique de l'Université de Padoue au XVIe siècle", *Revue d'histoire des sciences* 42 (1989) 3, 281–291 (287).

58 Belon, *Les remonstrances*, 44v. On the presence of plane trees in Italy, see also his *Observations de plusieurs singularitez et choses memorables, trouvees en Grece, Asie, Judee, Egypte, Arabie & autre pays estranges* [...] (Paris, Hierosme de Marnef – la Veuve Guillaume Cavellat: 1588) 358.

Pierre Belon, sent to Turkey by Francis I to Suleiman the Magnificent, brought back to France a specimen found on the island of Cos which he tried to cultivate ca. 1558 in the episcopal domain of Tourvoie, one of the first botanical gardens in France, near Le Mans.[59] François Rabelais, a physician like Belon, was interested in new plant specimens when he visited Italy in 1534 and described the great plane tree of Lake Nemi that Bembo had already mentioned in the *De Aetna*.[60] A letter from the Italian writer Giovanni Francesco Arrivabene about the same tree is preserved in the archives of the greatest Italian botanist of the 16th century, Ulisse Aldrovandi, who identifies it with the famous dining tree of Emperor Caligula mentioned by Pliny.[61] The tree was considered so valuable

59 Belon, *Les remonstrances*, 14v.

60 "Plantas autem nullas, sed nec animantia ulla habet Italia, quae non ante nobis et visa essent et nota. Unicam Platanum vidimus ad speculum Dianae Aricinae." François Rabelais, dedicatory letter to Jean du Bellay in Marliani Bartolomeo, *Topographia antiquae Romae* (Lyon, Sébastien Gryphe: 1534). See Cooper R., *Rabelais et l'Italie* (Geneva: 1991) 105.

61 Letter of Giovanni Francesco Arrivabene (1524–1575) to Camillo Paleotti (1520–1594), 2 May 1590 in Aldrovandi Ulisse, *Bologna, Biblioteca Universitaria, Ms. Aldrovandi 136*, vol. 13 (Bologna: 1588–1590) fols. 257–259: 'Dirò brevemente che stessimo sul lago un giorno e mezzo fermi, dico in doi castelli posti sul istesso lago lontani doi miglia, che sono cosi ben posti, che rendano una maravigliosa vaghezza, et avevano ragione quei antichi a godersi così fatta amenità di paesi, quali sono come Giardini, et sariano paradisi terrestri, si fossero ben coltivati dall habitatori, che ponto non hanno talento di fatigarsi. In la Riva di questo lago vi è il Platano celebrato da Plinio, andassimo a vederlo. Il Platino [sic.] ha il tronco terribilissimo, e manda fuori doi rami grossissimi, quali vanno in aria et da questi doi rami principali, se umagni [sic.?] una gran quantità di altri rami che fanno parere come se fossero otto o dieci albori insieme, et dicono gli habitatori che quando ha interamente le foglie/ che rende molto più, perché ancora non era finito di vestirsi delle sue foglie ben bene. Dentro il tronco io vi entrai et certamente sariano stati a sedere a una tavola rotonda otto a dieci huomini, parlo commodamente, et perché quelli piscatore talvolta per la pioggia solevano ridursi a coperto e farvi fuogo, ché ancora si vede guasto. Gli sig.ri del lago vi hanno provisto con un bando sotto pena della vita à chi vi portasse fuogo' ('I will briefly say that we spent a day and a half on the lake in two castles on the same lake, two miles apart, which are so well situated that they offer a marvellous sight, and the ancients were right to enjoy the amenities of such places, which are like gardens and seem to be earthly paradises when well cultivated by the inhabitants who have no need to toil. On the shores of this lake stands the plane tree celebrated by Pliny, which we visited. The tree has an enormous trunk, and it sends out two very large branches, which go up into the air, and from these two main branches shoot a great number of other branches, so that it seems as if there were eight or ten trees together, and the inhabitants say that when it has all its leaves it is much more impressive, because it had not yet finished putting on its leaves. I entered the trunk and eight or ten men could certainly have sat around a round table, I mean comfortably, and because those fishermen used to take shelter from the rain and make a fire there, the traces of which can still be seen, the local lords have issued a law forbidding anyone to make a fire there, on pain of death'). I am indebted to

FIGURE 7.13 Richard Wilson, *The Arbra Sacra on the Banks of Lake Nemi*, c.1754–1756. Drawing, 38.9 × 55.9 cm. New Haven, Yale Center for British Art
IMAGE © YALE CENTER FOR BRITISH ART

that the local lords threatened to condemn anyone who destroyed it to death. Known as "Arbra Sacra", it continued to fascinate travellers in the Eighteenth century and is the subject of a beautiful drawing by Richard Wilson [Fig. 7.13].[62]

The story of the planting of the first plane trees in the gardens of Renaissance Northern Europe is a complex and fascinating story. It is not clear whether the first plane trees in France were those planted at Tourvoie by Belon, at the Château de Madrid near Paris, or at Fontainebleau. In the entry for the plane tree in the great *Encyclopédie* of Diderot and d'Alembert, it is said that after its introduction to France, 'On prétend que cet arbre fut ensuite apporté en France, où les plus grands seigneurs faisoient un si grand cas de son ombre, qu'on exigeoit un tribut des gens qui vouloient s'y reposer' ('It is said that this tree was later brought to France, where Lords made such a great deal of its

Ada Vittorina Segre for bringing this letter to my attention and sending me a transcription of the manuscript. The letter cannot be dated 1590 since Arrivabene died in 1575. Aldrovandi and Arrivabene were in Rome during the same period c.1550.

62 See also "Memoirs of Thomas Jones Penkerring Radnorshire 1803", *The Volume of the Walpole Society* 32 (1946–1948) 58, April 30, 1777, who mentions Wilson's drawing.

shade, that a fee was demanded from people who wanted to rest there').[63] But this is, again, a deformation of one of Pliny's anecdotes (*N.H.*, 12.3). Scholars consider 1562, the date of the second part of William Turner's *New Herball*, as the first record of plane trees cultivated in England.[64] John Gerard, *The Herball or General Histories of Plantes* (1597) describes the plane as 'strange in Italy, it is nowhere seen in Germane, nor in the low countries [...] it growth in many places of Greece, and is found planted in some gardens of Italy, for pleasure rather than for profite'.[65] Carolus Clusius first introduced the tree in Vienna in 1576.[66] However, the plane tree had become so fundamental to the imagination of humanists that even where it could not grow, it could be invoked for the creation of an academy. Erasmus writes already in this sense in his *Antibarbari*, completed in 1494:

> But seriously, my friends, since we have happened by chance on such a splendid object, why should we not set up an academy here on the model of Plato's? No matter if we have not a plane tree like the Platonic or Ciceronian ones [...] I am offering you real trees instead [...], a whole orchard which you can see near our house on the left, bordered by noble oaks and the clearest of streams. What if we water this place too with our discussions, so that it will never dry up at any future time? And under that great pear tree, practically in the middle of the garden, there are some comfortable seats [...]. If so weighty a philosopher as Socrates could be allured (as in the *Phaedrus*) by the pleasantness of the place to lie down in the grass and converse beside the little spring, why should we not be tempted to sit down here, by these gardens which Epicurus himself might praise?[67]

---

63 Duval C., "Introduction du Platane en France", *Bulletin de la Société Botanique de France* 43 (1896) 194–198; Daubenton Pierre, "Platane", in Diderot D. – D'Alembert J.l.R. (eds.), *L'Encyclopédie*, vol. 12 (Paris: 1751) 733.

64 Fay M.F. – Rix M., "857. Platanus orientalis. Platanaceae", *Curtis's Botanical Magazine* 34.1 (2017) 36.

65 Gerard John, *The Herball or General Historie of Plantes*, ed. The English Experience, vol. 2 (n.p.: 1974), 1304.

66 Clusius Carolus, *Rariorum plantarum historia* (Antwerp, Joannem Moretum: 1601) 9.

67 *Collected Works of Erasmus*, in Thompson C.R. (ed.), *Literary and Educational Writings 1. Antibarbari / Parabolae*, vols. 23–24 (Toronto: 1978) 39. See Bietenholz P., "Felicitas (eudaimonia) ou les promenades d'Érasme dans le jardin d'Épicure", *Renaissance and Reformation* 30.1(2006) 69–70.

5 Conclusion

In his *Iconologia* (Siena, 1613), Cesare Ripa, following an invention of his friend and collaborator Giovanni Zaratino Castellini, describes the allegory, archetype, or idea of the academy as a woman of mature age with a golden crown holding in her right hand a file, around the handle of which is written *Detrahit atquet polit* [Fig. 7.14]. In her left hand, she holds a garland of laurel, ivy, and myrtle, with a pair of pomegranate fruits. Her seat is adorned with foliage and fruits of cedar, cypress, and oak, as well as olive branches. At her feet, lay many books and a baboon, the sacred baboon of the Egyptian god of wisdom Thot. Ripa gives every single detail of this scene a most precise meaning, a true demonstration, at the very beginning of his book, of his mastery of visual exegesis and textual erudition. The place where the woman stands is now familiar to us:

> Sederà in mezzo d'un cortile ombroso, o vero luogo boscareccio di villa con platani intorno [...] per memoria della prima academia, che fu principiata in villa da un nobil personaggio chiamato Academo, nella cui amena villa, non lungi d'Atene, si radunavano I platonici [...] a discorrere de studii dilettevoli.
>
> She will sit in the middle of a shady courtyard, or a wooded area in a villa with plane trees all around [...] in memory of the first academy, which was begun in the villa of a nobleman called Academus, in whose pleasant villa, not far from Athens, the Platonists gathered [...] to discourse on delightful studies.[68]

Although a less intellectual and more sensual approach to the garden developed in the Eighteenth century that may have eroded the academic and philosophical significance of the oriental plane tree in European gardens, Ripa's image will have a rich history until the nineteenth century at least.[69]

Recent studies in the field of biology indicate that a reconsideration of the cultural and historical importance of *Platanus orientalis* L., highlighted in the present study, can play a significant role in the preservation of veteran trees, which are particularly prone to disease, within historic gardens and

68 Ripa Cesare, *Iconologia* [...] (Florence, Bartolomeo Ruoti: 1613) 7.

69 On Ripa's personification of the Academy and its afterlife, see especially Zimmermann H.J., *Der akademische Affe. Die Geschichte einer Allegorie aus Cesare Ripas "Iconologia"* (Wiesbaden: 1991).

Parte Prima. 11

ACADEMIA.

dio nel libr. pr. de Ponto. acciò si emendino, e polischino.

Scilicet incipiam lima mordacius vti,
Vt sub iudicium singula verba vocem.

Onde Quintiliano lib. x. cap. iij. *opus poliat lima*, & non senza ragione si sdegna Horatio nella Poetica de i latini, che non poneuano al par de' Greci cura, e fatica, in limare, e polire l'opere loro.

Nec virtute foret clarisq; potentius armis,
Quàm lingua latium, si non offenderet vnum
Quenq; poetarum limæ labor, & mora vos;
Pompilius sanguis carmen reprehēdite, quod nō
Multa dies, & multa litura coercuit, atq;
Perfectum decies non castigauit ad vnguem.

Et il Petrarca Sonetto 18.

Ma trouo peso non de le mie braccie,
Ne opra di polir con la mia lima.

Quindi è, che molto accortamente dicesi, che ad vn'opera gli manca l'vltima lima, quando non è à bastanza tersa e pulita, veggasi ne gli Adagij. *Limam addere*: Da quali habbiamo cauato il motto, oue leggesi, circa l'emendatione de l'opere. *Lima detrahitur; atq; expolitur, quod redundat, quodq; incultum est, & limata dicuntur expolita*. La ghirlanda si tesse d'Alloro, Hedera, e Mirto, perche sono tutte tre piante poetiche, per le varie spetie di poesia, che ne l'Academie fioriscono, impercioche il Mirto è pertinente al Poeta melico amoroso, che con suauità, e piacere canta gli suoi amori, perche il Mirto, secondo Pierio Valeriano, è simbolo del piacere, & Venere madre de gli amori, anzi riferisce Nicandro, che Venere fù presente al giuditio di Paride incoronata di Mirto, tanto gli era grato, e però Virgilio in Melibeo.

Populus Alcidæ gratissima vitis, Iaccho,
Formosæ myrtus Veneri, sua laurea Phœbo.

Et Ouidio nel principio del 4. lib. de Fasti,

A 2 volen-

FIGURE 7.14 Allegory of the Academy. From: Ripa Cesare, *Iconologia* […] (Florence, Bartolomeo Ruoti: 1613) 2

landscapes.[70] This is not only true in Europe and Italy, but in many parts of the world, especially in Kashmir, India, where the tree has an exceptionally rich and ancient history and is being particularly threatened.[71] The necessity of recognising these trees as archaeophytes, i.e., non-native species introduced in ancient times (prior to 1500), participates in an effort to understand their history, their ecological impact on native species, and their symbolic status.[72] Preservation and study standards, the so-called COVE protocol, have recently been tested, notably in the garden of Villa Lante at Bagnaia, whose original planting dates to 1576.[73] The history of plane trees and their use in gardens also highlights the fundamentally sensory dimension of our relationship with trees. The history of plane trees is above all the history of the shade of plane trees, which protects from the heat of the sun and invites people to come and lie down at their feet. Rest and peace are experienced on a purely sensory level and create the conditions for meditation and philosophical dialogue, the possibility of feeling one with nature.[74] In Rome, the Accademia del Platano, founded in 1688 in the house of the abbot Rinaldi Cattani da Diacceto, chose *sub umbra* as its motto.[75]

It is in this simple appreciation of the protective dimension of the tree – different, for example, from the utilitarian dimension of fruit trees – that we must situate the many legends in which it appears as subject.[76] The garden

70 De Beer Z.W. – Santini A. – Tsopelas P. – Wingfield M.J., "Canker Stain: A Lethal Disease Destroying Iconic Plane Trees", *Plant Disease* 101 (2017) 645–658.

71 Haenraets J., "Save the Chinar: The Endangered Heritage of the Oriental Plane Trees of Kashmir", *Explore Earth. Landscapes, Heritage & Conservation*, https://explearth.org/save-the-chinar-kashmir/ [accessed 22.04.2023].

72 Giardini M. – Marignani M. – Masi A. – Rosati L., "Under the Shadow of a Big Plane Tree: Why Platanus orientalis should be considered an Archaeophyte in Italy", *Plant Biosystems* 149.1 (2015) 185–194.

73 Ciaffi M. – Alicandri E. – Vettraino A.M. – Paolacci A.R. – Tamantini M. – Tomao A. – Agrimi M. – Kuzminsky E., "Conservation of Veteran Trees within Historical Gardens (COVE): A Case Study applied to Platanus orientalis L. in Central Italy", *Urban Forestry & Urban Greening* 34 (2018) 336–347; Ciaffi M. – Alicandri E. – Vettraino A.M. – Adducci F. – Agrimi M. – Kuzminsky E., "Dimensional and Genetic Characterization of the Last Oriental Plane Trees (*Platanus orientalis* L.) of Historical Sites in Lazio (Central Italy)", *Urban Forestry & Urban Greening* 69 (2022), DOI: 10.1016/j.ufug.2022.127506.

74 On this topic, see especially Corbin A., "À l'ombre des arbres. Le répertoire des pratiques", in Corbin A., *La douceur de l'ombre* 291–344.

75 Piazza Carlo Bartolomeo, "Del Platano. In Casa dell'Abbate Rinaldi Cattani da Diaceto", in *Euseuologio romano overo delle opere pie di Roma* (Rome, Domenico Antonio Ercole: 1698) lxv–lxvi. The plane tree was represented with its trunk and roots, from which a small branch grows, with this verse by Claudian 'Revocat primordiae formae'.

76 This is the subject, for example, of Aesop's fable *The Plane Tree*, where two travellers resting under a large plane tree criticize it for being useless and bearing no fruit, while

is revealed here as the emblematic place where the senses and the mind are solicited in equal measure. In this respect too, our study could be added to the extensive list of scientific publications that demonstrate the intellectual and health benefits of outdoor living and education, a particularly important idea in ancient and Renaissance gardens centred on the concept of *otium*.[77] As so many outdoor schools around the world demonstrate, by creating a specific shade zone, the plane tree provides spaces that are particularly conducive to gathering, exchange, and the circulation of knowledge. They create the conditions for a privileged type of sociability not only between humans, but also between humans and their natural environments.[78] It is therefore important not only to remember how these trees have shaped our culture, but also to protect them as they have protected us for centuries with their benevolent shade. In this article I have only given a brief overview of the richness of the subject, but I hope that it may help some readers to take a closer look at the beautiful plane trees they encounter when visiting historic gardens and landscapes.

References

A'lam H., "*Čenār*, the Oriental plane (tree), *Platanus orientalis* L. (fam. Platanaceae)", *Encyclopædia Iranica* 5.2 (1990) 129–133.

Aldrovandi Ulisse, *Bologna, Biblioteca Universitaria, Ms. Aldrovandi 136*, 32 vols. (Bologna: 1588–1590).

enjoying its shade. The same topos is evoked also by another academy that used the plane tree as their emblem, the Accademia dei Trasformati in Milan, founded in 1546 and recreated again in 1743. The accompanying motto taken from Virgil, *Georgics*, 2,70 alludes to the fact that the plane tree bears no fruits: 'et steriles platani malos gessere valentes' ('barren plane trees will turn into vigorous apple trees'). See Maylender M., *Storia delle accademie d'Italia*,vol. 5 (Bologna: 1930) 338–342 and Maria Cristina Loi's essay in this book.

77 Harris M.H., "Growing among Trees: a 12–month process evaluation of school based outdoor learning interventions", *Journal of Adventure Education and Outdoor Learning* (2021) (with bibliography), accessible online. On the therapeutic dimension of gardens in the Renaissance, especially in Marsilio Ficino, see Brunon H., "Conjurer la mélancolie, ou la physiologie de l'allégresse dans l'Italie de la Renaissance", in Schmiele C. – Haberl H. – Lensing A. (eds.), *Jardin et mélancolie en Europe entre le XVIIIe siècle et l'époque contemporaine, Histoire culturelle de l'Europe 3* (2018). Katherine Bentz is preparing a book on this issue focusing on Popes and Roman cardinals: *Gardens as Green Medicine: Prelates and Health in Early Modern Rome*.

78 Zask J., *Quand la place devient publique* (Paris: 2018), 'La place du platane; écologie de la place publique'.

Alessi A., "La Palazzina Gambara: gli architetti e i pittori", in Frommel S. (ed.), *Villa Lante a Bagnaia* (Milan: 2005) 110–121.

Baltes M., "Plato's School, the Academy", *Hermathena* 155 (1993) 5–26.

Belon Pierre, *Les remonstrances sur le default du labour & culture des plantes, & de la cognoissance d'icelles* (Paris, Gilles Corrozet: 1558).

Belon Pierre, *Observations de plusieurs singularitez et choses memorables, trouvees en Grece, Asie, Judee, Egypte, Arabie & autre pays estranges* [...] (Paris, Hierosme de Marnef & la Veuve Guillaume Cavellat: 1588).

Bembo Pietro, *De Aetna* (Rome, Aldus Manutius: 1496).

Beneš M., "The Social Significance of Transforming the Landscape at the Villa Borghese, 1606–30: Territory, Trees and Agriculture in the Design of the First Roman Baroque Park", in Petruccioli A. (ed.), *Gardens in the Time of the Great Muslim Empires: Theory and Design* (Leiden: 1997) 1–31.

Biermann H., "War Leon Battista Alberti je in Urbino?", *Zeitschrift für Kunstgeschichte* 65.4 (2002) 493–521.

Bietenholz P., "Felicitas (eudaimonia) ou les promenades d'Érasme dans le jardin d'Épicure", *Renaissance and Reformation* 30.1 (2006) 37–86.

Boni G., "L'Arcadia sul Palatino", *Bollettino d'Arte* 8 (1914) 369–382.

Brown C.M., *Isabella d'Este in the Ducal Palace in Mantua* (Roma: 2005).

Bruce A. – Chandès H. – Couton P.E. – Gaudefroy I. (eds.), *Nous Les Arbres*, exh. cat., Fondation Cartier pour l'art contemporain (Paris: 2019).

Brunon H., "Amitiés respectueuses. Pour une archéologie de la relation jardinière", *Jardins* 6 (2015) 19–38.

Brunon H., "Comme par enchantement", in Lavigne E. (ed.), *Céleste Boursier-Mougenot: rêvolutions*, exh. cat., La Biennale di Venezia 2015, Pavillon français de la 56e exposition internationale d'arts visuels (Paris: 2015).

Brunon H., "*Locus secretus*: topique et topophilie", in Luciani D. – Mosser M. (eds.), *Petrarca e i suoi luoghi. Spazi reali e paesaggi poetici alle origini del moderno senso della natura* (Treviso: 2009) 41–55.

Brunon H. – Mosser M., "L'enclos comme parcelle et totalité du monde: pour une approche holistique de l'art des jardins", *Ligeia, dossiers sur l'art* 73.76 (2007) 59–75.

Buc'hoz P.J., *Dissertations sur le cèdre du Liban, le platane et le cytise* (Paris, Madame Buc'hoz: 1806).

Campitelli A., *Villa Borghese. Storia e gestione* (Rome: 2005).

Ciaffi M. – Alicandri E. – Vettraino A.M. – Paolacci A.R. – Tamantini M. – Tomao A. – Agrimi M. – Kuzminsky E., "Conservation of Veteran Trees within Historical Gardens (COVE): A Case Study applied to Platanus orientalis L. in Central Italy", *Urban Forestry & Urban Greening* 34 (2018) 336–347.

Ciaffi M. – Alicandri E. – Vettraino A.M. – Adducci F. – Agrimi M. – Kuzminsky E., "Dimensional and Genetic Characterization of the Last Oriental Plane Trees (*Platanus orientalis* L.) of Historical Sites in Lazio (Central Italy)", *Urban Forestry & Urban Greening* 69 (2022).

Clusius Carolus, *Rariorum plantarum historia* (Antwerp, Joannem Moretum: 1601).

Colonna Francesco, *Hypnerotomachia Poliphili* (Venice, Aldus Manutius: 1499).

Colonna Francesco, *Hypnerotomachia. The Strife of Loue in a Dreame*, trans. R. Dallington (London, Simon Waterson: 1592).

Collected Works of Erasmus, in Thompson C.R. (ed.), *Literary and Educational Writings 1. Antibarbari / Parabolae*, vol. 23–24, (Toronto: 1978).

Coffin D., *The Villa in the Life of Renaissance Rome* (Princeton: 1979).

Coleman K., "Melior's Plane Tree. An Introduction to the Ancient Garden", in Coleman K. (ed.), *Le jardin dans l'Antiquité*, Entretiens sur l'Antiquité classique de la Fondation Hardt 60 (Geneva: 2014) 1–26.

Comanducci R.N., "Gli Orti Oricellari", *Interpres* 15 (1995–1996) 302–355.

Cooper R., *Rabelais et l'Italie* (Geneva: 1991).

Corbin A., *La douceur de l'ombre. L'arbre, source d'émotions, de l'Antiquité à nos jours* (Paris: 2014).

Crescimbeni Giovanni Mario, *L'Arcadia* (Rome, Antonio de' Rossi: 1711).

Curtius E.R., *European Literature and the Latin Middle Ages* (Princeton – Oxford: 2013).

Daubenton Pierre, "Platane", in Diderot D. – D'Alembert J.l.R. (eds.), *L'Encyclopédie*, vol. 12 (Paris: 1751) 733–736.

De Beer Z.W. – Santini A. – Tsopelas P. – Wingfield M.J., "Canker Stain: A Lethal Disease Destroying Iconic Plane Trees", *Plant Disease* 101 (2017) 645–658.

De Meneses João Rodrigues de Sà, *Die Platane. Liber de platano (1537)*, eds. R. Friedlein – A. Lozar (Leiden: 2008).

Demandt A., *Über allen Wipfeln. Der Baum in der Kulturgeschichte* (Cologne: 2002).

Desnoyers G., *La Villa d'Este à Tivoli ou Le songe d'Hippolyte. Un rêve d'immortalité héliaque* (Paris: 2002).

Dixon S.M., *Between the Real and the Ideal. The Accademia Degli Arcadi and Its Garden in Eighteenth-Century Rome* (Newark: 2006) 56–57.

Dumargne L.M., "Jardiner à l'épreuve de la dualité sciences humaines / sciences naturelles," *Intermediality* 35 (2020).

Durante Castore, *Herbario Nuovo* (Rome, Bartolomeo Bonfadino: 1585).

Duval C., "Introduction du Platane en France", *Bulletin de la Société Botanique de France* 43 (1896) 194–198.

Fay M.F. – Rix M., "857. Platanus orientalis. Platanaceae", *Curtis's Botanical Magazine* 34.1 (2017) 29–40.

Ferrari G.R., "The Meaninglessness of Gardens", *The Journal of Aesthetics and Art Criticism* 68.1 (2010) 33–45.

Ficino Marsilio, *Commentaries on Plato, Phaedrus and Ion*, ed. M.J.B. Allen, 2 vols. (Cambridge MA – London: 2008).

Forster E., "Trees and Plants in Homer", *The Classical Review* 50.3 (1936), 97–104.

Gage F., "Exercise for Mind and Body: Giulio Mancini, Collecting, and the Beholding of Landscape Painting in the Seventeenth Century", *Renaissance Quarterly* 61.4 (2008) 1167–1207.

Galand-Hallyn P., "Aspects du discours humaniste sur la villa au XVIe siècle", in Galand-Hallyn P. – Lévy C. (eds.), *La villa et l'univers familial de l'Antiquité à la Renaissance* (Paris: 2008) 117–146.

Gatti H., *Ideas of Liberty in Early Modern Europe* (Princeton: 2015).

Gerard John, *The Herball or General Historie of Plantes*, ed. The English Experience, 2 vols. (n.p.: 1974).

Giardini M. – Marignani M. – Masi A. – Rosati L., "Under the Shadow of a Big Plane Tree: Why Platanus orientalis should be considered an Archaeophyte in Italy", *Plant Biosystems* 149.1 (2015) 185–194.

Grant K., "The Bosco Parrasio as a Site of Pleasure and of Sadness", *Histoire culturelle de l'Europe* 3 (2019).

Grimal P., *Les Jardins romains* (Paris: 1969).

Guerrieri Borsoi M.B., *Villa Sora at Frascati. Una villa romana fra '600 e '700* (Rome: 2000).

Hardie A., "Philitas and the Plane Tree," *Zeitschrift für Papyrologie und Epigraphik* 119 (1997) 21–36.

Kushner E., "Le rôle structurel du 'locus amoenus' dans les dialogues de la Renaissance", *Cahiers de l'AIEF* 34 (1982) 39–57.

Kuttner A.L., "Culture and History at Pompey's Museum", *Transactions of the American Philological Association* 129 (1999) 343–373.

Laird M., *A Natural History of English Gardening, 1650–1800* (New Haven – London: 2015).

Landino Cristoforo, *Disputationes Camaldulenses*, ed. P. Lohe (Florence: 1980).

Lauterbach I., "Gardens of the Milanese 'villeggiatura'", in Dixon Hunt J. (ed.), *The Italian Garden: Art, Design and Culture* (Cambridge: 1996) 127–159.

Lazzaro C., *The Italian Renaissance Garden* (New Haven – London: 1990).

Lee R.W., *Names on Trees: Ariosto Into Art* (Princeton: 1977).

Lemnius Levinus, *An Herbal for the Bible* (n.p: 1579).

Li H.L., *The Origin and Cultivation of Shade and Ornamental Trees* (Philadelphia: 1963).

Ligorio Pirro, *Enciclopedia del mondo antico* (Turin, Archivio di Stato).

Macaulay-Lewis E., "Imported Exotica: Approaches for the Study of the Ancient Plant Trade", in Macaulay-Lewis E. – Gleason K.L. (eds.), *The Gardens of the Ancient Mediterranean: Cultural Exchange Through Horticultural Design, Technology, and Plants* (Rome: 2010) 16–26.

Macaulay-Lewis E., "The Fruits of Victory; Generals, Plants, and Power in the Roman World", in Bragg E. – Hau L.I. – Macaulay-Lewis E. (eds.), *Beyond the Battlefields: New Perspectives on Warfare and Society in the Graeco-Roman World* (Newcastle: 2008) 205–225.

Marder M., "Plato's Plane Tree", in Marder M. (ed.), *The Philosopher's Plant* (New York: 2012) 1–18.

Marliani Bartolomeo, *Topographia antiquae Romae* (Lyon, Sébastien Gryphe: 1534).

Marzano A., "Roman Gardens, Military Conquests, and Elite Self-Representation", in Coleman K. (ed.), *Le jardin dans l'Antiquité*, Entretiens sur l'Antiquité classique de la Fondation Hardt 60 (Geneva: 2014) 195–244.

Mattioli Pietro Andrea, *I Discorsi* [...] (Venice, Vincenzo Valgrisi: 1568).

Maylender M., *Storia delle accademie d'Italia*, 5 vols. (Bologna: 1930).

Melikian Chirvani A.S. (ed.), *Architecture in Islamic Arts. Treasures of the Aga Khan Museum*, exh. cat., Aga Khan Trust for Culture (Geneva: 2012).

Melikian Chirvani A.S., *Le Chant du monde: L'art de l'Iran safavide, 1501–1736* (Paris: 2007).

"Memoirs of Thomas Jones Penkerring Radnorshire 1803", *The Volume of the Walpole Society* 32 (1946–1948) 1–143.

Mercuriale Girolamo, *De arte gymnastica libri sex* (Venice, Giunta: 1601).

Montelatici Domenico, *Villa Borghese fuori Porta Pinciana* (Rome, G.F. Guagni: 1700).

Moravus Augustinus, *Dialogus in defensionem poetices* (Venice: Paul Fridenperger, 1493).

Motte A., "Le pré sacré de Pan et des Nymphes dans le *Phèdre* de Platon", *L'antiquité classique* 32.2 (1963) 460–476.

Newlands C.E., "Martial, Epigrams 9.61 and Statius, Siluae 2.3: Two Branches from the Same Tree?", *Scholia* 20 (2011) 92–109.

Opsomer J.E., "Un botaniste trop peu connu Willem Quackelbeen (1527–1561)", *Bulletin de la Société Royale de Botanique de Belgique* 93 (avril 1961) 113–130.

Persio Sacconi Francesco, *Ristretto delle piante con suoi nomi antichi, e moderni, della terra, aria, e sito, ch'amano, consacrato alla S.C.R.M. Leopoldo I Imperatore* (Vienna, Andrea Heyinger: 1697).

Philandrier Guillaume, *M. Vitruvii Pollionis de Architectura Libri decem* (Lyon, Jean de Tournes: 1552).

Piana M., *Fallax Antiquitas: Gianfrancesco Pico della Mirandola's Critique of Antiquity* (Ph.D. diss., McGill University: 2017).

Piazza Carlo Bartolomeo, "Del Platano. In Casa dell'Abbate Rinaldi Cattani da Diaceto", in *Euseuologio romano overo delle opere pie di Roma* (Rome, Domenico Antonio Ercole: 1698).

Pico della Mirandola Gianfrancesco, *De Venere et Cupidine expellendis carmen* (Rome, Iacobus Mazochius: 1513).

Pignatti F., "Ricchieri, Lodovico Maria", *Dizionario biografico degli italiani* 87 (2016).

Piovene Pietro, *I Cesari in metallo mezzano e piccolo raccolti nel Museo Farnese*, 10 vols. (Parma, Nella Stamperia di S.A.S.: 1727).

Pona Francesco, *Il paradiso de' fiori overo Lo Archetipo de' Giardini* (Verona, Angelo Tamo: 1622).

Potts D.T., "Arboriculture in Ancient Iran: Walnut (*Juglans regia*), Plane (*Platanus orientalis*) and the '*Raddle dictum*'", *dabjïr* 6 (2018) 101–109.

Pontoppidan A., *Le platane* (Arles: 1996).

Ramade B., "Ex-Situ," in Bianchini S. – Quinz E.(eds.), *Behavioural Objects 1, A Case Study: Céleste Boursier-Mougenot* (London – Berlin: 2016).

Rashed M., "La mosaïque des philosophes de Naples: une représentation de l'académie platonicienne et son commanditaire", in Noirot C. – Ordine N. (eds.), *Omnia in uno. Hommage à Alain-Philippe Segonds* (Paris: 2013) 27–49.

Ranaldi A., *Pirro Ligorio e l'interpretazione delle ville antiche* (Rome: 2001).

Ribouillault D. – Weemans M., "Paysage sacré, Livre de la Nature, Exégèse: pour une reconception du paysage dans l'Europe de la première modernité", in Ribouillault D. – Weemans M. (eds.), *Sacred Landscape. Landscape as Exegesis in Early Modern Europe* (Florence: 2011).

Rice L., "*Pomis sua nomina servant*: The Emblematic Thesis Prints of the Roman Seminary", *Journal of the Warburg and Courtauld Institutes* 70 (2007) 195–246.

Ripa Cesare, *Iconologia* [...] (Florence, Bartolomeo Ruoti: 1613).

Rhodigini Ludovico Caelii, *Lectionum antiquarum* (Frankfurt – Leipzig, Danielis Fieveti: 1566).

Savonarola Girolamo, *Dialogus de veritate prophetica* (Venice, Bartolomeo de' libri: 1507).

Schievenin R., "Sotto il platano", *Incontri di filologia classica* 12 (2012–2013) 239–256.

Schimmel A., *Anvari's Divan: A Pocket Book for Akbar* (New York: 1983).

Soderini Giovanvettorio, "Il trattato degli arbori", ed. A. Bacchi Della Lega, *Opere*, 2 (Bologna: 1903).

Stubbings F.H., "Xerxes and the Plane-Tree", *Greece & Rome* 15.44 (1946) 63–67.

Stull W., "Deus ille noster: Platonic Precedent and the Construction of the Interlocutors in Cicero's 'De oratore'", *Transactions of the American Philological Association* 141.2 (2011) 247–263.

Tarrío A.M.S., *Paisagem e erudiçao no humanismo português. João Rodrigues de Sá de Meneses – De Platano (1527–1537)* (Lisbon: 2009).

Valéry A.C., *Voyages en Italie*, 3 vols. (Paris: 1831).

Williams G., *Pietro Bembo on Etna: The Ascent of a Venetian Humanist* (New York: 2017).
Wind E., *Pagan Mysteries in the Renaissance* (New Haven: 1958).
Zaslvsy R., "A Hitherto unremarked Pun in the *Phaedrus*", *Apeiron* 15 (1981) 115–116.
Zimmermann H.J., *Der akademische Affe. Die Geschichte einer Allegorie aus Cesare Ripas "Iconologia"* (Wiesbaden: 1991).
Zhong Mengual E., *Apprendre à voir: le point de vue du vivant* (Paris: 2021).

CHAPTER 8

Academies and Gardens of the Faubourg Saint-Marcel in Paris (1550–1578)

Laurent Paya

Abstract

In the 16th century, the Faubourg Saint-Marcel, beyond the fortified walls of Paris, was an area between town and country where three academies developed independently of the university authorities. The Académie de musique et de poésie, founded in 1571 by Jean-Antoine de Baïf, attracted King Charles IX for its musical performances. The *Lycium philosophicum*, an academy of natural philosophy founded in 1571 by Jacques Gohory, contributed to the development of botany. The Maison de la Charité chrétienne, a philanthropic academy founded in 1578 by the apothecary Nicolas Houel, included a garden dedicated to the cultivation of medicinal plants to provide free treatment for the poor and to teach orphans the art of apothecary. This paper discusses the history and function of the architecture and gardens of these academies. It also examines how they were integrated into the urban fabric and the links that existed between the various members of these ephemeral institutions.

Keywords

Paris in the 16th century – Académie de musique et de poésie – Lycium philosophal (academy of natural philosophy) – Maison de la Charité chrétienne (philanthropic academy) – Humanist networks

The garden, first in Greco-Roman antiquity and again during the *renovatio antiquitatis* of the early modern period, functioned as a *locus amœnus* favourable to the work of the academies.[1] Between 1570 and 1580, a period of civil war in France, the first French learned academies were founded in Paris *extra-muros* in and near the Faubourg Saint-Marcel [Figs. 8.1–8.2]. The Académie de musique et de poésie founded by Jean-Antoine de Baïf (1532–1589) was located

1 See the introduction.

 | DOI:10.1163/9789004517547_009

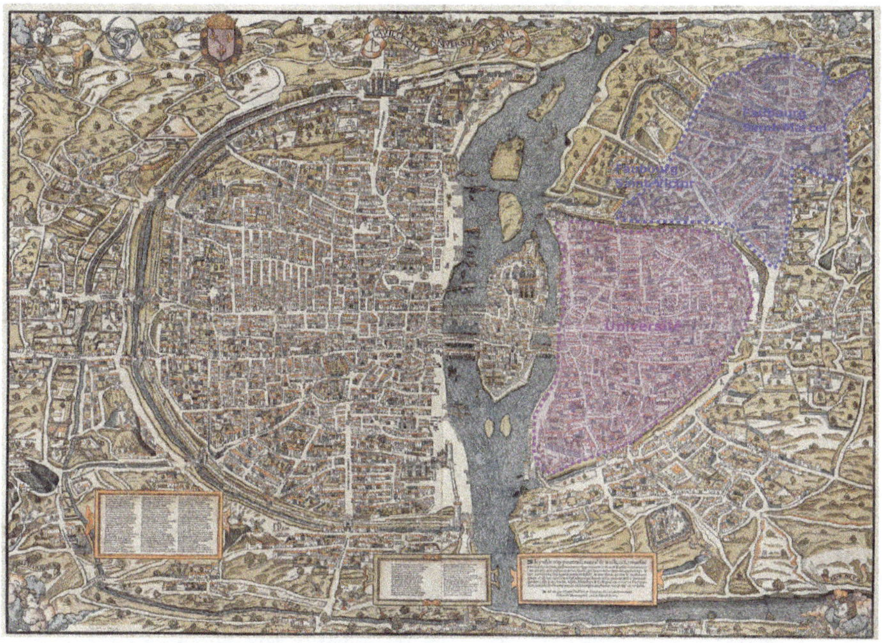

FIGURE 8.1 *The University of Paris and the Faubourg Saint-Marcel*, from the plan by Truschet and Hoyau, known as the Basel plan, *La ville, Cité et Université de Paris*, c.1550, facsimile of 1980, 1 map in 2 sheets, 94 × 120 cm. Amsterdam, Universiteitsbibliotheek Vrije Universiteit, LL.06979gk
IMAGE © UNIVERSITEITSBIBLIOTHEEK VRIJE UNIVERSITEIT

on rue des Fossés Saint-Victor,[2] the Académie paracelsienne established by Jacques Gohory (1520–1576) between rue Neuve Sainte-Geneviève and rue Saint-Marcel,[3] and the Académie philanthropique instituted by Nicolas Houel

2 Among the many references, Frémy E., *L'Académie des derniers Valois: académie de poésie et de musique 1570–1576, Académie du Palais 1576–1582* (Paris: 1887); Yates F., *Les Académies en France au XVIe siècle* (Paris: 1996); Vignes J., "Expérimentation musicale et pédagogie à la Renaissance: à propos du conflit entre l'Académie de Poésie et de Musique et le Parlement de Paris (1570)", *Les Académies dans l'Europe humaniste. Idéaux et pratiques* (Geneva: 2008) 493–506.

3 Hamy E.-T., *Un précurseur de Guy de La Brosse. Jacques Gohory et le Lycium philosophal de Saint-Marceau-lès-Paris (1571–1576)* (Paris: 1899); Bowen W.H., *Jacques Gohory (1520–1576)* (Ph.D., Harvard University: 1935); Idem, "The Earliest Treatise on Tobacco: Jacques Gohory's *Instruction on the Herb Petum*", *Isis* 28, 2 (1938) 349–363; Walker D.P., *Spiritual and Demonic Magic from Ficino to Campanella* (London: 1958) 101; Balmas E., *Saggi e Studi sul Rinascimento Francese* (Padua: 1982) 23–73; Kahn D., "Le paracelsisme de Jacques Gohory", *Aries* 19 (1995) 81–130; Idem, *Alchimie et Paracelsisme en France (1567–1625)* (Geneva: 2009); Gorris Camos R., "L'Hysope et la rose: le Lycium philosophal de Jacques Gohory", in D. Blocker (ed.), *Les Académies entre France et Italie* (Geneva: 2007) 549–590; idem, "Dans le labyrinthe de Gohory, lecteur et traducteur de Machiavelli", *Italian Laboratory* 8 (2008) 195–229.

FIGURE 8.2 1 – Abbaye Saint-Victor, 2 – location of the Académie de Poésie et de Musique, 3 – Abbaye Sainte-Geneviève, 4 – location of the house of André Thevet, 5 – location of the *Lycium Philosophal*, 6 – location of the Maison de la Charité chrétienne, in the plan of Truschet and Hoyau, known as the plan of Basel, *La ville, cité, université de Paris*, *c.*1550 (detail), facsimile of 1980, 1 map in 2 sheets, 94 × 120 cm. Amsterdam, Universiteitsbibliotheek Vrije Universiteit, LL.06979gk IMAGE © UNIVERSITEITSBIBLIOTHEEK VRIJE UNIVERSITEIT

(1524?–1587) nearby.[4] The latter two were notable for their medicinal plant gardens, representations or descriptions of which have survived.

4 Planchon G., "Le Jardin des Apothicaires", *Le journal de pharmacie et de chimie* 28 (1893) 250–258, 289–298, 342–349, 412–416; De Labord A., *Nicolas Houel, fondateur de la Maison de charité chrétienne* (Paris: 1937); Auclair V., "Nicolas Houel et Petrus Stephanus: un manuscrit enluminé du XVIe siècle, le *Traité de la Charité chrestienne*", *Revue de l'art* 132 (2001): 9–28; Idem, "Un logis pour l'âme des rois. Nicolas Houel (*c.*1520–*c.*1587) et les dessins de procession à la maison de la Charité chrétienne pour la famille royale", in *Henry III mécène des arts, des sciences et des lettres* (Paris: 2006) 39–54; Paya L., "Du Jardin des Simples de la Maison de la charité chrétienne au Jardin des Apothicaires (1578–1624)", in Vons J. (ed.), *La Médicine à la cour de France* (2008) http://cour-de-france.fr/article590.html; Broomhall S., "Hearts on Fire: Compassion and Love in Nicolas Houel's *Traité de la Charité chrestienne*", in Broomhall S. (ed.), *Ordering Emotions in Europe, 1100–1800* (Leiden: 2015) 121–160; Elliot L.K., "In Pursuit of Charity: Nicolas Houel and his *Maison de la Charité chrétienne* in Late Sixteenth-Century Paris", in Broomhall S. (ed.), *Experience of Charity, 1250–1650* (London: 2016) 149–170.

In the years around 1570, French academies were usually independent, elitist circles dedicated to the training of young talents and operated in competition with the university, where languages and ancient texts were traditionally studied.[5] Although some academies considered natural history, they were not yet specialised in this domain and remained more general in their approach.[6] The aforementioned institutions, which occupy a decisive place in the history of the arts, sciences and ideas of the sixteenth century, have all received in-depth study that has sought to define their particular academic projects.[7] However, as Frances A. Yates has already noted, a broader investigation into the humanist, historical and cultural relations that tie the institutions remains to be undertaken. Further unexamined still is the topic of their distribution across the Parisian urban fabric and the spatial relationships established therein. By outlining the epistemological contours and socio-relational dimensions of these specialist societies focused on the secrets of simples – a period term referring to medicinal or exotic plants – I will attempt to uncover the motivations for their establishment in this district of Paris. After outlining the physiognomy of this landscape, I will then turn my attention to the academies founded by Jacques Gohory (*c.*1520–1576) and Nicolas Houel (*c.*1524–1587), paying special interest to their gardens of simples.

1 The Faubourg Saint-Marcel: Suburban Landscape and Cultural Epicentre (1550–1574)

At the end of the eighteenth century, the Faubourg Saint-Marcel was a populous and rebellious district, one that had been recently integrated into the city of Paris. However, this had not always been the case.[8] Beginning in the twelfth century, the Faubourg developed southward alongside Charles V's fortifications and was a site of discontinuous urbanisation surrounded by fields, vineyards and gardens. In the fourteenth century, aristocratic residences were built along the Grand'Rue Saint-Marcel or rue Mouffetard and the river Bièvre; as early as 1290, King Louis IX's widow Marguerite of Provence (1221–1295) confirmed the attractiveness of this suburban landscape when she retired to a manor near the Cordelières convent. The eastern side of the rue de Saint-Marcel was then lined with private mansions as far as the church of Saint-Médard, the

5 Vignes, "Expérimentation musicale et pédagogie à la Renaissance" 454–455.

6 Boutroue M.-É., "Paolo Boccone et l'académie de Bourdelot", in Galand-Hallyn P. – Deramaix M. – Vagenheim G – Vignes J. (eds.), *Les Académies dans l'Europe humaniste, idéaux et pratiques* (Geneva: 2008) 663–675.

7 See notes 2, 3, and 4.

8 Mercier Louis-Sébastien, *Tableau de Paris* I (Hambourg, Virchaux: 1781) 160–163.

most important of which was the Maison des Patriarches, home of Bishop Guillaume de Chanac (*c.*1320–1383), Patriarch of Alexandria. The Hôtel des Carneaux was located on rue Daubenton (then rue des Bouliers) until 1388 when it was rechristened the Hôtel d'Orléans under the proprietorship of Duke Louis d'Orléans (1372–1407), brother to King Charles VI (1368–1422). Until the beginning of the sixteenth century, the towns of Saint-Médard and Saint-Marcel were a pleasant combination of fields, vineyards, gardens, artisanal factories and seigneurial residences. However, faced with the possibility of attack by Emperor Charles V's troops and the threat of religious unrest, Saint-Marcel seems to have quickly lost much of its prestige over the course of the sixteenth century. Aristocratic hotels were then built within the city walls in the Marais and around the Louvre.

One notable exception to this trend has survived; a private mansion decorated with terracotta medallions in the classicist style, built in the middle of a garden on an island in the River Bièvre and inhabited by Scipio Sardini (1526–1609), a financier and favourite of Catherine de Médicis. This noble residence was likely constructed *c.*1540–1565 prior to its acquisition and embellishment by Sardini between 1565 and 1596.[9] Its current condition and a 1639 deed of sale to the hospitals of Paris allow us to reconstitute its general design: 'a main building composed of several lower and upper rooms, attics, upper and lower galleries, going from the small dwelling to the large one, courtyard and garden, the whole enclosed by walls'.[10] In 1612, the hotel was transformed into a hospice and came to be known as the 'Hospital for the Shut-In Poor', tying it to the House of Christian Charity, an institution to which I will later turn my attention.[11]

A study of the toponymical descriptions of this district in the sixteenth century reveals its less attractive characteristics: the 'sentier aux tripes' and 'La Triperie', where the 'chantier aux tripes' (the 'tripe yard') rubbed shoulders with 'La Potence' (the gallows) and the 'Fosse aux larron' (the 'thieves' pit').[12] Activities deemed unsavoury and groups of inhabitants determined undesirable for health or moral reasons were relegated to live outside the city walls, settling in this area. The course of the Bièvre attracted tanneries and skinneries,

9 Record no. PA00088437, Mérimée database, French Ministry of Culture. Its construction is often attributed to Scipio Sardini.

10 *Code de l'hôpital-général de Paris, ou Recueil des principaux édits, arrêts, déclarations & réglements qui le concernent, ainsi que les maisons & hôpitaux réunis à son administration* (Paris, veuve Thiboust: 1728) 228.

11 Record no. PA00088437, Mérimée database, French Ministry of Culture.

12 Coyecque E., *Recueil d'actes notariés relatifs à l'histoire de Paris* I (Paris: 1923) 2492, 2586, 2731, 2811, II, 4376, 5519, 5567, 5903.

and the atmosphere of the adjacent streets was conducive to the establishment of cabarets and gambling houses.[13] The neighbouring Lourcine hospital took in syphilitic patients exiled from the urban centre and, contributing to the general unpleasantness of the area; the Bièvre was known to cause serious flooding and carry all sorts of 'filth'.[14]

As Yates has explained, the Abbey of Saint Victor [see Fig. 8.2], with its 'magnificent gardens', was a hotbed of medieval Platonism. Giordano Bruno frequented its library, and this contemplative space likely played a role in the emergence of academies in the Faubourg Saint-Marcel.[15] The proximity of the Faubourg Saint-Marcel to the city's southern quarter, today known as the University of Paris [Fig. 8.1], combined with the possibility of establishing gardens of simples in the shelter of the high walls, led important figures of the learned community to settle in the district.[16] This was the case for the famed anatomist, botanist and grammarian Jacques Dubois (1478–1555), often called Jacobus Sylvius; the exact location of his garden has not yet been determined and is only known through a later mention: 'Sylvius acquired two houses in Paris, one in the faubourg Saint Marceau where he had a garden in which he cultivated a large number of medicinal plants, and it was there that he went to demonstrate them to his students'.[17] Dubois's lectures there were quite successful, and another report mentions that he 'did the dissection, taught the preparation of Remedies, and demonstrated the Plants'.[18] In *La Pharmacopée* (first Latin edition 1542, French edition 1574), a treatise on the cultivation and medicinal preparation of simples, Dubois notes that 'on the highest spot of my garden' could be found a variety of iris native to Illyria and brought from Venice.[19] A man of diverse interests, he was prompted to address the plight of the poorest in his treatise *Conseil contre la famine* (*Advice against Famine*) (1546) after a series of food shortages befell Paris in the 1540s.[20]

13 de La Mare Nicolas, *Traité de la Police*, vol. 1 (Paris, Jean et Pierre Cot: 1710) 80; Antoine Fontanon, *Les Édicts et ordonnances des rois de France*, vol. 1 (Paris, n.d..: 1611) 842; Poëte M., *Une vie de cité. Paris, de sa naissance à nos jours*, vol. 2 (Paris: 1927) 159.

14 Fournier E., *Variétés historiques et littéraires*, vol. 2 (Paris: 1855) 234.

15 Yates, *The Academies*, 24–25.

16 Former name of the Latin Quarter.

17 Claude Pierre Goujet, *Mémoire historique et littéraire sur le Collège royal de France* (Paris: Martin Lottin, 1758) 4–8.

18 Niceron Jean-Pierre, *Mémoires pour servir à l'histoire des hommes*, vol. 29 (Paris, Briasson: 1733) 93.

19 The northwestern part of the Balkan Peninsula. Dubois Jacques, *La Pharmacopée* (Lyon, Loys Cloquemin: 1580) 57.

20 Baader G., "Jacques Dubois as a Practitioner", in French R.K. – Lonie I.M. (eds.), *The Medical Renaissance of the Sixteenth Century* (Cambridge: 1986) 152; Dupèbe J., "L'alimentation des

Dubois was not alone in Saint-Marcel during this period. A notarial archive dated to 1559 mentions that André Thevet (1516–1590), a famous cosmographer of the Valois royal family, acquired a 'tile-roofed house, courtyard and garden in the faubourg Saint-Marcel rue d'Ablon [...] at the sign of the golden hoof' [Fig. 8.2].[21] It is not known whether plants were cultivated there and Thevet makes no mention of an estate garden in his 1580 will.[22] Though he seems to have resided mainly in the central Parisian university district or the Latin Quarter, we know that he was responsible for the garden at the Hôpital Saint-Jacques as early as 1560.[23] In this position he was required

> to open the gardens and other customary places and open them on good days and festivals, when the people by devotion go to the hospital to gain indulgences, do their devotion and prayers. Item to maintain the vines and flowerbeds of the gardens of the paved alley and of the other small adjoining one giving onto the street.[24]

Thevet shared his contemporaries' infatuation with exotic plants, particularly those from the New World. He introduced hitherto unknown animals and plants and is credited with giving his name to a variety of manioc and the first depiction of a pineapple. He was also introduced to the practice of smoking tobacco by Indigenous people. Referred to as *petun*, a word from the Tupi *petyma, petyn*, Thevet cultivated the tobacco plant in his Angoulême garden. In his *Cosmographie Universelle* (1575), Thevet plagiarised other travellers' accounts, amplifying his discoveries and inventing journeys to the African continent and present-day Canada without ever having visited. These writings

pauvres selon Sylvius", in Margolin J.-C. – Sauzet R. (eds.), *Pratique et discours alimentaires à la Renaissance* (Paris: 1982) 41–56.

21 Archives Nationales, Minutes et répertoires du notaire François Cartault, 10 September 1539–12 August 1562 (study XI) MC/ET/XI/19f.65 vo. Note: At the end of the sixteenth century, this street became rue Neuve-Saint-Médard.

22 Archives Nationales, MC, XXIX, 29 published by Lestringant F., *André Thevet, cosmographe des derniers Valois* (Geneva: 1991) 364.

23 Lestringant F., "L'Histoire d'André Thevet, de deux voyages par luy faits dans les Indes Australes et Occidentales (v. 1588)", 4. http://editions-villegagnons.com/THEVET_Lestringant.pdf.

24 '[...] de faire ouverture des jardins et autres lieux accoutumés et ouvrir aux bons jours et fêtes, quand le peuple par dévotion va audit hôpital pour gagner les indulgences, faire leur dévotion et prières. Item d'entretenir les treilles et parterre des jardins de l'allée pavée et de l'autre petit joignant répondant sur la rue'. Archives Nationales, MC, LXXIII, 54.

were controversial, particularly among plant experts such as Pierre Belon, Jean Liebault and Jacques Gohory.[25]

Contributing to the rise of prestigious academies in the area, the Académie de musique et de poésie received its letters patent from Charles IX in November 1570. Founded by poet Jean-Antoine de Baïf (1532–1589) and musician Joachim Thibault de Courville (d. 1581), the Académie had little to do with herbology in its early years (1570–1574).[26] Baïf distinguished himself at the French Court and found loyal patrons in Catherine de Médicis, Charles IX and Henry III, with the latter sometimes travelling to Baïf's house between the Faubourg Saint-Victor and Faubourg Saint-Marcel to hear his and his friends' works. By analysing the area from existing sources, we know that by walking along the rue des Fossés (which runs parallel to Charles V's fortifications) from the Porte Saint-Victor to the Porte Saint-Marceau, one would arrive at the Maison de la Corne de Cerf from the Rue des Boulangers. And at the corner of this street, the Maison de l'Ange was followed by the Maison du Chapeau Rouge. Baïf occupied the latter which, in 1545, consisted of two main buildings linked by a gallery, two courtyards and a pleasure garden where he had to 'maintain the trellises, the cradle, the fruit trees and make the parterre'.[27] He founded a first version of his academy there in November 1570, which had already attracted Charles IX (1550–1574), the Duke of Anjou and future King Henry III (1551–1589), and the poet Torquato Tasso (1544–1595). The academy then moved to the neighbouring Maison de l'Ange, which had to be restored after the passage of soldiers in 1568. 'Enclosed by a wall', this large house contained 'two buildings, a courtyard in the middle, two gardens behind' as well as a 'small willow tree grove'.[28] It was jointly owned by one of Jean-Antoine de Baïf's old acquaintances, the apothecary Gérard or Girard Chassebras (?–1577), to whom Jean-Antoine's father, Lazare de Baïf (1496–1547), entrusted his servant as an apprentice in 1546.[29] Gardens occupied an important place in the district [Fig. 8.2], extending as far as the Rue des Boulangers; as Yates has observed, this *locus amœnus* is evoked in works by academy members. Among these, we count poems by Paulus Melissus (1539–1602) in *Schiedasmata* (1574) and 'Voici

25 Charles Estienne and Jean Liébault, *L'Agriculture et maison rustique* (Paris: Jacques Du Puy, 1583), fol. 126v. On the relationship between Gohory and Thevet, see Lestringant F., *Sous la leçon des vents. Le monde d'André Thevet, cosmographe de la Renaissance* (Paris: 2003) 237–243.

26 Yates, *The Academies* 27.

27 Coyecque, *Recueil d'actes notariés*, II, 3801.

28 'deux corps d'hôtel, cour au milieu, puys moicoien [mitoyen], deux jardins derrière' petite oseraie', Archives Nationales, MC/ET/LXXVIII/138.

29 Archives Nationales, MC/ET/XI/7.

le vert et beau Mai', one of the *Chansonnettes mesurées* (1586) by Jean-Antoine de Baïf, which Jacques Mauduit (1557–1627) set to music.[30] In 1575, Chassebras rented the Maison du Lion d'Or to Ronsard, which reveals the apothecary's proximity to this literate community.

2 Jacques Gohory's *Lycium philosophicum* (1571–1576)

Jacques Gohory was a polymath lawyer, translator, poet, mathematician, historian and follower of natural magic and the Great Work in alchemy. Although not a traveller, he was passionate about the simples, especially those from 'Antartic France'. This attraction brought him closer to André Thevet, sometimes via polemical writings. Gohory contributed to the poetics of pleasure gardens by helping to translate the *Hypnerotomachia poliphili*, which he presented to Jean Martin before 1546, as well as with his translations of several volumes of *Amadis de Gaule*.[31] With members of the Lycium philosophal, an academy with a garden of simples that he founded in 1571 in the Faubourg Saint-Marcel, he experimented with Paracelsian medicines, sometimes incorporating plants.[32] His translation of *Occultes merveilles et secretz de nature* (*Occult wonders and secrets of nature*, 1572) by Levin Lemne (1505–1568?) demonstrates his interest in the practices of magic and natural philosophy and their application to herbology. Through his varied activities, the Lycium paralleled two contemporary academies: the Platonic Academy in Florence frequented by Marsilio Ficino (1433–1499), a famed proponent of natural magic, and the Accademia dei Segreti, founded by Giambattista della Porta (1535–1615), author of *Magiae naturalis* (1558). The latter, located in Naples, is considered by some the first scientific society,[33] although it was forbidden by the Inquisition in 1574.[34] Perhaps inspired by the activities of his Italian peers, Gohory alludes to his ambition of 'making a happy academy' in his *Devis sur*

30 Yates F., *Astrée* (Paris: 1989) 25.

31 Jeannin Corbin M., "Jacques Gohory traducteur des *Amadis* ou l'annexion de l'altérité", in *Traduire le même, l'autre et le soi* (Aix-en-Provence: 2011) 155–161.

32 See Kahn, "Le Paracelsisme de Jacques Gohory" 81–130; idem, *Alchimie et Paracelsisme en France* 149–171. Walker, *Spiritual and Demonic Magic* 101.

33 Gohory referred to Marsilio Ficino and della Porta in Jacques Gohory, *Instruction sur l'herbe 'petum', ditte en France l'herbe de la royne, ou medicée, et sur racine mechiocan* (Paris: Galiot Du Pré: 1572) Epistre: 'Porta Neapolitinaus descrbit regulum Antomonii'; Gohory Jacques (Leo Suavius), *Theophrasti Paracelsi Philosophiae Et Medicinae Utriusque Universae, Compendium: ex optimis quibusque eius libris* (Basel, Peter Perna: 1568) 64; Magner L.N., *A History of the Life Sciences* (Boca Raton: 2002) 102.

34 Eamon W., *Science and the Secrets of Nature* (Princeton: 1994) 202.

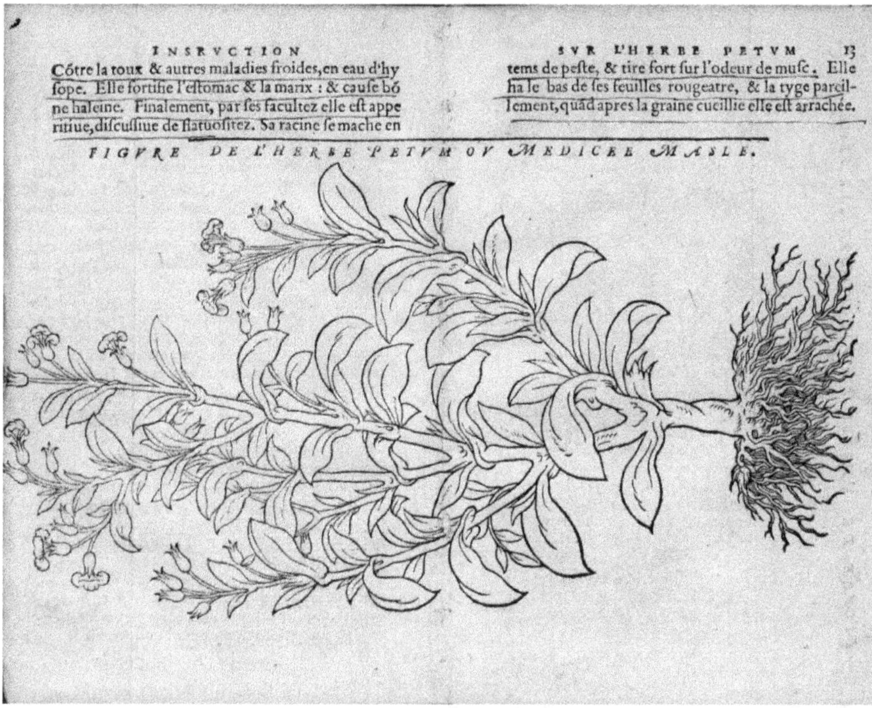

FIGURE 8.3 'Figure de l'herbe Petum ou médicée masle', *Instruction sur l'herbe petum ditte en France l'herbe de la Royne ou medicée*, Bibliothèque interuniversitaire de santé (Paris), Bibliothèque de pharmacie RES 19849
IMAGE © BIU SANTÉ

la vigne, published in 1549.[35] This dream became a reality in 1571 at his home in the Faubourg Saint-Marcel. Based on his May 1572 will, it is possible to mark the location of the Lycium between the rue Neuve Sainte-Geneviève and the rue Saint-Marceau [Fig. 8.2]: 'Also my house where the sign of Saint Estienne hangs in the grand rue of the faubourg saint marceau [...] the small house answering on the rue neuve sainte Geneviève'.[36]

In a pamphlet titled *Instruction sur l'herbe petum* (1574), Gohory writes that his garden is rich in 'rare and exquisite simples' that one must learn 'to handle philosophically' – an unequivocal reference to the field of natural philosophy [Fig. 8.3].[37] He was successful in cultivating plants of 'several new kinds, such

35 Gohory Jacques, *Devis sur la vigne, vin et vendanges* (Paris, Sertenas: 1550) fol. liiij v.

36 The will, dated 25 May 1572, is preserved in the Archives Nationales, Insinuations au Chatelet, Y 123, n. 3029, fols. 238r–39r and was transcribed by Bowen, *Jacques Gohory (1520–1576)* 81–83.

37 Gohory, *Instruction*, fol. 5r.

FIGURE 8.4
'Figure de l'herbe Petum ou médicée femelle', *Instruction sur l'herbe petum ditte en France l'herbe de la Royne ou medicée*, Bibliothèque interuniversitaire de santé (Paris), Bibliothèque de pharmacie RES 19849

as *pennaches* [plumes] with leaves of various colours, very pleasing to the eye, and flowering cabbages of excellent taste',[38] as well as the 'male and female *Petum* and several other rare Simples'.[39] In honour of Catherine de Médicis, Gohory stated that the *Petum* '[...] must rightly bear the name of Caterinaria, or Medicina: like the herb Arthemisia, of the Queen of Egypt: the gentian of Gentius King of Illyria: the Mithrydat of King Mithrydates'.[40] Here, the medicinal plant, the subject of an epidictic discourse, is erroneously considered a dioecious plant [Fig. 8.4–8.5]. In the chapter titled 'Consideration of the different

38 This appears to be a morphological characteristic and not a plant name. Pennache is also the name given to corals. See Du Tertre Jean Baptiste, *Histoire generale des Antilles habitées par les françois* (Paris, Thomas Jolly: 1667) 193–194.

39 Gohory, *Instruction*, fol. 5r.

40 Ibidem, fol. 2v.

FIGURE 8.5 Baptiste Pellerin, 'La Musique', *Liberal Arts*, pen and black ink on parchment, 19.5 × 28.5 cm. Paris, Musée du Louvre, Cabinet des Dessins (Inv. 774 RF)
IMAGE © RMN / MUSÉE DU LOUVRE

opinions on the previous preparations', Gohory questions some of the herbal preparations described in Charles Estienne (1504–1564)'s *La Maison rustique* as well as the description of certain plants by Pietro Andrea Mattioli, Leonhart Fuchs and other botanists.[41] No indication confirms whether the plants mentioned were grown in the *Lycium* garden. Gohory's demonstration names hyssop (*Hyssopus officinalis* L.), a Mediterranean medicinal plant frequently cited in garden literature and especially used in ornamental beds, 'Tuthie of Alexandria', which is not a plant but rather a zinc oxide used in ointments, and 'Euphrasia', which 'grow[s] in meadows' but not in gardens.[42] Gohory also cites well-known medicinal plants, some of which contributed to the ornamental and olfactory pleasures of gardens: rue (*Ruta graveolens* L.), verbena (*Verbena officinalis* L.), roses, chelidonia (*Chelidonium majus* L.), angelica (*Angelica archangelica* L.), 'two lunars', and the martagon (*Lilium martagon* L.) – a plant 'required by many curious people' and more a fragrant than medicinal

41 Ibidem, fols. 10r–11v.

42 Probably *Nigella damascena* L.

ornamental lily.[43] Gohory also mentions among the plants in his garden the 'Melanthium Alexandrin and the Romain, which bear their seeds in a round pod with some spikes like the head of a club' as well as the Scorpoides, Lagopus Alkali and Salicor.[44]

In 1571–1572, Gohory describes the lengths he took to protect his garden from the harsh winter weather:

> I have taken [...] great pains and care to preserve these noble plants with the flowering cabbages through the snows and frosts of this winter [plants that] the kind Choisnyn de Chastelleraud my good neighbour in this place gave me last year.[45]

From this note, we learn that at least one other simple garden was cultivated in the Faubourg Saint-Marcel near the Lycium: the garden of François Choisnin (*c*.1530–1584), physician to Jeanne d'Albret, Queen of Navarre, and Bernard Palissy's close friend.[46] He then gives what may be a surprising description of the layout of his Jardin de Simples:

> Now, I hope that in the spring, there will be rare and strange simples sown and planted to give contentment to people of spirit, who will delight as

43 Roses are probably *Rosa gallica* L. and *Rosa centifolia* L. The 'two lunars' are probably *Lunaria rediviva* L. The perennial lunaria, and *Lunaria annua* L. the annual lunaria, better known as the pope's coin. However, many plants have been named because of supposed relationships with lunar cycles. See Goupil T., "The moon and its relationship with early botanists", *Lejeunia, Revue de Botanique* 191 (2014), 1–18. Martagon, a plant 'required by many curious people', is more a fragrant, ornamental lily than medicinal. Dioscorides, *The Six Books* (Lyon, Balthazar Arnoullet: 1553) 408.

44 Gohory, *Instruction*, fol. 14r. 'Melanthium Romain' is *Nigella romana* L after Nicot Jean, *Thresor de la langue francoyse* (Paris, Douceur: 1606) 43. 'Melanthium Alexandrin' is probably *Nigella sativa* L. 'Scorpioides' is probably *Scorpiurus subvillosus* L. from the engraving illustrating this plant in Jacques Daléchamps, *L'histoire des plantes* (Lyon, Guillaume Rouillé: 1586) 1353. 'Lagopus' is probably the Harefoot *Plantago lagopus* L. after Matthiole Pierre André, *Commentaires de Dioscoride* (Lyon, Guillaume Rouillé: 1572) 538–539. 'Alkali' or 'Salicor' is the name given to the salicornia by several authors, which when reduced to ash releases salt, or alkali.

45 'J'ai pris [...] grand peine et soin à conserver ces nobles plantes avec les choux à fleurs par les neiges et gelées de cet hiver [des plantes que] me donna l'an passé le gentil Choisnyn de Chastelleraud mon bon voisin en ce lieu'. Gohory, *Instruction*, fol. 14.

46 Palissy Bernard, *Discours admirables* (Paris, Martin le Jeune: 1580) 202.

> well in the labyrinth of trees with its keep in the middle and its four turrets of elms bent to the four corners. Others will be near to the artificial fountain for which the pipes are lead. Others will admire the diversity of the fruit from the Grafts, planted in great number the length of the alleys and paths. Some with be at the edge of the two pavilions, one covered with plum trees, the other with cherry trees. Others will favour the ball or bowling games under a long and wide cradle of trellis. And when it will be time for them to leave, they will be able to look at the time on the horizontal compartment dial. Others, still, will make vocal and instrumental music under the historiated gallery (ornamented by fresco or sculptures).[47]

With its cradle-shaped paths and green pavilions, Gohory's quadrangular garden was characteristic of the period and, in his book, transformed into a labyrinth with 'turrets' and 'dungeons'. Emphatically evoking the mythical garden of Lucullus in ancient Rome, he compares the *Lycium* garden to the large gardens – about two hectares – of the nearby Abbaye Sainte-Geneviève, which was located on the other side of Charles v's wall [Fig. 8.2].[48] This description

47 'Or, j'espère, sur le printemps, qu'il n'y aura simple rare et étrange en ce pays qu'il n'y soit semé ou planté pour donner ce contentement aux gens d'esprit qui souvent se délectent au labyrinthe d'arbres garnis de son donjon au milieu et de quatre tourelles d'ormes courbez aux 4 coings. Les autres en la fontaine artificielle saillante par conduits de plomb. Les autres, ès fruits des Entes qui y sont de toutes sortes en grand nombre plantées à la ligne de deux côtés sur les allées et sentiers. Aucuns à l'orée des deux pavillons, l'un couvert de pruniers, l'autre de cerisiers. Autres a l'exercice de la boule ou quilles sous un long et large berceau de treillage. Et quand quelque assignation les presse de partir, regardent l'heure au cadran horizontal de compartiment. Autres s'adonnent à faire musique de voix et instruments en la galerie historiée'. Gohory, *Instruction*, fol. 14r. *Le Bon Mesnaiger. Au present volume des proffitz Champestres et ruraulx est traicté du labour des Champs, vignes, jardins, arbres de tous espèces, De leur nature et bonté, de la nature et vertu des herbes de la manière de nourrir toutes bestes, volailles et oyseaulx de proye* […] *Ledit livre compilé par Pierre des Crescens, jadis bourgeoys de Boulongne la grasse. Nouvellement corrigé, veu et amende sur les vieilz originaulx au par avant imprimez* […] (Paris. Charles Langelier: 1540) is the last edition of Crescenzi's *Opus ruralium commodorum, libri duodecim* (c.1320). On floral sundials and their use in Renaissance gardens, see further Ribouillault D., "Entre terre et ciel: les cadrans solaires floraux dans les jardins baroques", in Jakob M. (ed.), *Le Jardin de la Renaissance entre Arts et Sciences* (Pisa: 2022) 103–124.

48 'beautiful Lucullian garden of the magnificent abbot of Saincte-Geneviève Foullon, lord of this land'. ('beau jardin Lucullian du magnifique abbé de Saincte-Geneviève Foullon, seigneur de cette terre'.) Ibidem, fol. 14v.

conjures up the imaginary of the marvellous gardens in Pietro de Crescenzi's (1230–*c.*1320) *Le Bon Mesnaiger* (*The good husbandman*, 1540), Francesco Colonna's (1433–1527) *Poliphile* (1546), and *L'Onzième livre d'Amadis de Gaule* (1554), which Gohory translated. In this latter text, Phébus's castle has a garden '[...] furnished with all flowers, with alleys covered with cradles of musk-roses and jasmines, suave blooming and small fountains trickling by bushy places shaded by fruit trees [...]'.[49] We learn that '[this] beautiful garden provided herbs, gums and fruits: being the quarrels [beds] of it divided according to the various properties, one for sores, another for fevers, here the hot plants, here the cold ones and so on'.[50]

Despite calling himself 'solitary', Gohory cultivated not only medicinal plants and dreamlike gardens, but also friendships and scientific controversies. Staying away from the intrigues of the royal court and Parisian parliament, he remained close to natural philosophers. Referring to tobacco in his introduction to *L'Onzième*, Gohory writes:

> As for the cure of wounds, I have often experimented with the crushed leaf alone, of which the Sieur de la Brosse, the king's mathematician, a very learned man, provided me in abundance from his beautiful garden furnished with an infinity of rare plants and exquisite flowers.[51]

The de la Brosse to whom he refers and seems to have particularly liked was probably the father, grandfather, or uncle of Guy de la Brosse (1586–1641), the founder of the royal Jardin des Plantes (1626). The latter's father was a physician, astrologer, mathematician and Paracelsian known at the court of Henry IV.[52]

---

49 Gohory Jacques, *L'Onzième livre d'Amadis de Gaule* (Paris, Jean Longis: 1554) fol. 3v.

50 '[...] garni de toutes fleurs, avec allées couvertes de berceaux de roses muscades et jasmins, suave fleurant et petites fontaines ruisselant par lieux touffus ombragés d'arbres fruitiers [...] [ce] beau jardin fournissait les herbes, gommes et fruits: étant les quarreaux [parterres] d'icelui départis selon les propriétés diverses, l'un pour les plaies, l'autre pour les fièvres, ici les plantes chaudes, ici les froides et ainsi des autres'. Ibidem, fol. 4r.

51 'Quant à la cure des plaies, j'en ai fait souvent l'expérience de la feuille seule pilée, dont m'a fourni abondamment le sieur de la Brosse, mathématicien du roi, très docte, de son beau jardin garni d'une infinité de simples rares et de fleurs exquises'. Ibidem, fol. 7r.

52 Guerlac H., "Guy de La Brosse and the French Paracelsians", *Science, Medicine and Society in the Renaissance* (New York: 1972) 177–199; Howard R.C., "The Family of Guy de La Brosse, Founder of the Jardin des Plantes in Paris", *Proceedings of the XXVth International Congress in the History of Medicine* (Quebec: 1978) 771–789.

Gohory was not alone in his cultivation of tobacco in Paris. To this end, he mentions François Rasse (or Race Des Nœux, Neux, or Noux, still called Rasius), the surgeon to the king:[53] '[...] Race, surgeon, to plant a large quantity in a garden near the temple;[54] and of him I know the eldest, a very expert character in his art'.[55] Ambroise Paré, Palissy, and Jean Liébault, who continued the work of Charles Estienne, also reference Rasse.[56] Moving on from tobacco but remarking on Rasse's work, Gohory devotes the second part of the *Instruction* to the cultivation of mechoacan (*Convolvulus mechoacanna* Vitman.), a plant native to Mexico. He writes:

> you will be able to see it [the mechoacan plant] there in the medicinal gardens of Masters Nicolas Rasse, the learned and experienced surgeon, and Pierre Quthe, the knowledgeable and zealous apothecary, both of whom in Paris have enriched our France with an infinite number of rare and exquisite medicinal plants and gifted with singular virtues.[57]

Both Rasse and Gohory were close to Jacques Besson (1530–1572), who was known as a 'great mathematician, philosopher and engineer' as well as the author of a treatise on the distillation of simples.[58] This treatise so greatly marked Gohory that the 1580 edition of the *Instruction* included it in its first part.[59] Rasse and Gohory were known to frequent the apothecary Pierre Porret (*c.*1520–1596?), whom Gohory describes as 'a very ingenious man and

53 Veyrin-Forrer J., "Un collectionneur engagé: François Rasse des Neux, chirurgien parisien", in *La lettre et le texte* (Paris: 1987) 422–477; Idem, "Un collectionneur peu connu François Rasse des Neux, chirurgien parisien", *Studia in honorent Herman De La Fontaine verwey* (Amsterdam: 1968) 389–417.

54 A district of Paris.

55 Gohory, *Instruction*, fol. 6v.

56 Paré Ambroise, *Traicté de la peste* (Paris, André Wechel: 1568) 226; Palissy, *Discours admirables* 230.

57 'tu la pourras visiter [la plante de méchoacan] es jardins médicinaux de Messieurs, Maître Nicolas Rasse, le docte et bien expérimenté chirurgien, et Pierre Quthe le savant et soigneux Apothicaire, qui tous deux à Paris ont enrichi notre France d'une infinité de simples rares, exquis, et doués de singulières vertus'. Estienne, *L'Agriculture et maison rustique*, fol. 129v.

58 de La Croix du Maine François, *Les bibliothèques françoises* (Paris, Abel Angelier: 1584) 176.

59 Gohory Jacques – Besson Jacques, *Instruction de la connaissance des vertus et propriétés de l'herbe petum* [...] *Art et moyen parfaict de tirer hviles et eaux, de tous médicaments simples & oleagineux* (Paris, Jean Parant: 1580); Régnier-Roux D., "Notes sur les dernières années de Jacques Besson (1534?–1573?)", *Réforme, Humanisme, Renaissance* 77 (2013) 199–213.

a good simplist',[60] and La Croix du Maine, who was known as 'a good expert in distillations'.[61] An engineer for Charles IX, friend of Rasse des Neux, and agent of Christophe Plantin (1520–89), Porret gave lectures at the Lycium.[62] In Paris, Rasse and Porret were members of a controversial religious group with an Anabaptist background called the Famille de la Charité.[63] It would seem that the *Lycium philosophal* was not only the promoter of alchemy, an unorthodox science, but that it was also open to those with reformed religious ideas.

In addition to apothecaries and philosophers, Gohory maintained friendships with learned men of power like the magistrate Barnabé Brisson (1531–1591); as he wrote in *Instruction*, 'the famous lawyer Brisson, not only learned in his jurisprudence, but also in all antiquity and elegance of Greek and Latin letters'.[64] Important Italian lords also frequented the Lycium, such as Gian Francesco Carafa,[65] his friend Marcello Nonio Saia, and Filippo Strozzi, son of Pietro, 'maistre d'hostel du Roy: a character gifted with doctrine and good judgement'.[66] These gentlemen not only shared Gohory's scientific interests but also represented a privileged link to Catherine de Médicis. A man of diverse interests, Gohory was a known friend of the score editor Adrian Le Roy (1520–98);[67] Gohory wrote about his attraction to music and attested to its place in the garden in the prefaces he contributed to works on the subject. These observations make it possible to establish a link between Gohory and Baïf's academies, with the latter focusing on the relationships between music, alchemy, geometry and medicine. These disciplinary interrelations especially interested one of Baïf's prominent members, the musician Claude Lejeune (*c.*1530–1600), one of the preeminent secular composers of the time.[68]

60 Gohory, *Instruction*, fol. Aij.
61 La Croix du Maine, *Les bibliothèques françoises* 408–409.
62 Dolza L.M., "Reframing the Language of Inventions: The First Theatre of Machines", in Lefèvre W. – Renn J. – Schoepflin U. (eds.), *The Emergence of the Scientific Image, 1500–1700* (Basel: 2003) 120–156.
63 See Kirsop W., "The Family of Love in France", *Journal of Religious History* 3 (1964–1965) 103–118.
64 Gohory, *Instruction*, fol. 14v.
65 Ibidem, fol. aj r.
66 Ibidem, fol. 9r.
67 de Lassus Roland, *Secundus liber modulorum quinis vocibus constantium* (Paris, Adrian Le Roy & Robert Ballard: 1571); idem, *Moduli sex, septem et duodecim vocum* (Paris, Adrian Le Roy & Robert Ballard: 1573).
68 Bonniffet P., "Les sons et les mots"; Claude Lejeune, *Le Printemps*, by the ensemble Jacques Feuillie, 1995, booklet attached to the 2 CDs, 7–13; Brooks J., "Music as Erotic Magic in a Renaissance Romance", *Renaissance Quarterly* 60.4 (2007) 1207–1256.

In 1578, two years after Gohory's death, the Sorbonne rejected Paracelsianism during the trial of Roch Le Baillif (1540–1598).[69] Natural magic, which had already been the target of the Church in 1487 during the trial of Pico della Mirandola's *900 Theses*, was subject to witchcraft accusations in 1580 during the Jean Bodin's *De la demonomanie des sorciers* trial.[70] These condemnations likely had repercussions on the activities of other academies, even those that appear on first glance as very different to Gohory's. To this end, I will now turn our attention to another academy with a Jardin de Simples in the Faubourg Saint-Marcel.

3 Nicolas Houel's Academy of the 'Shameful Poor' (1578–1587)

Some authors have associated Nicolas Houel, the apothecary and master of the Maison de la Charité chrétienne philanthropic academy, with Paracelsian or familist thought.[71] However, this scholarly characterisation is more reflective of ideas circulating at the time than Houel's definite affiliation. Houel's intellectual environment seems to have been primarily oriented towards current and recently rediscovered medical and apothecary practices, which he subjected to a 'reduction into art' in his works *Traité de la Peste* (1573) and *Traité de la Theriaque et Mithridat* (1573).[72] At the same time, the Academy took on a charitable and devotional character linked to Catholicism, which was otherwise unheard of in France. This religious alignment was undoubtably influenced by Henry III's founding of the Order of the Holy Spirit in the same year, which brought together warriors in the service of God and their king and required they march in processions and perform charitable work.

Houel's commitment to academic pursuits seems to have begun at an early age, first in the company of apothecaries. In 1556, his name appears alongside Pierre Quthe in a lawsuit brought against former members of the corporation

69 Kahn, *Alchemy and Paracelsism in France* 278.

70 Walker, *Spiritual and Demonic Magic* 96–106. Bodin Jean, *De la démonomanie des sorciers* (Paris, Jacques du Puys: 1580) fol. 198v; Jacques-Chaquin N., *Jean Bodin, nature, histoire, droit et politique* (Paris: 1996) 43–70.

71 See Kahn D., "Paracelsisme et alchimie sous le règne de Henry III", in Poirier G. – Maillard J.-F. – De Conihout I. (eds.), *Henry III mécène des arts, des sciences et des lettres* (Paris: 2006) 91–103, and Yates, *Astrée* 192–193; Hamilton A., *The Family of Love* (Cambridge: 1981); Maillard J.-F., "Christophe Plantin et la Famille de la Charité en France: documents et hypothèses", *Mélanges sur la littérature de la Renaissance à la mémoire de R.-L. Saulnier* (Geneva: 1984) 235–253.

72 Vérin H., "Rédiger et réduire en art: un projet de rationalisation des pratiques", in Vérin H. – Dubourg Glatigny P. (eds.), *Réduire en art, La technologie de la Renaissance aux Lumières* (Paris: 2008) 17–58.

of apothecaries.[73] Throughout his life, Houel engaged in long, complex relationships of artistic patronage. These were defined by the production of pictorial and poetic epidictic works with a view to obtaining support from the royal court for the construction of his House of Christian Charity.[74] Such undertakings brought him closer to and even led to his integration in the artistic community of the time; in one case, he bore witness to the wedding of sculptor Germain Pilon. The Christ carried in the *Procession of Louise de Lorraine* (1583), the work to which we will turn more closely, includes a reproduction of a work by this sculptor.[75]

Between 1560 and 1574, Houel sought to found an academy that would combine characteristics of Baïf and Gohory's Academies under the patronage of Charles IX; his famous *Seven Liberal Arts* suite is believed to relate to this undertaking.[76] In the illustration of *La Musique* [Fig. 8.5], which clearly evokes the Académie de Poésie et Musique, muses perform a concert in a pleasure garden. Under a cradle of vines, elderly men and children – figures that symbolise the ages of life – study around a table bearing Houel's monogram. In the background, a deer, symbolising hearing, grazes on the edges of a labyrinthine flowerbed. In addition to this piece, another graphic work was produced with the intent of obtaining royal favours: the frieze portraying *The procession of Louise de Lorraine, wife of Henry III, going from the Louvre to the Faubourg Saint-Marceau to lay the stone of the New House known as the Maison Chrétienne, planned and even begun in 1584*. Depicting a royal procession from the Louvre to the Faubourg Saint-Marcel, this frieze illustrates the philanthropic academy imagined by Houel, which was at this time still called Maison de la Charité chrétienne. A series of eleven drawings, it shows a Jardin de Simples and an apothecary in five of them [Fig. 8.5, 8.6, 8.7, 8.8, 8.9 and 8.10].

---

73 Houel, Quthe, and other young apothecaries '[...] intentent un procès aux anciens de la corporation à propos des nombreux abus commis par ceux-ci'. Dorveaux P., *Pierre Quthe, maître apothicaire de Paris* (Paris: 1908) 12. Three years later, Pierre Quthe commissioned a portrait by François Clouet. The garden of Nicolas Rasse mentioned earlier was close to that of Pierre Quthe.

74 See in particular Auclair V., "Changement de Programme. De L'Histoire de la Reine Artémise de Nicolas Houel à la tenture d'Henri IV", in *À l'origine des Gobelins, la tenture d'Artémise* (Paris: 2007) 19–26.

75 It should be noted that the apothecary Chassebras, a close friend of Baïf, commissioned a sculpted decoration by Germain Pilon for the choir of the church of Saint-Étienne-du-Mont.

76 Toesca I., "Quelques dessins attribués à Etienne Delaune", *Revue des arts* 6 (1960) 255–259; Bjurström P., "Etienne Delaune and the Academy of Poetry and Music", *Master Drawings* 34.4 (1996) 351–364; Auclair V., "Etienne Delaune dessinateur? Un réexamen des attributions", in Zerner H. – Bayard M. (eds.), *Renaissance en France et renaissance française?* (Paris: 2009) 143–160.

FIGURE 8.6 *The procession of Louise de Lorraine, wife of Henry III, from the Louvre to the Faubourg Saint Marceau to lay the stone of the New House called Maison Chrétienne, planned and even begun in 1584*, 1583, 8th drawing, Paris, Bibliothèque nationale de France (Est Res pd-30-Fol)
IMAGE © BNF

FIGURE 8.7 *The procession of Louise de Lorraine, wife of Henry III, from the Louvre to the Faubourg Saint Marceau to lay the stone of the New House called Maison Chrétienne, planned and even begun in 1584*, 1583, 7th drawing, Paris, Bibliothèque nationale de France (Est Res pd-30-Fol)
IMAGE © BNF

FIGURE 8.8 *The procession of Louise de Lorraine, wife of Henry III, from the Louvre to the Faubourg Saint Marceau to lay the stone of the New House called Maison Chrétienne, planned and even begun in 1584*, 1583, 6th drawing, Paris, Bibliothèque nationale de France (Est Res pd-30-Fol)
IMAGE © BNF

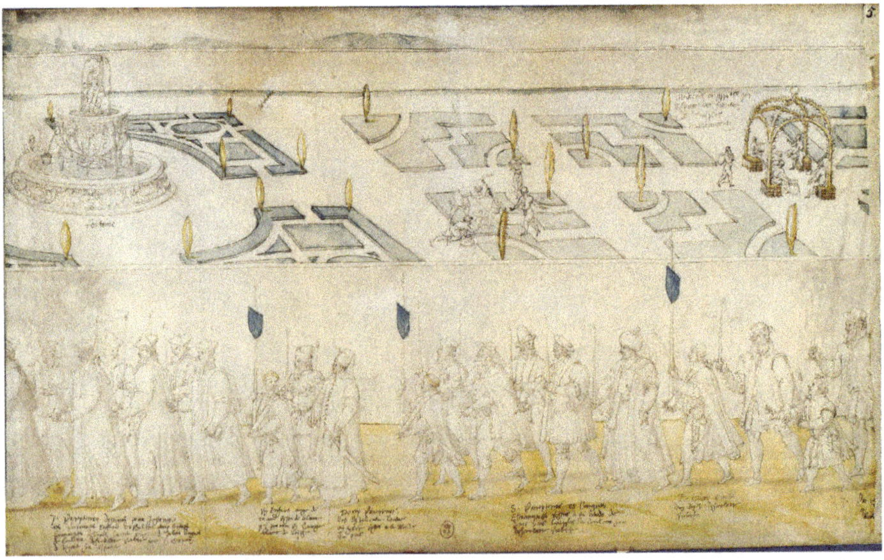

FIGURE 8.9 *The procession of Louise de Lorraine, wife of Henry III, from the Louvre to the Faubourg Saint Marceau to lay the stone of the New House called Maison Chrétienne, planned and even begun in 1584*, 1583, 5th drawing, Paris, Bibliothèque nationale de France (Est Res pd-30-Fol)
IMAGE © BNF

FIGURE 8.10 *The procession of Louise de Lorraine, wife of Henry III, from the Louvre to the Faubourg Saint Marceau to lay the stone of the New House called Maison Chrétienne, planned and even begun in 1584*, 1583, 4th drawing, Paris, Bibliothèque nationale de France (Est Res pd-30-Fol)
IMAGE © BNF

An official document from October 1576 – the year of Gohory's death and the transfer of Baïf's Academy to the Louvre – outlines a programme for the foundation of the future Maison de Charité:

> founded, built and endowed a house of charity, in which a certain number of orphan children will be fed and trained, first in piety and good letters, and then in the art of apothecary, and there will be prepared and supplied and administered free of charge all sorts of medicines and remedies suitable for the shameful poor of the city and suburbs of Paris.[77]

77 'fondé, édifié et doté une maison de charité, en laquelle soit nourri et institué certain nombre d'enfants orphelins, premièrement à la piété et bonnes lettres, et par après en l'art d'apothicairerie et y soient préparées et fournies et administrées gratuitement toutes sortes de médicaments et remèdes convenables aux pauvres honteux de la ville et faubourgs de Paris'. Félibien Michel, *Histoire de la ville de Paris*, vol. 3 (Paris, Guillaume Desprez and Michel Desessartz: 1725) 721.

The 'shameful poor' referenced are the destitute population reduced to begging and vagrancy for economic reasons. Their condition was perceived as an affront to French honour. Following the decree of 19 August 1577, Houel was attached to the Lourcine hospital and in 1578 appointed '[…] to the superintendence of the aforementioned house, chapel, apothecary and simple garden'.[78] The former Ourcine (or Saint-Marcel) hospital – then still called the Hôtel-Dieu du Patriarche – was thus rechristened Maison ou Hôpital de la Charité chrétienne.[79] A miniature of the manuscript *Traité de la Charité chrestienne* (*Treatise on Christian Charity*) suggests that the previous residents of the hospital suffering from syphilis were expelled and left to wander the streets, reflecting an updated mandate for transforming the poor rather than caring for the sick. Houel began work quickly, but in 1579: '[…] the apothecary and the garden of the Simples, which had been started and was well advanced […]' were seriously damaged by a sudden flood of the Bièvre.[80] In an *advertissement* printed in 1580, Houel describes his academy building project:

> When it will please the goodness and mercy of God to increase the good of this poor House, illuminate the King, Princes and Lords and other persons, filled with fervor for God and charity for their neighbour, to contribute large alms, in order to add to it the seven liberal arts and with the other disciplines and sciences up to the Greek and Hebrew languages, even foreign languages; so that it will be an Academy of all piety and science, all for the advancement of the Glory of God, profit and decoration of the Republic.[81]

The work and operations of the establishment are not well known. Exhausted by his reconstruction efforts, Houel died in 1587. In 1597, the mission of the House of Charity was again significantly modified with the requirement to

78 Ibidem, I, 5.

79 Letters Patent in the registers of the Parliament on the date of 9 March 1577, the day of their registration. Letters of provisions given in conformity by the King to Nicolas Houël on the same day, 20 January and similarly registered. Quoted by Babille Laurent-Jean, *Mémoire pour les Maitres et Gardes Apothicaires* (Paris, Michel Lambert: 1770) 14.

80 Houel Nicolas, *Advertissement* (Paris, Pierre Cheuillot: 1580) fol. Bij.

81 'Quand il plaira à la bonté et miséricorde de Dieu accroître le bien de cette pauvre Maison, allumer le Roy, les Princes et Seigneurs et autres personnes, remplies de zèle de Dieu et charité du prochain, à y aumôner de leurs biens, l'en y ajouter les sept arts libéraux et avec les autres disciplines et sciences jusques à la langue grecque et hébraïque, même les langues étrangères ; de sorte que ce sera une Académie de toute piété et science, le tout pour l'avancement de la Gloire de Dieu, profit et décoration de la République'. Houel Nicolas, *Traité de la Charité chrestienne*, fol. Bij v; illuminated manuscript, Princes Czartoryski Foundation, Kraków, MS 3092.

take in wounded soldiers.[82] In 1612, the Hotel Sardini – the former property of Scipio Sardini – received the "locked-in poor". In 1624, Houel's desired apothecary garden was finally established on the other side of the rue de Lourcine.[83]

Old plans show the location of the Maison de la Charité chrétienne on rue de Lourcine (now rue Broca) almost at the left entrance when arriving from rue Mouffetard. The adjoining land, quite large by urban standards, slopes down towards the Bièvre. Truschet and Hoyau's map of Paris (*c.*1552) mentions the location of the "hospital" [Fig. 8.2]. The architectural complex, like the actual gardens of the Christian Charity, appear modest in various plans produced between the sixteenth and eighteenth centuries.[84] However, the *Advertissement* specifies a garden project distinct in its ambitious scope:

> the Jardin des Simples, which, in imitation of the city of PADUA, will be filled with beautiful fruit trees, and rare and necessary fragrant plants and various species for the use of medicine for the relief of the sick, both rich and poor, which will bring great profit and decoration to the city of Paris.[85]

Perhaps following the advice of Pierre Belon, who testified in 1558 to the 'great recommendation of honour' earned by the Republic of Venice for the creation of its botanical garden in Padua, Houel imagined his garden as a source of prestige.[86] As indicated in *Traité de la thériaque et mithridat* (1573), he was in contact with Paduan officials who also sent seeds:

> As for tordylium (*Tordylium maximum* L.) or Sefeli de Candie, I once thought to have one in my garden, where I had sown some seed that had been brought to me, as Tordylium seeds, from the Medicinal Garden that is in Padua; but taking a closer look at the shape and taste of the said plant, I found it to be quite different from Tordylium.[87]

---

82 Félibien, *Histoire de la ville de Paris* III, 730.

83 Babille, *Mémoire pour les Maitres et Gardes Apothicaires* 19.

84 Paya, "Du Jardin des Simples".

85 '[…] le Jardin des Simples, lequel à l'imitation de la ville de PADOUE, sera rempli de beaux arbres fruictiers, et plantes odoriférantes rares et requises et de diverses espèces servant à l'usage de la médecine pour le secours des malades tant riches que pauvres, qui apportera un grand profit et une grande décoration à la ville de Paris'. Houel, *Advertissement*, fol. Bij r.

86 Belon Pierre, *Les Remontrances sur le défault de labour et culture des plantes et de la cognoissance d'icelles* (Paris, Gilles Corrozet: 1558) fols. 60r–65r.

87 'Quant à Tordylium où Sefeli de Candie, j'ai eu autrefois opinion d'en avoir en un mien jardin, où j'avais semé certaine graine qu'on m'avait apportée, pour graine de Tordylion, du Jardin Médicinal qui est à Padoue; mais prenant garde de plus près à la forme & au goût

Some of the sixty-four plants used in the manufacture of Thyriac, which also employed viper venom as one of its major ingredients, are probably among the plants grown in his Jardin des Simples.[88]

A plate from *The procession of Louise de Lorraine*, which depicts a snake hunt at the back of the garden, may allude to this medicinal preparation [Fig. 8.10]. In the image [Figs. 8.7, 8.8, 8.9, 8.10], the Jardin des Simples is represented as an idealised Christian charitable complex. The cultivated area of this exemplary garden is depicted as an oblong surface surrounded by a high wall, which is pierced by a gate surmounted by a pedimented portico on the Bièvre side. Although the walled garden follows a symmetrical axis, there is no relationship between the garden and the building. Modelled after the Belvedere courtyard designed by Bramante in 1505, the garden is arranged along the axis of a double staircase linking it to the terrace.[89] A group of figures contemplates an expanse of the plant carpet and the centre of the garden is occupied by a monumental fountain.[90] This vast basin is surmounted by a deep bowl from which the figure of Charity emerges. Water flows through the mouths of cherubic mascarons, and leaf scrolls, festoons and cartouches contribute to the lavish decoration. The depiction of the garden is treated in compartments of broken squares, while two cabinets are placed symmetrically, under which apothecaries converse.[91] In a manner comparable to the illustrations included in Gohory's *Lycium*, the characteristics of a pleasure garden are well represented.

In other plates of the *Queen's Procession*, we also find representations of the dwelling and school for the 'shameful poor'. A large building with a roof terrace, the complex evoked contemporary French palatial architecture. The draughtsman responsible for this illustration included detailed captions, amongst which we find: 'Various arts and crafts for crippled soldiers', the

---

de ladicte plante, je la trouvay bien différente du Tordylium'. Houel Nicolas, *Traité de la thériaque et mithridat, contenant plusieurs questions générales et particulières* (Paris, Jean de Bordeaux: 1573) fol. 93v. He also claims that Mathiole (Mattioli) received plants from the "jardin publique des simples" of Padua. Ibidem, fol. 96v.

88 See "The simple medicines entering the Theriach", Ibidem, n.p.

89 Paya, "Du Jardin des Simples".

90 Ibidem, for our restitution of the layout of the parterres. On Renaissance carpet gardens see Paya L., "Le *topos* du jardin comme 'tapis de Turquie' au temps des humanistes, une figure d'art et de mémoire", *Intermédialités* 35 (2020), https://www.erudit.org/fr/revues/im/2020-n35-im05945/1076368ar/.

91 The term "broken squares" refers to an ornamental pattern, translated in Hill Thomas, *The Gardener's Labyrinth* (London, H. Bynneman: 1577).

'French Academy of various kinds of craftsmen', the 'College of all disciplines and languages for poor students and poor orphan children', the 'School of Music', the 'Dwelling for poor widows', the 'School for poor girls to be married who will be instructed in various trades and sciences', and the 'Dwelling of the girls' mistresses'. By teaching the art of distillation in the apothecary [Figs. 8.6 and 8.7], this Academy promoted the utopian goal of transforming the most destitute into a new elite. In addition to learning music and arithmetic, the beneficiaries were also taught a variety of languages by teachers depicted wearing their national clothes in a procession [Fig. 8.9].[92] An adaptation of the *Seven Liberal Arts* [Fig. 8.5], *The procession of Louise de Lorraine* adopts the pictorial conventions of devotional and ceremonial representation in vogue under Henry III.[93] It portrays an ideal government devoted to the education of the poorest in the pursuit of peace and happiness and expresses hope for the return of royal visits to the Faubourg Saint-Marcel, as had been the case under Charles IX with his visit to the Académie de Baïf. This work, offered to Henry III, likely surpassed that which was offered to Charles IX and, with its dedication, also refers to the period during which Marguerite de Provence lived in the Faubourg Saint-Marcel.

Both of the academies with Jardins de Simples that were described and represented as pleasure gardens, the Lycium philosophal (1571–1578) and the House of Christian Charity (1578–1597), were the product of different, innovative projects. Their distinctiveness stems from their founders as well as the development of the history of ideas. The first Académie disappeared in a time when alchemy and natural magic were under attack in Paris by both the university and the Church; the establishment of a second Académie (1578–1580) coincided with a rising tide of religious and charitable courtly devotion. As a relatively independent researcher, Gohory allowed himself bold but controversial experiments. Houel, on the other hand, had recourse to the support of royal powers and thus had to tread more carefully by honouring his patrons with a project more unifying than polemic. These two academies are part of a diachronic history featuring elements of continuity that relate to the history of the urban periphery of the Faubourg Saint-Marcel from the thirteenth to the sixteenth centuries.

92 Paya, "Du Jardin des Simples".

93 Yates, *Astrée* 295–397.

As institutions seeking to train elites while maintaining independence from university authorities, early French academies were established in urban peripheries close to, but distinct from, the Parisian university district. This tendency extends beyond the sixteenth century to a longer history of prestigious, Parisian institutions. The emergence of academies can be connected to the legacy of scientific institutions such as the Jardin des Apothicaires, the Jardin des Plantes, the Muséum national d'Histoire naturelle, the Institut national d'Agronomie, the Hôpital de Broca, and the Faculté de Pharmacie, which all share a close location, or even direct proximity, to the original locations of the two sixteenth-century Jardins de Simples académiques, highlighting the understudied nature of their geographic distribution and its significance to the study of early modern scientific and artistic thought.

References

Auclair V., "Nicolas Houel et Petrus Stephanus: un manuscrit enluminé du XVIe siècle, le *Traité de la Charité chrestienne*", *Revue de l'art* 132 (2001) 9–28.

Auclair V., "Un logis pour l'âme des rois. Nicolas Houel (c.1520–c.1587) et les dessins de procession à la maison de la Charité chrétienne pour la famille royale", *Henry III mécène des arts, des sciences et des lettres* (Paris: 2006) 39–54.

Auclair V., "Changement de Programme. De L'Histoire de la Reine Artémise de Nicolas Houel à la tenture d'Henri IV", in *À l'origine des Gobelins, la tenture d'Artémise* (Paris: 2007) 19–26.

Auclair V., "Etienne Delaune dessinateur? Un réexamen des attributions", in Zerner H. – Bayard M. (eds.), *Renaissance en France et renaissance française?* (Paris: 2009) 143–160.

Baader G., "Jacques Dubois as a Practitioner", in Wear A. – French R. K. – Lonie I.M. (eds.), *The Medical Renaissance of the Sixteenth Century* (Cambridge: 1986) 146–154.

Babille Laurent-Jean, *Mémoire pour les Maitres et Gardes Apothicaires* (Paris, Michel Lambert: 1770).

Balmas E., *Saggi e Studi sul Rinascimento Francese* (Padua: Liviana, 1982) 23–73.

Belon Pierre, *Les Remontrances sur le défault de labour et culture des plantes et de la cognoissance d'icelles* (Paris, Gilles Corrozet: 1558).

Bodin Jean, *De la démonomanie des sorciers* (Paris, Jacques du Puys: 1580).

Bonniffet P., "Les sons et les mots. Geometry and alchemy in *Le Printemps*", in Claude Lejeune, *Le Printemps*, by the Jacques Feuillie ensemble, 1995, booklet attached to the 2 CDs, 7–13.

Boutroue M.-É., "Paolo Boccone and the Academy of Bourdelot", in P. Galand-Hallyn – M. Deramaix – G. Vagenheim – J. Vignes (eds.), *Les Académies dans l'Europe humaniste, idéaux et pratiques* (Geneva: Droz, 2008) 663–675.

Bowen W.H., *Jacques Gohory (1520–1576)* (Ph.D. dissertation, Harvard University: 1935).

Bowen W.H., "The Earliest Treatise on Tobacco: Jacques Gohory's *Instruction on the Herb Petum*", *Isis* 28.2 (1938) 349–363.

Brooks J., "Music as Erotic Magic in a Renaissance Romance", *Renaissance Quarterly* 60. 4 (2007) 1207–1256.

Broomhall S., "Hearts on Fire: Compassion and Love in Nicolas Houel's *Traité de la Charité chrestienne*", in Broomhall S. (ed.), *Ordering Emotions in Europe, 1100–1800*, (Leiden: 2015) 121–160.

Bjurström P., "Etienne Delaune and the Academy of Poetry and Music", *Master Drawings* 34.4 (1996) 351–364.

Code de l'hôpital-général de Paris, ou Recueil des principaux édits, arrêts, déclarations & réglements qui le concernent, ainsi que les maisons & hôpitaux réunis à son administration (Paris, veuve Thiboust: 1728).

Coyecque E., *Recueil d'actes notariés relatifs à l'histoire de Paris* (Paris: 1923).

Daléchamps Jacques, *L'histoire des plantes* (Lyon, Guillaume Rouillé: 1586).

De la mare (Delamare) Nicolas, *Traité de la Police*, vol. 1 (Paris, Jean et Pierre Cot: 1710).

Dioscorides, *The Six Books* (Lyon, Balthazar Arnoullet: 1553).

Dolza L.M., "Reframing the Language of Inventions: The First Theatre of Machines", in Lefèvre W. – Renn J. – Schoepflin U. (eds.), *The Emergence of the Scientific Image, 1500–1700* (Basel: 2003) 120–156.

Dubois Jacques, *La Pharmacopée* (Lyon, Loys Cloquemin, et Estienne Michel: 1574).

Dupèbe J., "L'alimentation des pauvres selon Sylvius", in Margolin J.-C. – Sauzet R. (eds.), *Pratique et discours alimentaires à la Renaissance* (Paris: 1982) 41–56.

Du Tertre Jean Baptiste, *Histoire generale des Antilles habitées par les françois* (Paris, Thomas Jolly: 1667).

Eamon W., *Science and the Secrets of Nature* (Princeton: 1994).

Estienne Charles – Liébault Jean, *L'Agriculture et maison rustique* (Paris, Jacques Du Puy: 1583).

Félibien Michel, *Histoire de la ville de Paris* vol. 3 (Paris, Guillaume Desprez and Michel Desessartz: 1725).

Fontanon Antoine, *Les Édicts et ordonnances des rois de France*, vol. 1 (Paris, s.n.: 1611).

Fournier E., *Variétés historiques et littéraires*, vol. 2 (Paris: 1855).

Frémy E., *L'Académie des derniers Valois: académie de poésie et de musique 1570–1576, Académie du Palais 1576–1582* (Paris: 1887).

Gohory Jacques, *Devis sur la vigne, vin et vendanges* (Paris, Sertenas: 1550).

Gohory Jacques, *L'Onzième livre d'Amadis de Gaule* (Paris, Jean Longis: 1554).

Gohory Jacques (Leo Suavius), *Theophrasti Paracelsi Philosophiae Et Medicinae Utriusque Universae, Compendium: ex optimis quibusque eius libris* (Basel, Peter Perna: 1568).

Gohory Jacques, *Instruction sur l'herbe petum, ditte en France l'herbe de la royne, ou Médicée; et sur la racine mechiocan* (Paris, Galiot Du Pré: 1572).

Gohory Jacques – Besson Jacques, *Instruction de la connaissance des vertus et propriétés de l'herbe petum* [...] *Art et moyen parfaitement de tirer huiles et eaux, de tous médicaments simples & oleagineux* (Paris, Jean Parant: 1580).

Gorris Camos R., "L'Hysope et la rose: le *Lycium philosophal* de Jacques Gohory", in Blocker D. (ed.), *Les Académies entre France et Italie* (Geneva: 2007) 549–590.

Gorris Camos R., "In the labyrinth of Gohory, reader and translator of Machiavelli", *Laboratoire italien* 8 (2008) 195–229.

Goujet Claude Pierre, *Mémoire historique et littéraire sur le Collège royal de France* (Paris, Martin Lottin: 1758).

Goupil T., "The moon and its relationship with the first botanists", *Lejeunia, Revue de Botanique* 191 (2014) 1–18.

Guerlac H., "Guy de La Brosse and the French Paracelsians", in Debus A.G. (ed.), *Science, Medicine and Society in the Renaissance* (New York: 1972) 177–199.

Hamy E.-T., *Un précurseur de Guy de La Brosse. Jacques Gohory et le Lycium philosophal de Saint-Marceau-lès-Paris (1571–1576)* (Paris: 1899).

Houel Nicolas, *Traité de la thériaque et mithridat, contenant plusieurs questions générales et particulières* (Paris, Jean de Bordeaux: 1573).

Houel Nicolas, *Advertissement et déclaration de l'institution de la maison de la Charité chrestienne* (Paris, Pierre Cheuillot: 1580).

Houel Nicolas, *Traité de la Charité chrestienne*, Princes Czartoryski Foundation, Kraków, MS 3092.

Howard R.C., "The Family of Guy de La Brosse, Founder of the Jardin des Plantes in Paris", in *Proceedings of the XXVth International Congress in the History of Medicine* (Quebec: 1978) 771–789.

Jeannin Corbin M., "Jacques Gohory translating *Amadis* or the annexation of otherness", in *Translating the same, the other and the self* (Aix-en-Provence: 2011) 155–161.

Kahn D., "Le paracelsisme de Jacques Gohory", *Aries* 19 (1995) 81–130.

Kahn D., "Paracelsisme et alchimie sous le règne de Henry III", in Poirier G – Maillard J.-F. – De Conihout I. (eds.), *Henry III mécène des arts, des sciences et des lettres* (Paris: 2006) 91–103.

Kahn D., *Alchimie et Paracelsisme en France* (1567–1625) (Geneva: 2009).

Elliot L.K., "In Pursuit of Charity: Nicolas Houel and his *Maison de la Charité chrétienne* in Late Sixteenth-Century Paris", in Broomhall I.S. (ed.), *Experience of Charity, 1250–1650* (London: 2016) 149–170.

Kirsop W., "The Family of Love in France", *Journal of Religious History* 3 (1964–1965) 103–118.

Laborde A., *Nicolas Houel, fondateur de la Maison de charité chrétienne* (Paris: 1937).

La Croix du Maine François de, *Les bibliothèques françoises* (Paris, Abel Angelier: 1584).

Lassus Roland de, *Secundus liber modulorum quinis vocibus constantium* (Paris, Adrian Le Roy and Robert Ballard: 1571).

Lassus Roland de, *Moduli sex, septem et duodecim vocum* (Paris, Adrian Le Roy and Robert Ballard: 1573).

Lestringant F., "L'Histoire d'André Thevet, de deux voyages par luy faits dans les Indes Australes et Occidentales (c. 1588)", 4, http://editions-villegagnons.com/THEVET_Lestringant.pdf.

Lestringant F., *André Thevet, cosmographe des derniers Valois* (Geneva: 1991).

Lestringant F., *Sous la leçon des vents. Le monde d'André Thevet, cosmographe de la Renaissance* (Paris: 2003).

Magner L.N., *A History of the Life Sciences* (Boca Raton: 2002).

Maillard J.-F., "Christophe Plantin et la Famille de la Charité en France: documents et hypotheses", *Mélanges sur la littérature de la Renaissance à la mémoire de R.-L. Saulnier* (Geneva: 1984) 235–253.

Matthiole Pierre André, *Commentaires de Dioscoride* (Lyon, Guillaume Rouillé: 1572).

Mercier Louis-Sébastien, *Tableau de Paris* I (Hamburg, Virchaux: 1781).

Niceron Jean-Pierre, *Mémoires pour servir à l'histoire des hommes*, 29 (Paris, Briasson: 1733).

Nicot Jean, *Thresor de la langue francoyse* (Paris, Douceur: 1606).

Palissy Bernard, *Discours admirables* (Paris, Martin le Jeune: 1580).

Paré Ambroise, *Traicté de la peste* (Paris, André Wechel: 1568).

Paya L., "Du Jardin des Simples de la Maison de la charité chrétienne au Jardin des Apothicaires (1578–1624)", in Vons J. (ed.), *La Médecine à la cour de France* (2008), http://cour-de-france.fr/article590.html.

Paya L., "Le *topos* du Jardin comme 'Tapis de Turquie' au temps des humanistes, une figure d'art et de mémoire", *Intermédialités* 35 (2020), https://www.erudit.org/fr/revues/im/2020-n35-im05945/1076368ar/.

Planchon G., "Le Jardin des Apothicaires", *Le journal de pharmacie et de chimie* 28 (1893) 250–258, 289–298, 342–349, 412–416.

Poëte M., *Une vie de cité. Paris, de sa naissance à nos jours*, vol. 2 (Paris: 1927).

Régnier-Roux D., "Notes on the last years of Jacques Besson (1534?–1573?)", *Reformation, Humanism, Renaissance* 77 (2013) 199–213.

Ribouillault D., "Hortus academicus: Les académies de la Renaissance et le jardin", in Jakob M. (ed.), *Des jardins et des livres* (Geneva: 2018) 23–34.

Ribouillault D., "Entre terre et ciel: les cadrans solaires floraux dans les jardins baroques", in Jakob M. – Pierre B. (eds.), *Le jardin de la Renaissance entre arts et sciences* (Pisa: 2022) 103–124.

Toesca I., "Quelques dessins attribués à Etienne Delaune", *Revue des arts* 6 (1960) 255–259.

Vérin H., "Rédiger et réduire en art: un projet de rationalisation des pratiques", in Vérin H. – Dubourg Glatigny P. (eds.), *Réduire en art. La technologie de la Renaissance aux Lumières* (Paris: 2008) 17–58.

Veyrin-Forrer J., "Un collectionneur engagé: François Rasse des Neux, chirurgien parisien", in *La lettre et le texte* (Paris: 1987) 422–477.

Veyrin-Forrer J., "Un collectionneur peu connu François Rasse des Neux, chirurgien parisien", in *Studia in honorent Herman De La Fontaine verwey* (Amsterdam: 1968) 389–417.

Vignes J., "Musical experimentation and pedagogy in the Renaissance: about the conflict between the Academy of Poetry and Music and the Parliament of Paris (1570)", in *Les Académies dans l'Europe humaniste. Idéaux et pratiques* (Droz: 2008) 493–506.

Walker D.P., *Spiritual and Demonic Magic from Ficino to Campanella* (London: 1958).

Yates F., *Astrée. Le symbolisme impérial au XVI*[e] *siècle* (Paris: 1989).

Yates F., *Les Académies en France au XVI*[e] *siècle* (Paris: 1996).

CHAPTER 9

Festive Erudition: the Garden of Laurentius Scholz von Rosenau in Breslau (Wrocław) 1588–1599

Christiane Lauterbach

Abstract

The garden of the physician and humanist Laurentius Scholz von Rosenau existed in Breslau (Wrocław) for only 12 years, from 1588 to 1599. It was both a place of pleasure and a botanical garden for the use of scholars. The Breslau 'collegium illustrium ingeniorum', which met in the Hortus Scholzianus, was a humanist community that cultivated erudite and cheerful exchanges. Its members met regularly in the garden for philosophical discussions and banquets. These meetings brought the tradition of ancient and Renaissance academies to Silesia.

Keywords

Breslau (Wrocław) – botanical garden – pleasure garden – late humanism – philosophical meetings

1 Breslau and Its 'collegium illustrium ingeniorum'

'Flos Sacer Europae, clarissimeque urbium ocelle' ('Sacred flower of Europe, glorious jewel among the cities'): with this verse in his poem *Ad Vratislaviam*, Valens Acidalius (1567–1595) launched into high praise of Breslau, today Wrocław in Poland.[1] In the fifteenth century, the powerful trading city experienced an educational boom that brought with it a blossoming of the school system and scientific study. As a result, Breslau became the capital of humanism and the Reformation movement in Silesia in the sixteenth century.

1 Acidalius Valens, *Valentini Acidalii Epigrammata. Ad Danielem Rindfleisch Bucretium, Vratislaviensem, fratrem carißimum certissimum* (Helmstedt, Jakob Lucius the Elder: 1589), fol. A 2v–A 3v, here fol. A 2v. On Acidalius, see IJsewijn J., "An admirer of Justus Lipsius: the German neo-Latin poet and philologist Valens Acidalius", *Mededelingen van de Koninklijke Academie voor Wetenschappen, Letteren en Schone Kunsten* 45 (1983) 183–206.

 | DOI:10.1163/9789004517547_010

The young, aspiring poet Valens Acidalius first published his praise of the city in 1589 in a collection of epigrams dedicated to his best friend and patron, physician Daniel Bucretius (1562–1621). Acidalius saw Breslau's fame at the end of the century as the 'flower of Europe' embodied in its most important citizens, those who had rendered outstanding services through their social endeavours and physical constructions. The eulogy gives us an overview of the late humanist bourgeois elite in Breslau in its most concise form. The first name Acidalius mentions is, of course, Bucretius, followed by the important Monaus and Rehdigers families – the latter represented by Thomas and Nicolaus –, Andreas Dudith, Johannes Crato von Krafftheim, Abraham von Jenckwitz, Zacharias Ursinus, Johann Matthäus Wacker von Wackenfels, Martin Schilling, and, finally, the physicians Johann Hermann and Laurentius Scholz. Many of these names could also be found in the learned circle that gathered in Breslau around Melanchthon's student, the physician and humanist Johannes Crato von Krafftheim (1519–1585) beginning in the middle of the sixeenth century.[2]

Under Crato's leadership, the circle quickly became the centre of the city's educated elite. Born in Breslau, he studied theology in Wittenberg and medicine in Leipzig and Padua before becoming a personal physician at the Viennese imperial court. Despite his extensive travels, Crato always returned to his hometown. In his 1620 *Vita Jakob Monau*, Melchior Adam describes the circle of late humanists, doctors, councillors and teachers that gathered around him:

> Erat tum Wratislaviae quasi collegium quoddam illustrium ingeniorum: quando uno et eodem tempore Nicolai Rhedingeri, pater et filius, Andreas Dudithius, Ioannes Crato, Petris et Iacobus Monavii fraters, Ribiscius, Engelhartus, Martinus Schillingus, Ioannes Hermannus, Petrus Vincentius, alii, urbem decorabant.[3]

---

2 Philip Melanchthon (1497–1560), humanist and reformer, made a significant contribution to making the Reformation an educational movement. He promoted the development of grammar schools and universities in the German territories as well as the education of theologians.

3 Melchior Adam, *Vitae Germanorum medicorum, qui seculo superiori et quod excurrit claruerunt* (Frankfurt am Main, Jonae Rosens Erben – Heidelberg, Johann Georg Geyder: 1620) 390. The Rehdigers – here represented by Nicolaus II Rehdiger (1525–1587) and Nicolaus III Rehdiger (1555–1616) – were considered the most important and wealthiest merchant and banker family in Breslau with trading offices in Antwerp and Danzig. While the father was active as a councillor for Breslau and a patron of the arts, the son devoted himself primarily to the sciences and arts. The humanist and former bishop convert to Protestantism, Andreas Dudith (1533–1589), moved to Breslau in 1579 and joined Crato's circle, in which he played a leading role until his death; see Ilić L., "Andreas Dudith und sein reformiertes Netzwerk

> At that time there was a kind of community of illustrious spirits in Breslau, when at the same time both Nicolaus Rehdigers, father and son, Andreas Dudith, Johannes Crato von Krafftheim, the brothers Peter and Jakob Monau, Seyfried Rybisch, Daniel Engelhardt, Martin Schilling, Johann Hermann, Petrus Vincentius and others ennobled the city.

Though the circle may have initially gathered around Crato, from 1579 to 1589 Andreas Dudith, his neighbour in Breslau, also occupied an important position in the community. Following Crato's death, the polymath and fellow Melanchthon student Jakob Monau (1546–1603) seems to have taken a leadership role in the circle in the last quarter of the century. Monau was a key figure in the flourishing late humanism in Breslau and lent a particular air of splendour to the circle. In turn, the illustrious group contributed to making the city the 'flower of Europe'.

The members of this 'community of illustrious spirits' were predominantly adherents of the Protestant faith. The fact that an inner circle

---

in Breslau am Ende des 16. Jahrhunderts", in Bahlcke J. – Dingel I. (eds.), *Die Reformierten in Schlesien. Vom 16. Jahrhundert bis zur Altpreußischen Union von 1817* (Göttingen: 2016) 53–63. Peter Monau (1551–1588), the younger brother of Jakob Monau, first settled as a physician in Breslau and later became the personal imperial physician in Prague on Crato's recommendation. Seyfried Rybisch (1530–1584) was an important historian and scholar of antiquity and worked as an imperial councillor and assessor of the imperial chamber court in Silesia. Daniel Engelhardt, a confidant of Dudith's, came from a merchant family in Breslau and studied in Padua and Siena in 1574–1575; see Zonta C.A., *Schlesische Studenten an italienischen Universitäten. Eine prosopographische Studie zur frühneuzeitlichen Bildungsgeschichte* (Ph.D. dissertation, University of Stuttgart: 1999) (Cologne: 2004) 210, no. 271. Martin Schilling was Laurentius Scholz's travelling and study companion in Italy; cf. Zonta, *Schlesische Studenten* 386, no. 1274. Johann Hermann (1527–1605) was also one of the men at the centre of the Breslau circle. He was the practising city physician in Breslau as well as the court physician in Liegnitz-Brieg and was a son-in-law of Lucas Cranach the Younger; cf. Fleischer, "Schlesischen Späthumanisten" 13; Fleischer M.P., "Der Garten des Laurentius Scholz", in Fleischer M.P., *Späthumanismus in Schlesien. Gesammelte Aufsätze*, Silesia 32 (Munich: 1984) 136–163, here 170, note 17; Zonta, *Schlesische Studenten* 260, no. 541. Petrus Vincentius (1519–1581) was the rector of the Elisabethgymnasium in Breslau from 1569 and inspector of the Breslau schools. Not mentioned by Melchior Adam, the legal scholar and diplomat Johann Matthäus Wacker von Wackenfels (1550–1619) also belonged to the circle. Wacker came from Crato's close circle and served as a courtier for the Rehdigers. He settled in Breslau in 1580 and later worked as the Breslau bishop's chancellor and imperial councillor in Prague. On the circle of scholars described by Adam, see Halsted D.G., *Poetry and Politics in the Silesian Baroque. Neo-Stoicism in the Work of Christophorus Colerus and His Circle*, Wolfenbütteler Arbeiten zur Barockforschung 26 (Wiesbaden: 1996) 47–51; Czarnecka M., "Breslau", in Adam W. (ed.), *Handbuch kultureller Zentren der Frühen Neuzeit. Städte und Residenzen im alten deutschen Sprachraum*, vol. 1 (Berlin: 2012) 197–238, here 206; Ilić, "Dudith" 59–61.

around Crato and Dudith harboured strong sympathies for the Reformed confession, which was particularly close to the humanist spirit, cannot be overlooked.[4] However, confessional boundaries or conflicts did not play a central role in this intellectual community. Besides their inclination towards Reformism, a spirit of Erasmian tolerance and scholarship characterised their meetings.

2 Laurentius Scholz von Rosenau

The Breslau city physician Laurentius Scholz von Rosenau (1552–1599) likely belonged to this 'collegium illustrium ingeniorum' under Monau's direction beginning in the 1580s. Though Adam does not mention him by name, he is listed in Acidalius's praise of the city, and not without reason. After Crato, Dudith and Monau's residences, Laurentius Scholz's garden was a regular meeting place for the distinguished group and the Hortus Scholzianus functioned as a most extraordinary social centre for the late sixteenth-century Breslau elite.

But who was this man who is today so little known? In this essay, I will endeavour to broaden current knowledge of his work and contribute to scholarship concerning his support of humanist activity in the period. Laurentius Scholz was born in 1552 in Breslau as the son of a doctor. After attending the Elisabeth gymnasium in his hometown, he began studying medicine in Wittenberg in 1572, which he followed with three years of study at the universities of Padua, Siena and Bologna between 1576 and 1579.[5] The time Scholz spent in Padua, which from the 1540s had emerged as a centre of innovation in the field of medical humanism in Europe, likely awakened his passion for botany. There, he took part in medical botanical courses given by Melchior Guilandinus (1520–89), prefect of the university's botanical garden. A drawing made by Scholz dated 16 February 1579 is preserved in the Bologna University library alongside Ulisse Aldrovandi's (1522–1605) manuscripts and shows a detailed

4 In 1584, the Breslau Prince-Bishop Martin von Gerstmann (1527–1585) suspected the circle of scholars around Crato including Andreas Dudith, Jakob Monau, Nicolaus Rehdiger, Johann Hermann, and Johann Matthäus Wacker von Wackenfels of crypto-Calvinism. The Breslau Council, which was entrusted with the investigation, played down the accusation, as no one publicly professed the forbidden Calvinist faith; see Garber, *Breslau* 53, 268–269; Czarnecka, "Breslau" 206.

5 For more on his biography, see Lauterbach C., *Der erzählte Garten des Laurentius Scholz. Bürgerliche Gartenkultur des Späthumanismus in Breslau*, Grüne Reihe 35 (Worms: 2018) 36–43.

FIGURE 9.1 Laurentius Scholz, *Hortus Patavinus*. 1579. Drawing, 56.5 × 42 cm
IMAGE © BIBLIOTECA UNIVERSITARIA DI BOLOGNA, MS. 124, NR. 40, *C*.82–83

ground and planting plan for the *Hortus Patavinus* for the year 1578–79, as well as a dedication to Aldrovandi [Fig. 9.1].[6]

Although Scholz was no pioneer of botany, he pursued this young science with passion and sought out contact with the leading scholars of his time.

6 Biblioteca Universitaria di Bologna, Fondo Aldrovandi, Ms. 124, vol. 40, cc. 82–85. Ubrizsy Savoia A., "L'horto di Padova all'epoca del Guilandino", in Minelli A. (ed.), *L'Orto botanico di Padova 1545–1995* (Venice: 1995) 173–195.

With the above-mentioned drawing and a written recommendation to Ulisse Aldrovandi from Guilandinus in his luggage, Scholz moved to the University of Bologna, which would prove to be a second important station during his *peregrinatio academica*. Aldrovandi's famous "Musaeum" quickly became a site frequently visited by Scholz, who mentions it in a letter to Joachim Camerarius the Younger (1534–1598) on 14 February 1579.[7] In March 1579, he took the opportunity to expand the scope of his knowledge and set off from Bologna on an educational trip through the sights of Italy with some of his Breslau compatriots.[8] Nicolaus III Rehdiger, Johann Matthäus Wacker and Martin Schilling belonged to this group and are representative figures of the social environment in which the young Scholz moved, one which eventually lead him into the illustrious circle around Monau.[9] The group first visited Rome, then Naples, and took in the sights of Florence and Milan on the way back. At this point, the group split up and Scholz and Schilling travelled via Basel to Valence, where the former received his doctorate in medicine and philosophy in 1580 before beginning his journey home to Breslau. In the year of his return, Scholz married Anna Maria Aurifaber, daughter of Johann Aurifaber (1517–68) and Sarah Heß. Aurifaber was the Breslau pastor at the Elisabeth Church and a reformer in Philipp Melanchthon's circle and Heß was the granddaughter of the important Silesian reformer Johann Heß (1490–1547). The latter was known for having helped the Reformation breakthrough in Breslau in 1524; for Scholz, a marriage to Heß's granddaughter would have undoubtedly been a significant gain in social status.

In 1585, a year of a severe plague in Breslau, Scholz accepted the office of city physician. He was at that time introduced to Johannes Crato von Krafftheim, whose German plague book he subsequently translated into Latin. Crato expresses his esteem for the young man in a letter to Joachim Camerarius the Younger dated 14 August 1585, in which he attests to his conscientiousness, sense of duty and modesty.[10] In the same year, in his function as Imperial

---

7 Erlangen-Nuremberg University Library, Trew letter collection; cf. Schmidt-Herrling E., *Die Briefsammlung des Nürnberger Arztes Christoph Jacob Trew (1695–1769) in der Universitätsbibliothek Erlangen*, Katalog der Handschriften der Universitätsbibliothek Erlangen 5 (Erlangen: 1940) 7, no. 3.

8 See Cohn F., "Dr. Laurentius Scholz von Rosenau, ein Arzt und Botaniker der Renaissance", *Deutsche Rundschau* 63 (1890) 109–126, here 115; Fleischer, "Garten Scholz" 140.

9 See Gillet J.F.A., *Crato von Crafftheim und seine Freunde. Ein Beitrag zur Kirchengeschichte* (Frankfurt am Main: 1860), vol. 2, 88–92; Cohn, "Scholz" 115–116; Fleischer, "Garten Scholz" 140.

10 Erlangen-Nuremberg University Library, Trew letter collection; Schmidt-Herrling, *Briefsammlung Trew* 134, no. 1239. The plague book mentioned is Crato's *Ordnung der*

Count Palatine, Crato awarded Scholz a coat of arms in recognition of his medical achievements in his function as Imperial Count Palatine. In choosing the motto 'Fac officium, Deus providebit' ('Fulfil your duty, God will take care'), Scholz declared piety and duty as the guiding principles of his life.[11] His social rise culminated in 1596 with the award of the 'von Rosenau' noble title. He continued to practice as a doctor in Breslau until he succumbed to tuberculosis three years later at the age of 47.

3 The Garden of Laurentius Scholz

Upon his return from Italy in the 1580s, Scholz began to lay out his garden in southeast Breslau [Fig. 9.2]. Located in an area behind St. Christopher's Church near the Ohle, a tributary of the Oder, it belonged to the so-called outer city, which stretched between the thirteenth- and fourteenth-century walls.[12] Here, between Weidenstraße and Taschenstraße, Scholz had inherited a plot of land, which he enlarged by purchasing neighbouring plots for the purpose of establishing his garden.[13] An inscription modelled after the Italian "Leges hortenses" – rules governing access to the gardens for visitors – that Scholz had carved into stone and placed above the entrance gate dates the completion of the garden to 1588:

> Hortum istum antea neglectum, et ruinosum, In Laudem Omnipotentis Dei, patriae honorem, Amicorum, ac rei herbariae studiosorum gratiam, Et in sui ipsius a laboribus, et studiis artis suae recreationem, Suis sumptibus instaurauit, ac inquilinis, et exoticis plantis exornavit

---

preseruation, wie man sich wider die erschreckliche seuche der pestilentz verwahren, auch rath wie die erkant vnd curiert werden sol (Wrocław, Crispin Scharffenberg: 1555) [*Order of preservation, how to guard against the terrible plague of pestilence, also advice on how it should be detected and cured*].

11 Cf. Cohn, "Scholz" 116; Fleischer, "Garten Scholz" 141.

12 See Mlynarska-Katetynowa M. (ed.), *Atlas historyczny miast polskich = Historical Atlas of Polish Towns. Vol. 4. Slask = Silesia. Dept. 4.1: Wrocław* (Torun: 2001) 19–21; Kurtzmann L., "Laurentius Scholtz und der erste botanische Garten in Breslau 1588–1599", *Schlesische Provinzialblätter* N.F. 5 (1866) 457–460 on the purchase contracts for the extension of the site; Fleischer, "Garten Scholz" 147. For a detailed description and classification of the garden, see Oszczanowski P., "Wrocławski ogród Laurentiusa Scholtza St. (1552–1599) – sceneria spotkań elity intelektualnej końca XVI wieku", in Hałub M. – Mańko-Matysiak A. (eds.), *Śląska Republika Uczonych. Silesian Scholars' Republic*, vol. 1 (Wrocław: 2004) 98–145 and Lauterbach, *Erzählte Garten* 63–80.

13 The garden had a base area of about 2,500–2,800 m^2; cf. Lauterbach, *Erzählte Garten* 42–43.

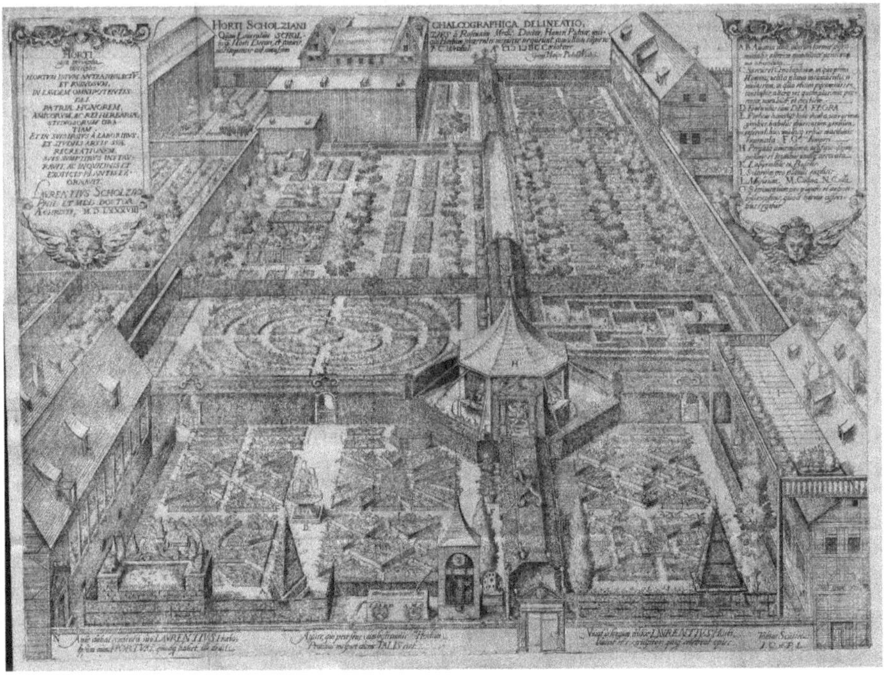

FIGURE 9.2 Georg Hayer (designer and engraver), *Horti Scholziani Chalcographica Delineatio*. 1598. Etching, 43.2 × 54.5 cm
IMAGE © BIBLIOTEKA UNIWERSYTECKA WE WROCŁAWIU, SHELFMARK 011185,1

Laurentius Scholtzius Phil. et Medic. Doctor. Anno Christi M.D. LXXXVIII.[14]

This previously completely neglected and desolate garden was restored at his own expense by Laurentius Scholz, Doctor of Philosophy and Medicine, in 1588, to the praise of Almighty God, to the glory of his hometown, for the benefit of the students of botany, as well as for his own recreation from the toils and studies of his art, and adorned with native and foreign plants [Fig. 9.3].

14 Scholz Laurentius, *In Laurentii Scholzii Medici Wratisl. Hortum Epigrammata Amicorum* (Wrocław, Georg Baumann: 1594), fol. A1v. On Leges Hortorum, see. Lauterbach C., "Gartengesetze des Späthumanismus. Der Garten als Raum gelehrter Geselligkeit", in Schweizer S. (ed.), *Gärten und Parks als Lebens- und Erlebnisraum. Funktions- und nutzungsgeschichtliche Aspekte der Gartenkunst in Früher Neuzeit und Moderne* (Worms: 2008) 21–32 and Coffin D., "The 'lex hortorum' and Access to Gardens of Latium during the Renaissance", *Journal of Garden History* 2 (1982) 201–232.

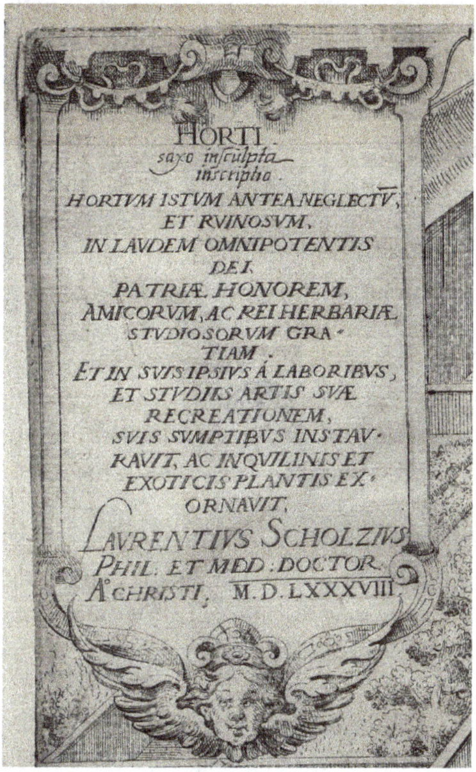

FIGURE 9.3
Georg Hayer (designer and engraver), Horti saxo insculpta inscriptio. Detail of *Horti Scholziani Chalcographica Delineatio*. 1598

The garden followed an elongated rectangular ground plan without any reference to the Scholz residence. When one passed through the primary portal, the main axis, which had been moved off centre and almost completely vaulted by arcades, led to the octagonal, wooden pleasure house [Fig. 9.4]. This pleasure house formed the centre of the main garden, which consisted of three flower and herb parterres and a hedge maze. The three geometric 'compartimenti' (compartments) were laid out, as two contemporary panegyrists described, in 'arte nova' ('novel artistry') and 'Belgarum Ausonidumque modo' ('in the manner of the Flemings and Italians'), that is in the style of Sebastiano Serlio and Hans Vredeman de Vries.[15] Scholz may have also been influenced by the botanical garden in Padua, which he knew in his youth; the Breslau grounds mirrored its four parterre fields ordered in a simple, formal geometry. The modern bed design and the furnishings of the parterre zone set this area apart

15 See Lauterbach, *Erzählte Garten* 67, notes 314 and 315.

from the rest of the grounds. Amidst the décor, we find a Mannerist style bowl fountain crowned by the goddess Flora with thin jets of water pouring from her breasts and head, several water basins and two aviaries festooned with garden honeysuckle and bittersweet nightshade, lending this section of the garden an opulence that appealed to all the senses.

Two corresponding buildings closed off the parterre zone to the neighbouring properties and afforded a view over the parterres from a slightly elevated position. The first structure was the gallery (E: "Porticus"), which, according to the legend, was dedicated to 'honestis ludis' ('virtuous games') and thus appealed to the tradition of princely ballrooms. The second served as the "Musaeum" (L) and housed part of Scholz's art collection. The gallery walls were decorated with murals showing the costumes of various peoples next to maritime scenes. This programme was complemented by depictions of exotic plants and the figure of Orpheus lamenting Eurydice's death with his music among listening birds and wild animals.[16]

The rear part of the grounds was occupied by two less aesthetically oriented areas, notably an orchard or arboretum and a zone containing low plant beds. In the orchard, fruit and ornamental trees were cultivated and shrubs grew in rows. A special quarter was dedicated to frost-sensitive plants and protected by slats in winter. Here, caretakers nurtured bitter orange, pomegranate, figs, pistachio and laurels for the enjoyment of Scholz and his friends and family. The eastern end of the complex was home to a garden with simple beds in which herbs and flowers were grown for wreaths that, following the ancient model, served to decorate both the table and the guests at festive gatherings.

Though moved out of the axial plan and only visible from the rear on the bird's-eye view, a particular building in the parterre zone and its adjacent recess carried important artistic and iconographic significance for the entire complex: the grotto and the cenotaph [Fig. 9.5]. The building, which was erected in a rustic style and gave an intentional impression of decay, was an early example of German grotto architecture; its interior was designed as a cave or natural hollow in the style of Italian nymphaea. A reclining male figure dominated the decoration of the room; an observer may have initially identified this figure as a river god. However, the inscriptions and epigrams of an associated etching point to a different reference. According to the etching, the figure extended beyond traditional grotto iconography to depict the sinful Adam with fruit from the tree of knowledge. According to this reading, the

16 See Calagius Andreas, *Hortus doct. Laurentii Scholzii medici et philosophi quem ille colit Vratislaviae, situm intra ipsa civitatis moenia, celebratus carmine* (Wrocław, Georg Baumann: 1592) fol. C 1r; on this text, see Lauterbach, *Erzählte Garten* 72–74.

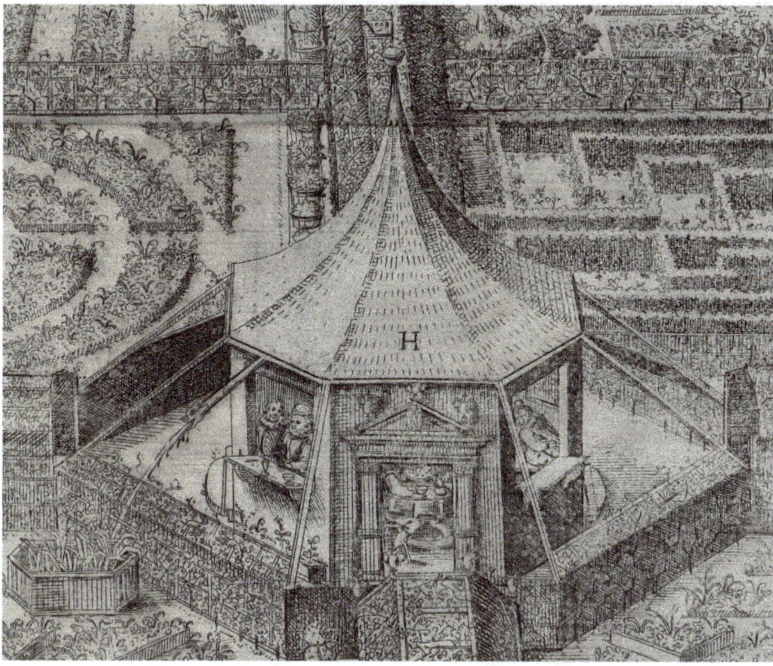

FIGURE 9.4 Georg Hayer (designer and engraver), pergula amoenissima, detail of *Horti Scholziani Chalcographica Delineatio*. 1598
IMAGE © BIBLIOTEKA UNIWERSYTECKA WE WROCŁAWIU

FIGURE 9.5 Georg Hayer (designer and engraver), *Chalcographica Delineatio Anterioris partis specus, sive cenotaphii*. 1598. Etching, 20 × 30.5 cm
IMAGE © BIBLIOTEKA UNIWERSYTECKA WE WROCŁAWIU, SHELFMARK 011185,2

cave suggests a causal link between Adam, who introduced death to humanity, and the burial cave of Joseph of Arimathea, in which Jesus's body lay until his resurrection. With its typological juxtaposition of the old and the new Adam, the Breslau cenotaph alluded to the concept of the "law and grace" allegory of faith, which enjoyed great popularity in sixteenth-century Protestant circles.

4 Literary Exaggeration of the Hortus Scholzianus

Sources reveal that the Hortus Scholzianus was both a space for pleasure and a botanical garden intended for use by scholars. An abundance of reference material, unique to the study of contemporary bourgeois gardens in sixteenth-century Germany, affords us the fortunate position of being able to reconstruct the use of the garden by the owner and his guests.[17] Among the available documentation, we find two plant inventories Scholz had printed to catalogue his collection and promote exchange with like-minded collectors.[18] The second inventory also included watercolour drawings of plants 'from nature', which he commissioned from the Breslau painter Georg Freiberger. These paintings were simultaneously intended to serve as illustrative material and teaching aids in the winter for his son and create an everlasting monument to his botanical collection.[19] In addition to the inventories and watercolours, the form of the garden is further documented in two etchings by the Breslau painter and engraver Georg Hayer (1559–1614). The first, a large-format bird's-eye view, depicts the entire complex, while the second offers a view of the garden's most extraordinary structure – the grotto and cenotaph[20] [Figs. 9.2, 9.5].

Not only contented with illustrating his garden in pictures, Scholz was also keen to exalt and immortalise it in literary texts. The Breslau author Andreas Calagius (1549–1609) penned the first work about the Hortus Scholzianus;

17 The pictorial and textual sources are presented in detail in Lauterbach, *Erzählte Garten* 44–62.

18 The first catalogue, which was probably published in a very small edition by Johann Scharffenberg's heirs in Breslau in 1587, is today considered lost. Scholz published the second catalogue with an extent of six sheets in 1594 with Georg Baumann under the title *Catalogus arborum, fruticum ac plantarum, tam indigenarum quam exoticarum horti Laurentii Scholzii, medici Vratislaviensis*. See Lauterbach, *Erzählte Garten* 48–52.

19 See Lauterbach, *Erzählte Garten* 52–53. Unfortunately, the watercolours have not survived.

20 The two leaves are enclosed in the copy of the epigrams kept by the University Library in Breslau. The view of the garden is no. 011185,1 and that of the grotto no. 011185,2.

FIGURE 9.6
Title page. 1594. Etching, 19.2 × 15.2 cm. From: Laurentius Scholz, *Hortum Epigrammata Amicorum* (Wrocław, Georg Baumann: 1594)
IMAGE © BIBLIOTEKA UNIWERSYTECKA WE WROCŁAWIU, SHELFMARK YE 601/1

his eighteen-page poem in praise of the garden appeared in print in 1592.[21] More ambitious and of greater interest for our research is the multi-author volume of epigrams published by Georg Baumann in Breslau in 1594 under the title *Laurentii Scholzii Medici Wratislaviensis Hortum Epigrammata Amicorum* [Fig. 9.6].[22] This volume is divided into three parts, the first containing the epigrams on the title and the others the *Janus Quadrifrons in Hortum Laurentii Scholtzii Medici Wratislaviensis* in iambic trimeters. Acidalius, a rising star in poetry of the late sixteenth century, composed this particularly notable

21 The title is *Hortus doct. Lavrentii Scholzii medici et philosophi qvem ille colit Vratislaviae, sitvm intra ipsa civitatis moenia, celebratus carmine* (Breslau, Georg Baumann: 1592).

22 Breslau University Library, shelfmark Ye 601/1. The Breslau University Library also holds under shelfmark Hs. R 2177 a paper manuscript owned by Scholz, which is a copy of the *Epigrammata Amicorum*, with blank pages throughout. It contains excerpts from letters as well as 24 additional epigrams on Scholz's garden, written between 1595 and 1597. Accordingly, an expanded new edition of the epigram collection was planned as early as 1595, but it was never published because Scholz died in 1599. The edition of this collection is a priority desideratum for future research. See Oszczanowski, "Wrocławski ogród" 102.

poem.[23] Having returned from a journey to Italy in 1593 and settled in Breslau, Acidalius's extraordinary talent made him a sought-after "poeta et orator publicus" among the city's humanist elites and he was already highly regarded beyond Breslau. The *Janus Quadrifrons* is an adaptation of the *Leges Hortenses* and *Leges Convivales*, which Scholz wrote in prose form for the 'Floralia Wratislaviensia' ('Breslau Flower Festivals') held in his garden. Both texts are printed in the volume and, taken together with Acidalius's work, are among the most informative sources on the use of the garden.

As a classic friendship book, the collection of epigrams reflects the late humanist, middle-class cult of sociability.[24] The forty-six epigrams by forty-one authors concern a single theme: the praise of the Hortus Scholzianus and its owner. As such, the work belongs to the genre of Neo-Latin occasional poetry, which was produced in enormous numbers in Silesia in the sixteenth century. Though stereotypical *topoi* and praise formulas make the reading somewhat difficult for contemporary scholars, the poems are still important academic sources. Sixteenth-century epigrams typically relate to their historical context and, despite their prescriptive nature, served the humanist project of self-definition and community formation.[25] This is also true of the poems in the Breslau epigram collection, which convey a meaningful image of Scholz and his garden through their direct reference to its objects.

Several of the panegyrists who contributed to the volume were members of the Breslau community of illustrious spirits. An epigram penned by Monau, the head of the circle at the time, occupies a special place in the collection in terms of content; some extant evidence suggests that he designed the ambitious pictorial programme for the cenotaph.[26] Among the other well-known members of the Breslau circle of scholars, we find Crato's student, jurist, diplomat, and poet Johann Matthäus Wacker von Wackenfels, as well as the

23 IJsewijn, "Admirer of Lipsius".

24 Trunz E., "Der deutsche Späthumanismus um 1600 als Standeskultur", in Alewyn R. (ed.), *Deutsche Barockforschung: Dokumentation einer Epoche* (Cologne: 1965) 167–171; Garber K., "Reformierte Mentalität und literarische Evolution. Aspekte kultureller Disposition der 'nobilitas literaria Silesiae' im europäischen Kontext", in Bahlcke J. – Dingel I. (eds.), *Die Reformierten in Schlesien. Vom 16. Jahrhundert bis zu Altpreußischen Union von 1817* (Göttingen: 2016) here 275–276.

25 Enenkel K., "Introduction. The neo-Latin epigram. Humanist self-definition in a learned and witty discourse", in De Beer S. (ed.), *The Neo-Latin Epigram. A Learned and Witty Genre*, Humanistica Lovaniensia – Supplementa 25 (Leuven: 2009) 1–23; on the genre definition of the epigram see also Garber, "Reformierte Mentalität" 293–294. On this topic, see also Denis Ribouillault's essay in this volume, *The Poetic Fortune of the Nymphaeum of Villa Giulia*.

26 For more details, see Lauterbach, *Erzählte Garten* 124–126.

figures of Nicolaus III Rehdiger, Daniel Engelhardt, and Johann Hermann – all represented with an epigram. Hermann's wife Barbara Cranach (*c.*1535–1601), daughter of Lucas Cranach the Younger, was also known to have taken part in the garden festivals. Among the other contributors of epigrams, we again find Acidalius and two of his most important friends and patrons, the Breslau physicians Johannes Fersius (1558–1611) and Daniel Bucretius. Several Breslau poets also contributed to the volume, who we can surmise were likely regular guests in the Scholz garden.[27] These include Kaspar Pridmann (1537–1598), rector of the Magdaleneum in Breslau from 1578, as well as his successor in office Johannes von Hoeckelshoven (1557–1618), who later made a name for himself as Martin Opitz's teacher. Also deserving of mention are Nicolaus Steinberg, rector of the Elisabeth Gymnasium and at the same time Breslau school inspector and imperial "poeta laureatus", Georg Seidel (1550–1626), teacher at the Elisabeth Gymnasium, the Breslau doctors Christoph Rumbaum (1555–1605), Martin Weinrich (1548–1609) and Andreas Büttner (1517–1603), who was a son-in-law of the reformer Johann Heß, the Breslau secretary Melchior Steinberg, and Georg Nostwitz (1542 – after 1620).

A diverse group of poets, scholars, doctors, and school and university teachers also contributed epigrams to the collection. Most of them would have come to know the garden as visitors, as Scholz and his friends endeavoured to solicit epigrams and spread knowledge of the space among notable humanists. A direct view of the garden was not necessarily a prerequisite for the composition of a poem, as in the 1594 correspondence between Justus Lipsius (1547–1606), one of Europe's leading late humanists, and Johannes Fersius and Valens Acidalius, who ask for an epigram by Lipsius on the Hortus Scholzianus.[28] The rank of the authors, whose fame would keep the garden renowned through the ages, was more important.[29]

5 Role Models

The Breslau "collegium illustrium ingeniorum" that congregated in the Hortus Scholzianus was a humanistic community that cultivated erudite and cheerful exchange. Its members met regularly for discussions in the garden and, together with their wives, daughters and sons, attended banquets on the "Floralia Wratislaviensia" held in the spring and summer. Scholz breathed new

27 For more details on this group of people, see Lauterbach, *Erzählte Garten* 58–59.

28 Ibid. 60–61.

29 Trunz, "Späthumanismus als Standeskultur" 167.

life into ancient festive customs through his celebration of the Breslau Floralia while avoiding directly copying the Roman Floralia, a popular festival held in spring to honour fertility. The exuberant folk character of the latter festival had little in common with the Breslau Floralia, which more closely resembled ancient feasts. First and foremost, Scholz claimed the patron goddess of blossoms for his garden and the festivals held there.

The Breslau meetings more closely corresponded to the schedule and content of the Italian academies, which adopted the tradition of the Platonic Academy and the Lyceum of the Peripatetics as well as other philosophical schools. In addition to the most famous example of the early Renaissance, the Florentine Academy, which formed around Marsilio Ficino (1433–1499) around 1460, countless other small academies, discussion circles, *sodalitates*, and *orti letterari* emerged based on the Ciceronian model in fifteenth- and sixteenth-century Italy, several of which are discussed in this book. Their meetings – like those of the Breslau circle of scholars – did not take place in a fixed, institutional framework. Rather, educated urban elitee met in a variety of locations for the purpose of exchanging on literature, philosophy and the arts, as well as to make music, recite poetry, play games and eat.[30] In the *Leges Convivales*, Scholz hints that ancient guest banquets served as models for the meetings in his garden and that he wished for cheerful conversations in the manner of the Platonists: 'Coena non Pythagorea, sed Platonica; non inter mutos, sed facundos esto: ne sileat aut langueat sine fabulis hilaritudo' ('Let the meal be not Pythagorean, but Platonic; let it be not among the silent, but among the loquacious: let not mirth be silent or idle without narratives').[31] With this statement, Scholz thus emphasises the function of his garden as a space of conversation and places it firmly in line with the tradition of the Greek philosophers' gardens.

6 The Tradition of the 'Leges hortorum' in Italy and the Netherlands

The culture of festivities and social visits in Breslau, for which the garden of Laurentius Scholz served as a kind of centre, is remarkably well known due to the three garden laws. These were published together with the "Epigrams of the Friends" and their purpose, announced in verse and prose, was to provide guests with rules for appropriate conduct in the garden and entertain

30 Chambers D.S. – Quiviger F. (eds.), *Italian Academies of the Sixteenth Century* (London: 1995) 10–13.

31 Scholz, *Epigrammata* fol. H 2r.

them through humanist scholarship.[32] In these texts, in a manner exceeding all other written and pictorial sources, the garden becomes tangible as a space that shaped the actions and communication of those who inhabited it.[33] It did so in two ways. First, it freed the visitors from the duties of everyday urban life and thus enabled them to spend carefree time together. Second, the publication of these laws also imposed new restrictions on behaviour despite a visitor's freedom to promenade leisurely. They thus offer a novel opportunity to apprehend the garden from the perspective of the actions that unfolded in it. However, we should be mindful that these instructions for conduct were indebted to a long literary tradition and also served a self-styling purpose. They should therefore not be understood as presenting an absolutely faithful image of a historical reality. Rather, they offer a narrative construction of their environment as fashioned by Scholz and Acidalius.

The tradition of the "Leges hortorum" in Italy can be traced back to the late fifteenth century; we owe our knowledge of the genre to David Coffin's scholarship.[34] An integral part of these early, usually very short laws included a dedication of the garden to friends who had been welcomed as guests in addition to the naming of the owner of the garden.[35] We can relate these characteristics of Italian laws in this early stage of development to the inscription on the main portal of the Hortus Scholzianus. [Fig. 9.3] Over time, garden laws became increasingly detailed and extended their invitation beyond the owner's narrow circle of friends to other citizens and travellers, reflecting the opening of private gardens to a growing public; this expansion compelled the production of concrete guidelines for use of the gardens. In the newly emerging public botanical gardens in Italy as well, an influx of students and visitors posed a danger to the precious plants and made it necessary to regulate visitor behaviour. The resulting laws are prosaically brief. By contrast, the laws of Italian villa gardens produced during the sixteenth-century are characterised by self-dramatisation in the guise of an ethical and learned code. The "Leges

32 Scholz, *Epigrammata* fols. G 1r–H 3v.

33 Schmid B. – Schmid R., "Die Allgegenwart des Raumes in den Kulturwissenschaften und die Ordnung der Dinge", in Kundert U. (ed.), *Ausmessen – Darstellen – Inszenieren. Raumkonzepte und die Wiedergabe von Räumen in Mittelalter und früher Neuzeit* (Zurich: 2007) 10–11.

34 Coffin, "Lex hortorum"; Coffin D., *Gardens and Gardening in Papal Rome* (Princeton: 1991) 244–247, here 268–269. The garden laws of Rome are the best researched, but one must assume that numerous garden laws existed outside Rome's villas as well.

35 The inscription plaque that Cardinal Oliviero Carafa (1430–1511) had placed in his villa in Rome in the last quarter of the fifteenth century is one such early example. See Coffin, "Lex hortorum" 202.

hortorum" of Italian villa gardens regularly included statements on the tone and moral conduct that the owner desired from his guests in the garden, which informed those produced in Breslau.

The reception of Italian Renaissance culture in the second half of the sixteenth-century resulted in an appropriation of Neo-Latin garden laws north of the Alps; the best-known and most influential among these are those penned by Justus Lipsius.[36] A first law was put forth in a dialogue in *Antiquae Lectiones* published in 1575, which Lipsius sets in his own, presumably fictitious, Louvain garden around 1568.[37] The exchange between Lipsius and three friends from his youth occurs during a banquet in his house, which is preceded by a tour of the garden. Biographical fragments from his years of study in Louvain are enriched by literary allusions to Rome's "Leges hortenses" – gleaned during his recent journey to Rome – and the 'imitatio' of classical models. Lipsius has the garden god Vertumnus proclaim the humourous verses in Plautine iambs, which were placed on a tablet above the garden gate. The length of the law contained in this poem suggests that it is purely fictional and speaks more to the literary ambition of its author. On the heels of the *Antiquae Lectiones*, Lipsius wrote a second law for his Leiden garden and attached it with a letter to the Flemish humanist Dominique Lampson (1532–1599) in 1587, 'Nam inserui, ut rideas' ('that you may laugh').[38] Lipsius has the law delivered from the protector of entrances, Janus bifrons:

> Quisquis es, sta, lege, pare. Janus ad hoc limen sic edico.[39]
>
> Whoever you may be, stop, read, obey.
> I, Janus, proclaim the following at this entrance.

---

36 Morford M., "The stoic garden", *Journal of Garden History* 7 (1987) 151–175; Lauterbach C., *Gärten der Musen und Grazien. Mensch und Natur im niederländischen Humanistengarten 1522–1655* (Ph.D. dissertation, University of Cologne: 2002) Kunstwissenschaftliche Studien 111 (Munich – Berlin: 2004) 72–85.

37 Lipsius Justus, *Antiquarum lectionum commentarius. Tributus in libros quinque, in quibus varia scriptorum loca, Plauti praecipue, illustrantur aut emendantur* (Antwerp, Christophe Plantin: 1575) 140–141. Lipsius dedicated the work to Thomas Rehdiger (1540–1576). See Halsted, *Poetry and Politics* 43; Lauterbach, *Gärten der Musen* 77–85.

38 Gerlo A. – Nauwelaerts M.A. – Vervliet H.D.L. (eds.), *Iusti Lipsi Epistolae* vol. 2 (Leuven – Paris – Bristol: 1978) 369, no. 546–87 06 19. The text is not printed in this edition. It is found in Lipsius Justus, *Opera Omnia, septem tomis distincta, postremum aucta et recognita*, vol. 2 (Antwerp, Jan Moretus: 1614) 140–142.

39 Cf. Morford, "The stoic garden" 151–152; Lauterbach, *Gärten der Musen* 73–84.

The extremely humorous verses were inscribed on a plaque at the gate of the garden and three of the laws are quoted:

> III. Sermones etiam ne exleges.
> Iocari licet:
> Narrare licet:
> Rogare licet:
> Sed nihil serium.
> Gratiarum hic locus est.
>
> IV. Siquid amoenius tamen in studiis;
> Inter ambulandum
> Dissere, doce, disce.
> Et Musarum hic locus est.
>
> V. Tu qui seriosus,
> Tu qui rixosus,
> Exeste.
> Clusius ego sum vobis.[40]
>
> III. Conversations are also bound by laws.
> It is permitted to joke.
> It is permitted to tell stories.
> It is permitted to ask questions:
> But nothing serious.
> This place is consecrated to the Graces.
>
> IV. But if something pleasant is found in your studies:
> While walking
> May you discuss it, teach, learn.
> This place is also consecrated to the Muses.
>
> V. You who are serious, you who are
> quarrelsome,
> stay outside.
> I am closed to you.

40 Lipsius, *Opera Omnia*, vol. 2, 140–142.

We again find this cheerful, scholarly tone in Scholz's garden. Both Lipsius's and Scholz's stand as gardens of the Muses in the ancient philosophical tradition and at the same time consecrate the Graces by becoming places of joy and grace, friendship and harmony. With their high degree of literary flair and stylisation, the Breslau garden laws further relate to the tradition of the Lipsian models, whose influence – especially on Acidalius – can be traced to individual turns of phrase.

7 An Illustrious Circle Sets Itself Apart

In the first verses of his *Leges Hortenses*, Scholz clarifies the higher power to which his garden is consecrated and those who may enjoy it with him:

> Hortus Laurentii Scholtzii Voluptati, vitae humanae sospitatrici, sed nec omni, nec omnium, sacer esto.
>
> Primo Heri ipsius inprimis: cui a laboribus et curis hic relevamen, animique exerciti restauramen. Proxime amicorum: quorum usu et fructu commune hoc cum Hero bonum. Deinde et hospitum peregrinorum: si quibus forte visere lubido. In his et illis tamen non omnium promiscam, sed aut eruditorum, aut eruditionis amantium. quique vel eorundem cum Hero sacrorum mystae, vel candidati, vel certe non prorsus alieni. Vulgus profanum procul abesto: locum sacrum nec oculis, nec pedibus violato.[41]

> Laurentius Scholz's garden is dedicated to pleasure, the healer of human life, but not to every pleasure, nor to all pleasures. Firstly, especially the pleasure of the owner of the house himself: he finds here a relief from his work and worries and a renewal for his weary spirit. Next, the pleasure of the friends: they share the use and the yield with the master of the house. Finally, of guests from afar: if they happen to look at it. Nevertheless, these are not common pleasures, but the pleasures of the educated or lovers of education: all these are either mystics of the same secrets with the master of the house, or suitors for them, but certainly not complete strangers. The common people keep far away: they do not violate the sacred place with their eyes or feet.

41 Scholz, *Epigrammata* fol. G 4r.

As a "locum sacrum", the Hortus Scholzianus was by no means without restriction. The garden laws speak in exactingly clear language on this point. Only the educated and lovers of education may enjoy themselves with the master of the house. Acidalius goes one step further in his garden proscriptions, having Janus Quadrifrons speak with contempt as the guardian of the garden:

> Exesse iubeo, liminique absistere.
> Odi profanum vulgus et procul arceo.[42]
>
> I command that all others stay outside and keep away from the threshold.
> I hate the profane people and keep far away from them.

The phrase 'Procul este profani', ('Keep away, you uninitiated') was already inscribed at the entrance to the statue garden in the Belvedere courtyard in Rome at the beginning of the sixteenth century and echoed Sybil of Cumae's warning call in Virgil's *Aeneid*.[43] In spite of this proclamation, the statue garden was actually freely accessible to visitors. Pronouncing a variation on this theme of access some seventy years later in his "Lex hortorum", Justus Lipsius demands,

> Quod promiscam omnes, servi sint an liberi,
> Dites inopesque ab aditu volt arcerier,
> Praeter Musarum qui initiati sint sacris.[44]
>
> That all without distinction, whether slave or free,
> rich or poor, shall be kept from entrance,
> except those who are initiated into the mysteries of the Muses.

These fixed *topoi* ritualise access to the gardens and attest to visitors' belonging to the 'respublica literaria' ('republic of letters') and the circle of 'studiosi rei herbariae' ('students of botany'). Scholz sets the bar correspondingly high for visitors to his garden, imploring that they be familiar with the secrets of the Muses, Phoebus and Asclepius, the Graces, Venus and Flora.[45]

---

42 Scholz, *Epigrammata* fol. G iv. With this, Acidalius refers almost literally to Horace: *Carmina* III, 1,1: 'Odi profanum vulgus et arceo'.

43 See Virgil, *Aeneid* VI, 258.

44 Lipsius, *Opera Omnia*, vol. 2, 140.

45 Scholz, *Epigrammata* fol. G iv.

Michel de Montaigne observed the contradiction between the opening of private gardens to visitors and the simultaneous restrictedness of a group of like-minded people formed around an owner on his trip to Rome in 1581. According to Montaigne, the gardens there only showed their 'public face' to visitors, while their true nature was reserved for an 'inner circle'.[46] With this in mind, we can thus extrapolate that the Breslau garden laws served several functions. On one hand, Scholz formulated both concrete rules of conduct and standards of value for spending time in his garden. On the other, they contributed to the self-fashioning and self-assurance of those who were members of the Breslau 'community of illustrious spirits' and functioned as important medium of social distinction.

8 Moderate Pleasure

By dedicating his garden to *voluptas* (pleasure), the 'healer of human life', Scholz situates himself in a long tradition. In his first sentence as the ruler of his garden and its guests, Scholz alludes to the life and pleasure-giving Venus from Lucretius' Venus Hymn – that inherent, all-moving force of nature.[47] He expresses that pleasure should be the driving motivation for visitors to his garden and articulates in great detail how this concept should be understood and through which activities it could be experienced.

> Quicunque admissi, Voluptatis caussa ineunto. Voluptatem autem, ut honesta sit et utilis, moderantor. Ergo vel aspectu florum et herbarum, cultusque omnis hortensis amoena contemplatione, vel alterni sermonis et colloquiorum iucunditate se oblectanto. Sed in utroque genere modeste se gerunto, et haec diligenter observanto.[48]
>
> Whoever is admitted shall enter for the sake of pleasure.
> But let them moderate their pleasure so that it may be decent and useful.

46 de Montaigne M. – Stilett H. (transl.), *Tagebuch der Reise nach Italien über die Schweiz und Deutschland von 1580 bis 1581* (Frankfurt: 2002) 189; Schweizer S., "Ethische und ästhetische Normierung. Zur Doppelfunktion von Garteninschriften", *Kritische Berichte* 36 (2008) 34–44, here 36–37.

47 'Aeneadum genetrix, hominum divumque voluptas, alma Venus'. Titus Lucretius Carus, *De rerum naturae*, I, 1–2; quoted from Diels H. (ed.), Lucretius, *Von der Natur. Latin – German* (Darmstadt: 1993) 6–7.

48 Scholz, *Epigrammata* fol. G 4v.

> Either they should be amused by the sight of flowers and plants, by the pleasant contemplation of every plant in the garden, or by the cheerfulness of conversation and amusement. But in every way they should behave moderately and observe this carefully.

Contemplation of plants and learned exchange with like-minded people are the key pursuits that Scholz's garden encouraged.

Scholz made pleasure, which he described as moderate, honourable, educated, and useful, the leitmotif of his garden ethos. This idea of *voluptas* is strikingly reminiscent of one of the most famous ancient examples – that of Epicurus, who lived and taught his philosophy revolving around pleasure from his gardens. The earliest "Lex hortorum" was found above the entrance to his garden, which exerted a major influence on the Renaissance tradition and proclaimed: 'Hospes, hic bene manebis, hic summum bonum voluptas est', ('Stranger, here you will do well to tarry, here our highest good is pleasure').[49] A predominantly Stoic reading of this pleasure-based ethos informed the approach of Renaissance gardens, rather than a faithfully Epicurean one. In his twenty-first letter to Lucilius, Seneca synthesised Epicurean and Stoic thought, an approach that would occupy an important place in discourses surrounding the development of Renaissance gardens. Seneca framed Epicurus's garden as the epitome of a simple life in harmony with surrounding nature – a life that derived its deepest pleasure from satisfying basic human needs. Seneca was convinced that pleasure, understood in this way, was entirely natural: 'Ista voluptas naturalis est'.[50] Through this reflexion, Seneca reconciled Stoic and Epicurean thought, championing the idea of a "pleasure garden" that provides its dwellers with both natural and moral enjoyment. This ethically aligned conceptualisation of pleasure became the core tenet of a discourse on the correct use and value of humanist gardens in the early modern period.[51] In this context, it is not surprising that the concept of *voluptas* had already taken hold in Italian garden laws from first half of the sixteenth century. Among these, Scholz's garden laws share a particular affinity to those of the Nymphaeum

49 Lucius Annaeus Seneca, *Epistulae morales ad Lucilium*, Epistula XXI, 10, cited in Apelt O. (ed.), Lucius Annaeus Seneca, *Philosophische Schriften*, vol. 3 (Hamburg: 1993) 76. Cf. Lauterbach, *Gärten der Musen* 59–60, 240–248.

50 Seneca, *Epistulae morales*, Epistula XXI, 11. Apelt, Seneca, *Philosophische Schriften* vol. 3, 77. Seneca considers the 'voluptas naturalis' to be natural, simple, and rational and therefore calls them good. Cf. Lauterbach, *Gärten der Musen* 240–242.

51 Lauterbach, *Gärten der Musen* for the Netherlands.

of the Villa Giulia in Rome, built in the 1550s for Pope Julius III, which begins with a dedication to the 'honesta voluptas'.[52]

However, Lipsius and his neo-Stoic garden ethos probably had an even greater influence on Scholz and Acidalius. Lipsius's primary moral philosophical work, *De Constantia in publicis malis*, was first published in 1584; its garden chapters focus on the concept of "modesta voluptas" as an ethical norm.[53] Alongside the two garden laws penned by Lipsius, this work is the most influential source for the three Breslau texts, especially Acidalius's.[54] Like Lipsius's garden the Hortus Scholzianus was by no means only a learned place of the Muses, but also a "Gratiarum locus" – a place of cheerful gatherings. Even in distant Silesia, urban elites found a connection to the guiding neo-humanist culture championed by Lipsius in the Netherlands. Both the banquets and time spent in the garden occurred under the sign of 'honourable and useful pleasure'.[55] Thus, visitors were advised in the 'Leges hortenses' not to behave licentiously or to challenge the purpose of the place during their conversations:

> Narrare, sed nihil serium; iocari, sed ne lascivius, liceto.
>
> Docti disputationibus amoenis ingenia exercento: de studio potissimum et re hortensi disserendo, interrogando; qui possunt, docento; qui volunt, discunto.[56]
>
> It should be permitted to tell stories, but nothing serious; to joke, but not too freely. Educated people should train their minds by nice conversations: best by discussing science and garden things and by asking questions; those who can, may instruct; those who want, shall learn.

52 The opening and closing verses have similar formulations and twists of content. Other laws dealing with pleasure are found in the garden house of Cardinal Cesarini in Rome and in the statue garden of Cardinal della Valle; Falk T., "Studien zur Topographie und Geschichte der Villa Giulia in Rom", *Römisches Jahrbuch für Kunstgeschichte* 13 (1971) 101–178, here 170; Coffin, "Lex hortorum" 204–206; Lauterbach, "Gartengesetze" 26; Holberton P., "'Honesta voluptas'. The Renaissance Justification for Enjoyment of the Natural World", in Goodchild K.H. – Oettinger A. – Prosperetti L. (eds.), *Green Worlds in Early Modern Italy: Art and the Verdant Earth* (Amsterdam: 2019) 69–86. See also Denis Ribouillault's essay in this book, *The Poetic Fortune of the Nymphaeum of Villa Giulia*.

53 See Lauterbach, *Gärten der Musen* 72–130, 240–248.

54 Cf. Lauterbach, "Gartengesetze" 26–27; Lauterbach, *Gärten der Musen* 72–85; IJsewijn, "Admirer of Lipsius".

55 'Hortus hic Voluptati honestae et utili sacer, etiam Conviviis'. Scholz, *Epigrammata* fol. H 1r.

56 Scholz, *Epigrammata* fols. G 4v–H 1r.

The garden not only provided the setting for gatherings, but also significantly influenced the topics of conversation among guests. As we learn from these garden laws, a carefully controlled tone of cheerful erudition made the garden a place of pleasure for the humanistically educated elites of the city.

9 The Owner's Use of the Garden

On the inscription above the portal, Scholz announced to those entering the garden his vision regarding its function: 'Et in sui ipsius a laboribus, et studiis artis suae recreationem' ('rest from the labours and studies of his art'), meaning medicine.[57] Scholz explained this function in greater detail in the preface to *Janus Quadrifrons*, stating 'quem curarum ac molestiarum mearum, in quotidianis artis et vocationis meae laboribus, et occupationibus interdum propemodum nimijs, quasi asylum, ac secessum quendam liberalem, vel museum potius suavißimum nominare soleo […]' ('which I am in the habit of calling an asylum and a noble retreat for my worries and troubles, for the daily toil of my art and vocation, and for the demands that are sometimes almost too great, or rather a very pleasant "museum"').[58] With this formulation, Scholz cites a passage from the *Hortus Medicus et Philosophicus* by his friend Joachim Camerarius the Younger almost verbatim. He too called his garden 'quasi refugium quoddam et secessum liberalem vel potius museum iucundissimum ab ipsa Natura' ('a refuge and a free retreat or rather the most pleasant museum of nature itself').[59]

Building on the overall ambition of the gardens, the Acidalius law explains how the octagonal pleasure house served as the centrepiece of the garden [Fig. 9.4]:

> Haec pergula aedes est, sacellum est: hic herus
> Studiorum habet domicilium, habet sapientiae:
> Naturae in hoc arcana scrutari solens,

57 Scholz *Epigrammata* fol. A 1v.

58 Scholz *Epigrammata* fol. F 3v.

59 Joachim Camerarius the Younger, *Hortus medicus et philosophicus. In quo plurimarum stirpium breves descripiones locorum natalium, novae icons non paucae indicationes locorum natalium, observationes de cultura earum peculiares, atque insuper nonnulla remedia euporista, nec non philologica quaedam continetur. Item Sylva Hercynia. Sive catalogus plantarum sponte nascentium in montibus & locis plerisque Hercyniae syluae quae respicit Saxoniam, conscriptus singulari studio à Ioanne Thalio* (Frankfurt: Johann Feyerabend, 1588) fol. B 2r.

Nova, mira quaerens, ante nulli cognita.
Sed et otij sedem hic, sed et requietis hic,
Atque hilaritatis sedem habet laetißimam.
Hic suetus animum solus, hic cum aequalibus,
Hic cumque amicis, cumque longinquo hospite
Laxare doctis musicis sermonibus.[60]

This pleasure house is a temple, a small sanctuary. Here the owner of the house has a domicile for his studies, a domicile for wisdom:
by exploring the secrets of nature, by searching for the
new, the wonderful and for things previously unknown to anyone.
Here he has both a place for leisure and for rest,
and here he also has the most cheerful place for gaiety.
Here he cultivates his spirit alone, here with like-minded people,
here with friends, and here with a guest from afar, in
learned conversation consecrated to the Muses.

As a place of study consecrated to the Muses, the Scholz garden continues a tradition dating back to Plato's Academy. Plato and his students walked through a sacred grove with an altar dedicated to the Muses, the first philosopher's garden in Greece and a green place of teaching and learning. Aristotle's Lyceum and Epicurus's garden in Athens followed his example and were also environments that provided a solid institutionalised framework for philosophical discussions.[61] As a place of leisure, Scholz's garden also follows in the Roman convention of "otium philosophicum" established by figures like Cicero, Horace and Seneca. Like his predecessors who hosted in villa gardens at the gates of the city, Scholz cultivated philosophical conversations among like-minded people as well as the study of natural mysteries.

10 Rules for Visitors to the Garden and Guests of the Breslau Floralia

The *Leges Hortenses* and *Leges Convivales* penned by Scholz, as well as Acidalius's elaborate treatment of the topic in the *Janus Quadrifrons*, set out guidelines for visitors on correct conduct and morality and offer us deep insight into the activities Scholz and his guests pursued in the garden. Scholz welcomed respectable and learned visitors to his garden; access was not only

60 Scholz, *Epigrammata* fol. G 1v.

61 Lauterbach, *Gärten der Musen* 47–49. See also Ribouillault's introductory essay.

granted to men, but also to women and children accompanied by an adult 'guardian'.[62] They were permitted to walk through all parts of the estate but were forbidden from entering the flower beds so as not to damage exotic and precious plants.[63] All visitors were encouraged to stroll along the paths,

> Sed pendula planta atque sollicito gradu,
> Ne forte temere hic quod satum aut natum obteras.
> Nimio labore et sumtibus nimiis hero
> Ipsae herbulae ac cultura earum constitit.
> Nam quod vides cunque, omne paene exoticum,
> Advectum ab oris orbis amoti procul,
> Procul orbe nostro, procul et hoc sole et solo:
> Ex Italorum transditum viridariis,
> E Graeciaeque, Hispaniaeque, et Galliae,
> Ex Indiae penitißimae, atque ex Arabiae.[64]
>
> But with a hovering foot and a careful step,
> so that you don't accidentally trample down what has been sown or grown here.
> The owner of the house has put a lot
> of effort and a lot of money into the
> care of even the smallest plants.
> For what you see everywhere is almost all foreign,
> imported from the shores of a distant world,
> far from our land, far from this sun and this soil:
> acquired from the parks of Italy, from
> those of Greece, Spain, and France,
> from those of the interior of India and Arabia.

Women were prohibited from wearing robes with a long, hanging train, as these could cause damage to the garden and were deemed unsuitable in bad weather.[65] Only plant experts were permitted to touch the plants while others could only possess them with their eyes. If they wished to take seeds, a branch or cuttings, they were entreated to ask the master of the house, the gardener

62 Scholz, *Epigrammata* fols. G 4r–G 4v.
63 Scholz, *Epigrammata* fol. G 4v.
64 Scholz, *Epigrammata* fol. G 2v.
65 Cf. Scholz, *Epigrammata* fol. G 2r.

('curatore'), or expert picker ('carpatore') for help.[66] Guests were expected to limit the duration of their visit to the garden to a few hours, depending on the time of year and the time that the host could spare for them.[67]

Scholz dedicates the *Leges Convivales* to the banquets held in the *Lusthaus* or pleasure house, underlining the importance of these festivities to the author. Unlike access to the garden, these festivities were reserved for his inner circle of friends and scholars:

> Convivia nomine et tutulo Floralium Wratslaviensium celebrantor. In haec sacra et hanc voluptatem idonei conviviae, viri boni, elegantes, eruditi, faciles, hilares vocantor: nec matronae castae, aut formosae honestaeque virgines excluduntor.[68]
>
> The guest banquets shall be celebrated under the name and the honorary title 'Breslauer Floralia'. Suitable guests, honourable, witty, educated, skilful and cheerful men shall be invited to this celebration and to this pleasure: morally strict wives or beautiful and honourable girls shall not be excluded.

Scholz limited the number of banquet guests between three and nine, given that the octagonal garden house could not comfortably accommodate a larger group. Conversely, an uncomfortable emptiness would be felt if the circle were too small.[69] With respect to age and social rank, a certain equality was ensured by the choice of guests, who were selected in part to guarantee an atmosphere of free exchange:

> Venientes, domi curas et sollicitudines omnes relinquunto: Lubentiam, Risum, Iocum comites adducunto: ad Floralia, sed honesta et modesta, quibus et Cato intersit, ire se cogitanto.[70]
>
> Let the guests leave all their cares and worries at home: let them bring merriment, laughter and jesting as companions: let them remember that

66 Scholz, *Epigrammata* fol. G 4v.
67 Scholz, *Epigrammata* fols. G 2r; H 1r.
68 Scholz, *Epigrammata* fol. H 1r.
69 Scholz, *Epigrammata* fol. H 1v.
70 Scholz, *Epigrammata* fol. H 1v. Marcus Porcius Cato, also known as Cato the Censor or the Elder, known for his virtuous and simple life.

> they are going to the festivals of Flora, but honourable and moderate ones, in which even Cato should be able to take part.

Scholz presided over the celebrations as 'rex et modimperator' ('king and president'). The welcoming and introduction of each individual was followed by a walk through the garden, where all could be inspected.[71] After washing their hands, guests took their places at the table, which were dictated according to order of precedence and the will of the host.[72] They could expect to be presented with wreaths of flowers from the garden and seated amidst wine cups decorated with garlands and a table adorned with flowers.[73] According to the laws, the food served at the meal should neither be too luxurious nor too expensive while the costs should not exceed a middle-class level.[74] Scholz desired that offerings should primarily be prepared from the yields of the garden in alignment with the classical topos of the "dapes inemptae" (i.e., unbought food homegrown produce) and belief that the food should be morally pure.[75] Wine – served mixed with water – could be consumed in moderation so as to promote, and not destroy, happiness.[76] Toasts to the welfare of the master of the house, to the garden, to friends present or absent, or in honour of important men were all welcome. With these edicts, Scholz hoped that the festivities at the Floralia would run with light-hearted cheerfulness, regulated in right measure.[77]

> Igitur Ioca et Dicta, quae sal et sol etiam reliquae vitae proprie in convivatione habitant, largiter adhibentor. Lascivia tamen in illis, in his acerbitas vitantor: utraque intra licentiam et convitia sunto.[78]

> Therefore, jokes and sayings, which are also common as salt and sunshine in the rest of life, but especially at communal meals, should be used liberally. Nevertheless, licentiousness and severity should be avoided; both should be below insolence and blasphemy.

71 Scholz, *Epigrammata* fol. H 1v.
72 Scholz, *Epigrammata* fol. H 1v.
73 Scholz, *Epigrammata* fols. H 2v–H 3r.
74 Scholz, *Epigrammata* fol. H 2r.
75 Scholz, *Epigrammata* fol. H 2r.
76 Scholz, *Epigrammata* fol. H 2v.
77 Scholz, *Epigrammata* fol. H 2r.
78 Scholz, *Epigrammata* fol. H 2r.

Though convivial in nature, banquets were to be held 'Convivium ipsum Philosophice et Musice instruitor' ('in a philosophical and learned manner').[79] At the dining table, Scholz therefore favoured conversations on topics consecrated to the Muses, philosophy and medicine:

> Sed idipsum ne nimis arduum et anxium, sed nec obvium aut tritum, et amoenum vero et facetum sit, vidento. Praecipue ex cultu hortorum, reque herbariam, et naturam rerum universam ZHTHMATA lepidiora et loco cognata sumunto.[80]

> But be careful that even these are not too sublime and too tense, but unobtrusive and not worn out and truly graceful and elegant. Preferably, one should take from the field of the care of gardens, botany and the whole nature of the material world more pleasing questions and those connected with the place.

For Scholz, Apollo and the Muses were welcome guests at the banquet in the pleasure house. Bacchus and his son Comus were also accepted, as long as they remained sober or intoxicated within decent limits. If guests should go overboard, however, and behave in an unrestrained manner by drinking too much or becoming too boisterous, the guardian of the garden, the four-faced Janus, would immediately order them out the door, as Acidalius warned with a wink.[81] If conversations lost their liveliness, guests might enjoy the singing of the birds in the aviaries or make music themselves. Instruments were available for this purpose in the *Lusthaus*. In this way, banquets could go on deep into the night.[82]

The final lines of the *Leges Convivales* again reference the exclusivity of the inner circle around Scholz, in which Scholz forbids the guests to say anything about what happened or was said in the garden upon their return to town. 'Qui hasce leges frausus erit, e consortio bonorum eicitor, ex amicorum numero deletor' ('Whoever transgresses these laws shall be banished from the community of the honourable and erased from the number of friends'), Scholz sternly warns, underscoring the delicate balance of pleasure and philosophical refinement.[83]

79 Scholz, *Epigrammata* fol. H 2r.
80 Scholz, *Epigrammata* fol. H 2r.
81 Scholz, *Epigrammata* fol. G 1v.
82 Scholz, *Epigrammata* fol. H 3r.
83 Scholz, *Epigrammata* fol. H 3v.

11 Regulated Freedom of the Mind

Scholz's garden ascribed a high priority to pleasure and laughter. Guests were encouraged to be merry within the bounds of moderation and propriety dictated by the laws. This 'taming of laughter', a term which Gernot Böhme aptly coined for the Christian West, is precisely the aim of Scholz's instructive texts.[84] The 'community of the honourable' that Scholz and Acidalius evoke in their works can be characterised as an educated, late humanist public with an established meeting place in the Hortus Scholzianus. Their prescribed gatherings and literary focus notwithstanding, the group was also unified through laughter, and thus were driven by humour as much as by a desire for learning.[85] In cheerful company, visitors could leave behind the constraints and worries of everyday life and reassure themselves of their common moral standards and educational horizons.

Scholz was concerned, however, with cultivating a civilised form of laughter; a form of laughter that could on the one hand liberate, and on the other hand tame. Laughter and conviviality were in no way a rejection of the social norms of the time and should not be misunderstood as a kind of escape from them. Rather, the Breslau laws precisely reflect the highly cultivated and deeply refined nature of the guests, who engaged in moderate joking and laughter. In the ordered space of the garden, the polished guest found a place of pleasure consecrated to the Muses and Graces – a space of moderation, restrained affect, and intellectual exchange.

12 Another Room

The Hortus Scholzianus is firmly rooted in the history of Breslau and its urban culture, even if Scholz restricted access to the site as though it were a sacred place. Aleida Assmann describes those places where one can experience the presence of the gods as sacred and, indeed, the visitor to the Breslau garden

84 Böhme G., "Es gibt vielleicht auch für das Lachen noch eine Zukunft", in Beil R. (ed.), *Mathilda is calling. Erinnerung als Zukunft*, exh. cat., Mathildenhöhe Darmstadt (Stuttgart: 2006) 44–55, here 46.

85 On the term, see Röcke W. – Velten H.R., "Einleitung", in Röcke W. – Velten H.R. (eds.), *Lachgemeinschaften. Kulturelle Inszenierungen und soziale Wirkungen von Gelächter im Mittelalter und in der frühen Neuzeit*, Trends in medieval philology 4 (Berlin: 2005) 9–31; Kuhn C. – Bießenecker S. (eds), *Valenzen des Lachens in der Vormoderne (1250–1750)*, Bamberger Historische Studien 8 (Bamberg: 2012) 19, 165–172.

encountered a cast of ancient gods.[86] This pantheon was led by Flora and Venus, the protectors of the garden, and followed by the four-faced guardian Janus.[87] If Venus and Flora embodied the creative power of nature, Janus, the god born out of primordial chaos, represented the order of the cosmos. With his four faces, Janus stands for the cyclical division of the day and seasons into four parts, as well as the cardinal directions and primary elements. Placed in the middle of the garden in the pleasure house and positioned to regulate natural and human behaviour alike, Janus exemplifies the order of the world par excellence.

The murals in the gallery complement the round dance of ancient gods with their depiction of Orpheus making music amidst exotic fauna and flora and lamenting the death of Eurydice. The singer, who captivated wild animals and even trees, here represents the human who can tame nature through artistry. Orpheus's beguilingly beautiful music symbolises the dream of humans as rulers of a nature with which they live in harmony. The blossoming and fruit-bearing plants in the garden, the music-making of the guests, the accompanying cooing of the pigeons in the aviaries, and not least the overall order of the garden reflect this desired harmony. The musical, creative and scientifically curious man of the late sixteenth century could here reflect on and formulate his place in the cosmic whole. Thus, the combined tenets of antiquity and humanism formed a fecund ground for the development of one's perspective towards life and view of nature, as in the festive culture of the Breslau garden. Here, the gods of antiquity were transposed to sixteenth-century experience with a view of shaping cultural and natural identity by narrative and artistic means.

In *Of Other Spaces*, Michel Foucault describes the 'space of the outside ('espace du dehors')', 'in which we live and in which the erosion of our life, our time and our history takes place'.[88] For Laurentius Scholz, this space was the city of Breslau, where he lived and worked as a doctor. Through his garden, Scholz managed to create a 'different space', situated in the city but at the same time removed from it. As a 'counter-space' in the Foucauldian sense, Scholz's garden was both a real and mythologised negation of the urban environment. It withdrew from the hustle and bustle of the city by way of walls, gates, and

86 See Assmann A., *Erinnerungsräume. Formen und Wandlungen des kulturellen Gedächtnisses* (Munich: 1999) 303.

87 See Scholz, *Epigrammata* fol. G 4r.

88 Foucault M., "Andere Räume" in Barck K. (ed.), *Aisthesis. Wahrnehmung heute oder Perspektiven einer anderen Ästhetik. Essais* (Leipzig: 1992) 34–46, here 38. See also Ribouillault's introductory essay.

restricted access, and offered Scholz and his guests an escape from the corrosive nature of everyday life. Here, the host and his friends could immerse themselves in a counter-world of erudition and festive gaiety that revived ancient customs. In so doing, an urban elite gathered based on their shared connoisseurship of ancient art and culture, creating a sacred space for scholarship, plant research, and festivity.

References

Calagius Andreas, *Hortus doct. Laurentii Scholzii medici et philosophi quem ille colit Vratislaviae, situm intra ipsa civitatis moenia, celebratus carmine* (Wrocław, Georg Bauman: 1592).

Coffin D., "The 'Lex Hortorum' and Access to Gardens of Latium during the Renaissance", *Journal of Garden History* 2 (1982) 201–232.

Coffin D., *The Villa in the Life of Renaissance Rome* (Princeton: 1988).

Coffin D., *Gardens and Gardening in Papal Rome* (Princeton: 1991).

Cohn F., "Dr. Laurentius Scholz von Rosenau, ein Arzt und Botaniker der Renaissance", *Deutsche Rundschau* 63 (1890) 109–126.

Eysymontt K., "Ogród Laurentiusa Scholza we Wrocławiu i jego europejskie parantele", *Biuletyn historii sztuki* 51.1 (1989) 3–12.

Fleischer M.P., "Der Garten des Laurentius Scholz", in Fleischer M.P., *Späthumanismus in Schlesien. Gesammelte Aufsätze*, Silesia 32 (Munich: 1984) 136–163.

IJsewijn J., "An admirer of Justus Lipsius: the German neo-Latin poet and philologist Valens Acidalius", *Mededelingen van de Koninklijke Academie voor Wetenschappen, Letteren en Schone Kunsten* 45 (1983) 183–206.

Ilić L., "Andreas Dudith und sein reformiertes Netzwerk in Breslau am Ende des 16. Jahrhunderts", in Bahlcke J. – Dingel I. (eds.), *Die Reformierten in Schlesien. Vom 16. Jahrhundert bis zu Altpreußischen Union von 1817* (Goettingen: 2016) 53–63.

Lauterbach C., *Gärten der Musen und Grazien. Mensch und Natur im niederländischen Humanistengarten 1522–1655* (Ph.D. dissertation, University of Cologne: 2002) Kunstwissenschaftliche Studien 111 (Munich – Berlin: 2004).

Lauterbach C., "Gartengesetze des Späthumanismus. Der Garten als Raum gelehrter Geselligkeit", in Schweizer S. (ed.), *Gärten und Parks als Lebens- und Erlebnisraum. Funktions- und nutzungsgeschichtliche Aspekte der Gartenkunst in Früher Neuzeit und Moderne* (Worms: 2008) 21–32.

Lauterbach C., *Der erzählte Garten des Laurentius Scholz. Bürgerliche Gartenkultur des Späthumanismus in Breslau*, Grüne Reihe 35 (Worms: 2018).

Oszczanowski P. – Gromadzki J., *Theatrum vitae et mortis. Graphik, Zeichnung und Buchmalerei in Schlesien 1550–1650* (Wrocław: 1995).

Oszczanowski P., "Wrocławski ogród Laurentiusa Scholtza St. (1552–1599) – sceneria spotkań elity intelektualnej końca XVI wieku", in Hałub M. – Mańko-Matysiak A. (eds.), *Śląska Republika Uczonych. Schlesische Gelehrtenrepublik*. Bd. 1 (Wrocław: 2004) 98–145.

Oszczanowski P., "The Art Collection at the Garden of Laurentius Scholz the Elder in Wrocław", *Studia Rudolphina* 6 (2006) 36–44.

Scholz Laurentius, *In Laurentii Scholzii Medici Wratisl. Hortum Epigrammata Amicorum* (Wrocław, Georg Baumann: 1594).

Scholz Laurentius, *Catalogus arborum, fruticum ac plantarum, tam indigenarum quam exoticarum horti Laurentii Scholzii, medici Vratisl* (Wrocław, Georg Baumann: 1594).

CHAPTER 10

Botany and Academy: the Gardens of Cardinal del Monte in Early Seventeenth-Century Rome

Alessandro Spila

Abstract

Thanks to his great cultural eclecticism, Cardinal Francesco Maria Del Monte (1549–1626) pioneered a new culture based on the observation of natural and artistic phenomena. His residence at Palazzo Madama can be considered a true academy, as can the garden of his last residence on the banks of the Tiber. Among many other disciplines, medicine and botany were central to the Cardinal's interests, and his circle included figures of the caliber of Fabio Colonna (1567–1640), member of the Academia dei Lincei, and Tobia Aldini, director and curator of the famous Farnese Gardens. This chapter focuses on Cardinal Del Monte's three private gardens in Rome, all of which can be considered exemplary of the fundamental links between gardens and academic research.

Keywords

Cardinal Francesco Maria del Monte – Botany – Alchemy – Tobia Aldini – Accademia dei Lincei – Palazzo Madama – Rome – Gardens

1 Introduction

Between the end of the sixteenth and beginning of the seventeenth century, Cardinal Francesco Maria Bourbon del Monte Santa Maria (1549–1626) [Fig. 10.1] was undoubtedly one of the figures most closely linked to the future concept of the academy. His academy was understood both as an elite cultural coterie and a place for the dissemination of knowledge by the young scholars and artists who were fortunate enough to be welcomed into his large entourage. My focus on Francesco Maria del Monte in relation to a still little-studied aspect of his otherwise well-known patronage – namely, the gardens he established – emerges from a small archival note discovered during my research on the Barberini family a few years ago. From the accounts of Cardinal Francesco Barberini (1597–1679), we learn that he employed the celebrated

 | DOI:10.1163/9789004517547_011

FIGURE 10.1
Ottavio Leoni, *Cardinal Francesco Del Monte*, 1616. Black chalk heightened with white chalk, 22.9 × 16.5 cm. Sarasota, The John and Mable Ringling Museum of Art
IMAGE © WIKIMEDIA COMMONS

physician and botanist from Cesena, Tobia Aldini, in 1625. In the same year, Aldini co-authored the famous printed description of the Farnese Gardens on the Palatine Hill and, in 1626, he was paid for the transport of 'mortari di bronzo e di pietra per stillare l'acqua dal palazzo di piazza Madama; leoni altre robe dal palazzo del già Card. Farnese a Monte Cavallo' ('bronze and stone mortars to drain water from the palazzo in Piazza Madama, lions and other objects from the palazzo of the former Cardinal Farnese at Monte Cavallo'), according to a specification from an expense note, detailed 'Nota della spesa fatta da me Tobia Aldini per far andar diverse robbe al Palazzo dell'ill.mo sig. card. Barberino alle Quattro Fontane dal giardino degli heredi del sig. card. Del Monte di Ripetta e da Madama' ('Expenses note created by me, Tobia Aldini, to get several things moved to the palazzo of the Most Eminent Cardinal Barberino at Quattro Fontane from the gardens of the heirs of Cardinal del Monte from Ripetta and Madama').[1] These short passages provides a great deal of new information on

1 Vatican Library, Barberini Archive, Justification 1, Cardinal Francesco Barberini *juniore*, portfolio 46, number 696. Document cited in Spila A., "Feste e tornei alla corte dei Barberini:

del Monte's patronage. Notably, we learn of the probable existence of a garden at Cardinal del Monte's primary residence – the Palazzo Madama near Piazza Navona – in addition to the one already known in Via di Ripetta, as well as Aldini's presence among Cardinal del Monte's protégés.[2]

Aldini was best known for his role as superintendent of the Farnese Gardens in the service of Cardinal Odoardo Farnese (1573–1626). First and foremost a chemist, pharmacist, and doctor of medicine, his knowledge of botany and the natural sciences was inherently connected to his practice of these professions. He was associated with numerous members of the newly founded Accademia dei Lincei and signed the famous treatise *Exactissima descriptio rariorum quarundam Plantarum quae continentur Romae in Horto Farnesiano* [Fig. 10.2].[3] However, his friend and esteemed colleague Pietro Castelli (1570–1661) was the true author of this text, who worked in those same years as the superintendent of the Vatican gardens surrounding Pius IV's villa.[4] The Farnese gardens managed by Aldini on the Palatine Hill were one of the most attractive examples in the first decades of the seventeenth century in Rome and the subject of study and publications by leading botanists of the time including Giovanni Battista Ferrari (1584–1655) and Fabio Colonna (1567–1640).[5]

Bernini e l'"ordigno del cavallo' nel giardino alle Quattro Fontane", *Miscellanea Bibliothecae Apostolicae Vaticanae* 24 (2018) 665.

2 Insufficient references were found to identify the palazzo belonging to Cardinal Odoardo Farnese on the Quirinal Hill. Although there has been past speculation about the Villa Aldobrandini in Magnanapoli, which somehow came to the Farnese family through the marriage of Duke Ranuccio and Margherita Aldobrandini, it must be acknowledged that the actual owner at the time was the elder Olimpia Aldobrandini, who gave it in 1626 to her son, Cardinal Ippolito. See Benocci C., *Villa Aldobrandini a Roma* (Rome: 1992) 59.

3 Aldini Tobia, *Exactissima descriptio rariorum quarundam Plantarum quae continentur Romae in Horto Farnesiano* (Rome, Mascardi Giacomo: 1625). On Aldini, see Coggiatti S., "Gli orti farnesiani, la Gaggia e Pietro Castelli alias Tobia Aldini", *Strenna dei Romanisti* 37 (1976) 465–468; Nocchi P.B. – Pellegrini E., "La collezione botanica del Cardinale Odoardo", in Morganti G. (ed.), *Gli Orti farnesiani sul Palatino* (Rome: 1990) 413–429. He was also in close contact with other members of the Accademia dei Lincei such as Giovanni Faber. See Cortesi F., "Una lettera inedita di Tobia Aldini a Giovanni Faber", *Annali di Botanica* 7 (1908) 403–405.

4 On Castelli, see Dollo C., *Modelli scientifici e filosofici nella Sicilia spagnola* (Naples: 1984) 148–155; Trabucco O., "La corrispondenza tra Pietro Castelli e Marco Aurelio Severino (con un'appendice di lettere inedite)", in Dollo C. (ed.), *Filosofia e scienze nella Sicilia dei secoli XVI e XVII* (Catania: 1996) 109–137; Ottaviani A., "Da Fabio Colonna a Paolo Boccone: momenti della storia della botanica fra Napoli e la Sicilia (con un'appendice di lettere inedite)", in Dollo (ed.), *Filosofia e scienze nella Sicilia dei secoli XVI e XVII* 137–168; De Ferrari A., "Castelli, Pietro", *Dizionario biografico degli italiani* 21 (1978) 747–750.

5 Among the extensive literature on Botanica at the time, see, for example: Ottaviani A., *Natura ed esattezza all'alba della scienza galileiana: le 'Observationes' di Fabio Colonna*, Letture di pensiero e d'arte 128 (Rome: 2021); Idem, "L'immagine di Galileo nella Roma dotta",

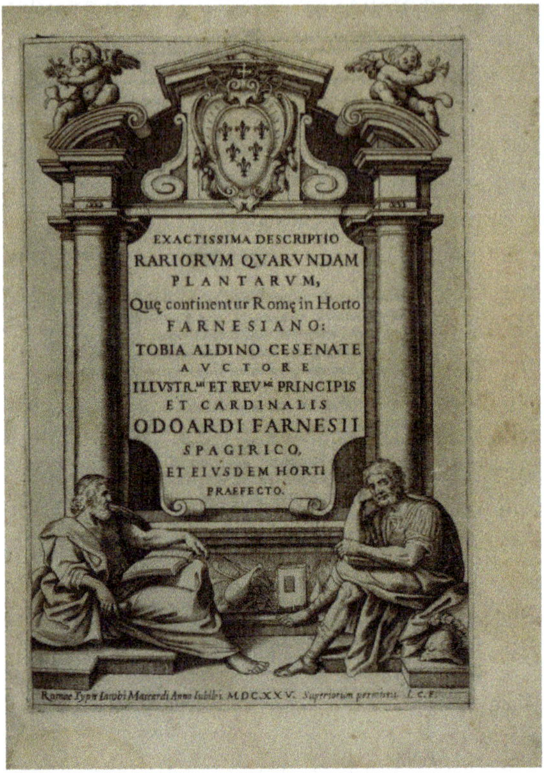

FIGURE 10.2
Luca Ciamberlano (engraver), frontispiece with Theophrastus and Pedanius Dioscorides. Copper engraving, 28 × 16 cm. From: Tobia Aldini/Pietro Castelli, *Exactissima descriptio rariorum quarundam Plantarum quae continentur Romae in Horto Farnesiano* (Rome, Pietro Mascardi: 1625)
IMAGE © INTERNET ARCHIVE

2 Cardinal del Monte – Patron of Alchemy, Science, and Art

A diplomat, politician, and connoisseur of the arts, sciences and music, Cardinal del Monte was known as a culturally eclectic man. With interests spanning diverse fields of human knowledge, he pioneered a new cultural attitude based on the observation and examination of natural phenomena. Better known as the first patron of Michelangelo Merisi da Caravaggio (1571–1610) and for his support of Galileo Galilei (1564–1642), del Monte was also a long-time

in Montesinos J. – Solís C. (eds.), *Largo campo di filosofare. Eurosymposium Galileo 2001* (Orotava: 2001) 915–923; on Fabio Colonna, among many, see Faraglia N.F., "Fabio Colonna linceo", *Archivio storico per le province napoletane* 10 (1885) 665–749; Gabrieli G., "Il 'Liceo' di Napoli. Lincei e linceabili – amici e corrispondenti della vecchia Accademia dei Lincei nel Mezzogiorno d'Italia", *Rendiconti della R. Accademia nazionale dei Lincei, classe di scienze morali, storiche e filolologiche*, XIV (1939) 499–565; Ottaviani A., "La natura senza inventario: aspetti della ricerca naturalistica del linceo Fabio Colonna", *Physis* 34 (1997) 31–70.

patron of the Accademia di San Luca and the Congregazione dei Musici, which would become one future Accademia Nazionale di Santa Cecilia, one of the oldest music institutions in the world. Protected by the Cardinal and eventually the Grand Duke of Tuscany, Ferdinando de' Medici (1549–1609), del Monte established his residence in Rome in the Palazzo Medici, better known as Palazzo Madama, in 1589. Under his influence, this site became a driving force behind the cultural and scientific activities of the time. In the years it was occupied by Del Monte and his entourage, Palazzo Madama was a multifunctional, strategic place of high standing. It was home to collections of paintings, sculpture and *objets d'art*, and became a haven of sorts for young artists from all walks of life eager to demonstrate their talent and exhibit their work. In addition to Galileo Galilei, Caravaggio and Federico Cesi (1585–1630), many politicians, diplomats, poets, mathematicians, as well as musicians and scientists frequented the palazzo.[6] Over the years, the Palazzo Madama and the Cardinal's other summer residences at Porta Pinciana, then Via di Ripetta were home to a veritable cultural and artistic coterie. They served as one of the most important artistic salons in Rome and Europe and were a place of excellence and cross-pollination among diverse cultures and disciplines of knowledge. Del Monte's residences were forerunners of the type of institution that Federico Cesi would create shortly thereafter with the 1603 establishment of the Accademia dei Lincei, which focused on scientific knowledge.

---

6 Among the vast bibliography on this topic, specific reference should be made to D'Alessandro D.A., "Caravaggio's 'Musical' Paintings for Cardinal Del Monte", in Christian K.W. – D'Alessandro D.A. – Terzaghi M.C. (eds.), *Caravaggio – i musici* (Venice: 2017) 49–65; Loskoutoff Y., "The 'Cupid Affair' (1596): Cardinal Francesco Maria Del Monte as a Collector of Antiquities", *Journal of the History of Collections* 25 (2013) 19–27; Carofano P., *Francesco Maria del Monte e Caravaggio: Roma, Siena, Bologna; opera biografia, documenti*, Libera Accademia di Studi Caravaggeschi (Pontedera: 2011); Lorizzo L., *La collezione del cardinale Ascanio Filomarino: pittura, scultura e mercato dell'arte tra Roma e Napoli nel Seicento ; con una nota sulla vendita dei beni del cardinal Del Monte*, Biblioteca Electa Napoli. Studi (Naples: 2006); Sickel L., "Ein 'Heiliger Franziskus' für Sixtus v. Kardinal Francesco Maria del Monte, Eleonora Orsini und die Bedeutung eines Gemäldegeschenks im Januar 1589", *Mitteilungen des Kunsthistorischen Institutes in Florenz* 43 (1999) 239–241; Waźbiński Z., *Il cardinale Francesco Maria del Monte, 1549–1626: 1. Mecenate di artisti, consigliere di politici e di sovrani; 2. Il 'dossier' di lavoro di un prelato*, Accademia Toscana di Scienze e Lettere La Colombaria: Studi 137 (Florence: 1994); Idem, "The Cardinal Francesco Maria del Monte and the New Patronage", *National Gallery of Art, Center for Advanced Study in the Visual Arts* 10 (1989/1990) 75–76; Spezzaferro L., "La cultura del cardinal del Monte e il primo tempo del Caravaggio", *Storia dell'arte* 9–10 (1971) 57–92; Sutherland Harris A., "Andrea Sacchi's 'Portrait of a Cardinal'", *Bulletin of The National Gallery of Canada* 14 (1969) 9–14.

3 Botany and Alchemy

Primarily due to Zygmut Waźbiński's studies, we know the vast extent of the Cardinal's interests in art and science. In Waźbiński's opinion, discussion of the Cardinal's involvement in the fields of botany and alchemy, the latter in particular, should be limited to an analysis of scientific interests, notably medicine and pharmacology.[7] For Maurizio Calvesi and others, however, his study of botany and alchemy might be significantly linked to the esoteric and astrological traditions championed by numerous members of the Medici family, including Ferdinando.[8] Ultimately, it is a fairly widely held opinion that the Cardinal intended to adapt and transpose the magical, esoteric, and alchemical into experimental, scientific fields.[9] His remarkable passion for botany, revealed in his possession of numerous herbaria, must have been closely connected to the interests described above. Until now, the name of the botanist directly employed by del Monte was unknown, though del Monte was understood to have had a series of indirect relationships with Ulisse Aldrovandi and the learned Jesuit naturalists of the Collegio Romano. The latter were among the most up to date in their knowledge of botany and were acquainted with the most recent discoveries from European missions to the Americas. These important relationships were enabled by the Cardinal's close friendship with Roberto Bellarmino. but their significance is dwarfed by the relationship he maintained with Fabio Colonna, who was recommended by del Monte first to the Florentine court, then to Federico Cesi's academy, and later became a leading exponent of the Accademia dei Lincei.[10] A direct relationship between Colonna and the Cardinal cannot be ruled out in view of their production of

7 Waźbiński, *Il 'dossier' di lavoro di un prelato* 433–434. To the series of doctors listed in his service we can now add the name of the Jesuit Giulio Filippelli 'medico', author of, inter alia, 'Iulii Philippelli eugubini [...] Epistola apologetica ad praeclarissimos almae Vrbis medicos Romae' (Rome, heirs of Zannetti Bartolomeo: 1624), see Archivio di Stato, Roma: Trenta Notai Capitolini, Paulus Vespignanus, Office 28, Volume 138, fol. 595v.

8 Calvesi M., *Le realtà del Caravaggio* (Turin: 1990) 176; Idem, "Trasparente l'autore come il soggetto: ancora su Caravaggio del Casino Ludovisi a Roma", *Art e dossier* 6 (1991) 22–26.

9 Benocci C., *Villa Ludovisi* (Rome: 2010) 94. With Pietro Castelli and the founding of the Accademia dei Lincei, alchemical theories and practices became a central reference for the emerging chemistry and therapeutic sciences. See Clericuzio A., "Chemical Medicine and Paracelsianism in Italy (1550–1650)", in di Pelling M. – Mandelbrote S. (eds.), *The Practice of Reform in Health, Medicine, and Science. 1500–2000* (Aldershot: 2005) 59–79.

10 Waźbiński, *Il 'dossier' di lavoro di un prelato* 443–446. On the importance of Bellarmino and the Jesuits in general as valuable commentators on botanical disciplines and the art of gardens, see also Witte A., *The Artful Hermitage: the Palazzetto Farnese as a Counter-Reformation 'diaeta'*, LermArte 2 (Rome: 2008) 89–123.

the famous naturalistic prints that found their way into Cassiano dal Pozzo's collections.[11] The connection between botany and alchemy in del Monte's patronage also had repercussions for the field of art, as is apparent in the descriptive details and symbolism of plants and flowers that figure in the work of the artists he protected, above all Caravaggio and his early works such as the *Rest on the Flight into Egypt* (*c.*1597).[12]

4 The Garden at Porta Pinciana

Already in possession of several vineyards in areas such as Genzano and Albano near Rome, del Monte assume a perpetual lease from 1596 on Cavalier Francesco del Nero's vineyard and associated hunting lodge near Porta Pinciana and made it his primary suburban residence.[13] Along with the neighbouring Villa Orsini, this residence formed the initial core of the future Villa Ludovisi. Close to his patron Ferdinando de' Medici's villa, the chosen site and famous *stilleria* constructed shortly thereafter became a sanctum for del Monte's alchemical and medicinal passions. This laboratory was used for chemical and pharmacological experiments and is known, above all, for the famous frescoes painted by Caravaggio depicting the alchemical triad – Jupiter, Neptune, and Pluto – around the spherical universe [Fig. 10.3].[14] This extensive property was

11 Waźbiński, *Il 'dossier' di lavoro di un prelato* 445; Solinas F., "L'erbario miniato e altri fogli di iconografia botanica appartenuti a Cassiano dal Pozzo", in Haskell F. – McBurney H. – Freedberg D. – Solinas F. – Napoleone C. (eds.), *Il museo Cartaceo di Cassiano dal Pozzo. Cassiano naturalista*, Quaderni puteani 1 (Milan: 1989) 52–76.

12 On this painting, see Calvesi M., "Riposo durante la fuga in Egitto", in Strinati C. (ed.), *Caravaggio*, exh. cat., Scuderie del Quirinale di Roma (Rome: 2010) 34–41. One typical feature is the meticulous depiction of plants, such as the Barbasso Yew immediately below the Virgin. This is a medicinal herb portrayed in many of the painter's other works, recurring, for example, in the three representations of St. John the Baptist preserved respectively in the Capitoline Museums in Rome, the Museo Tesoro Catedralicio in Toledo and the Nelson-Atkins Museum in Kansas City.

13 Frommel C.L., "Caravaggios Frühwerk und der Kardinal Francesco Maria del Monte", *Storia dell'arte* 9–10 (1971) 5–52; Benocci, *Villa Ludovisi* 101.

14 Pascucci I., "Gli enigmi del Camerino del Monte: scoperte e ipotesi ricostruttive del programma iconografico", in Lapucci R. (ed.), *Evidenza Caravaggio* (Rome: 2020) 295–317; Whitfield C., "The 'camerino' of Cardinal del Monte", *Paragone. Arte* 59 (2008) 3–38; Camiz F., "'Luogo molto vago et delitioso […]': il casino del cardinale del Monte ed un suo soffitto dipinto da Caravaggio", *Ricerche di storia dell'arte* 47 (1992) 81–88. For a summary of the many interpretations by scholars such as Maurizio Calvesi, Christoph Luitpold Frommel, Nancy Cole Wallach, and Matthias Winner, see Bernardini M.G., "Il dipinto del

FIGURE 10.3
Caravaggio, *Jupiter, Neptune and Pluto*, Rome, Villa Boncompagni Ludovisi; palazzetto (formerly Del Monte), *c.*1597. Oil on plastered ceiling, 300 × 180 cm

the ideal place for del Monte's circle to pursue literary and social advancement and also served as a place to grow plants, flowers and medicinal herbs destined for use in the distillery.[15] Thanks to our knowledge of the purchase and installation of water pipes, we can estimate that the Cardinal likely commenced the renovation of a garden surrounding the site around 1600. As Carla Benocci points out, Matthäus Greuter's 1618 map of Rome is the most faithful visual representation of the garden [Fig. 10.4], which shows how the Cardinal chose the area along the Via di Porta Pinciana to arrange the varied succession of parterres in his garden. Bordering the Medici property, they are laid out using different shapes and according to a regular, geometric grid. This design was in line with the dictates of sixteenth-century texts and any particular emphasis on originality is absent. The sector towards the hunting lodge was bordered by a first row of cypress trees, while a second row marks the boundary with the

---

Caravaggio nel casino Ludovisi: i risultati del restauro", in Gregori M. – Acanfora E. (eds.), *Come dipingeva il Caravaggio*, exh. cat., Galleria Palatina di Firenze (Milan: 1996) 102–105.

15 Whitfield, "The 'camerino'" 12.

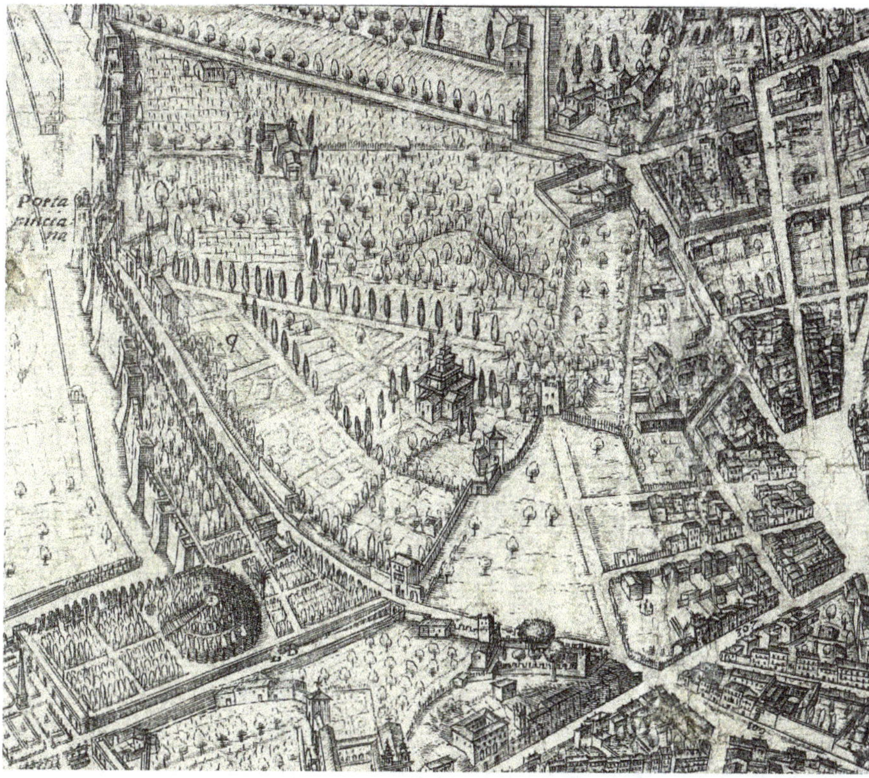

FIGURE 10.4 Matthäus Greuter (engraver and designer), the casino and vineyard near Porta Pinciana. Copper engraving, 15.02 × 16.96 cm. From: idem, *Disegno nvovo di Roma moderna con le sve strade, siti et edifitii in pianta esatta cosi come sta al presente sotto il fel.mo pont.to di N.S. Paolo P.P.V.* (Rome, Matthäus Greuter: 1618). Etching and engraving, 1 print (16 folios), 130 × 217 cm IMAGE © CENTRO DI STUDI SULLA CULTURA E L'IMMAGINE DI ROMA

Orsini property. Other successive rows of cypresses indicate the main paths that cross the various terraces and further emphasise the axes of the hunting lodge. Cultivated and wooded areas of the most common types were also indicated as present on the property. The nearby Villa Medici garden appears to have served as a model for del Monte's garden, as well as that of the neighbouring Palazzo Sforza, which was also being renovated by the architect Giovanni Fontana in those same years and was where the Cardinal spent his youth.

The garden at Porta Pinciana was the only large garden cultivated by Cardinal del Monte in Rome and was soon fought over by higher ranking figures supported by wealthier families. Del Monte was initially forced to cede the villa to the powerful Cardinal Pietro Aldobrandini (1571–1621), nephew of Pope Clement VIII, for two years, which was followed by its later occupation by

Cardinal Ludovico Ludovisi (1595–1632), nephew of Pope Gregory XV elected in 1621. In that same year, Del Monte was eventually forced to sell the villa to the latter. Despite the villa's fine embellishments and famous dressing room, Del Monte's limited financial resources did not allow him to fully exploit the vast estate or create a monumental garden on par with Ferdinando de' Medici's or that of the future Villa Ludovisi.

5 The Garden at the Palazzo Madama

In addition to confirming the employment of the botanist, the aforementioned note from the Barberini archive also attests to the presence of instruments belonging to Aldini at the Cardinal's primary urban residence, the Palazzo Madama. This might suggest that, like many of the Cardinal's other employees, Aldini resided there or had his own workshop linked to the garden. The inventory drawn up on 7 October 1626 on behalf of his second heir, Alessandro del Monte, confirms the existence of a *stilleria* and garden at the Palazzo Madama. This was ordered following the death of the first heir, Uguccione del Monte, just eight days (7 September 1626) after the Cardinal's own death.[16] This inventory was published by Christoph L. Frommel, but the first and as yet unpublished part of this inventory can be found in the previous volume of the same notary's office.[17] It reads:

> Inventarium omnium et singulororum bonorum repertorum in hereditate ch: mem. Ill.mi D. Uguccioni a Monte factum a Instantiam Ill.mi Ecc.mi Alexandri a Monte Episcopi Eucubinensis Germani fratris et eredis ex testamento q. Uguccioni que bona sequuntur et sunt infrascripta [...][18]

> Inventory of all the assets found in the estate of the clear memory of the most illustrious Don Uguccione del Monte, made at the request of the most illustrious and most excellent Alessandro del Monte, Bishop of Gubbio, his full brother and heir, as per the will of the late Uguccione, whose assets are as follows [...]

16 On the Cardinal's heirs, multiple wills, and related inventories, see Waźbiński, *Il 'dossier' di lavoro di un prelato* 567–570 and 633–642.

17 Frommel, "Caravaggios Frühwerk" 30–49.

18 Archivio storico di Roma (ASR), Trenta Notai Capitolini, Office 28 (Notary Paolo Vespignani), Volume 137, fols. 59r–82v. The inventory was drawn up on several occasions between October 1626 and May 1627, as can also be seen from the opening words of the inventory published by Frommel: 'Continuatur fuit [...]' (Continued).

From the second, already known portion of the will, we understand that most of del Monte's art collections had been transferred to the residence on Via di Ripetta by the time of his death.[19] The unpublished part of the will is especially important in terms of its description of the goods belonging to Cardinal del Monte that remained in the Palazzo Madama. Mainly a list of clothing, fabrics, sacred vestments and minor furnishings, it also includes a few interesting items such as 'Un baldacchino con tre cascate di damaschetto rosso, che serve per il giardino' ('A canopy with three cascades of red damask, which is used for the garden'),[20] and 'tutte le robbe che sono nella stillaria del Palazzo' ('all the things that are in the palazzo's *stilleria*').[21] Moreover, Alò Giovannoli's 1616 engraving depicting the remains of the 'thermae Neronianae in aedibus card. de Monte' ('Nero's baths in the palazzo of Cardinal del Monte') [Fig. 10.5] provides a small visual record of a secret garden – or, rather, a viridarium – in the

FIGURE 10.5 Aloisio Giovannoli (engraver and designer), *Thermae Neronianae in aedibus card. de Monte*. Copper engraving, 22.3 × 37.3 cm. From: idem, *Roma antica*, vol. 3 (Roma, Mascardi: 1619) fol. 9
IMAGE © BiASA

19 Ibidem, fol. 82v: '[...] in Palatio apud Platea Madama [...]' (in the palazzo at Piazza Madama).

20 Ibidem, fol. 62r.

21 Ibidem, fol. 83r. On the paucity of information about the possessions remaining at the Palazzo Madama and the transfer of the collections to Ripetta, at least from 1610 onwards (the year in which the Medici seem to have wanted to regain possession of the building), see Waźbiński, *Mecenate di Artisti* 116–117.

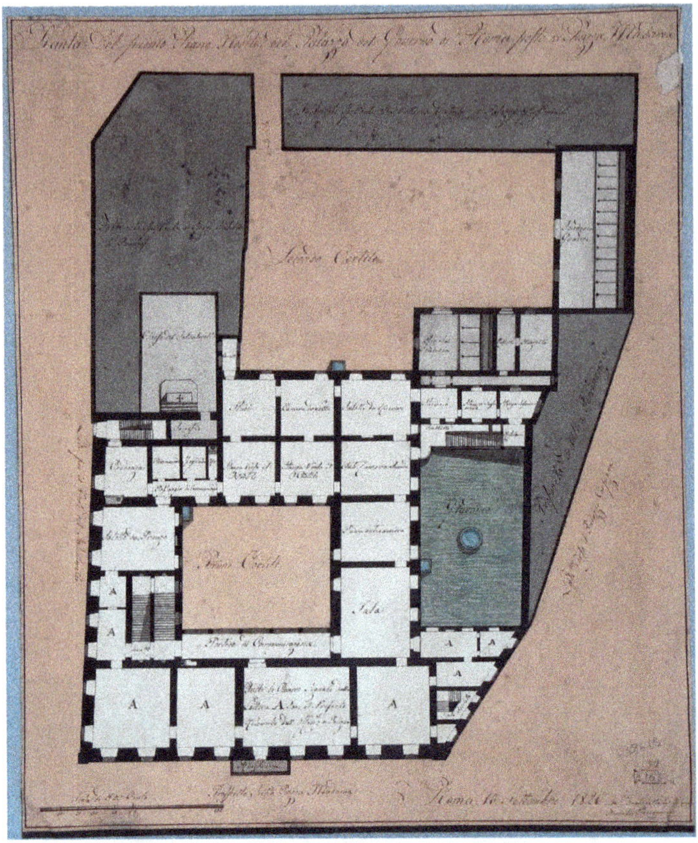

FIGURE 10.6 Francesco Paccagnini, plan of the first floor of Palazzo Madama in Rome, 1826. Pen and ink drawings, wash, 42.7 × 59.2 cm. Rome, Biblioteca dell'Istituto di Archeologia e Storia dell'Arte. Inv. Roma XI.52.163 IMAGE © BiASA

palazzo's second courtyard.[22] This partial view complements the information contained in the will and reveals the presence of flowerbeds arranged in parterres set against the backdrop of the ruins of the ancient Alexandrine baths, which have since been lost. These engravings refer to numerous earlier and contemporary references to various figures who were very close to Cardinal del Monte. This engraving also shows the view from inside the courtyard, which bordered other houses along the rear elevation facing the Palazzo Giustiniani

22 Giovannoli A., *Roma antica*, vol. 3 (Rome, Mascardi: 1619) fol. 9. On del Monte's lengthy residencc at the Palazzo Madama, see Waźbiński, *Mecenate di artisti* 113–118 and 185–217.

and was oriented towards the Palazzo Carpegna.[23] The viridarium would have been located in the second large courtyard and, as can be seen in a drawing from the early nineteenth century [Fig. 10.6], where another open space could be found to its side prior to the transformation of the palazzo into the Italian Senate. This space, albeit enlarged, still exists today under the name of the "orange garden". The small garden at the Palazzo Madama, distinct from its predecessor in both size and function, would have carried particular significance for its proprietor. Unlike the vineyard at Porta Pinciana, it was not only a secluded refuge in which to pursue one's passions. Rather, it also connected to the public, political, diplomatic, and above all academic role that the Cardinal conferred to his city residence. The presence of the garden-distillery and records of Aldini's personal activities within it might suggest an additional, intimate character for this part of the palazzo. Indicating an overlap between del Monte's public and private roles, the distillery served as a hideaway for rest and meditation where strictly personal passions could be cultivated alongside the primary function of the palazzo as a site of public affairs.

6 The Garden at Ripetta

Demonstrating an academic, encyclopedic approach to collecting, Cardinal del Monte's ambitions in the field of science were also revealed in a more spacious garden including an area dedicated to the accomodation of artists. These were realised in the later part of his life from 1621 onwards, when the large vineyard at Porta Pinciana was finally sold to Cardinal Ludovisi. Del Monte had begun to seek more stable accommodation as early as 1610, when his removal from the Palazzo Madama by the Medici became an increasingly real possibility because of their renewed use of the property. The 1615 purchase of the small Palazzo Avogadro near Via di Ripetta marked the construction of a new residence; this property, bordering Amerigo Capponi's palazzo, was conveniently

23 On the history of the building, see Borsi F., *Palazzo Madama, I palazzi del Senato* (Rome: 1999); Fumagalli E., "Committenza e iconografia medicea a Roma nel Seicento: il ciclo di affreschi di Palazzo Madama", *Mitteilungen des Kunsthistorischen Institutes in Florenz* 41 (1997) 314–347; Idem, "La facciata quattrocentesca del palazzo Medici in piazza Madama: un disegno e alcune considerazioni", *Annali di architettura* 3 (1991) 23–31; Miarelli Mariani G., "Il Palazzo Medici a Piazza Navona: un'utopia urbana di Giuliano da Sangallo", in Garfagnini G.C. (ed.), *Firenze e la Toscana dei Medici nell'Europa del '500, 3. Relazioni artistiche. Il linguaggio architettonico* (Florence: 1983) 977–993.

adjacent to a vineyard already owned by the Cardinal.[24] The building was best known for its famous loggia, which Giovanni Pietro Bellori recorded as having been decorated by Andrea Sacchi in 1620:

> Aveva il cardinal Del Monte adornato un giardino per suo diporto vicino al Tevere su la via di Ripetta, oggi del signor principe Borghese, edificatavi una loggia di ricreazione, nella quale volendo in tempo di estate fare l'invito di una cena, ordinò ad Andrea che quanto prima la dipingesse, ristrettogli il termine di alquanti giorni.[25]
>
> Cardinal Del Monte had adorned a garden for his pleasure near the Tiber on the Via di Ripetta, now belonging to Prince Borghese, and had built a recreational loggia, which, wishing to host a dinner party in summer, he ordered Andrea to paint as soon as possible, giving him only a few days to do so.

The paintings, long believed to have been lost and known only through a few sketches, were rediscovered in 2011 during restoration work in an apartment.[26] The large fresco and three lunettes that were uncovered reveal mythological scenes inflected with references to alchemy, with Venus and Mars, Proserpina and a Naiad, and Vulcan and Apollo portrayed on the lunettes. The painting depicts the four seasons as represented by Bacchus (autumn), a farmer (winter), a woman with a mirror (spring), and Venus (summer). The Cardinal's passion for alchemy is further confirmed by the presence of a laboratory or *stilleria* – in this case perfectly described with a list of all its equipment in the 1627 inventory.[27] In addition to a series of chemical and pharmacological instruments, the room was decorated with portraits of famous personalities linked to the history of alchemy and esotericism. Already a topic of analysis

---

24 Ważbiński, *Mecenate di artisti* 310–316. In the same year, the Florentine Amerigo Capponi purchased the larger, adjacent building and began the creation of a garden through extensive hydraulic work. This also involved the property owned by Cardinal del Monte. See Papini M.G., *Palazzo Capponi a Roma: casa vicino al Popolo, a man manca per la strada di Ripetta* (Rome: 2003) 222–223.

25 Bellori Giovanni Pietro, *Le vite de' Pittori, Scultori et Architetti moderni* (reprint Turin: 1976) 539–541; Idem, *The Lives of the Modern Painters, Sculptors and Architects: a new translation and critical edition*, trans. Sedgwick Wohl A. – Wohl H. – Montanari T. (Cambridge: 2005) 377.

26 Sutherland Harris A., *Andrea Sacchi. A Complete Edition of the Paintings* (Oxford: 1977) 1–5. See Fidanza G.B., *Andrea Sacchi and Cardinal del Monte: the Rediscovered Frescoes in the Palazzo di Ripetta in Rome* (London: 2022).

27 Frommel C.L., "Caravaggios Frühwerk" 45–46.

by Maurizio Calvesi in 1990, these figures include Hermes Trismegistus, Geber Arabs, Ramon Llull, Paracelsus, Petri Murien and Isabella Cortese, among others, lending the room a true aura of an alchemical *sancta sanctorum*.[28]

The discovery of Sacchi's fresco makes it possible to identify the exact location of the villa, which we now know extended upwards from the banks of the Tiber on the plot between Vicolo del Vantaggio and Vicolo delle Scalette, now the Palazzo San Martino Valperga on Passeggiata di Ripetta, between the current Via del Vantaggio and Via Angelo Brunetti. The elevation of the loggia overlooking the garden, characterised by an elegant Palladian window surrounded by later additions, also appears to have been preserved. The garden can be discerned in various seventeenth-century maps of Rome [Figs. 10.7–10.8]. Some additional details about the garden are provided in an urban land registry map from the early nineteenth century, which shows a large, central parterre with a fountain and hedges [Fig. 10.9]. The garden is also briefly described in the 1627 inventory, which identifies three sectors of differing heights in addition to a courtyard: a first terrace, a second area characterised by the presence of a fishpond (basin), and a third area referred to as the large garden. Both the courtyard and the three terraces were marked by the presence

FIGURE 10.7 Giovanni Battista Falda (engraver and designer), the casino and vineyard Del Monte near Ripetta. Copper engraving, 6.4 × 7 cm. From: idem, *Nuova pianta et alzata della città di Roma* […] (Rome, Giovanni Giacomo de' Rossi: 1676). Etching and engraving,
157 × 155 cm (12 folios, 39 × 51 cm)
IMAGE © CENTRO DI STUDI SULLA CULTURA E L'IMMAGINE DI ROMA

28 Ibidem 47; Calvesi M., *Le realtà del Caravaggio* 177–178.

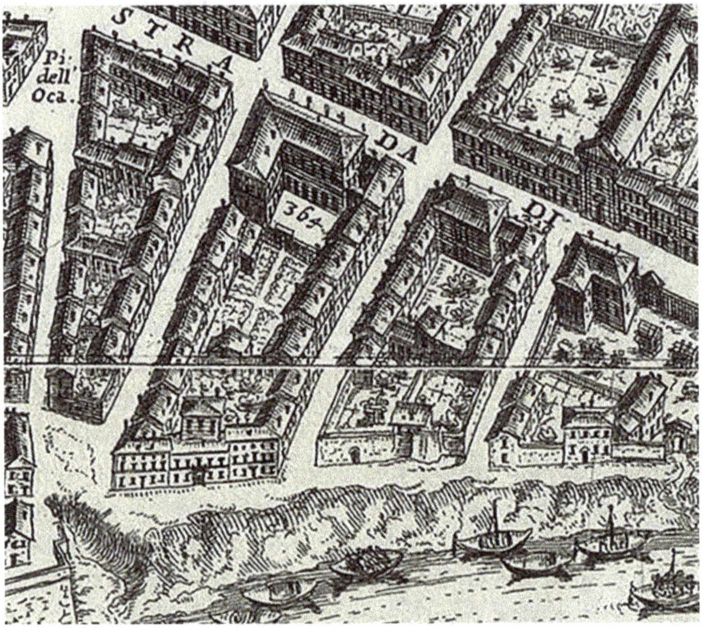

FIGURE 10.8 Matthäus Greuter (engraver and designer), the casino and vineyard Del Monte near Ripetta. Copper engraving 9.6 × 15.52 cm. From: idem, *Disegno nvovo di Roma moderna con le sve strade, siti et edifitii in pianta esatta cosi come sta al presente sotto il fel.mo pont.to di N.S. Paolo P.P.V.* (Rome, Matthäus Greuter: 1618), 1 print (16 folios), 130 × 217 cm

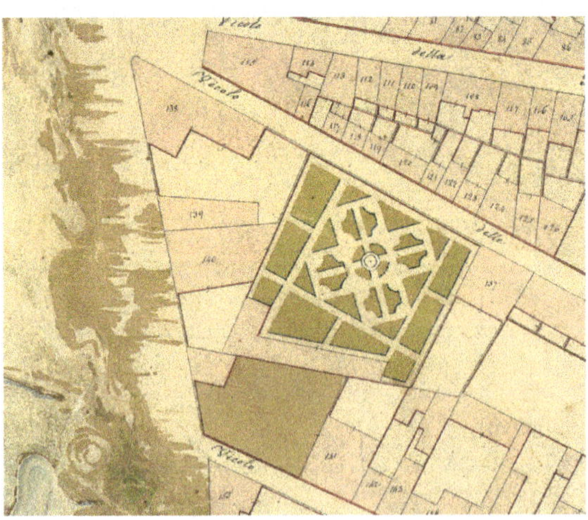

FIGURE 10.9 Giacomo Palazzi, Gaspare Salvi, detail of the block corresponding to the Del Monte vineyard in Ripetta in 1824. Pen and ink drawings, wash, 11.73 × 13.73 cm. Part of *Catasto Urbano di Roma, Rione IV, Campo Marzio*, fol. 1 (64 × 89.1 cm), Rome, Archivio di Stato

of numerous ancient statues and bas-reliefs, testifying to the Cardinal's other passion for archaeological collecting. The inventory describes various putti and fauns, along with figures of Apollo (first sector), Spinarius and Bacchus, as well as a goat, a leopard, a bas-relief with consuls and female figures, and a bas-relief with a woman on a seahorse (second sector, around the fishpond). Along with many unidentified statues, we find mentions of various statues of Venus, Adonis, Eros, a bas-relief with two Harpies, and diverse representations of Hermes in the large garden.[29] Waźbiński's studies of the Cardinal's correspondence mention a theatre where plays were often performed for the guests during sumptuous receptions. Such a theatre was likely located outdoors, since the detailed inventory of the interior does not mention such a room.[30] The hydraulic clock was another notable element of the garden; it not only showed the time but, connected to a pipe organ, also played the hours. Cassiano dal Pozzo, who frequented the del Monte garden from the time of his arrival in Rome in 1613, described this curiosity. Dal Pozzo reports that this mechanical jewel was similar to those found at the Villa Aldobrandini at Frascati, Villa d'Este at Tivoli, and the Quirinal Gardens, although probably more monumental. [31]

As further evidence of the cardinal's investment in, and not just appreciation of, the arts, a building on the property was used as a studio for the restoration of ancient statues. This place was made available to the numerous artists who frequented the villa, among them not only sculptors but also painters and other artists.[32] The building was enlarged over the years and survived until at least the mid-nineteenth century, according to a document drawn up by the owners of the villa at that time. Its presence in the villa allows us to piece together the truly academic vocation the villa served for the Cardinal, and which persisted for many years despite the adverse events that prevented del Monte's wishes from being fulfilled after his death [Fig. 10.10].[33] According

29 Frommel C.L., "Caravaggios Frühwerk" 38–39.

30 Waźbiński, *Mecenate di artisti* 117.

31 Idem, *Il 'dossier' di lavoro di un prelato* 457; Ribouillault D., "Hermeneutics and the Early Modern Garden: Ingenuity, Sociability, Education", in Enenkel K.A.E. – Melion W.S. (eds.), *Landscape and the visual Hermeneutics of Place, 1500–1700*" (Leiden – Boston: 2021) 311.

32 Waźbiński, *Il 'dossier' di lavoro di un prelato* 465.

33 Archivio di stato di Roma, Trenta Notai Capitolini, Office 02, Volume 837, fol. 121bis: 'Pianta dimostrativa della proprietà di Alessandro e Laura Aducci e sue adiacenze' ('Demonstrative plan of the Alessandro and Laura Aducci property and its surroundings'), consisting of (trad.) 'an entire house with garden located in Rome at Vicolo del Vantaggio, 1, with another lodge or building corresponding to Passeggiata di Ripetta, or Ripa di Fiume, 14–19, with perennial water and sculpture studios'. On that date (1854), we can see how the garden and the adjacent cottages partly used as original sculpture studios

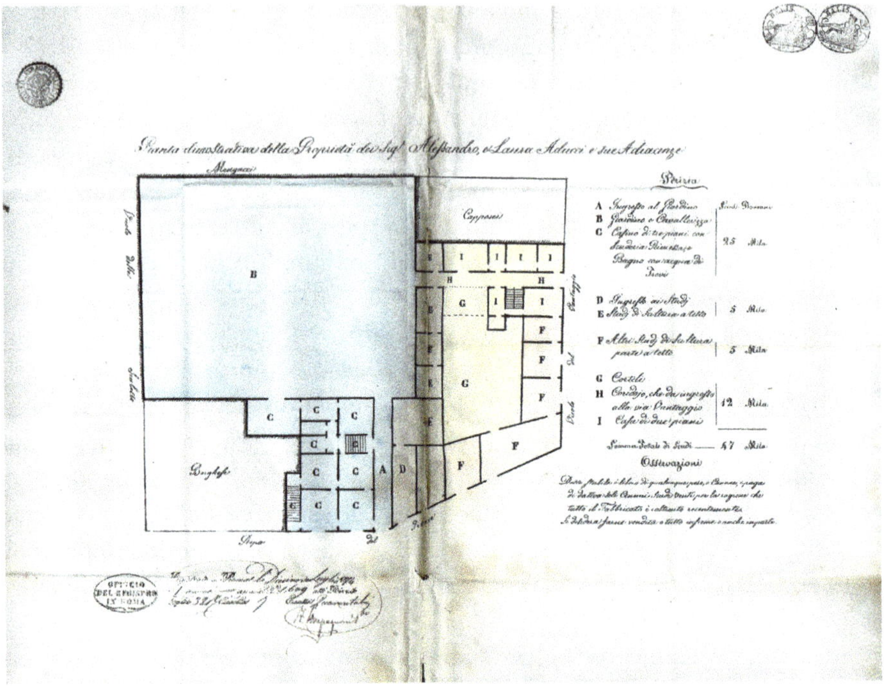

FIGURE 10.10 Carlo Nicola Carnevali, floor plan of the Borghese casino (formerly Del Monte) and adjacent sculpture studios (letters C, E, F) belonging to Alessandro and Laura Aducci in 1854. The drawing was made for a mortgage valuation and is attached to the notarial deed (Notary: Antonio Blasi). Pen and ink drawings, wash, 43.9 × 56 cm. Rome, Archivio di Stato
IMAGE © ARCHIVIO DI STATO DI ROMA

to Waźbiński, the sculpture studio, garden, library, and other spaces like the villa's galleries and cabinets of curiosities were all established for the purpose of educational advancement. They were fundamental to the scientific and artistic academy that del Monte hoped to found, which would have been itself a means of keeping the extraordinary cultural collections and experiences he had built in his decades at the Palazzo Madama alive. The academy was to be a 'collective monument', namely a public art institute according to Waźbiński. Explicitly declared in del Monte's first will, its mandate was to be upheld under the aegis of various religious institutions, most notably the Society of Jesus and its General.[34] Unfortunately, this ambitious project did not come to

had been detached from the lodge proper (and the famous loggia) and still belonged to the Borghese princes, as they had in Bellori's time.

34 Waźbiński, *Mecenate di artisti* 310–316; Idem, *Il 'dossier' di lavoro di un prelato* 633–642.

fruition, mainly due to the hostility of the Barberini family who stopped it from developping when Pope Urban VIII came to power in 1623. Envisioning the centralisation of the arts and total domination of the papacy across political and cultural domains, the Barberini established a set of radical reforms of the Accademia di San Luca, which adversely affected the realisation of del Monte's vision for his own academy.[35]

The role of the garden in the context of del Monte's academy plans should not, therefore, be treated as secondary to his better-known collections. On the contrary, a pleasant place for contemplation and reflection was essential to the study of art, nature, and their intersections, which was a central interest for academics in the period. In addition to its profusion of statues, the garden must have been home to a substantial collection of rare plants and flower essences as well as medicinal herbs. Owing to the Cardinal's close relationship with the Jesuits, these likely would have included samples from the Americas. Like the contents of del Monte's library, little is currently known about his botanical collection, but his will attests to its importance, as it stipulated that a topographical map of his property in Ripetta (the lodge and adjoining gardens) be drawn up after his death. A detailed inventory of everything in the garden, including plants and statues, was to be numbered and catalogued in the same way as the volumes in his library:[36]

> voglio che subito seguita la mia morte si facciano quattro accurati inventari: uno dello stabile, con tutte le qualità annesse et opportune, valuta, pianta, misure, confini a giudizio dell'esecutore, e di persone intelligenti da pigliarsi da detto esecutore, al qual inventario infrascritto poi s'aggiungano li beni stabili da co denari o delle cose vendute, o dell'entrate accumulate, o altri in altra occasione. Secondo delle cose proprie del giardino, le quali gli siano diligentemente e con ogni diligenza numerate, o mancando per casi fortuiti, o per altre ragioni in vantaggio contro cambiati, o risarciti, come io particolarmente mi raccomandai nel fidecommesso fatto della libreria sui libri, casse, e scansie [...] delli beni mobili siano nel palazzo nella Casa di Ripetta, nella vigna [...] Circa questi inventarij de tutti l'herede pigli appresso di se una copia del primo che è degli

35 Idem, *Mecenate di Artisti* 329–331.

36 Ibidem 312. Waźbiński points out that the inventory of books kept at Ripetta (over 3,000 volumes), made shortly afterwards and found by him, is rather general. A few shelves of books, perhaps of minor importance, had remained at the Palazzo Madama, see Archivio di stato di Roma (ASR), Trenta Notai Capitolini, Office 28 (Notary Paolo Vespignani), Volume 137, fol. 82r: 'various shelves of books consigned [signed?]'.

stabili, ne dia una per uno al Presente Generale del Gesu, alle monache di Sant'Urbano, alli Padri di Sant'Onofrio, aggiungendovi per tutto lo stabile che dalle cose vendute o dalle spese rimesse con l'entrate, o in altro modo all'heredità s'unisca, del secondo che è delle cose proprie del giardino e del terzo e quarto ne dia uno alla Compagnia di Gesù, del terzo delli libri ne dia uno anco al Vescovo sempre in detta copia autentica.[37]

I want four detailed inventories to be made immediately after my death: one of the buildings, with all the annexed and appropriate qualities, values, plans, measurements, boundaries in the judgement of the executor, and of intelligent persons to be taken by that executor, to which inventory below should then be added the immovable property from the money or things sold, or from the accumulated revenue, or those from other sources. According to the things belonging to the garden, which are to be diligently and with all diligence numbered, or missing by chance, or for other reasons held reciprocally, or compensated, as I particularly recommended in the *fideicommissum* [a deed forbidding the heirs from splitting up the collection] made for the library for the books, boxes, and shelves [...] of the movable property to be in the palazzo in the Ripetta house, in the vineyard [...] With regard to these inventories of everything, the heir should take a copy of the first, which is of the buildings, and give one to each of the present General of the Society of Jesus, the nuns of Saint Urban and the Fathers of Saint Onofrio, adding to the buildings any revenue from the things sold or from the expenses remitted with income, or in any other way added to the estate; for the second, which is of the things belonging to the garden, and for the third and fourth, the heir should give one copy to the Society of Jesus; for the third, relating to the books, the heir should also give one copy to the Bishop, always in the form of that authentic copy.

It is very significant that all the artistic and scientific collections gathered in the Ripetta residence were de facto entrusted to the aforementioned religious institutions. As a result of this directive, the Cardinal's nephew and official heir, Ranieri Marquis of Monte Baroccio, was only able to sell the goods remaining in the Palazzo Madama, which had evidently been left there for this

37 State Archive of Rome, Trenta Notai Capitolini, Office 11 (Notary Angelus Justinianus), wills, 1624, fol. 825r. These unpublished passages are taken from the will entered in the Urban Archive register on 5 September 1625, probably a copy of the one dated 27 June 1625 and partially transcribed in: Waźbiński *Il 'dossier' di lavoro di un prelato* 639–640.

intended purpose. Thought they were valuable objects, they were not considered part of the museum collection.[38]

7 Considerations

While awaiting the discovery of additional sources – both written and, especially, graphic – that might better establish the layout of the Cardinal's gardens and the architect(s) involved and thus add to the limited information currently available, I will attempt to offer some preliminary considerations.[39] In so doing, I will keep in mind the peculiar qualities of the Cardinal's persona rather than the more general aspects of Roman garden tradition. As one of the most multifaceted patrons of his time, Cardinal del Monte had a wide range of interests that make it impossible to frame our observations in terms of a single vision. Rather, they must be viewed in accordance with his wide-ranging contributions to a variety of fields.

In each of the three cases I have here considered, the Cardinal's passion for alchemy is a constant factor. This interest was closely linked to the study of medicine and botany but also carried symbolic implications, which were repeatedly demonstrated by way of painted decoration, and can be connected to hermeticism. The Cardinal's knowledge of the hermetic tradition seems to be confirmed by the series of portraits that appeared in the *stilleria*, starting with Hermes Trismegistus. We should therefore not rule out the possibility that the design of the gardens might have incorporated complex, symbolic

38 Archivio di stato di Roma, Trenta Notai Capitolini, Office 11 (Notary Angelus Justinianus), wills 1624, fols. 829r–v (trans): 'All movable things, large or small, that at my death are to be found in the palazzo where I live, or in the vineyard, or elsewhere, or that in the judgement of the executor may belong to the garden, which are and may be quite a few, I want to be assigned to him and written down in his inventory, and while the heir is not living in Rome they should be stored reasonably, and closed according to the judgement of the executor. Because there may be many things in the furniture from the palazzo that are not needed by the heir, and it would be better to make money from them, I declare that I want them to be sold, silks, carpets, and similar furnishings for the walls of the rooms or also many, many pieces of silver will be placed before all things, and much before. I say the same of similar things that are in the garden and his house, with which also I wish to do so, at the judgement of the executor, and sell them, investing the money as above.'

39 No information on the architects is found in the vast bibliography on the topic. In the list of creditors drawn up after his death, some workers generally linked to the figure of Carlo Maderno are mentioned, such as 'Pompeo Pagliari, muratore' and 'Giovanni Volpetta Falegname, for the Porta Salara site', see Archivio di stato di Roma, Trenta Notai Capitolini, Paulus Vespignanus, Office 28, Volume 138, fol. 595v.

meanings linked to esotericism. Interpretations based on alchemical considerations of the Boboli Gardens in Florence as well as those at the Villa Medici on the Pincian Hill have been suggested on several occasions.[40] The Cardinal's very close ties to the Medici family, from whom his passion for alchemy almost certainly derived, could then lead us to deduce that these gardens served as a point of reference and departure when building his own. Although they differ in size and monumentality, the analogous presence of small terraces in both the Medici Garden and the del Monte Garden could also be indicative of an ascending, initiatory symbolism. While certainly related to the orographical constraints of the Roman region, this conceptual orientation would have aligned with the academic features of these gardens and the association of different physical levels with the knowledge acquired by academics, followers, or those belonging to a close circle of compatriots. Such an interpretation could only be advanced after a careful examination of the symbolism expressed in the statuary and botany of the various sectors; current scholarship to this end is still too general. In any case, the number three seems to be related, in principle, to the three degrees of alchemical transmutation (namely the three states of matter) and potentially related to the '*Magnum opus*: *nigredo, albedo, rubedo*', i.e. degrees of spiritual metamorphosis respectively associated with earth, water, and fire. The presence of the fishpond in the second terrace, as well as the presence of statues depicting mostly human figures, as opposed to the deities in the more elevated sector, seem to support this hypothesis.[41] Finally, there is another garden with similar formal and cultural characteristics: the gardens of the Palazzo dei Lincei in Acquasparta, with three sloping terraces. The Accademia dei Lincei was born, as we know, from the esoteric and initiatory tradition.[42]

---

40 See, for example, Maresca P., *Alchimia, magia e astrologia nella Firenze dei Medici*, Giardini e dimore simboliche (Florence: 2012); Riva C., *Boboli: il giardino alchemico* (Chianciano Terme: 2010); Idem, *La Grotta Grande di Boboli: laboratorio di meraviglie. Il giardino alchemico* (Florence: 2011).

41 Another significant factor is the presence of the bas-relief with a Nereid on a sea monster, a recurring iconographic image from ancient Greece symbolising the long journey to be made after death, in this case understood as the passage from the earthly realm to the ethereal. A famous large sculptural group on the same subject, now in the Uffizi Gallery in Florence, was in the Villa Medici on the Pincian Hill, and before that in the collections of Angelo Colocci, see Bober P.P. – Rubinstein R., *Renaissance Artists & Antique Sculpture: A Handbook of Sources* (London: 2010) 145.

42 Benocci C., "La natura come sfida e passione: i giardini Cesi di Roma e Acquasparta; XVI–XVIII secolo", in de Petra G. – Monacchia P. (eds.), *I Cesi di Acquasparta, la dimora di Federico il Linceo e le accademie in Umbria nell'età moderna* (Perugia: 2017) 585–627. On the Lincei, see, for example, Ricci S., "I Lincei: l'invenzione della mediazione accademica.

One aspect of the garden that might at first appear antithetical to its alchemical character is its devotional dimension. Not entirely irreconcilable with the public and academic components of the property, Del Monte's garden suggested the incorporation of a hermitage as a retreat for spiritual contemplation. This element is typical of the gardens of high-ranking churchmen in the Counter-Reformation; the hermitage belonging to the Cardinal's friend Cardinal Odoardo Farnese, the focus of Arnold Witte's recent studies, is a particularly fitting and comparable example to the Ripetta garden because of its size, configuration and proximity to the river.[43] In addition to Cardinal Pietro Aldobrandini's garden at Monte Magnanapoli, many models of this type could be found in Rome. According to Witte, Cardinal Roberto Bellarmino's *Buen Retiro* at the famous garden of the Jesuit novitiate between the Quirinal Hill and the church of San Vitale was perhaps the most emblematic of the type.[44] Owing to its terraced structure, the richness of its botanical gardens, and the general didactic dimension of the site used for the training of the religious order's novices, this model may have appeared to del Monte as an archetype. He had always been very close to Bellarmino and considered the Society of Jesus as the highest authority in the doctrine of the Holy Roman Church, exemplary in both the scientific and artistic fields. The garden of the novitiate also features as many as two clocks, in this case sundials, described by the Jesuit Louis Richeome in his famous *La peinture spirituelle*, further illustrating the probable influence of the former on the development of the Ripetta garden.[45]

A third aspect that I would like to underline is the antiquarian dimension of the garden, which, although not new for a Roman garden, is fundamental both in terms of possible symbolic references and didactic dimensions; it would have therefore attracted artists who were allowed to study the statuary collections. Courtyards and gardens adorned with ancient statues were common in Rome from the Middle Ages, with the first evidence of this practice dating to the fifteenth century, with the Orti della Valle, Savelli Gardens and

Nuova scienza, religione, vita civile", in *Sciences et religions. De Copernic à Galilée (1540–1610)*, Proceedings of the International Conference in Rome 12–14 December 1996, Collection de l'École française de Rome 260 (Rome: 1999) 205–234.

43 Witte, *The Artful Hermitage* 38–41.

44 Ibidem 102–108.

45 Richeome Louis, *La peinture spirituelle ou L'art d'admirer aimer et louer Dieu en toutes ses oeuures, et tirer de toutes profit salutere* (Lyons: Rigaud Pierre, 1611) 472; Ribouillault D., "Sundials on the Quirinal: Astronomy and the Early Modern Garden", in Fischer H. et al. (eds.), *Gardens, Knowledge and the Sciences in the Early Modern Period*, Trends in the History of Science (Basel 2016) 114–117.

Vatican Belvedere among the most famous examples.[46] The first attempts to re-introduce classical gardens based on Pliny's described villas were recorded on the Vatican Hill by Bramante and at the Villa Madama on Monte Mario by Raphael. The terracing system of the latter example is most remarkable and follows the same model as that proposed in the Horti Farnesiani on the Palatine Hill. These employ a system of hanging gardens, enclosures, structures and belvederes in a project first suggested by Giacomo Barozzi da Vignola and then by Girolamo Rainaldi.[47] The Villa Farnese was also characterised by its engagement with ancient ruins, similar to that found in del Monte's viridarium at the Palazzo Madama. This peculiar interest was widespread in Rome; we need only mention Cardinal Du Bellay's larger gardens at the Baths of Diocletian[48] and those planned by Baldassarre Peruzzi at Orsini's Baths of Agrippa.[49] The most fitting examples are perhaps those created by del Monte's close contemporaries such as the garden described above at the Palazzo Sforza on the Quirinal Hill – formerly the antiquarian garden of Cardinal Rodolfo Pio da Carpi[50] – and the garden built by Cardinal Alessandro de' Medici before he became Pope Leo XI, also on the Quirinal Hill before the residence of Carlo Borromeo.[51] These spaces are all characterised by ruins, statues, fountains and botanical gardens in a setting similar to the Cesi garden in the Vatican, where, along with the Acquasparta garden, Federico Cesi established the first seat

46 See, for example, Belli Barsali I., "I giardini di Statue antiche nella Roma del '500", in Morganti, *Gli Orti farnesiani sul Palatino* (Rome: 1990) 341–372. Among the most recent studies, see Christian K.W., *Empire Without End: Antiquities Collections in Renaissance Rome, c. 1350–1527* (New Haven – London: 2010).

47 Fagiolo M., "Arche-tipologie degli Orti farnesiani", 245–251.

48 See, most recently, Bardati F., "Jean du Bellay et les Horti Bellaiani: bâtir à l'antique, sur l'antique et pour l'antique", in Lurin E. – Burlot D.M. (eds.), *L'artiste et l'antiquaire* (Paris: 2017) 231.

49 Hülsen C., *Die Thermen des Agrippa: ein Beitrag zur Topogaphie des Marsfeldes in Rom* (Rome: 1910) 14–16; Würm H., *Baldassare Peruzzi: Architekturzeichnungen. Tafelband* (Tübingen: 1984) 147; Bentivoglio E., "Il progetto di Baldassarre Peruzzi per gli Orsini di Pitigliano e Nola nell'area delle terme di Agrippa: una rilettura archeologico-topografica e architettonica", *Palladio* 58 (2016) 5–24. For a general overview see Ribouillault D., *Rome en ses jardins. Paysage et pouvoir au XVIe siècle* (Paris: 2013) 33–92.

50 Bilancia F., "Giovanni Fontana per la committenza degli Sforza di Santa Fiora: il palazzo alle Quattro Fontane e altre opere", *Palladio* 23 (2010) 105–136; Eiche S., "Cardinal Giulio della Rovere and the Vigna Carpi", *Journal of the Society of Architectural Historians* 45 (1986) 115–133.

51 Checchi T., "I Giardini Colonna al tempo del cardinale Ascanio", in Picozzi (ed.), *Palazzo Colonna, giardini* 119–136; Spila A., "The Double Stairway of the Temple of Serapis on the Quirinal Hill: Fame and Oblivion in the Early Modern Period", *Pegasus* 18–19 (2018) 39–71.

of the Accademia dei Lincei.[52] In light of this analysis, Cardinal Del Monte's importance as a patron of gardens becomes clear, especially when considering the relationship between gardens and the nascent concept of the academy at the beginning of the seventeenth century.

References

Aldini T., *Exactissima descriptio rariorum quarundam Plantarum quae continentur Romae in Horto Farnesiano* (Rome, Mascardi Giacomo: 1625).

Bardati F., "Jean du Bellay et les Horti Bellaiani: bâtir à l'antique, sur l'antique et pour l'antique", in Lurin E. – Burlot D.M. (eds.), *L'artiste et l'antiquaire* (Paris: 2017) 171–189.

Belli Barsali I., "I giardini di Statue antiche nella Roma del '500", in Morganti G. (ed.), *Gli Orti farnesiani sul Palatino* (Rome: 1990) 341–372.

Benocci C., *Villa Aldobrandini a Roma* (Rome: 1992).

Benocci C., *Villa Ludovisi* (Rome: 2010).

Benocci C., "La natura come sfida e passione: i giardini Cesi di Roma e Acquasparta; XVI–XVIII secolo", in de Petra G. – Monacchia P. (eds.), *I Cesi di Acquasparta, la dimora di Federico il Linceo e le accademie in Umbria nell'età moderna* (Perugia: 2017) 585–627.

Bentivoglio E., "Il progetto di Baldassarre Peruzzi per gli Orsini di Pitigliano e Nola nell'area delle terme di Agrippa: una rilettura archeologico-topografica e architettonica", *Palladio* 58 (2016) 5–24.

Bellori Giovanni Pietro, *Le vite de' Pittori, Scultori et Architetti moderni* (reprint Turin: 1976).

Bellori Giovanni Pietro, *The Lives of the Modern Painters, Sculptors and Architects: a new translation and critical edition*, trans. Sedgwick Wohl A. – Wohl H. – Montanari T. (Cambridge: 2005).

Bilancia F., "Giovanni Fontana per la committenza degli Sforza di Santa Fiora: il palazzo alle Quattro Fontane e altre opere", *Palladio* 23 (2010) 105–136.

Bober P.P. – Rubinstein R., *Renaissance Artists & Antique Sculpture: A Handbook of Sources* (London: 2010).

Borsi F., *Palazzo Madama*, I palazzi del Senato (Rome: 1999).

Calvesi M., *Le realtà del Caravaggio* (Turin: 1990).

Calvesi M., "Trasparente l'autore come il soggetto: ancora su Caravaggio del casino Ludovisi a Roma", *Art e dossier* 6 (1991) 22–26.

52 Benocci, "La natura come sfida e passione" 585–627; Eiche S., "On the Layout of the Cesi Palace and Gardens in the Vatican Borgo", *Mitteilungen des Kunsthistorischen Institutes in Florenz* 39 (1995) 258–281.

Calvesi M., "Riposo durante la fuga in Egitto", in Strinati C. (ed.), *Caravaggio*, exh. cat., Scuderie del Quirinale di Roma (Rome: 2010) 34–41.

Camiz F., "'Luogo molto vago et delitioso […]': il casino del cardinale Del Monte ed un suo soffitto dipinto da Caravaggio", *Ricerche di storia dell'arte* 47 (1992) 81–88.

Carofano P., *Francesco Maria del Monte e Caravaggio: Roma, Siena, Bologna; opera biografia, documenti*, Libera Accademia di Studi Caravaggeschi (Pontedera: 2011).

Christian K.W., *Empire Without End: Antiquities Collections in Renaissance Rome, c. 1350–1527* (New Haven – London: 2010).

Clericuzio A., "Chemical Medicine and Paracelsianism in Italy (1550–1650)", in di Pelling M. – Mandelbrote S. (eds.), *The Practice of Reform in Health, Medicine, and Science. 1500–2000* (Aldershot: 2005) 59–79.

Coggiatti S., "Gli orti farnesiani, la Gaggia e Pietro Castelli alias Tobia Aldini", *Strenna dei Romanisti* 37 (1976) 465–468.

Cortesi F., "Una lettera inedita di Tobia Aldini a Giovanni Faber", *Annali di Botanica* 7 (1908) 403–405.

De Ferrari A., "Castelli, Pietro", *Dizionario biografico degli italiani* 21 (1978) 747–750.

de Petra G. – Monacchia P. (eds.), *I Cesi di Acquasparta, la dimora di Federico il Linceo e le accademie in Umbria nell'età moderna* (Perugia: 2017).

di Pelling M. – Mandelbrote S. (eds), *The Practice of Reform in Health, Medicine, and Science. 1500–2000* (Aldershot: 2005).

D'Alessandro D.A., "Caravaggio's 'Musical' Paintings for Cardinal Del Monte", in Christian K.W. – D'Alessandro D.A. – Terzaghi M.C. (eds.), *Caravaggio – i musici* (Venice: 2017) 49–65.

Dollo C., *Modelli scientifici e filosofici nella Sicilia spagnola* (Naples: 1984).

Dollo C. (ed.), *Filosofia e Scienze nella Sicilia dei Secoli XVI e XVII* (Catania: 1996).

Eiche S., "Cardinal Giulio della Rovere and the Vigna Carpi", *Journal of the Society of Architectural Historians* 45 (1986) 115–133.

Eiche S., "On the layout of the Cesi Palace and Gardens in the Vatican Borgo", *Mitteilungen des Kunsthistorischen Institutes in Florenz* 39 (1995) 258–281.

Fagiolo M., "Arche-tipologie degli Orti farnesiani", in Morganti G. (ed), *Gli Orti farnesiani sul Palatino* (Rome: 1990) 245–251.

Faraglia N.F., "Fabio Colonna linceo", *Archivio storico per le province napoletane* 10 (1885) 665–749.

Fidanza G.B., *Andrea Sacchi and Cardinal del Monte: the Rediscovered Frescoes in the Palazzo di Ripetta in Rome* (London: 2022).

Filippelli Giulio, *Iulii Philippelli eugubini* […] *Epistola apologetica ad praeclarissimos almae Vrbis medicos Romae* (Rome, Zannetti Bartolomeo: 1624).

Frommel C.L., "Caravaggios Frühwerk und der Kardinal Francesco Maria del Monte", *Storia dell'arte* 9–10 (1971) 5–52.

Fumagalli E., "Committenza e iconografia medicea a Roma nel Seicento: il ciclo di affreschi di Palazzo Madama", *Mitteilungen des Kunsthistorischen Institutes in Florenz* 41 (1997) 314–347.

Fumagalli E., "La facciata quattrocentesca del palazzo Medici in piazza Madama: un disegno e alcune considerazioni", *Annali di architettura* 3 (1991) 23–31.

Gabrieli G., "Il 'Liceo' di Napoli. Lincei e linceabili – amici e corrispondenti della vecchia Accademia dei Lincei nel Mezzogiorno d' Italia", *Rendiconti della R. Accademia nazionale dei Lincei, classe di scienze morali, storiche e filolologiche* 14 (1939) 499–565.

Garfagnini G.C. (ed.), *Firenze e la Toscana dei Medici nell'Europa del '500, 3. Relazioni artistiche. Il linguaggio architettonico* (Florence: 1983).

Gregori M. – Acanfora E. (eds), *Come dipingeva il Caravaggio*, exh. cat., Galleria Palatina di Firenze (Milano: 1996).

Giovannoli A., *Roma antica* (Rome, Mascardi: 1619).

Hülsen C., *Die Thermen des Agrippa: ein Beitrag zur Topogaphie des Marsfeldes in Rom* (Rome: 1910).

Lapucci R. (ed), *Evidenza Caravaggio* (Roma: 2020) 295–317.

Lorizzo L., *La collezione del cardinale Ascanio Filomarino: pittura, scultura e mercato dell'arte tra Roma e Napoli nel Seicento; con una nota sulla vendita dei beni del cardinal Del Monte*, Biblioteca Electa Napoli. Studi (Naples: 2006).

Loskoutoff Y., "The 'Cupid Affair' (1596): Cardinal Francesco Maria Del Monte as a Collector of Antiquities", *Journal of the History of Collections* 25 (2013) 19–27.

Maresca P., *Alchimia, magia e astrologia nella Firenze dei Medici*, Giardini e dimore simboliche (Florence: 2012).

Miarelli Mariani G., "Il Palazzo Medici a Piazza Navona: un'utopia urbana di Giuliano da Sangallo", in Garfagnini G.C. (ed.), *Firenze e la Toscana dei Medici nell'Europa del '500, 3. Relazioni artistiche. Il linguaggio architettonico* (Florence: 1983) 977–993.

Montesinos J. – Solís C. (eds.), *Largo campo di filosofare. Eurosymposium Galileo 2001* (Orotava: 2001).

Nocchi P.B. – Pellegrini E., "La collezione botanica del Cardinale Odoardo", in Morganti G. (ed), *Gli Orti farnesiani sul Palatino* (Rome: 1990) 413–429.

Ottaviani A., "Da Fabio Colonna a Paolo Boccone: momenti della storia della botanica fra Napoli e la Sicilia (con un'appendice di lettere inedite)", in Dollo (ed.), *Filosofia e scienze nella Sicilia dei secoli XVI e XVII* (Catania: 1996) 137–168.

Ottaviani A., "La natura senza inventario: aspetti della ricerca naturalistica del linceo Fabio Colonna", *Physis* 34 (1997) 31–70.

Ottaviani A., "L'immagine di Galileo nella Roma dotta", in Montesinos J. – Solís C. (eds.), *Largo campo di filosofare. Eurosymposium Galileo 2001* (Orotava: 2001) 915–923.

Ottaviani A., *Natura ed esattezza all'alba della scienza galileiana: le* 'Observationes' *di Fabio Colonna*, Letture di pensiero e d'arte 128 (Rome: 2021).

Papini M.G., *Palazzo Capponi a Roma: casa vicino al Popolo, a man manca per la strada di Ripetta* (Rome: 2003).

Pascucci I., "Gli enigmi del Camerino del Monte: scoperte e ipotesi ricostruttive del programma iconografico", in Lapucci R. (ed.), *Evidenza Caravaggio* (Rome: 2020) 295–317.

Ribouillault D., *Rome en ses jardins. Paysage et pouvoir au XVIe siècle* (Paris: 2013).

Ribouillault D., "Sundials on the Quirinal: Astronomy and the Early Modern Garden", in Fischer H. et al. (eds.), *Gardens, Knowledge and the Sciences in the Early Modern Period*, Trends in the History of Science (Basel 2016) 103–134.

Ribouillault D., "Hermeneutics and the Early Modern Garden: Ingenuity, Sociability, Education", in Enenkel K.A.E. – Melion W.S. (eds.), *Landscape and the visual Hermeneutics of Place, 1500–1700*" (Leiden – Boston: 2021) 291–325.

Ricci S., "I Lincei: l'invenzione della mediazione accademica. Nuova scienza, religione, vita civile", in *Sciences et religions. De Copernic à Galilée (1540–1610)*, Proceedings of the International Conference in Rome 12–14 December 1996, Collection de l'École française de Rome 260 (Rome: 1999) 205–234.

Richeome Louis, *La peinture spirituelle ou L'art d'admirer aimer et louer Dieu en toutes ses oeuures, et tirer de toutes profit salutere* (Lyons: Rigaud Pierre, 1611).

Riva C., *Boboli: il giardino alchemico* (Chianciano Terme: 2010).

Riva C., *La Grotta Grande di Boboli: laboratorio di meraviglie. Il giardino alchemico* (Florence: 2011).

Sickel L., "Ein 'Heiliger Franziskus' für Sixtus v. Kardinal Francesco Maria Del Monte, Eleonora Orsini und die Bedeutung eines Gemäldegeschenks im Januar 1589", *Mitteilungen des Kunsthistorischen Institutes in Florenz* 43 (1999) 239–241.

Spezzaferro L., "La cultura del cardinal Del Monte e il primo tempo del Caravaggio", *Storia dell'arte* 9–10 (1971) 57–92.

Solinas F., "L'erbario miniato e altri fogli di iconografia botanica appartenuti a Cassiano dal Pozzo", in Haskell F. – McBurney H. – Freedberg D. – Solinas F. – Napoleone C. (eds.), *Il museo Cartaceo di Cassiano dal Pozzo. Cassiano naturalista*, Quaderni puteani 1 (Milan: 1989) 52–76.

Spila A., "Feste e tornei alla corte dei Barberini: Bernini e l''ordigno del cavallo' nel giardino alle Quattro Fontane", *Miscellanea Bibliothecae Apostolicae Vaticanae* 24 (2018) 659–689.

Spila A., "The Double Stairway of the Temple of Serapis on the Quirinal Hill: Fame and Oblivion in the Early Modern Period", *Pegasus* 18–19 (2018) 39–71.

Strinati C. (ed.), *Caravaggio*, exh. cat., Scuderie del Quirinale di Roma (Rome: 2010).

Sutherland Harris A., "Andrea Sacchi's 'Portrait of a Cardinal'", *Bulletin of The National Gallery of Canada* 14 (1969) 9–14.

Sutherland Harris A., *Andrea Sacchi. A Complete Edition of the Paintings* (Oxford: 1977).

Trabucco O., "La corrispondenza tra Pietro Castelli e Marco Aurelio Severino (con un'appendice di lettere inedite)", in Dollo C. (ed.), *Filosofia e scienze nella Sicilia dei secoli XVI e XVII* (Catania: 1996) 109–137.

Waźbiński Z., "The Cardinal Francesco Maria del Monte and the New Patronage", *National Gallery of Art, Center for Advanced Study in the Visual Arts* 10 (1989/1990) 75–76.

Waźbiński Z., *Il cardinale Francesco Maria Del Monte, 1549–1626: 1. Mecenate di artisti, consigliere di politici e di sovrani; 2. Il "dossier" di lavoro di un prelato*, Accademia Toscana di Scienze e Lettere La Colombaria: Studi 137 (Florence: 1994).

Whitfield C., "The 'camerino' of Cardinal Del Monte", *Paragone. Arte* 59 (2008) 3–38.

Witte A., *The Artful Hermitage: the Palazzetto Farnese as a Counter-Reformation 'diaeta'*, LermArte 2 (Rome: 2008).

Würm H., *Baldassare Peruzzi: Architekturzeichnungen. Tafelband* (Tübingen: 1984).

CHAPTER 11

Vitaliano Borromeo's Academy of the Isola Bella

Sergio Monferrini

Abstract

For forty years, Vitaliano VI Borromeo (1620–1690) transformed a "bare rock" into a "miracle of art and nature", giving life to what is now Isola Bella, one of the islands of Lake Maggiore in northern Italy. On his garden island, Vitaliano welcomed distinguished guests such as the writer Carlo Maria Maggi, the mathematician Pietro Paolo Caravaggio, the Jesuit Fathers Barella and Edera, and gave life to an academy particularly interested in moral philosophy. Although not institutionalised, this "Accademia dell'Isola" met in the gardens, where musical performances were staged on the themes of pleasure and usefulness, in the wake of the ethical discussions that developed there and which also found an echo in the "Accademia dei Faticosi" in Milan. The decoration of the garden was also conceived in relation to its philosophical and "academic" function, which this article attempts to describe and explain.

Keywords

Vitaliano VI Borromeo – Isola Bella – academy and gardens – moral philosophy

Recent studies on the Isola Bella in Lake Maggiore, Verbano, have made it possible to deepen research into the Borromeo family and their artistic commissions in Lombardy.[1] Through these studies, the multifaceted figure of Count

---

1 Morandotti A. – Natale M. (eds.), *Collezione Borromeo. La Galleria dei Quadri dell'Isola Bella*, (Cinisello Balsamo: 2011); Galli A.E., "Borromei Libertini. Renato II e Giberto III tra ideali epicurei e letture satiriche", in Spiriti A. (ed.), *Libertinismo erudito. Cultura lombarda tra Cinque e Seicento* (Milan: 2011) 171–185; Carpani R., *Le feste e il teatro all'Isola Bella: allegrezze familiari e immagine politica della famiglia Borromeo fra Sei e Settecento* (2011), https://drammaturgia.fupress.net/saggi/saggio.php?id=4950; Galli A.E. – Monferrini S., *I Borromeo d'Angera. Mecenati e collezionisti nella Milano del Seicento* (Milan: 2012); Galli A.E., "'Un Eremo proprio per contemplar il Cielo'. La villa Borromeo di Senago tra XVII e XVIII secolo", in Spiriti A. (ed.), *Lo spazio del collezionismo nello Stato di Milano (secoli XVII–XVIII)* (Rome: 2013) 17–59; Galli A.E., "I Borromeo d'Angera e l'isola di San Vittore: 'luogo delitioso, e da Prencipe'. Nuovi elementi per la storia del palazzo e dei giardini dell'Isola Madre", in Dell'Omo M. – Monferrini S.

 | DOI:10.1163/9789004517547_012

FIGURE 11.1
Salomon Adler, *Vitaliano VI Borromeo*, 1680, oil on canvas, 119.5 × 93.5 cm. Isola Bella, Borromeo Palace
IMAGE © BORROMEO

Vitaliano VI Borromeo (1620–1690) has emerged not only as a skilled and intelligent soldier and statesman, but also as a refined and cultured patron [Fig. 11.1].[2] Borromeo was responsible for the construction of the palace and gardens on Isola Bella,[3] that 'meraviglioso Artificio' ('marvelous Artifice') where Flora and Pomona lavished their wealth among 'Viali, Habituri, Teatri Boscarecci, altissime Piante, Statue, Spalliere, Spruzzi d'Acque abbondanti, ed altre delitie'

(eds.), *Forme che volano. 1630–1738. Il barocco nelle province di Novara e del Verbano Cusio Ossola*, atti delle giornate di studio, Archivio di Stato di Novara (Borgomanero: 2014) 277–282; Morandotti A. (ed.), *Le isole incantate. Vedute dei domini Borromeo da Gaspar Van Wittel a Luigi Ashton* (Milan: 2015).

2 Morandotti A. – Natale M. (eds.), *Vitaliano VI Borromeo. L'invenzione dell'Isola Bella*, exhibition catalogue, Isola Bella, Palazzo Borromeo (Milan: 2020).

3 For a brief history of the Borromean Islands, see Natale M., *Le Isole Borromeo e la Rocca di Angera. Guida storico artistica* (Cinisello Balsamo: 2000); for a first reading of the construction events of Isola Bella see Azzi Visentini M., "Vitaliano VI Borromeo e l'Isola Bella: un'eccezionale celebrazione barocca", in Fagiolo M. (ed.), *Atlante tematico del Barocco in Italia. Il sistema delle residenze nobiliari. Italia settentrionale* (Rome: 2009) 194–204; Galli A.E. – Monferrini S., "Un cavaliere e la sua 'dama'. Vitaliano VI Borromeo tra politica, filosofia e amore per l'Isola Bella", in Morandotti – Natale (eds.), *Vitaliano VI Borromeo* 34–40.

('avenues, dwellings and woodland theatres, tall trees, statues, espaliers, jets of abundant water, and other delights'), as described by the painter Luigi Scaramuccia in 1674.[4]

From 1650 to 1690, Vitaliano VI transformed a 'nudo scoglio' ('bare rock') into a miracle 'd'Arte e di Natura' ('of Art and Nature'), an ambitious and difficult task that entailed creating terraces, crushing rock, filling hollows and bringing in soil to cultivate plants. Construction was uninterrupted until he died in 1690.[5] The project also involved building of foundations in the lake itself to give a more regular shape to the island, as well as the erection of a palace, theatres, and gardens. This was a long and constantly evolving task increasing in sophistication over time according to 'le idee più magnifiche et amene' ('the most magnificent and pleasant ideas').[6] The design precisely aimed to avoid 'dal solito et dalle cose viste e fatte in altre parti' ('usual and things seen and done elsewhere'). The island served as an extraordinary tool of self-promotion for Borromeo and his family, used 'per farsi amici, e stima' ('to make friends and obtain esteem') and to secure 'merito, e dimestichezza con signori grandi' ('merit, and familiarity with great lords').[7] In both the development of the palace architecture and gardens and their interior and exterior decoration, the making of the grandiose "scenic machine" occupied a large part of Vitaliano VI's life.[8] A charming site in which to welcome guests, it became a place to amaze and to impress. Scaramuccia pointed out the 'superbissimi arredi', ('very superb furnishings'), among which 'la pittura [...] con gran pregio in molto buon numero pomposamente risiede' ('painting [...] with great value in a very good number pompously resides').[9] The building contained over a thousand canvases at the time of the death of its founder.[10]

4 Scaramuccia Luigi, *Le finezze de pennelli italiani* (Pavia, Giovanni Andrea Magri: 1674; reprint, Milano 1965) 148.

5 For the letter from Vitaliano to the architect Carlo Fontana dated 28 December 1686, see Crespi C., *Vitaliano VI Borromeo e l'Isola Bella: un collezionista e mecenate milanese nella seconda metà del Seicento*, vol. 2 (Ph.D. diss., Università degli studi di Milano: 1986–1987) 236.

6 Morandotti A., "La formazione della Galleria e la sua storia tra la seconda metà del Seicento e la fine del Settecento", in Morandotti – Natale (eds.), *Collezione Borromeo* 9. Magnificence was one of the elements sought by Vitaliano because it was a fundamental attribute of aristocratic nature; for Plato it is the philosopher's main quality; *Republic* 487 a2–8.

7 Vitaliano himself remembers this in his last will, 19 January 1682 (Archivio storico civico del Comune di Milano – Biblioteca Trivulziana, *Famiglie*, b. 240).

8 On celebrations and entertainments on the island, see Carpani R., *Scritture in festa. Studi sul teatro tra Seicento e Settecento* (Pisa – Roma: 2008).

9 Scaramuccia, *Le finezze de pennelli italiani* 147–148.

10 Morandotti A., "La formazione della Galleria e la sua storia tra la seconda metà del Seicento e la fine del Settecento", in Morandotti – Natale (eds.), *Collezione Borromeo* 13.

Vitaliano's skilful use and advertising of the island's multimedia environment (through word of mouth, letters, drawings, paintings, engravings, etc.) was instrumental in promoting his reputation as a patron and his recognition as an expert in the fine arts. A figure whom Scaramuccia qualified as 'accorto non meno che bellissimo ingegno e l'animo splendididissimo' ('shrewd with no less than beautiful wit and a splendid soul'), Vitaliano showed a particular predilection for the arts; a visit to the island was a fundamental rite of passage for Milanese and foreign nobility, as well as for members of the Lombard artistic milieu.[11]

Isola Bella became a point of reference for this former group, a fact that recent studies increasingly confirm.[12] The vast network of relationships and connections that Vitaliano carefully built allowed him admission into the most important Italian and European courts, particularly via key figures from the local nobility and state officials. The Isola was a useful tool in obtaining a significant role in the great "theatre" of the world, serving to enhance the 'reputation' and 'good opinions' – the fundamental basis of social capital in the early modern period – as well as the 'applause' and 'affection' that he sought.[13]

The distinguished guests Vitaliano welcomed to his garden-island included the writer Carlo Maria Maggi, the mathematician Pietro Paolo Caravaggio, the Jesuit Fathers Giovanni Battista Barella and Pietro Giuseppe Ederi, Francesco Bondicchi and Angelo Bon, who resided in Milan as representatives of the Grand Duke of Tuscany and of the Republic of Venice, respectively.[14] Thus, the

On Vitaliano's taste, see Morandotti A., "La ricchezza degli interni. 1. I criteri di allestimento e il ruolo della pittura", in Morandotti – Natale (eds.), *Vitaliano VI Borromeo* 44–63; Zanuso S., "La ricchezza degli interni. 2. Scultura e arredi", ibidem 64–79.

11 Scaramuccia, *Le finezze de pennelli italiani* 147.

12 Monferrini S., "'Raccomandazioni' di Vitaliano VI Borromeo per alcuni artisti. Note d'archivio su Cignaroli, De Groot, Garzia, Lanzani, Saglier, Tempesta, Vismara, Volò", *Arte lombarda* 186–187 (2019) 180–189.

13 The terms mentioned here are the same ones that Vitaliano used in indicating to his nephew Carlo IV Borromeo Arese the behaviour to be followed on his embassy visit in Rome in 1686 (Archivio Borromeo Isola Bella ABIB, *Famiglia Borromeo, Carlo IV Borromeo Arese, Ambasciatore presso la S. Sede*); I am presently completing a study on this topic.

14 The bibliography on Maggi is very extensive. See Isella D., *I Lombardi in rivolta. Da Carlo Maria Maggi a Carlo Emilio Gadda* (Turin: 1984); Zardin D., "Carlo Maria Maggi e la tradizione culturale milanese tra Sei e Settecento", *Annali di storia moderna e contemporanea* 3.3 (1997) 9–50; Carpani R., *Drammaturgia del comico. I libretti per musica di Carlo Maria Maggi nei 'theatri di Lombardia'* (Milan: 1998); Frare P., "La sincerità degli affetti: sulle Rime varie di Carlo Maria Maggi", *Testo. Studi di teoria e storia della letteratura e della critica* 26.36 (1998) 45–74. Father Edera later became confessor of the Empress Eleonora Magdalena of the Palatinate-Neuburg and then of the emperor.

Accademia dell'Isola, as it came to be called, was born. Without precise rules or procedures for membership, it was not a formal institution but rather a casual congregation accessible to guests who visited the island. A variety of topics were open to discussion and those covered depended on the occasion and natural development of dialogue, though a particular emphasis was placed on philosophical issues.[15] By referring to his community as an "academy", Vitaliano meant a conversation dealing with points or issues to be debated in turns. Participants were encouraged to discuss concepts and problems of note and exchange knowledge.

Vitaliano had a special interest in moral philosophy, especially Plato's articulation of the topic, and was regarded as an expert in this domain, often walking 'a spasso nelle miniere di Platone' ('through Plato's mines'), as he wrote.[16] He read the classical Greek and Roman philosophers from a young age and was drawn to Senecan philosophy, whom he called the 'maestro di vera felicità' ('master of true happiness').[17] During his education in Rome in the 1640s, he

---

15 The first certified documents of the Academy date back to the early 1680s, but there is no doubt that the academies, as Vitaliano understood them, had already been held on the island for many years.

16 Muratori reports that Borromeo was 'perhaps the most respectable [among the Milanese knights] both for the theory and for the practice of this study' ('forse il più stimabile [fra i cavalieri milanesi] sì per la Teorica, sì per la pratica di tale studio'); Muratori L.A., *Vita di Carlo Maria Maggi* (Malatesta – Milano: 1700) 38. Muratori described him as 'a lover of Platonic doctrines, on which he spent no small amount of time' ('amantissimo delle dottrine Platoniche, dietro alle quali spendeva non poco tempo') (Ibidem). For the studies about Plato, kwown especially through Marsilio Ficino, and Platonism, also in relation to faith, see Kristeller P.O., *Il pensiero filosofico di Marsilio Ficino* (Florence: 1998); Taliaferro C., *Evidence and Faith* (Cambridge: 2005); Douglas H. – Hutton S. (eds.), *Platonism at the Origins of Modernity* (Dordrecht: 2008); Bonazzi M., *Il Platonismo* (Turin: 2015). For an important contribution to Platonism and modern thought, see Schmitt A., *Modernity and Plato. Two Paradigms of Rationality* (Rochester: 2012).

17 ABIB, Famiglia Borromeo, Carlo III, 29 January 1642. The study of Platonic philosophy showed him the way to happiness. This idea is significantly summarized by Marsilio Ficino: 'Tutti gli huomini naturalmente desiderano far bene, cioè viver bene. Costoro viveranno bene se eglino haranno de i beni assai, et li beni son questi ricchezze, sanità, bellezza, gagliardia, nobiltà, sangue, honori, potenza, prudenza, oltra ciò, Giustitia, Fortezza, Temperanza, et piú di tutti gli altri la Sapienza, la quale senza dubbio alcuno, contiene in sé, et abbraccia tutta la forza di questa felicità la quale consiste in un prospero, et favorevole acquisto del desiderato fine, et questo acquisto in ogni facultà o arte lo dona la Sapienza' ('All men naturally wish to do well, that is, to live well. They will live well if they have a lot of goods, and these goods are riches, health, beauty, vigor, nobility, blood, honors, power, prudence, besides that, Justice, Fortitude, Temperance, and more than all the others Wisdom, which without any doubt, contains in itself and embraces all the strength of this happiness which consists in a prosperous and favourable acquisition

recorded his progress in the study of morality: 'et il strepito della corte mi ha fermato l'animo; dal vedere la fortuna et i suoi sforzi più da vicino, sono venuto in cognitione più certa che non vi è altro in questo mondo che si accosti alla beatitudine salvo che una solitudine erudita' ('and the clamour of the Court strengthened my soul; from observing Fortune and its efforts more closely, I have come to a more certain knowledge that there is nothing in this world that bring us closer to bliss than an erudite loneliness').[18] Vitaliano's period of Roman residency deeply influenced him; he remembered not only the memory of the places he had inhabited but also the knowledge he acquired at the most important court in the world, not to mention the impact of visits to palaces, villas and gardens as well as his participation in ceremonies, celebrations and performances.[19] Upon his return to Milan, Vitaliano first pursued a career in the military before moving into politics, obtaining the rank of general before becoming a member of the Milanese Privy Council and commissioner to the Imperial fiefs in Italy. Despite countless tasks, which undoubtedly left him little leisure time, he continued to advance his studies.[20] This achievement was partly due to the advice of the Jesuits, by whom he always wished to be visited on Isola Bella, and for whom he asked the Superior General for a special licence so that they could leave their colleges and come to the island.[21] The ambitious Vitaliano's access to important theologians and leading figures of the Order who resided in the Professed House of Milan and the House of Probation in Arona (near the island) in the heart of the family fiefs offered him a distinct intellectual advantage.[22]

---

of the desired end, and this acquisition in every faculty or art is given by Wisdom'); Ficino Marsilio, *Tomo primo delle divine lettere del gran Marsilio Ficino tradotte in lingua thoscana per M. Felice Figliuccii senese* (Venice, Gabriele Giolito de Ferrari: 1549) 2r.

18 ABIB, *Famiglia Borromeo, Vitaliano VI Borromeo, Corrispondenza 1642*, letter from Vitaliano to his father Carlo III, 28 June 1642.

19 For the Roman gardens of the time and their political relevance, see: Ribouillault D., *Rome en ses jardins: Paysage et pouvoir au XVIe siècle* (Paris: 2013).

20 Galli – Monferrini, *Un cavaliere e la sua 'dama'* 27–33.

21 'Il tratenermi io in questo paese alcune volte mi fa desiderare il non rimanere privo de' Padri della Compagnia di Vostra Paternità Reverendissima e vorrei poterne havere in questa Isola talhora due di Milano o di Arona per poterne sempre apprendere dal loro zelo e virtù' '(Staying in this country sometimes makes me wish not to be deprived of the Fathers of the Company of Your Most Reverend Paternity and I would like to be able to have two of them in this island from Milan or Arona in order to always be able to learn from their zeal and virtue') (ABIB, *Famiglia Borromeo, Vitaliano VI, Corrispondenza 1680*, letter draft from Vitaliano to Father Giovanni Paolo Oliva, Superior General of the Jesuits, Cesano Maderno, undated).

22 For Milan, see Dossi L., *I Gesuiti a Milano. 1563–1970* (Ponte Sesto di Rozzano: 1970); Zanlonghi G., *Teatri di formazione. Actio, parola e immagine nella scena gesuitica del*

Vitaliano's Academy gathered in the three-level Teatro Massimo amphitheatre, which was the focal point of the gardens of the Isola Bella. The amphitheatre incorporated arches, niches, pillars, balustrades, and ornament in stone and tuff and featured statues of Art and Nature, originally bound with golden chains to a statue of Honour who rode on a rearing Unicorn – the emblem of the Borromeo family.[23] These were surrounded by statues of the four seasons and the four elements in the terraced pyramid and complemented by innumerable putti holding what Vitaliano described as hieroglyphs, symbolic elements linked to the family. Obelisks, called spires or pyramids, were interspersed throughout the space. Below, statues of Lake Maggiore and the Ticino and Po rivers set amongst large shells and dolphins amplified the "aquatic" identity of the island, which Borromeo called 'la mia dama' ('my lady'), as if it were a personification of Venus rising from the Greek seas [Figs. 11.2–11.3].

Revealingly, Vitaliano wrote to the Marquis Giulio Spinola in Genoa on 8 March, 1672, thanking him for his 'finezze' ('kindness') in providing plants for the island, which 'cadono sopra una parte, che mi è di gran premura e piacere mentre miro e galanteo quella Isola come dama mia e di molta mia satisfatione' ('fall on a part, which is of great concern and pleasure to me, while I enjoy and serve that island as my lady and with great satisfaction').[24] Vitaliano was deeply enamoured with the island and declared himself indebted to those who favoured it with gifts, showed their appreciation, or even simply visited: 'chi honora la dama honora l'inamorato' ('whoever honours the lady, honours the lover').[25] Asking the architect Andrea Biffi, who was responsible for the building plans, to hasten his arrival to Verbano, Vitaliano jokingly wrote that the island was more beautiful than any widow – likely referencing a specific woman to whom Biffi was sentimentally tied – and, prepared to testify on this with Plato's help, argued that 'le bellezze che si fanno e si godono sono migliori

Sei-Settecento a Milano (Milano: 2002). On Arona see Papale A., "La Casa di probazione di Arona", in Signorelli B. – Uscello P. (eds.), *La Compagnia di Gesù e la società piemontese. Le fondazioni del Piemonte Orientale*, atti del convegno, Vercelli (Vercelli: 1995) 249–267. Among the most important Jesuits who were on the Island were Father Tommaso Ceva (1648–1737), who was a teacher of rhetoric and mathematics at the Braidense College in Milan for over forty years; see Gronda G., "Ceva, Tommaso", *Dizionario biografico degli italiani* 24 (1980).

23 For the Academies in gardens, see Denis Ribouillault's introductory essay.

24 ABIB, *Famiglia Borromeo, Vitaliano VI, Minute 1672–1674*.

25 ABIB, *Famiglia Borromeo, Vitaliano VI, Minute 1675–1676*, letter's draft from Vitaliano to Angelo Bon, 1675 September 11.

FIGURE 11.2 Teatro Massimo, elaborated by Sergio Monferrini
IMAGE © SERGIO MONFERRINI

FIGURE 11.3 Borromeo's Garden, elaborated by Sergio Monferrini
IMAGE © SERGIO MONFERRINI

FIGURE 11.4 Statue of Love, Isola Bella, Borromeo Gardens
IMAGE © SERGIO MONFERRINI

di quelle che si usano' ('the beauties that are made and enjoyed are better than those that are used').[26]

In front of the pyramid of stepped terraces, an effigy of the god Love, who united the four seasons and elements to create the island, dominates a small, square garden decorated with coloured mosaics.[27] This statue of a winged child does not feature Cupid's typical attributes (bow, quiver, arrows), but rather holds a bunch of flowers, originally gilded, as a tribute to the island where they grew luxuriantly [Fig. 11.4]. For Plato, Love-Eros is a philosopher because of his love for beauty – and wisdom is one of the most beautiful things. Vitaliano

26 Vitaliano was however certain that, due to Biffi's affection for him, he would arrive immediately without having to 'ricorrere a quell'autore, et alla comprovatione del vero' ('resort to that author, and to prove the truth'). The architect ensured that 'bastevoli vidove o zitelle a trattenere il desiderio mio sempre a tutti superiore di servire a Vostra Signoria illustrissima et all'Isola' ('widows or spinsters would not be enough to hold back my desire always superior to all to serve your illustrious Lordship and the Island') (ABIB, *Stabili in Isola Bella, Fabbriche e riparazioni 1654–1665*, 1680 June 15 and 17).

27 ABIB, *Stabili in Isola Bella, Fabbriche e riparazioni, 1630–1665*, Count Vitaliano's memory for the wedding of his nephew Carlo IV Borromeo Arese, 1677.

was aware of Plato's argument that virtue is worthless when separated from wisdom and that wisdom involves moderation, justice and courage. By following the virtuous path, man frees his soul and leads it towards perfection: 'l'operare con virtù è bene e perciò ragione, derivante dal supremo regolatore e della suprema regola che è Dio' ('operating virtuously is good and therefore rational, and this generates from the supreme Ordainer and the supreme rule which is God'). He was also conscious that 'pochi uomini sono quelli nei quali la natura sia in totale perfezione, che è il mezzo della virtù' ('few men are those in which nature is entirely perfect, which is the standard of virtue'); and that 'la natura non è altro che la disposizione stessa della mente di Dio' ('nature is nothing more than the very disposition of the mind of God').[28] In light of these statements, the statues of Nature and Art linked together in the most visible place on the island imparts a precise and highly effective meaning: Vitaliano, a philosopher because he tries to achieve the perfection of his soul through the pursuit of virtue, has shaped the nature of the island by transforming it through art. Contemplating such a "paradise", as the Jesuit Barella wrote, did not distract from celestial contemplation but rather encouraged it, because 'il fine e la perfetta felicità dell'uomo, che si cerca dalla morale, consiste nella contemplazione dell'ottimo' ('the purpose and perfect happiness of man, is sought from morality, consisting in the contemplation of excellence').[29]

The "Teatro delle acque" (Water Theatre, then Hercules' Theatre) near the palace, the first built by Borromeo, consists of a large exedra housing a colossal statue of Hercules' 'in atto d'appoggiarsi alla mazza' ('in the act of leaning on the club') with the lion's skin on his shoulder. Over four metres high, the work, taking inspiration from the Farnese Hercules, was placed on a pedestal about two metres high in 1671 [Fig. 11.5].[30] In the niches stood 'le quatro Virtù, per fingere che Ercole colà ripara fra le virtù, e due altre figure con i trofei di Ercole' ('the four Virtues, to pretend that Hercules shelters there among the virtues, and two other figures with the trophies of Hercules').[31] Vitaliano intended

28 ABIB, *Famiglia Borromeo, Vitaliano VI, Produzioni letterarie e scientifiche*, vol. 1.

29 As taught 'not only by Christian theologians, but also by the best philosophers, Plato, Aristotle, and their followers' ('non solo i teologi cristiani, ma anche i megliori filosofi, Platone, Aristotele, e i loro seguaci') (ABIB, *Isola Bella, Addobbi e mobili*, letter from Father Giovanni Battista Barella to Vitaliano, 24 November 1684).

30 On the diffusion of the image of the Farnese Hercules, see Gasparri C., "The Making of an Icon. The Farnese Hercules and the Power of Place", in Settis S. – Anguissola A. – Gasparotto D. (eds.), *Serial Classic. Multiplying Art in Greece and Rome*, exhibition catalogue, "Serial Classic", Milan, and "Portable Classic", Venice (Milan: 2015) 171–179.

31 ABIB, *Stabili in Isola Bella, Chiesa 1670*, letter from Vitaliano to his brother cardinal Giberto III, 27 October 1671.

FIGURE 11.5 Hercules's Theatre on an old postcard. Private collection
IMAGE © SERGIO MONFERRINI

for the large statue to remain behind a canvas of water falling from above, so that the hero would appear 'riposarsi' ('to rest') as though having 'eletto il porsi nella niccia dietro le cascate stesse' ('chosen to place himself in the niche behind the falls themselves').[32] Commenting on an ancient medal in 1559, Sebastiano Erizzo wrote that Hercules was 'la Idea di tutte le virtù' ('the Idea of all virtues'), his club was a 'simolacro di valore e di virtù' ('symbol of value and virtue'), and the lion's body 'dimostra una generosa fortezza d'animo' ('demonstrates a generous strength of soul').[33] Placed at a crossroads in Vitaliano's theatre to symbolise humanity's eternal conflict between good and evil, between *virtus* and *voluptas*, Hercules was depicted as having made his choice: he rests among the virtues and, through earthly labours to fulfil Jupiter's will, he awaits the call to Olympus. Current scholarship does not offer insight into Borromeo's knowledge of the importance ascribed to Hercules in Tivoli as a 'reclaimer of

32 ABIB, *Stabili in Isola Bella, Fabbriche e riparazioni 1666–1685*, letter from Vitaliano to his brother cardinal Giberto III, 13 January 1672.

33 Erizzo Sebastiano, *Discorso di Sebastiano Erizzo sopra le medaglie antiche* (Venice, Bottega Valgrisiana: 1559) 277.

uncultivated lands and marshy places'.[34] However, Hercules was symbolically present at the Villa d'Este, which Vitaliano certainly visited during his years spent in Rome; this theme was particularly well suited to the extraordinary transformation later carried out on the island.[35]

Like the mythical hero, Vitaliano was fully immersed in the civil life of his time and chose the path of virtue. On the island, he found an ideal place to restore his spirit from political and military commitments, a place where he could leave the *negotia* for the *otia* not as an end in and of itself, but for 'honeste sodisfationi' ('honest satisfactions') granted by God.[36] The pleasures he enjoyed in this place were accompanied by a greater profit, because, as Agostino Gallo wrote in his book on the pleasures of the villa – a text present in the count's library –, 'vera utilità è quella che porta honesto contento all'animo nostro, e massimamente quando ci leva l'intelletto a contemplare quelle cose, che fanno l'uomo eternamente felice' ('true usefulness is that which brings honest contentment to the soul and intellect is best served in lifting us up to contemplate those things which make man eternally happy').[37] An interesting example of the island's ability to evoke joy and ward off misfortune is an intermezzo performed in the recently completed theatre in 1666. Hypochondria was 'scacciata dall'Allegria, che non le permette dimora nella buona stanza

---

34 Pierattini C., "Aspetti, funzioni, e fortuna dell'Ercole tiburtino", *Atti e Memorie della società Tiburtina di Storia e Arte* 54 (1981) 7–40; Barisi I. – Fagiolo M. – Madonna M.L. (eds.), *Villa d'Este* (Rome: 2003) 88–89.

35 In the 'Philosophers' Hall' (today on the stairs) a wooden sculpture of Atlas supporting a mirror is still connected to the myth of the hero and his daily commitment to the good of the world. This object arrived on the island in 1678 (Zanuso, *La ricchezza degli interni* 68). The possible connection between Atlas and Hercules is very interesting and had philosophical implications, as can be found, for example, in Villa Aldobrandini's gardens at Frascati, which Vitaliano certainly saw. See Ribouillault D., "Atlas and Hercules in the Garden. Scientific Culture and Literary Imagination at the Villa Aldobrandini at Frascati", *Nuncius* 30 (2015) 124–160.

36 An illness in 1668 had warned him 'of our fragility, and that it is madness not to enjoy here the good that God has granted' (della nostra fragilità, e che è pazzia il non godere pur qua del bene che Dio ne ha concesso) (ABIB, Isola Bella in generale, Cen-Des, letter from Vitaliano VI to Aron Cuchino, 17 August 1688). Maggi dictated an inscription for him, placed in 1671 at the bottom of the gardens, in which Vitaliano is remembered: 'Informibus scopulis /Substruens, et extruens, / Dignitatem otiis, maiestatem deliciis / Comparabat' (Muratori, *Vita di Carlo Maria Maggi* 41).

37 Gallo Agostino, *Le dieci giornate della vera agricoltura e piaceri della villa* (Venice, Domenico Farri: 1565) 224v.

dell'Isola' ('banished by Joy, which did not allow her to stay in the pleasant site of the island').[38]

Musical comedies were another of Vitaliano's passions, showcasing two main themes – pleasure and usefulness – that aligned with the ethical discussions developed during his stays on the island. With Vitaliano as promoter and Maggi as poet, the musical comedies he commissioned consolidated a connection of 'soave amistà' ('sweet friendship')[39] between the two that continued throughout their lives.[40] The count referred to Maggi as 'uno de più virtuosi sogetti di questa città possedendo la lingua latina e scrivendo eminentemente, et è bravissimo poeta et ripieno d'infinite virtù, et il maggiore amico secondo la vera morale che io qui habbi' ('one of the most virtuous subjects of this city [Milan] mastering the Latin language and writing eminently, and is a very good poet and full of infinite virtues, and the greatest friend according to the true morality that I had here').[41] This passion for performance led Borromeo to build a real "Teatro delle commedie" (Theatre of Comedies), equipped with stages and capable of welcoming guests in three hundred seats. The theatre was built in 'imitatione de i teatri grandi' ('imitation of large theatres'),[42] since 'un teatro grande e bello recita sempre, et ancorché non vi si faccino comedie' ('a great and beautiful theatre always plays, even if comedies are not performed there').[43] As with most of the dramatic texts performed in Italian theatres, love occupied the central narrative focus. Maggi, however, interpreted this theme in light of the considerations that figured prominently in the erudite conversations of Vitaliano and the intellectual circle he frequented. Coherent with Jesuit teachings, he referred to love as the 'prima gloria del bene' ('first glory of good') and his was a moralising conceptualisation.[44] At its core were a set

38 ABIB, *Famiglia Borromeo*, Card. Giberto III, Corrispondenza 1666, letter from Vitaliano VI, 1666 december; Carpani, *Drammaturgia del comico* 76–78. Maggi's musical comedy consisted of a prologue and two interludes with seven main characters and up to twenty actors. It was performed twice to an audience of over four hundred people.

39 Muratori, *Vita di Carlo Maria Maggi* 38.

40 Borromeo and Maggi were also involved in the performances of the Milan theatre: Carpani, *Drammaturgia del comico* 35–117; Monferrini S., "Nuovi documenti sulla gestione impresariale del teatro di Milano fra Sei e Settecento. Il Teatro Ducale milanese (1642–1716) e la rappresentazione de *La Floridea* (Novara, 1674)", *Fonti Musicali Italiane* 23 (2018) 38–39.

41 ABIB, *Famiglia Borromeo, Stabili in Isola Bella, Fabbriche e riparazioni 1666–1685*, letter from Vitaliano to his brother cardinal Giberto III, 10 december 1671.

42 Carpani, *Drammaturgia del comico* 51.

43 ABIB, lettera Barale, 1669 novembre 20.

44 Maggi Carlo Maria, *Il trionfo d'Augusto in Egitto* (Milan, Marc'Antonio Pandolfo Malatesta: 1672) 4. The reading of Cardinal Pietro Sforza Pallavicino's *Del bene* (Sforza Pallavicino

of values intended to serve as 'ammaestramento' ('training') as well as 'diletto' ('pleasure'), this in full awareness of the didactic and recreational functions of the theatre.[45] By placing the deep bond that united love with good at the centre of his storytelling, Maggi exalted the creative power of the virtuous, highlighting its capacity to generate positive thoughts and actions and to guide in the right direction. On stage, the characters were judged according to this value schema and rewarded or punished based on their existential conduct.

Vitaliano shared Maggi's understanding of theatre and music and was aware of how their approach compared to performances presented elsewhere. He wrote to his brother concerning this topic, assuring him that the island comedies had 'ogni circostanza. Et in quanto al verso et al costume et all'intreccio superano di gran longa quelle di Venetia, e forse quelle di costì [cioè di Roma]' ('every circumstance. And in terms of verse, costume and narrative frame, they far exceed those of Venice, and perhaps those of there [Rome]').[46] In 1686, Vitaliano's nephew Charles IV reaffirmed the importance ascribed to the ethical value of theatre on the island when he asked his brother-in-law Livio Odescalchi for some musical comedies by Alessandro Scarlatti to please ladies and knights and to 'mettere in credito il Teatro dell'Isola' ('position the island theatre well'). Charles had had the opportunity to listen to *L'Aldimiro* at the home of Countess Maria Taverna, a work 'ch'ha havuto grand'applauso non solo per l'eccellenza della musica, quanto per esservi maggiore proprietà e modestia di quelle di Venetia, che sono quelle che si sentono nel nostro Teatro di Milano' ('that received great applause not only for the excellence of the music, but also for having greater capacity and modesty than those of Venice,

Pietro, *Del bene* [Rome: eredi di Francesco Corbelletti, 1644]) was fundamental. According to Muratori, Maggi found in it 'un gran fondo di vera Filosofia' ('great grounds of true Philosophy'): Muratori, *Vita di Carlo Maria Maggi* 118. Vitaliano also often consulted it together with the *Arte della perfezione Cristiana*, probably in the Milan edition (Milan: Lodovico Monza, 1666): 'sono infatti ripieni di gran scienza morale, e si è il signor cardinale con gran ordine con gran perspicacia inoltrato in essa' ('they are in fact filled with great moral science, and the cardinal with great order with great perspicacity has advanced into it') (ABIB, *Stabili in Isola Bella, Chiesa a tutto il 1669*, letter from Vitaliano to his brother cardinal Giberto III, 7 april 1666).

45 Maggi Carlo Maria, *Affari ed amori* (Milan, Marc'Antonio Pandolfo Malatesta: 1675) 8. On the theme of morality in Maggi, Muratori's interpretation, and an evaluation in relation to the theatre between the seventeenth and eighteenth centuries, see Apollonio M., *Carlo Maria Maggi e il moralismo Lombardo* (Florence: 1932); Taviani F., "Il teatro dalla morale alla cultura", in Guccini G. (ed.), *Il teatro italiano nel Settecento* (Bologna: 1988) 101–120.

46 ABIB, lettera Barale, 20 november 1669.

which are those that can be heard in our Milan Theatre').[47] Maggi emphasised that he had avoided 'tutti i motti men che onesti' ('all less than honest words'), and affirmed pleasure as 'il più sano e 'l più soave è il dolce dell'onestà' ('the sweetness of honesty is the healthiest and most pleasant).[48] The formal qualities of the dramaturgical texts are well expressed by the voice of Comedy, an allegorical character from the drama *Il Girello*:

> Maestra d'allegrezza
> festosa accortezza
> da me si comparte.
> Sanar con la gioia
> è il meglio de l'arte.
> Se il vizio, se noia
> v'offende, vi tedia
> d'ambi vi curerò. Son la Comedia.[49]
>
> Teacher of joy
> festive care
> is granted by me.
> To heal with joy
> is the best of art.
> If the vice, if tedium
> offends you, bores you
> I will heal you from both. I am Comedy.

During these theatrical performances, the audience certainly derived pleasure, but it also received training in a manner like the instruction that took place through conversations on the island. These conversations were never an end in themselves, but rather contributed to the development of the participants' knowledge, supported them on the path of virtue, and intended to bring them

---

47 Monferrini S., "Carlo IV Borromeo Arese, Alessandro Scarlatti e la Cappella Reale di Napoli", in Della Libera L. – Maione P. (eds.), *Devozione e passione: Alessandro Scarlatti nella Napoli e Roma barocca* (Naples: 2014) 68–69. For Milan theatre in those years, see at least: Cascetta A. – Carpani R. (eds.), *La scena della gloria. Drammaturgia e spettacolo a Milano in età spagnola* (Milan: 1995).

48 Maggi Carlo Maria, *La Bianca di Castiglia* (Milan, Marc'Antonio Pandolfo Malatesta: 1674) 3.

49 Idem, *Il Girello drama faceto*, ([Milan], Marc'Antonio Pandolfo Malatesta: 1674) 5.

closer to the supreme good of God: 'solo è bello a gran'Alma immenso Dio' ('the great soul of the Immense God is solely beautiful').[50]

Academies like Vitaliano's presented an opportunity for moral and spiritual growth; Father Barella wrote with satisfaction: 'O che bel filosofare su la morale all'Isola' ('Oh, how beautiful it is, philosophising on morality on the island').[51] The twenty phrases dictated by Maggi in 1680 that were to be inscribed on marble slabs on the island, described by Charles IV as 'piene di tanto sapere' ('full of so much knowledge'), have yet to be found. Vitaliano saw in the inscriptions 'la morale sì ben dimostrata e distinta' ('the morality so well demonstrated and distinct') that was valued, and which could be attained, on the island. The poet's intention was not to teach these ethical concepts. His intention was to describe them and to inspire them by virtuous examples 'a forza di virtù' ('by force of virtue').[52] In these brief reflections, Vitaliano affirmed the power of teaching by example, which he considered superior to direct teaching. Maggi's comedies worked in much the same way, offering a contrast between virtue and vice that was clear and unambiguous and did not allow the latter to offend or flatter. Thus, the prize of love obtained by the protagonists in his musical comedies through their sincerity and faith resulted in their happy ending.[53] On *The Island Day*, a cantata by Maggi dedicated to the Borromeos, the poet exclaims:

> Isoletta in cui si giace
> l'alma sciolta dagli affanni:
> ond'esilio hanno gl'Inganni,
> ove ha patria la mia Pace.
>
> Tu più belle, e più propizie
> fai le stelle al cuor giocondo,
> la maggior di tue delizie
> è il dividermi dal Mondo.
>
> Qui dal Mondo son distinto,
> per quest'Isola, in cui giaccio,

50 The verse is taken from one of Vitaliano's poems published by Muratori, *Vita di Carlo Maria Maggi* 39.

51 ABIB, *Famiglia Borromeo, Vitaliano VI, Corrispondenza 1676*, 18 November 1676.

52 Crespi, *Vitaliano VI Borromeo e l'Isola Bella* 109–110 (letters, 12 and 19 June 1680).

53 On Maggi's conception of theatre, see Carpani, *Drammaturgia del comico* 129–138; Petrella A., "Carlo Maria Maggi's Theory of Comedy", *Italian Quarterly* 12 (1969) 223–237.

qui mi par dall'onde cinto
esser tutto al Cielo in braccio.

Dalle voglie più moleste
l'Alma qui non sente guerra.
Qui provò d'esser celeste
con spiccarsi dalla Terra.

In quest'Isola depose
l'Alma stanca ogni sua cura,
qui disgiunta dalle cose
sente il dolce d'esser pura.

Sovra lei grazie divine
di bei lumi il Cielo spande:
e ristretta in tal confine,
più s'accorge d'esser grande.[54]

Islet in which lies
the soul free from worries:
Whereby Deceptions have exile,
where my Peace has a homeland.

You more beautiful, and more propitious,
make the stars for a joyful heart,
the greatest of your delights
is dividing myself from the World.

Here I am distinct from the World,
for this Island, where I lie,
here it seems to me surrounded by waves
to be completely in the arms of Heaven.

From the most persistent cravings
the Soul here does not feel war.
Here it proved to be heavenly
Taking flight from the Earth.

---

54 Maggi Carlo Maria, *Rime varie*, vol. 1 (Milan, Giuseppe Malatesta: 1700) 151.

On this island laid down
the tired Soul in all its care,
here detached from things,
it feels the sweetness of being pure.

Above it, the divine graces of
Heaven spread beautiful lights:
and confined in this border,
the more she realises that she is great.

For Maggi, the island was a place to distance himself from the thoughts and worries of the world, an environment in which the soul felt pure, grand, and could aspire to Heaven. In this respect, Vitaliano's "lady" fulfilled its aspiration of promoting healing from earthly tribulations while at the same time assisting its patrons in their orientation towards God.

The Accademia dei Faticosi in Milan, established in the Theatine Father's college, also echoed the dichotomy of delight and usefulness that seemed constantly present on the Isola Bella; Vitaliano served as "prince" and dictated its constitutive rules. The foundation of this academy in 1660 resulted from the initiative of the Theatine fathers Giovanni Battista Rabbia and Celso Quattrocase. Its motto was "Firmabitur", represented by the symbol of an obelisk raised by machines, which referred to the restoration of virtue lost in the previous war.[55] By emphasising discussion, the practice of virtue, and the study of sciences and skills useful to practitioners and the public good, the Academy foregrounded learning directed to the glory of God, service of the king, and honour of Milan. Discussions of ethics were central to its institutional focus and largely derived from Aristotle's moral philosophy. Literary considerations focused on the achievement of virtue, which was inextricably linked to fortitude, prudence, magnanimity, piety and the exaltation of friendship and patriotism, in contrast to vices like ambition, anger, envy, ignorance, falsehood and duplicity. During public academy sessions, two "arguments" were presented: the first was dedicated to moral or political themes, and the second was more open. The latter argument, delivered 'all'improvviso' ('improvised'), responded to a theme proposed by officials such as a historic tale, part of a poem, lecture on a composition by an illustrious author, an instruction on an important matter, or even a debate between several people 'ad imitatione delle controversie

55 Carpani, *Scritture in festa* 52–53. On the "Accademia dei Faticosi", see Carpani, *Drammaturgia del comico* 9–18; Carpani, *Scritture in festa* 49–77.

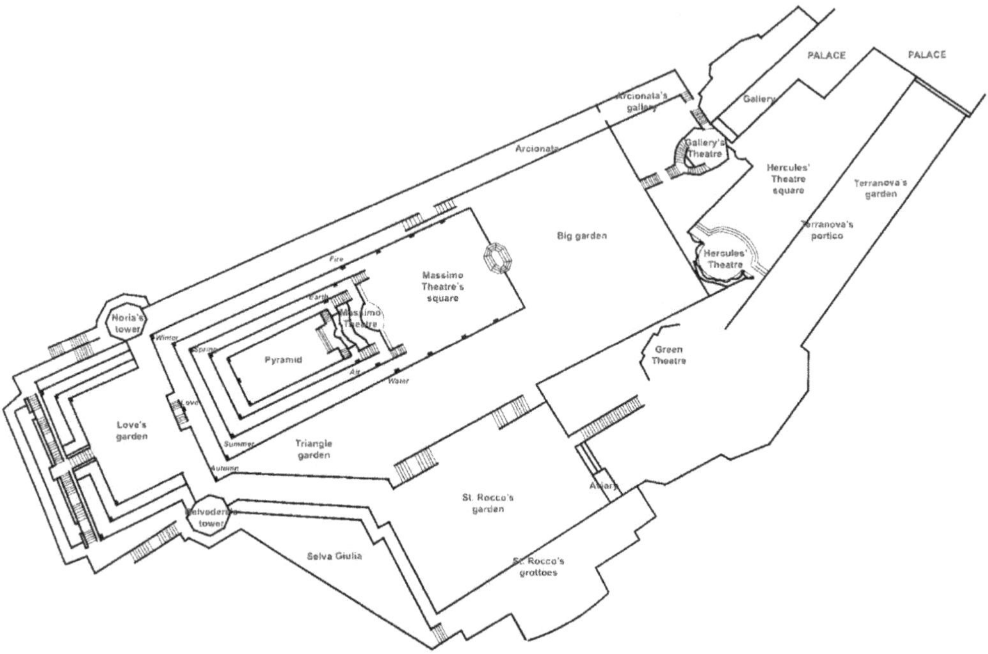

FIGURE 11.6 Plan of Isola Bella gardens, designed and elaborated by Sergio Monferrini
IMAGE © SERGIO MONFERRINI

di Seneca' ('in imitation of Seneca's controversies').[56] The academy was a civic institution and its public orientation contributed to municipal cultural activity. It was closely linked to the Ambrosian church and the power structures of the State, with the governor and the archbishop serving as its special protectors. The official nature of the Academy's activities placed it on a level quite distinct from the island, which was far from the city and court etiquette and, particularly in the villa residence, allowed for a familiarity that would have otherwise been impossible in another setting.

The long walks along the avenues of the Isola Bella, designed in imitation of those suggested by Plato, induced the "academics" to speculate while enjoying views of the lake, mountain landscape, flower beds, and citrus espaliers. Stopping in the cool grottoes or enjoying the fountains, they were totally immersed in a Nature reworked and recreated by humans to make it more beautiful, more desirable, and more like Paradise [Fig. 11.6].[57] From the house, a

56 Carpani, *Drammaturgia del comico* 23.

57 Father Ceva also compared the Island to Paradise: 'Paradisi sed tamen hortus, / Divinas aedes gyro qui plurimus ambit, / Paulo aliter positus. Verbana haud discolor illi / Insula, ad eorum qua formisissima solem / Vergit opaca cedris, Borromeosque penates / Desuper aspectat'. Tommaso Ceva, *Iesus puer* (Milan, Carlo Antonio Malatesta: 1690) 99.

large avenue with arches covered by plants ("arcionata") led to a decorated gallery. A series of niches, tufa walls and statues of Jupiter, Apollo and Daphne – the mythological figure transformed into laurel, a fragrant plant abundant on the island and symbolic of the renunciation of earthly love – led to the gardens. On the left of the avenue, a series of porticos partially decorated as grottos permitted a walk almost entirely protected from the sun. This environment was meant to evoke the setting where Aristotle held his lectures, the *peripatum* in Athens near the wood of the Muses. In reference to Maggi, Ludovico Antonio Muratori wrote that he had learned the first principles of moral philosophy from Plato and then developed them 'con la scorta più sicura e chiara d'Aristotele' ('with the safer and clear companionship of Aristotle').[58] Plato was also an essential reference point for Vitaliano, who often lingered on his philosophy, mediated by the writings of Fabio Albergati.[59] A letter from Bon, who recalls walks on the island and its 'sontuose, delitiose maraviglie' ('sumptuous, delicious marvels'), explicitly mentions the 'portico della Morale' ('portico of morality') and 'arco del disinganno' ('the arch of disillusionment') as striking points on the route, both from a physical and philosophical perspective.[60]

According to Vitaliano's description, the Isola Bella gardens constituted an ideal place to study that aspect of philosophy 'che abilita ad intendere secondo il vero' ('that enables us to understand according to truth'), as well to understand morality that allows us to 'conoscere il ben et il male, et il maneggio delli animi' ('know good and evil, and the management of souls').[61] There, one could also practice love of 'perfettissima Sapienza' ('the most perfect Wisdom') while engaging in research on the three aspects of philosophy.[62] In his *Iconologia*, Cesare Ripa describes the three aspects of philosophy, the active one 'che compone l'animo nelli buoni costumi, la contemplativa che investiga i segreti della natura e la razionale in cui consiste la ragione con la quale disputando si discerne il vero dal falso' ('that composes the soul in good habits, the contemplative one that investigates the secrets of nature, and the rational one,

58 Muratori, *Vita di Carlo Maria Maggi* 14.

59 In the small library that he kept on the island were: 'seven books, bound in goatskin parchment, of the Morale by Fabio Albergatti' and 'Discorsi politici by Fabio Albergatti, bound in goatskin parchment, in octavo' (Crespi, *Vitaliano VI Borromeo e l'Isola Bella*, inventory, 26 December 1690).

60 ABIB, *FB*, *Vitaliano VI*, *Corrispondenza 1680/1137*, letter from Angelo Bon to Vitaliano, Venice, 22 June 1680.

61 Archivio storico civico del Comune di Milano – Biblioteca Trivulziana, *Famiglie*, b. 240, Count Vitaliano VI Borromeo's last will, 19 January 1681. For the quotes, see Ripa Cesare, *Iconologia* (Siena, Heredi di Matteo Florimi: 1613) 269.

62 Ibidem.

FIGURE 11.7 Isola Bella, aerial view
IMAGE © BORROMEO

that allows us to distinguish what is true from what is false').[63] In the palace's "Philosophers' Hall", seven large wooden telamons argued with each other in a nod to the ancient symposia of the Seven Sages. The garden, on the other hand, was a place to meditate and contemplate the gifts that God had given in an optimistic reappraisal of nature and creation [Fig. 11.7].

According to the Jesuits' worldview, this contemplative attitude was also meant to guide social life, interpersonal relationships and the performance of duties related to one's social status. A realistic, concrete, positive and optimistic attitude allowed an individual to face the world while maintaining a delicate equilibrium between emotion and reason. St. Francis de Sales and St. Philip Neri initially promoted this practice. In later years, Charles IV Borromeo Arese would refer to St. Francis de Sales, recognising him as a guide in spirituality like Aristotle's service as a guide in politics. He suggested that his son Federico apply himself to the study of morality, which would allow him to see 'gli abiti e le regole della virtù della magnanimità' ('the habits and rules of the virtue of magnanimity') and to discover that 'anche il Paradiso si compiace delle

63 Ibidem.

virtù morali' ('even Heaven delights in moral virtues').[64] He thus demonstrated his strong understanding of the principles debated in the academic conversations of the island and that he was indeed a worthy descendant of his uncle. So accomplished was he in this pursuit that Muratori came to qualify him by 'Sangue Borromeo, professore e maestro della buona morale' ('Borromean blood, professor and teacher of good moral').[65]

Vitaliano's letters frequently reference divine Providence, which, even in adversity, allows us to see the positive side of all opportunities and face trials stoically so that they may temper the spirit, strengthen the character, and improve us. Maggi indicated how the 'Isoletta fortunata' ('lucky islet') enabled the banishment of all pain, a serenity of the soul, and understanding:

> Quell'amor di Providenza,
> il cui studio orna le piaggie,
> così bello all'alme saggie
> così dolce all'innocenza.[66]
>
> That love of Providence,
> whose study adorns the beaches,
> so beautiful to wise souls
> so sweet to innocence

Realising this ambition required a 'umor candido' ('a candid mood'), aided by the Aristotelian virtue of 'eutrapelia' (friendliness, ease, and humour), which was also studied and practiced at the Accademia dell'Isola. The transformed "rock" was 'albergo del godimento e dell'allegrezza' ('the harbour of pleasure and joy'),[67] as Barella wrote, because, according to the logic of *docere et delectare*, studies of morality that result in an improved spirit should be infused with playful moments of games, laughter, song and music. Thus, according to Muratori, an environment such as the one created by Vitaliano was a 'vero pascolo de gli animi grandi' ('real pasture for great souls').[68]

---

64 ABIB, Famiglia Borromeo, c. Federico VI, Corrispondenza, letter from Carlo IV Borromeo Arese, Arona, 5 May 1722.

65 Monferrini S., *Scene di corte all'inizio del Settecento. Cerimoniali barocchi per Carlo IV Borromeo Arese viceré di Napoli (1710–1713)* (Milan: 2021) 40–42.

66 ABIB, *Stabili in Isola Bella, Visite, 1590–1707*, Maggi's poem in a letter from Vitaliano to his brother Cardinal Giberto III, 22 May 1699.

67 ABIB, *Famiglia Borromeo, Vitaliano VI, Corrispondenza 1680*, letter from Giovanni Battista Barella to Vitaliano, 21 June 1680.

68 Muratori, *Vita di Carlo Maria Maggi* 39.

Vitaliano Borromeo's Academy of the Isola Bella is an episode of particular interest in Lombard culture. The initiative of a singular person, its realisation was intimately linked to its creator; its informal character, removed from urban centres and their institutions, makes it remarkable. The academy cannot be separated from the physical location in which it was organised, which was inextricably tied to its built and natural surroundings. Early modern academies and gardens are closely linked and interdependent. If gardens were borne out of a desire to enjoy the beauty of the landscape, purity of the air, quiet away from the worries of the city, and simple pleasure, their value nonetheless shifted over time.[69] They became an important element in Vitaliano's plan to relaunch his family's political agenda as well as a place to project his identity on a personal and familial level. Here, Borromeo attempted to "compose" a unique place where art and nature could coexist in harmony. Contemplation in this environment allowed visitors to appreciate the joys of earthly life, and meditate and reflect on the heavenly one, bringing the soul closer to God with the help of moral philosophy.

Acknowledgements

Special thanks to: Lorena Barale (Archivio Borromeo Isola Bella), Maurizio Bettoja, princes Borromeo Arese, Serena Sogno (Borromeo's art collection Curator), Sandra Spoto.

References

Apollonio M., *Carlo Maria Maggi e il moralismo Lombardo* (Florence: 1932).

Azzi Visentini M., "Vitaliano VI Borromeo e l'Isola Bella: un'eccezionale celebrazione barocca", in Fagiolo M. (ed.), *Atlante tematico del Barocco in Italia. Il sistema delle residenze nobiliari. Italia settentrionale* (Rome: 2009) 194–204.

Barisi I. – Fagiolo M. – Madonna M.L. (eds.), *Villa d'Este* (Rome: 2003).

Bonazzi M., *Il Platonismo* (Turin: 2015).

Carpani R., *Drammaturgia del comico. I libretti per musica di Carlo Maria Maggi nei 'theatri di Lombardia'* (Milan: 1998).

Carpani R., *Scritture in festa. Studi sul teatro tra Seicento e Settecento* (Pisa – Roma: 2008).

69 Archivio storico civico del Comune di Milano – Biblioteca Trivulziana, *Famiglie*, b. 240, Count Vitaliano VI Borromeo's last will, 19 January 1681.

Carpani R., *Le feste e il teatro all'Isola Bella: allegrezze familiari e immagine politica della famiglia Borromeo fra Sei e Settecento* (2011), https://drammaturgia.fupress.net/saggi/saggio.php?id=4950.

Cascetta A. – Carpani R. (eds.), *La scena della gloria. Drammaturgia e spettacolo a Milano in età spagnola* (Milan: 1995).

Ceva Tommaso, *Iesus puer* (Milan: Carlo Antonio Malatesta, 1690).

Crespi C., *Vitaliano VI Borromeo e l'Isola Bella: un collezionista e mecenate milanese nella seconda metà del Seicento*, II, (Ph.D. diss., Università degli Studi di Milano: 1986–1987).

Dossi L., *I Gesuiti a Milano. 1563–1970* (Ponte Sesto di Rozzano: 1970).

Douglas H. – Hutton S. (eds.), *Platonism at the Origins of Modernity* (Dordrecht: 2008).

Erizzo Sebastiano, *Discorso di Sebastiano Erizzo sopra le medaglie antiche* (Venice, Bottega Valgrisiana: 1559).

Ficino Marsilio, *Tomo primo delle divine lettere del gran Marsilio Ficino tradotte in lingua thoscana per M. Felice Figliuccii senese* (Venice, Gabriele Giolito de Ferrari: 1549).

Frare P., "La sincerità degli affetti: sulle Rime varie di Carlo Maria Maggi", *Testo. Studi di teoria e storia della letteratura e della critica* 26.36 (1998) 45–74.

Galli A.E., "Borromei Libertini. Renato II e Giberto III tra ideali epicurei e letture satiriche", in Spiriti A. (ed.), *Libertinismo erudito. Cultura lombarda tra Cinque e Seicento* (Milan: 2011) 171–185.

Galli A.E., "'Un Eremo proprio per contemplar il Cielo'. La villa Borromeo di Senago tra XVII e XVIII secolo", in Spiriti A. (ed.), *Lo spazio del collezionismo nello Stato di Milano (secoli XVII–XVIII)* (Rome: 2013) 17–59.

Galli A.E., "I Borromeo d'Angera e l'isola di San Vittore: 'luogo delitioso, e da Prencipe'. Nuovi elementi per la storia del palazzo e dei giardini dell'Isola Madre", in Dell'Omo M. – Monferrini S. (eds.), *Forme che volano. 1630–1738. Il barocco nelle province di Novara e del Verbano Cusio Ossola*, atti delle giornate di studio, Archivio di Stato di Novara (Borgomanero: 2014) 277–282.

Galli A.E. – Monferrini S., *I Borromeo d'Angera. Mecenati e collezionisti nella Milano del Seicento* (Milan: 2012).

Galli A.E. – Monferrini S., "Un cavaliere e la sua 'dama'. Vitaliano VI Borromeo tra politica, filosofia e amore per l'Isola Bella", in Morandotti A. – Natale M. (eds.), *Vitaliano VI Borromeo. L'invenzione dell'Isola Bella*, exhibition catalogue, Isola Bella, Palazzo Borromeo (Milan: 2020) 34–40.

Gallo Agostino, *Le dieci giornate della vera agricoltura e piaceri della villa* (Venice, Domenico Farri: 1565).

Gasparri C., "The Making of an Icon. The Farnese Hercules and the Power of Place", in Settis S. – Anguissola A. – Gasparotto D. (eds.), *Serial Classic. Multiplying Art in Greece and Rome*, exhibition catalogue, "Serial Classic", Milan, and "Portable Classic", Venice (Milan: 2015) 171–179.

Gronda G., "Ceva, Tommaso", *Dizionario biografico degli italiani* 24 (1980).
Kristeller P.O., *Il pensiero filosofico di Marsilio Ficino* (Florence: 1998).
Isella D., *I Lombardi in rivolta. Da Carlo Maria Maggi a Carlo Emilio Gadda* (Turin: 1984).
Maggi Carlo Maria, *Il trionfo d'Augusto in Egitto* (Milan, Marc'Antonio Pandolfo Malatesta: 1672).
Maggi Carlo Maria, *Il Girello drama faceto*, (Milan, Marc'Antonio Pandolfo Malatesta: 1674).
Maggi Carlo Maria, *La Bianca di Castiglia* (Milan, Marc'Antonio Pandolfo Malatesta: 1674).
Maggi Carlo Maria, *Affari ed amori* (Milan, Marc'Antonio Pandolfo Malatesta: 1675).
Maggi Carlo Maria, *Rime varie*, I (Milan, Giuseppe Malatesta: 1700).
Monferrini S., "Carlo IV Borromeo Arese, Alessandro Scarlatti e la Cappella Reale di Napoli", in Della Libera L. – Maione P. (eds.), *Devozione e passione: Alessandro Scarlatti nella Napoli e Roma barocca* (Naples: 2014) 67–118.
Monferrini S., "Nuovi documenti sulla gestione impresariale del teatro di Milano fra Sei e Settecento. Il Teatro Ducale milanese (1642–1716) e la rappresentazione de *La Floridea* (Novara, 1674)", *Fonti Musicali Italiane* 23 (2018) 27–66.
Monferrini S., "'Raccomandazioni' di Vitaliano VI Borromeo per alcuni artisti. Note d'archivio su Cignaroli, De Groot, Garzia, Lanzani, Saglier, Tempesta, Vismara, Volò", *Arte lombarda* 186–187 (2019) 180–189.
Monferrini S., *Scene di corte all'inizio del Settecento. Cerimoniali barocchi per Carlo IV Borromeo Arese viceré di Napoli (1710–1713)* (Milan: 2021).
Morandotti A. (ed.), *Le isole incantate. Vedute dei domini Borromeo da Gaspar Van Wittel a Luigi Ashton* (Milan: 2015).
Morandotti A., "La formazione della Galleria e la sua storia tra la seconda metà del Seicento e la fine del Settecento", in Morandotti A. – Natale M. (eds.), *Collezione Borromeo. La Galleria dei Quadri dell'Isola Bella* (Cinisello Balsamo: 2021) 9–45.
Morandotti A., "La ricchezza degli interni. 1. I criteri di allestimento e il ruolo della pittura", in Morandotti A. – Natale M. (eds.), *Vitaliano VI Borromeo. L'invenzione dell'Isola Bella*, exhibition catalogue, Isola Bella, Palazzo Borromeo (Milan: 2020) 44–63.
Morandotti A. – Natale M. (eds.), *Collezione Borromeo. La Galleria dei Quadri dell'Isola Bella* (Cinisello Balsamo: 2011).
Morandotti A. – Natale M. (eds.), *Vitaliano VI Borromeo. L'invenzione dell'Isola Bella*, exhibition catalogue, Isola Bella, Palazzo Borromeo (Milan: 2020).
Muratori L.A., *Vita di Carlo Maria Maggi* (Malatesta – Milan: 1700).
Natale M., *Le Isole Borromeo e la Rocca di Angera. Guida storico artistica* (Cinisello Balsamo: 2000).

Papale A., "La Casa di probazione di Arona", in Signorelli B. – Uscello P. (eds.), *La Compagnia di Gesù e la società piemontese. Le fondazioni del Piemonte Orientale*, atti del convegno, Vercelli (Vercelli: 1995) 249–267.

Petrella A., "Carlo Maria Maggi's Theory of Comedy", *Italian Quarterly* 12 (1969) 223–237.

Pierattini C., "Aspetti, funzioni, e fortuna dell'Ercole tiburtino", *Atti e Memorie della società Tiburtina di Storia e Arte* 54 (1981) 7–40.

Ribouillault D., *Rome en ses jardins: Paysage et pouvoir au XVIe siècle* (Paris: 2013).

Ribouillault D., "Atlas and Hercules in the Garden. Scientific Culture and Literary Imagination at the Villa Aldobrandini at Frascati", *Nuncius* 30 (2015) 124–160.

Ripa C., *Iconologia* (Siena: Heredi di Matteo Florimi, 1613).

Scaramuccia Luigi, *Le finezze de pennelli italiani* (Pavia, Giovanni Andrea Magri: 1674; anastatic reprint, Milano: 1965).

Schmitt A., *Modernity and Plato. Two Paradigms of Rationality* (Rochester: 2012).

Sforza Pallavicino Pietro, *Arte della perfezione cristiana* (Milan, Lodovico Monza: 1666).

Sforza Pallavicino Pietro, *Del bene* (Rome, eredi di Francesco Corbelletti: 1644).

Taliaferro C., *Evidence and Faith* (Cambridge: 2005).

Taviani F., "Il teatro dalla morale alla cultura", in Guccini G. (ed.), *Il teatro italiano nel Settecento* (Bologna: 1988) 101–120.

Zanlonghi G., *Teatri di formazione. Actio, parola e immagine nella scena gesuitica del Sei – Settecento a Milano* (Milano: 2002).

Zanuso S., "La ricchezza degli interni. 2. Scultura e arredi", in Morandotti A. – Natale M. (eds.), *Vitaliano VI Borromeo. L'invenzione dell'Isola Bella*, exhibition catalogue, Isola Bella, Palazzo Borromeo (Milano: 2020) 64–79.

Zardin D., "Carlo Maria Maggi e la tradizione culturale milanese tra Sei e Settecento", *Annali di storia moderna e contemporanea* 3.3 (1997) 9–50.

CHAPTER 12

Academicians and Their Gardens in Genoa in the First Half of the 17th Century

Lauro Magnani

Abstract

Giovan Vincenzo Imperiale (1582–1648), a leading figure in the Italian academic milieu of the early seventeenth century, expressed his poetic project in the poem *Lo stato rustico*, offering an extraordinary verbal reconstruction of the family garden at Sampierdarena. The interweaving of the space of the academy and the fabric of the garden is revealed in an itinerary that takes the poet and his readers from the villa of Sampierdarena, in Genoa, to the mythical Mount Helicon. This ability to combine academic activities with the creation of new garden spaces would be adopted by other Genoese aristocrats, such as the Di Negro family in their villa Dello Scoglietto. In the middle of the century, Anton Giulio Brignole Sale, also a man of letters and a leading figure of the local aristocracy, revived this ideal relationship between literature and the garden, critically comparing the ideas of freedom and civic engagement with the artificial nature of the Italian garden.

Keywords

Genoa – gardens – academies – Giovan Vincenzo Imperiale – Lo Stato rustico

Academies were an extraordinary network of cultural and social relationships and a fundamental space of exchange for the leading figures of the early seventeenth century. Beyond Cesi's 'scientific' critiques and the crisis reported by Trajano Boccalini,[1] this becomes particularly clear when we examine cases

1 Ricci S., "I Lincei: l'invenzione della mediazione accademica. Nuova scienza, religione, vita civile", in *Sciences et religions. De Copernic à Galilée*, Actes du colloque international – Rome 12–14 décembre 1996 (Rome, Publications de l'École Française de Rome: 1999) 205–234, 260; for Boccalini in particular, see 231. Federico Cesi was the founder of the Accademia dei Lincei and Trajano Boccalini the author of the *Ragguagli di Parnaso*. Both are well-known figures in the scholarly discourse on the development of academies.

 | DOI:10.1163/9789004517547_013

like that of the Genoese aristocrat and politician Giovan Vincenzo Imperiale (1582–1648).[2] For Imperiale, belonging to and being embraced by several Academies factored significantly in his travel plans, affected his career as a man-of-letters, and influenced his political activity. A member and possibly 'founder and patron' of the Accademia dei Mutoli, as noted by Martinoni, Imperiale was a wealthy, busy scion of one of the ruling oligarchic families in Genoa at the beginning of the century.[3] His academy name "Il Desioso" was emblematic of his manner, the rigour of his research, and his desire for recognition. This recognition came from the most prominent figures of his contemporary aristocratic and literary milieus, including Giovanni Battista Marino, the most important poet of his time. Furthermore, the network that supported Imperiale participated in the Academies to which he belonged.[4] Among these we count the "Addormentati" (following the "Mutoli") in Genoa and the "Intrepidi" in Ferrara, where he took the name "Ripercosso" in 1612. Introduced by Francesco Paleotti, a member of the illustrious Bolognese family, to the "Gelati" between the 1620s and 1630s,[5] Imperiale officially joined the group in 1637 as the "Attempato".[6] By this time, he was also already a member of the "Umoristi" of Rome as the "Agitato", as well as the "Oziosi" of Naples, which he joined in 1632 with the name "Fermo". In *Le glorie degli Incogniti* (1647), a text listing the members of the Venetian Academy, he is described as a man 'così famoso al mondo per la grandezza non meno delle sue virtù, e delle sue fortune' ('as famous for his greatness as for his virtues and his fortune').[7] In these academies he found social and cultural recognition as well as a network of protectors, which was confirmed during the greatest crisis of his political career.

In 1635, Imperiale was labelled "rebellious", accused of murder and, as a result, condemned to two years in exile. He was, in reality, a victim of tensions with Spain at a time when high-ranking figures like himself seem to have been

2 The fundamental source on Imperiale remains Martinoni R., *Gian Vincenzo Imperiale politico, letterato e collezionista genovese del Seicento* (Padua: 1983).

3 Idem 15. For the Accademia dei Mutoli, see also Morando S., "Tra 'muta lingua' e 'cor loquace': echi neoplatonici nell'Accademia dei Mutoli di Genova", in Beltrami L. – Chichiriccò E. – Morando S. (eds.), *I diversi fuochi della letteratura barocca. Ricerche in corso*, Atti del Convegno di Genova, 29–30 ottobre 2015 (Genoa: 2018) 63–79.

4 On the network of academies and Giovan Vincenzo Imperiale's place in them, see the Database of Italian Academies, funded by the Arts and Humanities Research Council UK and created in collaboration with the British Library, Royal Holloway University of London and the University of Reading; https://www.bl.uk/catalogues/ItalianAcademies.

5 On Imperiale and the "Gelati", see Beltrami L. (ed.), *Gian Vincenzo Imperiale, Il ritratto del Casalino* (Lecce: 2009) 18–20.

6 Ibidem 19.

7 *Le glorie degli Incogniti o vero gli huomini illustri dell'Accademia de' Signori Incogniti di Venezia*, (Venice, Francesco Valvasense: 1647) 261.

trying to establish a distance from Spanish influence and secure autonomy for the Republic.[8] At this critical moment, Giovan Francesco Loredan, founder of the "Incogniti" in Venice, gave Imperiale his support. He found further backing and protection for his defence among the "Gelati" of Bologna, an affiliation aided by his friendship with intellectuals like Virgilio Malvezzi and the powerful Paleotti family, particularly Galeazzo and his son Francesco.[9] The social role of the Academy took concrete form at these critical political junctures.

Notably, Imperiale evoked a symbolic experience of his hosts' and own gardens in his writings. Gardens marked stops in his journey of exile, such as the Farnese garden in Parma and the country residence of the Paleotti family.[10] The latter, known as the "Arienti" and the Casalino dei Paleotti, was a charming site between the Reno and Setta rivers, as well as the seat of the Accademia dei Gelati. A place of simple pleasures, this was the refuge offered by the Bolognese family to Imperiale, who extolled its virtues in a 1635 volume titled *Il ritratto del Casalino* and published in Bologna.

Imperiale explores the relationship between written text, description of place and the interdependence of owner, patron and property in the introduction of *Il ritratto del Casalino*. After noting his country house birth in Sampierdarena and his later "rebirth" in the Casalino country residence, he declares that 'Chi è padrone della fabrica, sia padrone del dissegno' ('The master of the house is also the master of the design').[11] Imperiale described the site and the activities that took place there as the 'sketch', referring to the proprietor's ability to construct 'literally' and to 'restore', through text, the design of his residence. In his moment of crisis, Imperiale emphasises the contrast between his sumptuous villa in Sampierdarena and the studied and refined 'medietas' of the aristocratic Bolognese family's residence. His description of the Bolognese villa in Part 2 of *Il ritratto del Casalino* recalls, by contrast, his description of his villa in Part 10 of *Lo stato rustico*, which he had written almost thirty years earlier [Fig. 12.1]. According to Luca Beltrami, the simplicity and purity of the Bolognese residence highlight the possibility of an 'economically self-sufficient' villa model, approaching 'a natural idyll guaranteed by a Petrarchan solitude', as opposed to the 'private and illusory Arcadia' of the Genoese model.[12] Imperiale's texts offer two alternative models of dwelling: the 'otium' of the countryside in contrast to the 'negotium' of the city. The opening descriptions of the two models seem to underline the contrast

8 Beltrami, *Gian Vincenzo Imperiale* 9–10.

9 Ibidem 14.

10 Ibidem 11.

11 Imperiale Gio. Vincenzo, *Il ritratto del Casalino Componimento dell'Ill. Sig.* [...] (Bologna, Herede di Vittorio Benacci: 1635), proemio. See Beltrami, *Gian Vincenzo Imperiale* 73.

12 Ibidem 221.

FIGURE 12.1 Frontispiece, from Giovanni Vincenzo Imperiale, *Lo stato rustico* (Genoa, Pavoni: 1611), private collection
IMAGE © LAURO MAGNANI

between the perceived artifice of the monumental building and large garden of the Genoese villa and the deconstructed character of the Bolognese site. The 'arrogant' façade of the Genoese villa, with its classical orders from 'del Corinto lavor l'arte più illustre [...] del dorico ornamento i precetti osservati' ('the illustrious Corinthian order to Doric ornamentation carefully observed'), are opposed to the Casalino façade where the 'Corinzio fregio e dorico rilievo qui non macchiano il dosso a calci intatte' ('Corinthian frieze and Doric relief do not stain the simply plastered façade').[13] These diverging models also distinguished the gardens, with the ascent to Helicon in the case of Genoa and the rural idyll symbolised by Casalino.

Imperiale's personality is most vividly revealed and the close link between the Academy and the garden is best illustrated in a related description of the Genoese villa. In *Le grazie rivali. Declamazioni accademiche del cav. Gio. Battista Manzini*, printed in Bologna in 1637, the author celebrates Imperiale and discusses the clash between the 'moderns' and the champions of a more moderate 'Ciceronian' style during talks by Academy members at the Genoese poet's villa. The introduction evokes Imperiale's garden in the 'alpestre' area stretching from the beaches of Sampierdarena to the mountains as a foreword to the scholarly talks that follow. The author's description starts with the highest point on the hill and ends at the palazzo that rises on the shore, tracing the course of the water that runs from the summit. Following the stream through the 'amenità di un colle dedicato ad Apollo, et alle Muse' ('delights of a hill dedicated to Apollo and the Muses'), woods, fields of flowers, fishponds, waterfalls, and fountains the author eventually arrives at the 'royal residence'.[14] There, among one thousand delights, he again conjures the imagery of the woods in the garden 'posciache l'occhio, fascinato dale postergate delizie del bosco, non saprebbe privarsi di sì godevole oggetto' ('since the eye, fascinated by the neglected delights of the forest, could no longer do without such an enjoyable thing').[15] In this beloved place, Imperiale had gathered noble intellects, including more than sixty knights and ladies, to listen to Manzini's three "Declamations" and to decide which should receive 'il voto dell'Accademia' ('the Academy's vote').[16]

In his best-known text, *Lo stato rustico*, Imperiale suggests interpreting the garden as the image of his Academy. Marino confirmed the success of this

13 *Il ritratto del Casalino*, Second part, IV. See Beltrami, *Gian Vincenzo Imperiale* 97. For the relationship between the façades of the two villas at Casalino and Sampierdarena, 221.

14 *Le grazie rivali. Declamazioni Accademiche del Cavalier Gio. Battista Manzini* (Bologna, Niccolò Tebaldini: 1637), Introduzione n.p.

15 Ibidem.

16 Ibidem.

work, citing Imperiale's poem in the opening of his famous and extraordinary *Adone*. Here, he identifies Clizio, who initiates Adonis into the 'delizie di Cipro' ('pleasures of Cyprus'), with 'il Sig. Gio. Vincenzo Imperiali, gentilhuomo Genovese di belle lettere' ('Giovan Vincenzo Imperiale, Genoese gentleman, man-of-letters').[17] Marino has Clizio/Imperiale praise the pastoral life, identify the rustic state as the golden age and exalt the natural state over the civil one.[18] Numerous academy members similarly affirmed Imperiale's prominence in his contemporary literary milieu, as evidenced by their compositions that accompanied the first edition of *Lo stato rustico* in 1607, the second in 1611, and the final version in 1613. The centrality of the villa and garden in Imperiale's poem directly reflect his perception and experience of the garden in parallel to his literary and academic work. This relationship has not yet been sufficiently emphasised in scholarly literature, even in Ottavio Besomi, Augusta López-Bernasocchi and Giovanni Sopranzi's formidable critical edition on the text.[19] The verbal play and praise of Claudio Achillini – an eminent member of many academies, not least among them the Lincei – suggests a poetic engagement with the physical form of the villa:

> il signor Giovanni Vincenzo ha tolto i pregi alle città e n'ha arricchito le ville; ma se gli alberghi civili portano invidia ai boschi, i boschi all'incontro si dolgono d'essere stati rubati in queste carte e communicati alle città: così cotesto signore non ha saputo involarsi alla città per donarsi alla villa, senza involar la villa alla villa e farne parte alla città.[20]
>
> Giovanni Vincenzo has robbed the beauties of the cities and adorned the villas; but while the city residences envy the woods, the woods for their part complain that they have been robbed in these writings and communicated to the cities: and so this man was unable to steal himself from the city to give himself to the villa without stealing the villa from the villa and introducing it to the city.

17 Marino Giovan Battista, *Adone* (Venice, Giacomo Sarzina: n.d. [1623–1626]), Canto Primo, Allegoria, 4.

18 Besomi O. – López Bernasocchi A. – Sopranzi G. (eds.), Giovan Vincenzo Imperiale, *Lo Stato rustico* (Rome: 2015) I, 9–10.

19 After the first two Genoese editions printed in 1607 and 1611 for Giuseppe Pavoni, the work was reprinted in Venice in 1613 by Evangelista Deuchino. See Imperiale, *Lo Stato rustico* (2015).

20 Achillini C., "Epistolario", in Marino G., *Epistolario: seguito da lettere di altri scrittori del seicento*, eds. Borzelli A. – Nicolini F. (Bari: 1912) II, 118–119; Imperiale, *Lo Stato rustico* (2015) 13.

A play on words, this imagery evokes the overlaps between arcadia, academy and villa that Imperiale made concrete. Amidst endless digressions, dialogues, descriptions, pastoral and amorous adventures, myths, discussions of hunting, fishing, and bird-catching, revelations of iconography and symbology and celebrations of ancient and modern poets, the poem narrates Clizio's travels from Genoa across the Italian peninsula.[21] Accompanied by the muse Euterpe, they journey away from the splendid city, discovering the rustic, pastoral beauty of landscapes in Liguria, Piedmont, the Po Valley, the Veneto, Tuscany, Lazio, and Campania, finally to arrive in Helicon, Greece. The poem is made up of sixteen parts and includes the revelation of Apollo and his concert with the Muses and choir of poets in Saturn's wood. In the middle of this narrative (part 10), a precise, detailed report of Imperiale's villa and his garden is provided. Although embellished in Baroque poetry, the description of its parts holds up a mirror to the narrative: a microcosm in a macrocosm, a world described and passed through by the poet and his Muse to get to Helicon and experience the revelation of Apollo's concert.

The description of the garden visit in Part 10 begins with Clizio's dream of an Eden-like Arcadian setting, a dream of pastoral beauty and love. Euterpe reveals that she too has dreamed, but her dream is capable of elevating the aristocrat's dream to a higher level. The Muse's dream starts with the 'nobil cerchio' ('noble circle') of 'gentil, spiriti ingenosi' ('gentle, ingenious spirits') gathered at Imperiale's villa who 'Mutoli ad arte e si nomaro e Mutoli si fero' ('wished to call themselves Mute and Mute they made themselves').[22] Their chants could ring from the 'Ligustici lidi' ('Ligurian shores') to the ('rive d'Helicona') 'banks of Helicon'.[23] Thus begins the descriptive journey of a dream villa – in reality, Imperiale's – through the itinerary of the imagination, describing the actual elements of the garden. Embedded in the middle of the text, the poem's passages describing the ideal journey to Helicon align with the visit to the garden described in Part 10. Euterpe accompanies the aristocrat in a model garden like Imperiale's, taking on the role of the actual owner as he guides members of the Academy through his reconstructed Parnassus. The narrative focuses on the description of the Imperiale villa and garden, one of the most important in Genoa and emblematic of the design trends of the second half of the sixteenth century. The Genoese aristocracy were the drivers behind

21 On the literary references and models for Imperiale's poem, see Selmi E., "Pastorale in romanzo: un contributo per lo 'Stato rustico' di Gian Vincenzo Imperiali", in Beniscelli A. – Chiarla M. – Morando S. (eds.), *La tradizione della favola pastorale in Italia: modelli e percorsi*, atti del Convegno di studi, Genoa, 29 November–1 December 2012 (Bologna: 2013) 243–279.

22 Gio. Vincenzo Imperiale, *Lo stato rustico* (Genoa, Giuseppe Pavoni: 1611) 422.

23 Ibidem 423.

FIGURE 12.2 Cristoph Friedrich Krieger, *View of Sampierdarena*, watercolor engraving, after Johann Cristoph Volkamer, *Nürbergische Hesperides, oder, Gründliche Beschreibung der edlen Citronat-Citronen-und Pomerantzen-Früchte* [...], 1708. Private collection
IMAGE © LAURO MAGNANI

developments of the villa in this period, not to mention the urban palazzi so successfully celebrated by Peter Paul Rubens.[24]

With a prime location bordering the city [Fig. 12.2], the described garden sat between the sea and the hills and was situated on the first riviera at the end of the surrounding Polcevera and Bisagno torrent valleys.[25] Teeming with elements that signalled their refined and innovative nature, the gardens housed internationally renowned aviaries and featured artificial grottoes.[26]

24 See *I palazzi di Genova* (Antwerp, Christophe Plantin: 1622) and *Palazzi moderni di Genova raccolti e designati da Pietro Paolo Rubens* (Antwerp, Giacomo Merusio: 1652).

25 On the gardens of Genoa, see Magnani L., *Il tempio di Venere. Giardino e villa nella cultura Genovese* (Genoa: 1987); on the Villa Imperiale in particular, see 125–140. The cultural and social aspects of the phenomenon are also analysed in Magnani L., "The Rise and Fall of Gardens in the Republic of Genoa, 1528–1797", in *Bourgeois and Aristocratic Cultural Encounters in Garden Art 1550–1850*, Conan M. (ed.), Dumbarton Oaks Colloquium on the History of Landscape Architecture 23, Dumbarton Oaks, (Washington, D.C.: 2002) 43–76.

26 Magnani L., "L'uso d'ornare i fonti: Galeazzo Alessi and the Construction of Grottoes in Genoese Gardens", *Journal of Garden History* 5 (1985) 135–153; Magnani L., "Introduzione alle grotte genovesi", in Cazzato V. – Fagiolo M. – Giusti M.A. (eds.), *Atlante delle grotte e ninfei in Italia; Italia settentrionale, Umbria, Marche* (Milan: 2002) 38–47 and the catalogue entries p. 53–84; Hanke S., *Zwischen Fels und Wasser. Grottenanlagen des 16. und*

FIGURE 12.3 Martin Pierre Gauthier, *Villa Impériale à Sampierdarena, Vue générale*, in *Les plus beaux édifices de la ville de Gênes*, 2 vols. (Paris: 1832), II, pl. 50. Genoa, Collection Galleria San Lorenzo
IMAGE © GALLERIA SAN LORENZO

The description outlines the architectural structure of the country palazzo as well as its interior and exterior decoration. It then turns to the garden, moving through the geometric sections adjacent to the building [Fig. 12.3]. The poem speaks of flowerbeds with low hedges most likely located on its east and west sides, cut 'col gemino coltel, che fabro accorto uniò nel pié, di scema Luna in guisa, E fabricò arrendevole, e tagliente' ('with a double knife that an able blacksmith joined together at the bottom to form a kind of waning moon, making it easy to use and sharp').[27] These were possibly accompanied by circular labyrinths, fountains with high jets of water, a mosaic floor featuring animal motifs, and marble statues of Diana and Actaeon in the centre. These elements were destroyed when the villa was rebuilt but they can be seen on nineteenth-century plans and maps [Fig. 12.4]. Behind the building was an

17. *Jahrhunderts in Genua* (Münster: 2008). As for aviaries, the monumental one at the Villa Doria del Principe was particularly admired; Camurati M., "L'uccelliera di Palazzo del principe Doria a Fassolo in Genova", *Archeologia dell'architettura* 8 (2003) 89–110, as was the one cited by John Evelyn in 1647 at the villa of Orazio Di Negro at Fassolo.

27 Imperiale, *Lo Stato rustico* (1611) 440.

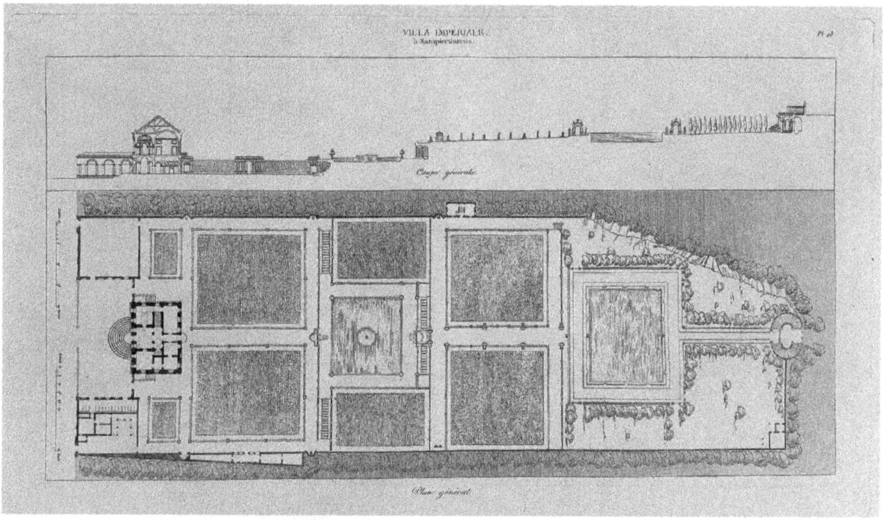

FIGURE 12.4 Martin Pierre Gauthier, *Villa Impériale à Sampierdarena, Coupe générale*, in *Les plus beaux édifices de la ville de Gênes*, 2 vols. (Paris: 1832), II, pl. 48. Genoa, Collection Galleria San Lorenzo

'ampio quadro' ('large square field') surrounded by laurel trees. Then, 'Nel più sublime là dove poi termina il prato' [...] ad erto in mezo, e a polito muro, in prospettiva gratiosa [...] appare da termini marmorei, arco di marmo ad alto nicchio eretto' ('at the highest point where the meadow ends [...] in a pleasant perspective, in the center of a high wall [...] a marble arch [framing] a high niche appears, supported by marble terms').[28] Next, Euterpe describes a fountain with some marble ornament and rustication: one with stones, shells and limescale ('conche, chiocciole, coralli') decorated with two fighting putti and embellished by the sound of the water that spouted from the horn blown by one figure. The opening of the via Cantore in 1932–1933 destroyed the first level of the garden [Figs. 12.6–12.7], but the late sixteenth-century tripartite nymphaeum with marble herms is still well preserved [Fig. 12.8].[29] The garden moved up from that level 'dove il muro, qual fabricato ponte, ambe le braccia, in largo giro ne piega' ('where the wall, like a fabricated bridge, bends its arms in a wide circle') via two ramps decorated with balusters made of 'corte e spesse colonne alabastrine' ('short and thick alabastrine columns').[30]

28 Ibidem 448.
29 Poleggi E. – Cevini P., *Genova* (Bari: 1981) 219–231.
30 Imperiale, *Lo stato rustico* (1611) 451.

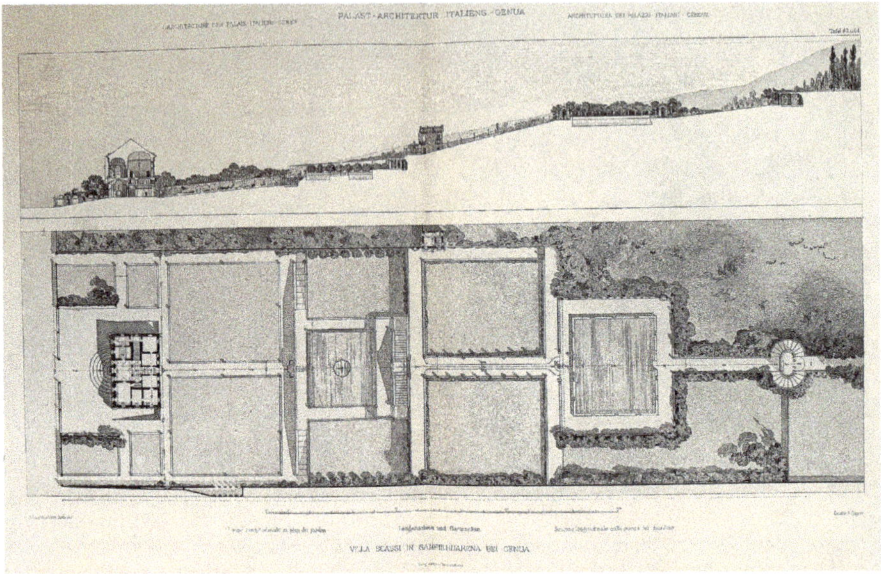

FIGURE 12.5 Longitudinal section and plan of the Imperiale garden, from Robert Reinhardt, *Villa Scassi in Sampierdarena bei Genua*, in *Palastarchitektur von Ober-Italien und Toscana, vom 13. bis 18. Jahrhundert*, vol. I: *Genua* (Berlin: 1886), pl. 83–84. Università degli Studi di Genova
IMAGE © LAURA GUIDA, UNIVERSITY OF GENOA

At the top of the first ramp was a small, flat area with a statue of the Goddess of Abundance on one side and a statue of the Goddess of Peace on the other, both sculpted out of 'Africana pietra' ('African stone'). In this case as well, one can still make out the structure as it appears in historical plans [Fig. 12.5]. The second ramp led to a 'gran piano di tronchi più pregiati e più fecondi' ('large flat area of fine, flourishing trees'), divided into 'quasi in fertili sale in quadro inchiuse' ('what are almost like fertile rooms enclosed by trees').[31] The ground level was surrounded by a wall of cedars and divided into individual gardens with covered paths. Here, the man-of-letters, guided by his muse, engages in a masterful description of a long list of plants and flowers.[32] The second terrace can still be easily made out, although it was transformed by the planting of different tree species when converted into a public park. At the top of this large area of level ground, Euterpe exclaimed 'mi volgo, ove gran parte e più sublime de l'ameno giardin posta à la fronte' ('I turn, where a great and most sublime part of the lovely garden is placed before me'). There stands a grotto or 'spring'

31 Ibidem 452.
32 Ibidem 454–466.

FIGURE 12.6 View of the Imperiale garden, from Robert Reinhardt, *Villa Scassi in Sampierdarena bei Genua, Ansicht des Gartens*, in *Palastarchitektur von Ober-Italien und Toscana, vom 13. bis 18. Jahrhundert*, vol. 1: *Genua* (Berlin: 1886), pl. 86. Università degli Studi di Genova
IMAGE © LAURA GUIDA, UNIVERSITY OF GENOA

made up of three parts: 'sorge l'alta fontana in tre spelunche ripartita ugualmente' ('the tall fountain rises in three equally divided caves'), where Art and Nature compete with each other.[33] A sculpture of a 'winged snake' was placed in the larger cave. She goes on:

> Indi rimiri nel suo doppio fianco e l'una grotto, e l'altra avvivar quasi, e far quasi spirar marmorei corpi [...] da l'una parte Amor, dall'altra Apollo [...] di pioggia saettar il suolo e il tetto; Anzi ferir di colpi inaspettati, gl'incauti spettatori e spensierati

33 Ibidem 467–468.

FIGURE 12.7 *The Imperiale-Scassi Garden at Sampierdarena from the terrace of the Villa's Palace*
IMAGE © MARCO STUCCHI

FIGURE 12.8 *The fountain-nymphaeum in the wall that supports the second terrace*, Villa Imperiale Scassi, Sampierdarena (Genoa)
IMAGE © MARCO STUCCHI

FIGURE 12.9 *The second terrace with the fountain and the grotto*, Villa Imperiale Scassi, Sampierdarena (Genoa)
IMAGE © MARCO STUCCHI

> then you see on its two sides two grottos in which the marble figures of Cupid and Apollo seem to live and breathe, while water shoots from the floor and ceiling. This way, the unsuspecting and carefree visitor is unexpectedly struck by the water [as if by Cupid's arrows].[34]

The interior was decorated with mixed media mosaic: 'Ne le volte di mezo e ne i pilastri, sono per opra, ò di Mosaico impressi, o di dorato, ricamato intaglio' ('the middle vaults and pillars are made either of mosaic, or of gilded, embroidered carving'). In this way, art and nature became inseparable: 'se miri, con tal arte osservi, l'arte osservata, e l'arte nascosta ad arte, ch'à natura n'ascrivi il don de l'arte' ('if you look carefully, you see how, with this artifice, art is artfully hidden, so that you attribute to nature the gift of this artistic creation').[35] Although it was somewhat transformed in the nineteenth century, the grotto still exists and preserved statues from the Imperiale antiquarian collection until the 1980s, today in the deposits of the museum of Sant'Agostino in Genoa [Fig. 12.9].

34 Ibidem 470.
35 Ibidem 471.

Next in the described itinerary came a climb up a 'soave ascesa' ('gentle slope') to the upper level 'di mille balaustri adorno il giogo che fa cupola al fonte' '(with a thousand balusters adorned the yoke that dome the fountain'). Then a vast slope full of flowers 'qual dolce si sollevi a poco a poco' ('whose sweetness rose little by little'), 'con ordine lungo' ('in long order'), i.e. that it was difficult to see its extent from one end to the other. Here, one looks back to admire the path taken thus far and the broad view extending down to the sea.[36] A large oak formed the focal point of this expansive, precipitous ascent with the mountain as its backdrop and where farmers cultivated rows of grapevines: 'Di sì bel calle à la felice meta vedi posta sublime immensa quercia' ('You can see how an immense oak is the happy destination of such a beautiful path'), evoking the oaks of the Golden Age and the Age of Saturn.[37] This part of the garden is now crossed by a wide brick path. The large oak can be seen in a painting of the Imperiale family likely made in the last decade of Giovan Vincenzo's life [Fig. 12.10]. Behind the oak, where the path becomes flat again, we find 'at its foot' a fishpond 'in quadro muro' ('enclosed in a square-shaped wall') reflecting its green fronds [Fig. 12.11]. The banks on either side of the fishpond and opposite the oak tree were decorated with statues of sea divinities: 'da cento statue Protei, Glauchi, Tritoni, Nereidi misti ha quanto ha il mare Ninfe e Sirene' ('by a hundred statues Proteus, Glauchi, Tritons, Nereids mixed as the sea has Nymphs and Sirens') and on the opposite shore water spurts from the trident of Neptune, flanked by Thetis and Amphitrite.[38] This fountain stands opposite the oak tree. Around the fishpond, 'ogni stagion tu vedi star più d'un Cigno candido, canoro' ('you can see more than one singing, white swan in any season'), referencing Plato's famous dying swan and its perennial song by noting how 'Canta il Cigno se muor; canta se vive' ('the swan sings if it dies, it sings if it lives'), accompanied by the babbling water.

The garden composition appears as a sacred, enchanted place amidst dense woods: 'a le mura, a le statue, a l'acque, ai cigni, cerchio onorato et immortal corona, fanno co' i tronchi lor folti boschetti' ('the thick groves with their trunks make an honoured circle and immortal crown to the walls, to the statues, to the waters, to the swans').[39] Though it can be orographically reconstructed, this section was entirely lost to urban development, which spared only the part of the garden on the same axis as the building. Walking along this lost part that turned slightly westwards from the axis of the garden, Euterpe

36 Ibidem 472.

37 Ibidem 474.

38 Ibidem 475.

39 Ibidem 476–477.

FIGURE 12.10 Worshop of Domenico Fiasella, *Giovan Vincenzo Imperiale with his family on the terrace of the villa of Sampierdarena*, oil on canvas, 270 × 314 cm, 1642, Musei di Strada Nuova, Palazzo Bianco (currently on view at Palazzo Imperiale, Genoa)
IMAGE © LAURO MAGNANI

FIGURE 12.11 Giovanni Battista Derchi (1879–1912), *The great fishpond in the garden of the Villa Imperiale Scassi at Sampierdarena* (*Genoa*), oil on canvas, 38 × 66 cm. Private collection
IMAGE © MUSEI DI STRADA NUOVA

shows the reader a rustic path that opens to pristine natural setting and ends where 'dentro à gran nicchio gran fontana adorna' ('a large ornate fountain stands in a large niche').[40] This is an artificial hill, an 'alto monte alpestre di ruvidi, scogliosi monticelli, di gran pomixce un sovra l'altra composti ad arte in ordine incomposto' ('high alpine mountain of rough, rocky mounds, of great pumice stone one above the other artfully composed in a disordered order'). Striking the highest stone upon it, Pegasus unleashes a flow of water 'in cristalline in argentee vene, Di novo fium far novo Hippocrene' ('in crystalline silver veins, making a new Hippocrene from a new stream'). Flowing through the rocks, this water creates a stream 'giù diroccando un musico ruscello' ('down a musical brook'), 'forma soave un suono; onde forse anco apprendon l'armonia de i canti loro quei cigni, che tutt'hor da la peschiera, quì volano e rivolano; bevendo con l'onda armoniosa in un' istante, quell'armonia canora ' ('producing a sweet sound from which those swans might have learned the harmony of their song, still today flying to and from the fishpond, drinking from that harmonious water that is at once both song and undulating wave').[41]

Gauthier's and Robert Reinhardt's plans [Figs. 12.4–12.5] merely illustrate the main axis of the garden with a belvedere at its end [Fig. 12.12], but earlier eighteenth-century maps show that the property followed a west-facing valley [Fig. 12.13]. Coming to the top of the garden, the path is interrupted by a structure that might have coincided with the artificial Helicon. Marble statues of wild animals are still found in the garden, confirming the description of a 'marble tiger' that guards the fountain. Still climbing in the same westward direction, one encounters the wild part of the park: the woods described by Euterpe as 'lunga selva, e ampia' ('a forest tall and vast').[42] Emerging from this dark forest that rises towards the mountain, which leads towards the sky and incites one to admire Berenice's constellation 'fin de la solitaria, alta foresta, anzi pur de la villa altera, augusta' ('to the end of the lonely, high forest, or

40 Ibidem 478.

41 Ibidem 479. For a reconstruction of aspects of the garden, see Magnani L. (ed.), *Grotte e giardini ai tempi di Rubens. 'Delizie' e 'meraviglie' a Genova all'alba del Seicento*, exhibition catalogue, Palazzo della Meridiana (October 2022–January 2023) (Genoa: 2022), 85–96.

42 Imperiale, *Lo stato rustico* (1611) 480. Since Imperiale says 'Sta quinci, e quindi' ('it is on one side and on the other'), it is likely that there was more than one statue of tigers. They were moved to other parts of the garden and appear in a painting by Giovanni Battista Derchi (Genoa 1879–1912) (Collection of the Civic Hospitals of Genoa – Sampierdarena). The statues were later placed on the main, surviving axis of the garden, only to be replaced by copies.

FIGURE 12.12 Giovanni Battista Derchi (1879–1912), *The fishpond seen from the Belvedere in the garden of the villa Imperiale Scassi at Sampierdarena (Genoa)*, pencil drawing, 21.5 × 23 cm. Private collection
IMAGE © LAURO MAGNANI

rather of the haughty, Augustan villa'), one encounters a large lake surrounded by tall trees [Fig. 12.13].

The end description in Part 10 stresses the exact overlap between contrived verbal constructions and Imperiale's villa. In Baroque ekphrasis rather than fantastical expression, Euterpe exclaims: 'hor chi vide giamai, Villa più bella' ('now who has ever seen a more beautiful villa'), playfully inviting Clizio/Imperiale to build something 'almeno a lei simile [...] nell'alta patria tua' ('at least similar to it [...] in your homeland').[43] A parallel reading of Imperiale's text and the remains of his grand villa together with existing visual sources seems to confirm his desire to create his villa and garden in monumental form as an image of a journey to Helicon. This ideal place made real through his Academy,

43 Ibidem 482–483.

FIGURE 12.13 Matteo Vinzoni, *Pianta del Soborgo di Sampierdarena in cui sono segnate tutte le Strade Palazzi Case Giardini* […], 1757, detail with the Imperiale property. Genoa, Centro DocSAI, Archivio Topografico del Comune di Genova
IMAGE © CENTRO DocSAI

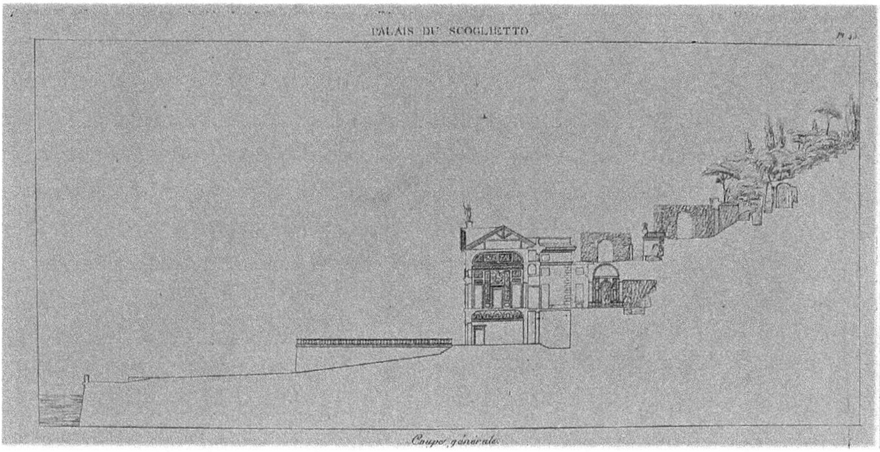

FIGURE 12.14 Martin Pierre Gauthier, *Palais du Scoglietto* (Villa Di Negro): *Coupe générale*, in *Les plus beaux édifices de la ville de Gênes*, 2 vols. (Paris: 1832), II, pl. 45. Genoa, Collection Galleria San Lorenzo
IMAGE © GALLERIA SAN LORENZO

which became the location for the meetings because of Imperiale's patronage, concretely realised the rustic state longed for in the poetic gatherings.

Scholars have studied other similarly important garden projects undertaken by leading figures in Academy circles. These early-seventeenth century Genoese patrons combined gardens with ornament designed to shape the specific symbolic meaning of their spaces. One such case is the Villa Di Negro, among the most cited by non-Italian visitors including Joseph Furttenbach and John Evelyn.[44] Ambrogio Di Negro and his son Orazio designed the space, which was monumental in scale and extended between the seashore and the hill west of the city [Fig. 12.14]. A wealthy financier and collector associated with Academy circles in Flanders, Ambrogio Di Negro was also a poet. Orazio followed his example and was one of the four founders of the Accademia degli Addormentati in 1587, the most influential civil and literary academy in Genoa between the end of the sixteenth century and the first half of the seventeenth.[45] The German architect Joseph Furttenbach noted the grotto in

44 On this villa and its garden, see Magnani, *Il Tempio di Venere* 141–150 and Santamaria R., "Venere a Genova. La collezione di Orazio Di Negro nella villa dello Scoglietto: dipinti, statue e 'altre galanterie'", in Orlando A. (ed.), *La Sacra famiglia di Van Dyck e le collezioni Di Negro Doria a Genova* (Genoa: 2018) 52–65.

45 Beltrami L. (ed.), *Leggi e ordini dell'Accademia degli Addormentati di Genova (1587)* (Rome: 2016) 7.

the Di Negro garden, which he remembered as 'the most charming in Italy' and built 'in the modern style', the monumental aviary and lake and the presence of statues.[46] Numerous ancient or *all'antica* sculpture groups were reported in the collection; John Evelyn remarked in 1647 that 'there is a grove of stately trees, among which are sheep, shepherds, and wild beasts, cut very artificially in a grey stone'.[47] An inventory dated 1667 notes the presence of Roman travertine in the garden, though this was more likely stone from Finale.[48] Together with 'fountains, rocks and ponds for fish', the sculptures created a 'silent, wild' environment within the garden space.[49] Combined with its extremely fortunate location, the Di Negro garden created an artificial Arcadia, a 'Bosco Parrasio' and so perhaps a Helicon. This symbolic connection is further underscored by the presence of a statue of a shepherd with his sheep in the garden, evoking Hesiod's meeting with the Muses: 'And they once taught Hesiod the art of singing verse, While he pastured his lambs on holy Helicon's slopes'.[50] Similarly, in 1674, Raffaelle Soprani recalled Orazio's desire to commission Agostino Tassi to fresco a 'little house dedicated to the Muses' in the garden in 1610, the 'cazoto nel bosco' as it was called in the seventeenth-century inventory (1618).[51] These elements might logically be connected with the activities of the Di Negro Academy. Beyond the Neoclassical alterations, we can still see a clear, structural break in the garden between the geometric terracing at the beginning and the wilder part – the woods – enclosed within high walls and accessed through an arch with a railing.

The orography typical of the Ponente Genovese villa region seems to evoke a climb to a mountain, as in the case of Imperiale's project. In Di Negro's garden, a hill rises from woods cut through by water and gives way to a hallowed clearing inhabited by shepherd statues like the Muses' sacred Helicon in the villa at Sampierdarena. Referencing a different garden, Gabriello Chiabrera, one of the best-known poets of the new classicism of the first half of the seventeenth century, observed that the garden is the ideal setting for the Arcadian scene.[52] One such example is the villa Negrone at Prà, where we find the remains of a curious, rectangular outdoor nymphaeum located at the north end of the garden above the country palazzo. Here, a large marble basin datable to the

46 Furttenbach Joseph, *Newes Itinerarium Italiae* (Ulm, J. Saur: 1627) 217–218.
47 De Beer E.S. (ed.) *The Diary of John Evelyn* (London: 1959) 97–98.
48 See Santamaria, *Venere a Genova* 58.
49 De Beer, *The Diary* 97–98.
50 Hesiod, *Theogony* 23–24.
51 Santamaria, *Venere a Genova* 58–59.
52 See Magnani L., "Residenze di villa e di giardino tra realtà e mito", in Bianchi F. – Russo P. (eds.), *La scelta della misura. Gabriello Chiabrera: l'altro fuoco del barocco italiano* (Genoa: 1993) 471–475.

FIGURE 12.15 *The villa of Anton Giulio Brignole Sale in Albaro (Genoa)*
Photograph by Paolo Monti 1963

first decades of the seventeenth century is topped by a wall with two rusticated pilasters flanking a niche and supporting a triangular gable. Two other niches can be found on its sides, which would have originally been adorned with statues. Between the pilasters, the deepest niche still preserves its green stone and stalactite decoration and hosts a stucco statue of Pegasus striking Helicon with his hoof and creating the Hippocrene spring. The overall effect of the nymphaeum is to simulate an outdoor space, making it the perfect place for Academy meetings at Bendinelli Negrone's residence.[53]

Highlighting the fluid relationship of academy, garden and attempts to enact civic ideals, I turn to a final example. Sharing a similar background and active in the same milieu, though a generation younger than Orazio Di Negro and Imperiale, the aristocrat, politician, poet and later Jesuit, Anton Giulio Brignole Sale stressed the importance of the garden as a site for intellectual discussion. To this end, he chose his Albaro garden and villa [Fig. 12.15] as the setting for *Instabilità dell'Ingegno*, presenting the main characters, four young

53 Lissoni A., "Genova Prà, Ninfeo di villa Negrone", catalogue entry in *Atlante delle grotte* 81.

noblewomen and four young noblemen, as members of the Accademia degli Addormentati to which he had belonged since 1628. Brignole seems to have been a reformer of the Genoese Academy, as suggested in his *Tacito abburrattato* (published for the first time in 1643), in which he analyses the faults of the Academy and his hopes for its concrete change:[54] 'Ecco dunque convenirsi alle Accademie per natura il far de' fatti' ('This, therefore, is what it would befit the Academies to do by nature the making of facts').[55] Between the 1630s and 1640s, Brignole developed a different position, one mindful of countering the supposed call to rustic nature at a time when aristocratic Genoese patrons, who were often prominent members of the Academies, were driven by 'quell'istinto, nemico della natura e naturale delle ricchezze, posto in non prezzare [apprezzare] ciò che non costa' ('that instinct for wealth, enemy of nature, and lack of appreciation for that which cannot be bought') to such an extent that they:

> recidon l'ombre degli alberi, per inalzarvi sopra quelle de' sassi, e fabrican palagi, […] che mostran che gli habitatori non per lasciar la Città, ma per portarcela habbian amato la Villa. Qualhor s'affacciano alle finestre dominatrici […], mirano da lor tesori seminate per man di rusticane fatiche, essere prodotti da qualunque parti dell'amena collina giardini, dove l'aure Ligustiche si profuman con odori Fiaminghi: boschetti dove falce legislatrice nelle chiome frondose non permette soperchierie, praticelli dove vanno carolando schiere di Ninfe, uccelliere ove Musici volatori con le licenze della voce i confini prescritti alle lor penne compensano, fontane, dove l'acque nell'angustia di canali martirij purificatori provando, uscite finalmente alla luce entro à limpidissimi scherzi se stesse con se stesse per la libertà ricuperata festeggiano.

> severed the shade of the trees, to raise up stone walls, and built so many palazzi […] that it is clear that they love living in the countryside not to

54 Brignole Sale Anton Giulio, *Le instabilità dell'Ingegno divise in otto giornate* (Bologna, Giacomo Monti: 1637). The first edition was published in Bologna in 1635. Another edition was published in Venice in 1641. Brignole wrote *Le instabilità* when he was working to take control of the management and reform of the Accademia degli Addormentati. On Brignole's cultural and political project, see Marini Q., "Per una nuova edizione delle 'Instabilità dell'ingegno' di Anton Giulio Brignole Sale", in Beltrami L. – Chichiriccò E. – Morando S. (eds.), *I diversi fuochi della letteratura barocca. Ricerche in corso*, Atti del convegno di studi, Genoa, 29–30 ottobre 2015 (Genoa: 2017) 188–212.

55 Brignole Sale Anton Giulio, *Tacito abburattato discorsi politici, e morali* (Venice, Combi: 1646) 17.

> get away from the city but to bring the city to the country. And if you look out the windows (of their dominantly-situated country palazzi), you see gardens everywhere, admire their treasures sown by the hand of rustic toil, as products of any part of the pleasant hill-gardens, where the Ligurian air is perfumed with Flemish scents: little woods where the trees have been trimmed into geometric shapes, small fields with multitudes of dancing Nymphs, aviaries where song birds compensate for the confinement of their plumes with the freedom to sing, fountains where the waters that suffer every purifying martyrdom in the narrowness of the canals, finally come out into the light in limpid plays of water and celebrate their regained freedom.[56]

The constraints of the Italian garden are thus perceived as the limits imposed on academic research; the literary artifice, like the artificial forms of the garden, contrasts with the desired reform, aimed at the idea of an academy entirely dedicated to civic engagement and the formation of the citizen of the Republic. Ironic confirmation of the relationship between the Academy and the garden: the reference to a conception of the garden that, by moving away from artificial forms, is better suited to the representation of nature, and corresponds to a distancing from an illusory evocation of the Muses and Helicon. This suggests, even through the idealised forms of an Arcadian society, the need to reform both the Academy and politics.

References

Achillini C., "Epistolario", in Marino G., *Epistolario: seguito da lettere di altri scrittori del seicento*, eds. Borzelli A. – Nicolini F., vol. II (Bari: 1912).

Beltrami L. (ed.), Gian Vincenzo Imperiale, *Il ritratto del Casalino* (Lecce: 2009).

Beltrami L. (ed.), *Leggi e ordini dell'Accademia degli Addormentati di Genova (1587)* (Rome: 2016).

Besomi O. – López Bernasocchi A. – Sopranzi G. (eds.), Giovan Vincenzo Imperiale, *Lo stato rustico*, 2 vols. (Rome: 2015).

Brignole Sale Anton Giulio, *Le instabilità dell'Ingegno divise in otto giornate* (Bologna, Giacomo Monti: 1637).

Brignole Sale Anton Giulio, *Tacito abburattato discorsi politici e morali* (Venice, Combi: 1646).

56 Brignole Sale Anton Giulio, *Le instabilità dell'Ingegno divise in otto giornate* (Bologna, Giacomo Monti: 1637) 1–2.

Camurati M., "L'uccelliera di Palazzo del principe Doria a Fassolo in Genova", *Archeologia dell'architettura* 8 (2003) 89–110.

Database of Italian Academies, funded by the Arts and Humanities Research Council UK and created in collaboration with the British Library, Royal Holloway University of London and the University of Reading; https://www.bl.uk/catalogues/ItalianAcademies.

De Beer E.S. (ed.), *The Diary of John Evelyn* (London: 1959).

Furttenbach Joseph, *Newes Itinerarium Italiae* (Ulm, J. Saur: 1627).

Hanke S., *Zwischen Fels und Wasser. Grottenanlagen des 16. und 17. Jahrhunderts in Genua* (Münster: 2008).

Imperiale Gio Vincenzo, *Il ritratto del casalino componimento dell'Ill. Sig.* [...] (Bologna, Herede di Vittorio Benacci: 1635).

Imperiale Gio Vincenzo, *Lo stato rustico* (Genoa, Giuseppe Pavoni: 1611).

Le glorie degli Incogniti o vero gli huomini illustri de l'Accademia [...] (Venice, Francesco Valvasense: 1647).

Lissoni A., "Genova Prà, Ninfeo di villa Negrone", in Cazzato V. – Fagiolo M. – Giusti M.A. (eds.), *Atlante delle grotte e ninfei in Italia; Italia settentrionale, Umbria, Marche* (Milan: 2002).

Magnani L., "'L'uso d'ornare i fonti': Galeazzo Alessi and the Construction of Grottoes in Genoese Gardens", *Journal of Garden History* 5 (1985) 135–153.

Magnani L., *Il Tempio di Venere. Giardini e ville nella cultura genovese* (Genoa: 1987; reprint Genoa: 2005).

Magnani L., "Residenze di villa e di giardino tra realtà e mito", in Bianchi F. – Russo P. (eds.), *La scelta della misura. Gabriello Chiabrera: l'altro fuoco del barocco italiano* (Genoa: 1993) 471–475.

Magnani L., "Introduzione alle grotte genovesi", in Cazzato V. – Fagiolo M. – Giusti M.A. (eds.), *Atlante delle grotte e ninfei in Italia; Italia settentrionale, Umbria, Marche* (Milan: 2002) 38–47.

Magnani L., "The Rise and Fall of Gardens in the Republic of Genoa, 1528–1797", in Conan M. (ed.), *Bourgeois and Aristocratic Cultural Encounters in Garden Art 1550–1850*, Dumbarton Oaks Colloquium on the History of Landscape Architecture 22 (Washington, D.C.: 2002) 43–76.

Magnani L., *Grotte e giardini ai tempi di Rubens. 'Delizie' e 'meraviglia' a Genova all'alba del Seicento*, exh. cat, Palazzo della Meridiana (Genoa: 2022).

Manzini Giovan Battista, *Le grazie rivali. Declamazioni accademiche del cav. Gio. Battista Manzini* [...] (Bologna, Nicolò Tebaldini: 1637).

Marini Q., "Per una nuova edizione delle 'Instabilità dell'ingegno' di Anton Giulio Brignole Sale", in Beltrami L. – Chichiriccò E. – Morando S. (eds.) *I diversi fuochi della letteratura barocca. Ricerche in corso*, Atti del Convegno di studi, Genoa, 29–30 ottobre 2015 (Genoa: 2017) 188–212.

Marino Giovan Battista, *Adone* (Venice, Giacomo Sarzina: n.d. [1623–1626]).

Martinoni R., *Gian Vincenzo Imperiale politico, letterato e collezionista genovese del Seicento* (Padua: 1983).

Morando S., "Tra 'muta lingua' e 'cor loquace': echi neoplatonici nell'Accademia dei Mutoli di Genova", in Beltrami L. – Chichiriccò E. – Morando S. (eds.), *I diversi fuochi della letteratura barocca. Ricerche in corso*, Atti del Convegno di Genova, 29–30 ottobre 2015 (Genoa: 2018) 63–79.

Poleggi E. – Cevini P., *Genova* (Bari: 1981).

Ricci S., "I Lincei: l'invenzione della mediazione accademica. Nuova scienza, religione, vita civile", in *Sciences et religions. De Copernic à Galilée*, Actes du colloque international – Rome 12–14 décembre 1996 (Rome: 1999) 260, 205–234.

Rubens Pietro Paolo, *I palazzi di Genova* (Antwerp, Christophe Plantin: 1622).

Rubens Pietro Paolo, *Palazzi moderni di Genova raccolti e designati da Pietro Paolo Rubens* (Antwerp, Giacomo Merusio: 1652).

Selmi E., "Pastorale in romanzo: un contributo per lo 'Stato Rustico' di Gian Vincenzo Imperiali", in Beniscelli A. – Chiarla M. – Morando S. (eds.), *La tradizione della favola pastorale in Italia: modelli e percorsi*, atti del Convegno di studi, Genoa, 29 November–1 December 2012 (Bologna: 2013) 243–279.

Santamaria R., "Venere a Genova. La collezione di Orazio Di Negro nella villa dello Scoglietto: dipinti, statue e 'altre galanterie'", in Orlando A. (ed.), *La Sacra famiglia di Van Dyck e le collezioni Di Negro Doria a Genova* (Genoa: 2018) 52–65.

CHAPTER 13

A Place Both Real and Imagined: Play, Performance and Narrative in the Gardens of the Arcadian Academy in Rome

Katrina Grant

Abstract

The Accademia degli Arcadia (Arcadian Academy), founded in 1690 in Rome, regularly met in gardens and described their meeting place as the 'Bosco Parrasio' (Parrhasian Woods). For the first thirty years of the academy's existence, they met across many different sites in Rome until a permanent home was built for the Academy on the Janiculum Hill in the 1720s. Nature, landscape and gardens are a constant theme in the writings of the Academy as they reimagine Rome as a new Arcadia. The textual sources, including poems, discourses, published histories, longer prose texts and the proceedings of the *Giuochi Olimpici* (the Arcadians revival of the Olympic Games) all convey how nature, landscape and garden form part of the foundations for the Academy's intellectual endeavours. This chapter looks how the physical sites reflected Arcadian ideals of simplicity, egalitarianism, memorialisation and a nostalgic yearning for the classical landscape.

Keywords

Rome – Arcadia – Poetry – Myth – Debate

1 The Idea of Arcadia in the Gardens of Rome

The garden played an important role in courtly and academic life in early modern Italy; it was an intimate and restricted space compared to the public streets of a city, allowing for more informal activities. Visitors often shared a tacit acceptance that etiquette could be relaxed and that individuals could take on different identities in the garden. As such, gardens offered a setting for conversation and debate that was different to a palace or a university. Beyond this practical function, gardens also operated on a symbolic level, with members

 | DOI:10.1163/9789004517547_014

drawing associations between early modern academies and Plato's Grove of Academe or the gathering of the Muses on Mount Parnassus.[1] From the fifteenth century onwards, groups of poets and humanists in Rome embraced the symbolism of meeting amongst the ruins and landscape of the ancient city.[2]

The literary Accademia degli Arcadi, founded in Rome in the late seventeenth century, built on these various traditions and actively promoted the idea that pastoral nature as represented by the garden was the ultimate setting for intellectual debate and the performance of poetry. The ideals of the Academy, captured in their laws, histories, poems, and meeting records, were bound up with a nostalgia for the perceived simplicity of a pastoral life, an embrace of egalitarianism and the rejection of class structures, and a belief that a pastoral setting was conducive to the composition of literature and other intellectual activities. The meeting place for their gatherings was a central part of their identity. Acceptance to the Academy (*acclamato*) required the new member to adopt a pseudonym as a shepherd or shepherdess (*pastore* or *pastorella*) and an association with a region of Ancient Arcadia. This "disguise" extended to their meeting place, with each *orti*, *prato* or *giardino* taking on the guise of the Bosco Parrasio, the Parrhasian Woods. Evoking this ideal pastoral place rather than a specific, modern garden or landscape was important to the Arcadians. The names "Arcadia" and Bosco Parrasio recalled the place in ancient Greece described by poets like Virgil and the ideal vision of Arcadia popularised by Jacopo Sannazaro in his sixteenth century text of the same name.[3]

2 'Some Pleasant Place': Accounts of the Bosco Parrasio in the Histories of the Arcadians

The Academy did not have a permanent home until the 1720s. For the first thirty years the peripatetic academy carried their meeting place with them, from the fields near the Castel Sant'Angelo to the slopes of the Aventine Hill. These *boschi* were usually sections of larger gardens of palaces or suburban

1 MacDougall E., "The Sleeping Nymph: Origins of a Humanist Fountain Type" *The Art Bulletin* 57 (1975) 361–363. Brown A., "Defining the Place of Academies in Florentine Culture and Politics", in Everson J.E. – Reidy D.V. – Sampson L. (eds.), *The Italian Academies 1525–1700: Networks of Culture, Innovation and Dissent* (Abingdon – New York: 2016) 22.

2 Christian K.W., "Landscapes of Ruin and the Imagination in the Antiquarian Gardens of Renaissance Rome", in Conan M. (ed.), *Gardens and Imagination: Cultural History and Agency* (Washington D.C.: 2006) 122–123.

3 On Sannazaro and Neapolitan academies and gardens, see George Brocklehurst's essay in this book.

villas belonging to cardinals or noble Roman families that were adapted to the needs of the Academy. Some sites were only used once or twice, while others were occupied for several years. Although these sites are mostly lost, they can be pieced together from written descriptions and a few sketches and engravings.[4] Earlier studies have looked at several of these sites in detail, but it is worth following the Arcadians through their meeting places as they were recorded in the official histories written by the first *custode*, Giovanni Mario Crescimbeni (1663–1728), and the third, Michele Giuseppe Morei (1695–1767).[5] These texts, interspersed with sonnets and verse, describe the physical locations and appearance of the Academy's meeting sites and offer insights into how they were conceptualised by the academicians.

In volume four of Crescimbeni's *L'istoria della volgar poesia* (1730), he describes the role that pastoral sites played in the establishment of the academy. Explaining the origins of the academy, he writes that it had become the custom for himself and others, including the poet Vincenzo Leonio (1650–1720), to go to 'qualche ameno luogo' ('some pleasant place') to share their poetic compositions, listen to each other, and offer 'pronto, e sano giudizio' ('prompt and sound judgement') on the poems presented.[6] The idea for an Academy was established from these informal meetings. Crescimbeni records that this informal group was gathered in a delightful place on the edge of Rome near the Castel Sant'Angelo on a summer's day in 1690 when one of the members wittily suggested that they had 'risorta arcadia' ('resurrected Arcadia').[7] This spontaneous statement inspired Crescimbeni and Leonio to act on the idea for a new Academy that they had been contemplating for some time. It inspired them to associate it closely with the idea of Arcadia and to maintain the custom of meeting out of doors in natural surroundings. The academy would revive the

4 Crescimbeni and Morei often repeat the same descriptions; see for example Crescimbeni Giovanni Mario, *L'istoria della volgar poesia*, vol. 6 (Venice, Lorenzo Basegio: 1730) 218–219; Morei Michele Giuseppe, *Memorie Istoriche dell'adunanza degli Arcadi* (Rome, Antonio de' Rossi: 1761) 63–68. The sketches, possibly by Crescimbeni himself, are in the Biblioteca Angelica, Arcadia MS. 16, folio 393v, 402r.

5 Malfatt S., "Boschi e mecenati: Nuovi documenti sull'Arcadia delle origini (1690–1707)", in Campanelli M. – Petteruti Pellegrino P. – Russo E., *Le accademie a Roma nel Seicento* (Rome: 2020) 239–257; Dixon S.M., *Between the Real and the Ideal. The Accademia degli Arcadi and Its Garden in Eighteenth-century Rome* (Newark: 2006); Minor V.H., "Ideology and Interpretation in Rome's Parrhasian Grove: The Arcadian Garden and Taste," *Memoirs of the American Academy in Rome* 46 (2001) 183–228; Rinaldi A., "Viaggio al Bosco Parrasio. Arcadia Attraverso i Giardini Romani", in Morganti G. (ed.), *Gli Orti Farnesiani sul Palatino* (Rome: 1990) 485–504; Predieri D., *Bosco Parrasio: un giardino per l'Arcadia* (Modena: 1990); Alhaique Pettinelli R., *L'Accademia dell'Arcadia e il Bosco Parrasio* (Rome: 2016).

6 Crescimbeni, *L'istoria della volgar poesia* 218.

7 Ibidem 219.

'dolci studi' ('sweet studies') and 'innocenti costumi' ('innocent customs') of the Arcadians.[8] In his history of the academy written in 1712, *Storia dell'Accademia degli Arcadi istituta in Roma l'anno 1690*, Crescimbeni records that:

> Sogliono ragunarsi gli Arcadi pubblicamente in qualche bosco o prato, sette volte l'anno nella state, cioè dal 1 di Maggio, che il Custode apre, siccome si finge, il Bosco Parrasio, luogo immutabilmente destinato per le Ragunanze, invitando tutti i Pastori sparsi per l'Arcadia a concorrervi, fina a 7 d'Ottobre, che il medesimo Custode chiude il Bosco, e licenzia i Pastori che vi sono concorsi.[9]
>
> The Arcadians gather publicly in some wood or meadow, seven times a year in the summer, that is, from the 1st of May, when the Custodian opens, as is the fiction, the Parrhasian Wood, the place immutably destined for the Gatherings, inviting all the Shepherds scattered throughout Arcadia to attend, until the 7th of October, when the same Custodian closes the Wood, and dismisses the Shepherds who have attended.

The first official meeting took place in the woodland (*selva*) next to San Pietro in Montorio on the Janiculum. Though only used for a brief time as membership quickly grew too large to fit, this first site closely aligned with the ideal of the Bosco Parrasio. It was more of a small woodland than a formal garden and the members had to sit upon the hard earth. Crescimbeni describes this place as a 'nuovo Parnaso' ('new Parnassus') and Morei later recorded the origins of the Academy in a sonnet:

> Qui'nacque Arcadia, in questo Colle, in questa
> Selva trà i sacri boscherecci orrori;
> Il dicon questi Fonti, e questi Allori,
> Che ancor dolce memoria in lor ne resta.
> Quì dove l'erba umìl Teatro appresta,
> Si assisero quei primi almi Pastori.
>
> Here Arcadia was born, on this hill, in this
> Forest, among the sacred, terrible, woods;
> These Fountains and these Laurels say,

8 Ibidem 218–219.

9 Crescimbeni Giovanni Mario, *Storia dell'Accademia degli Arcadi istituita in Roma l'anno 1690 per la coltivazione delle scienze, delle lettere umane e dela poesia* (London, T. Becket: 1804) 6. My translations unless otherwise noted.

> That still sweet memory remains in them.
> Here, where the humble grass Theatre presents itself,
> Those first noble Shepherds sat.[10]

Writing approximately fifty years after the Academy was founded, Morei makes it clear that this first meeting site had itself taken on mythical qualities, reminding the Arcadians of their 'humble' origins. The ideal is not so much one of a formal garden, but rather one of a wild, pastoral landscape with a garden standing in for the 'fiction' of the woods. The academy was also declared a 'repubblica democratica' in the woods next to San Pietro in Montorio.[11] This idea of the academy as a place where all members were equal is key to understanding the importance of the natural setting. Crescimbeni records that the members were attracted by the requirement to disguise themselves as Arcadian shepherds because this enjoyable 'fiction' removed status and put all members on an equal footing. He writes, again in *L'istoria della volgar poesia*, that:

> Come anco allettati dall'amenità, e novità d'andar tutti mascherati sotto la finzione de' Pastori dell'antica Arcadia, dalla quale la conversazione prese il nome; e i Soggetti, che la compongono, Pastori Arcadi si appellarono, e si appellano, con che si venne a togliere ogni riguardo di preminenza, e precedenza tra i Personaggi, che la dovevano formare.[12]

> As well as being allured by the amenity and novelty of being disguised as Shepherds of ancient Arcadia, from which the conversation took its name; and the Subjects, who compose it, called themselves Arcadian Shepherds, and still call themselves that, so that all regard of pre-eminence and precedence was removed from the Personages, who were to form it.

These ideals are also conveyed in a *Discorso* by Benedetto Menzini (his name "in Arcadia" was *Euganio Libade*), which was presented at an early meeting in 1692. His long text addresses the aims of the Arcadians and the significance of their association with the mythical site of Arcadia in ancient Greece. He writes:

> Thus by the name Arcadia, we mean to signify an honourable and civil assembly that, bearing us away occasionally from the popular bustle,

10 Morei Michele Giuseppe, *Poesie* (Rome, Antonio de' Rossi: 1745) 37. 'Orrori' as used in this poem may intend to describe the untamed and wild aspect of the woods rather than indicate that they were simply 'horrible'.

11 Crescimbeni, *L'istoria della volgar poesia* 219.

12 Ibidem 220.

> turns the wooded groves into an academy and puts their innocent pleasures to good use by exercising our minds. Man by his nature cannot remain for long away from pleasure, and whoever does find pleasure in the Good ends up crossing over to the opposite side, deluded by false affections.[13]

Menzini's description alludes to the idea of gardens as a place of retreat away from busy, urban life, a common conceit in the period. He also depicts this retreat into nature for intellectual activities as a type of beneficial pleasure, as opposed to other more indulgent human pleasures. He goes on to liken 'good discourse' to a 'strong plant that truly extends its spacious and diverse branches while connecting them by a single trunk'.[14] Menzini also celebrates the *bassezza* (baseness) of the Arcadians and the way in which they invited people of different backgrounds or simple birth to participate in intellectual debates and composition of poetry.[15] This observation reflected Menzini's own experience, whom Crescimbeni records as having been born to parents that were 'poveri sì, ma onesti' ('poor, yes, but honest'). Crescimbeni presents this condition as an important ingredient in the reform and promotion of poetry and writing, which introduces an interesting role for the garden in an intellectual academy.[16] Traditionally, the aristocratic gardens that had been used by academies in Rome were elite places, more relaxed, perhaps, than aristocratic palaces or the papal court, but not intended to draw in a mix of classes.[17] Both the garden setting and the conceit that it represented rustic Arcadia appear to have allowed a relaxation of strict social protocols and promoted egalitarian principles. This social openness also played a part in allowing the admittance of women to the Academy, first as part of the audience, but increasingly as participants in its intellectual activities.[18]

---

13 Menzini as translated in Dooley B.D., *Italy in the Baroque: Selected Readings* (New York: 1995) 609. The original is in Crescimbeni Giovanni Mario, *Prose degli Arcadi*, vol. 1 (Rome, Antonio de' Rossi: 1718) 112–125.

14 Ibidem.

15 Crescimbeni, *Prose degli Arcadi* 107.

16 Crescimbeni Giovanni Mario, *Le Vite degli Arcadi illustri, Scritte da diversi autori, e pubblicate d'ordine* vol. 1 (Rome, Antonio de' Rossi: 1708) 169.

17 Ehrlich T.L. "Pastoral Landscape and Social Politics in Baroque Rome", in Conan M. (ed.), *Baroque Garden Cultures: Emulation, Sublimation, Subversion* (Washington, D.C.: 2005) 162–3.

18 Norlander Eliasson, S., *Portraiture and Social Identity in Eighteenth-Century Rome* (Manchester: 2009), 112–124 and Giuli, P. "Prospero Lambertini and the Accademia Degli Arcadi (1694–1708)", in Messbarger R. – Johns C.M.S. – Gavitt P. (eds.), *Benedict XIV and the Enlightenment* (Toronto: 2016) 315–316.

The *Leges Arcadum* or Arcadian Laws, thought to be authored by Gianvincenzo Gravina, underlined this ideal of equality. The *leggi* were displayed at the sites used as the Bosco Parrasio and informed the members of their obligations and the standards of behaviour that they were expected to observe both at meetings and in their daily lives. In his exploration of Arcadianism and language, Nicholas Temple observes that these tablets served as a visible reminder of the relationship 'between *otium* and *negotium* cultivated during this time'.[19] The laws reiterate the centrality of pastoral customs and the importance of the locale to maintaining them. The first law 'Pese commune summa potestas esto. Ad idem cuilibet provcare just esto' ('The supreme power shall belong to the community. Every individual shall have a right to resort to the same') reiterates the ideal of egalitarianism. The sixth law, which sets out the conditions of the meetings, emphasises that they will meet in spring and summer ('in the Parrhasian grove in the open air').[20]

In 1696, when the Arcadians were based at the Orti Farnesiani on the Palatine Hill, the Farnese Duke gave the *pastori* two great marble slabs engraved with the Academy's laws. The inauguration of the *leggi* was a major occasion for the Academy during its time at the Orti Palatini. The members gathered for a special event on 20 May 1696 to witness the presentation of these laws, inscribed on marble tablets. The *leggi* were composed in archaic Latin and their inscription in marble was clearly intended to mirror the style of ancient Roman inscriptions (with which they were surrounded on the Palatine).[21]

The presence of a space that could be used as a theatre was another key element of these gardens. Crescimbeni's *Stato della basilica diaconale, collegiata e parrocchiale di S. Maria in Cosmedin di Roma nel presente anno 1719*, which offers a substantial discussion of the Academy, includes descriptions of the meeting places. These descriptions tend to emphasise performances of poetry and composition that took place in the theatres. Although theatres were a common feature of seventeenth-century gardens and had often been associated with academies or academic activities, the first Arcadian theatres were simple

---

19 Temple N., *Architecture and the Language Debate: Artistic and Linguistic Exchanges in Early Modern Italy* (Abingdon – New York: 2020) 194.

20 Anderson James, *The Bee, Or Literary Intelligencer* (London, James Anderson: 1792) 197–199.

21 They are recorded by Gianvincenzo Gravina, who had made the translation and marked the occasion with an oration, also in Latin; see Gravina G., *Orationes et opuscula: quorum series conspicitur post praefalionem* (Utrecht: Gulielmum vande Water, 1713), 234–246. His translation caused a friction between him and Crescimbeni over the ownership of the original composition of the laws; see Minor, "Ideology and Interpretation in Rome's Parrhasian Grove" 185.

spaces 'provided by nature'; the descriptions make a virtue of the need for the members to sit upon the earth.[22] However, over these first few decades, more substantial structures were built specifically to host the Arcadian meetings. In the chapter titled 'Del Teatro degli Arcadi, col ristretto dell'Istoria di questa celebre Adunanza' ('On the Theatre of the Arcadians, with a short history of this celebrated Gathering'), Crescimbeni describes the site for a meeting in August 1699 held in the domestic gardens of Duke Antonio Maria Salviati as 'ove nel pendio d'una Collinetta, fece scavare il Teatro di forma ovale, che nella parte superior aveva tre ordini di sedili, e nell'inferiore uno' ('where on the side of a small hill was carved out an oval-shaped Theatre, that had in the upper section three rows of seats and in the lower section one').[23] The setting at the Giardino Ginnasi, where the academy members were hosted by Prince Francesco Maria Ruspoli from 1712–20, is described as:

> un Anfiteatro in forma di scalea murata di tre ordini di sedili, il più elevato de'quali ha un comodo appoggio, o come suoi dirsi, paraschiena. Il diametro della platea di messo è di palmi sessantotto.[24]

> an Amphitheatre in the form of a walled staircase with three rows of seats, the highest of which has a comfortable backing, or as it is called, a back protector. The diameter of the orchestra is sixty-eight *palmi* [about 15–16 metres].

Crescimbeni goes on to record that the theatre allowed the voices to be heard 'senza stento' ('without difficulty') and had a capacity of three hundred seats – twelve reserved for Cardinals – with space for another hundred to stand. It featured a five-metre-high marble statue of Apollo flanked by a lyre and the pan pipes – the symbol of Arcadia – made of golden metal. The stage was covered with thyme (*serpollo*) to make it seem like a grassy field and the amphitheatre was enclosed by espaliered laurel, presumably to create the sense of a green forest.[25] An engraving from one of Crescimbeni's books shows this site with a substantial stone theatre designed by Girolamo Odam [Fig. 13.1].[26]

---

22 Grant K., "Hedge Theatres of Lucca", in Marshall D.R. (ed.), *Art, Site and Spectacle* (Melbourne: 2007).

23 Crescimbeni G.M., *Stato della basilica diaconale, collegiata e parrocchiale di s. Maria in Cosmedin di Roma nel presente anno 1719* (Rome, Antonio de' Rossi: 1719) 128.

24 Crescimbeni, *Stato della basilica diaconale* 130.

25 Crescimbeni, *Stato della basilica diaconale* 129–131.

26 The engraving is in Crescimbeni, *Stato della basilica diaconale* 131.

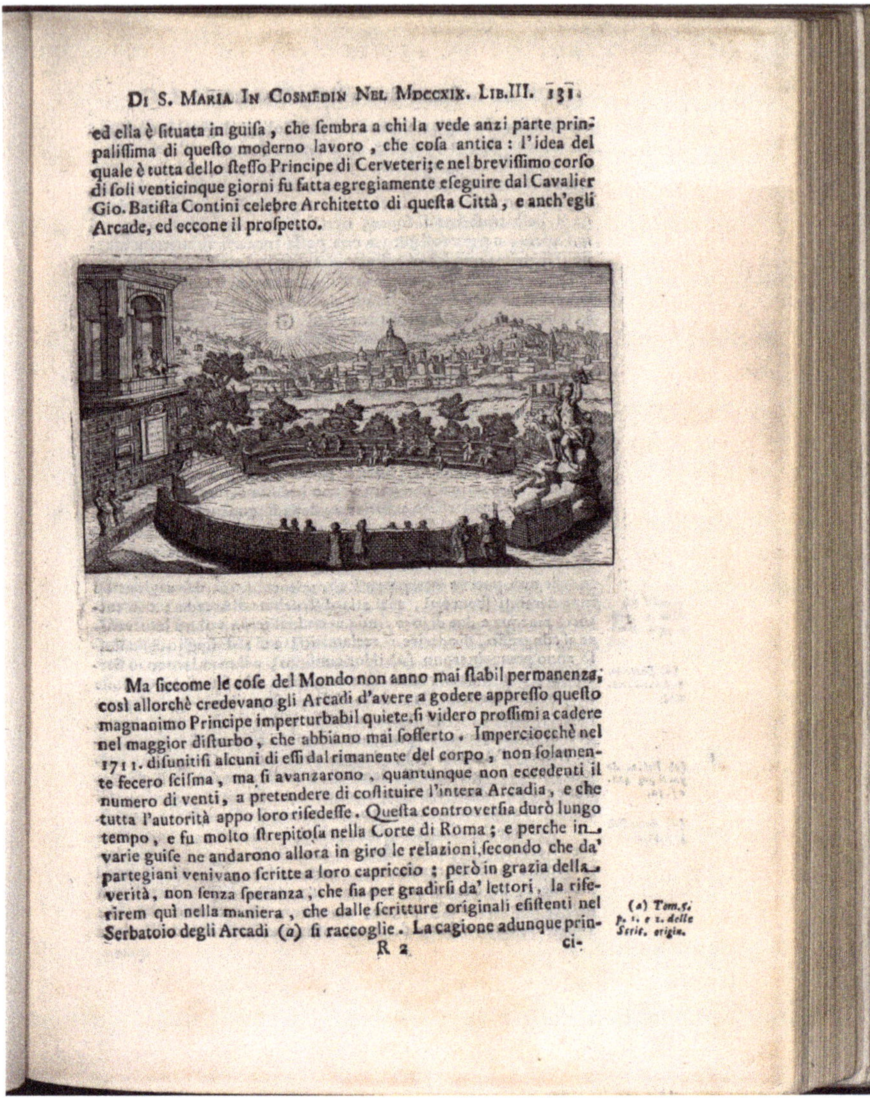

Di S. Maria In Cosmedin Nel Mdccxix. Lib.III. 131

ed ella è ſituata in guiſa, che ſembra a chi la vede anzi parte principaliſſima di queſto moderno lavoro, che coſa antica: l'idea del quale è tutta dello ſteſſo Principe di Cerveteri; e nel breviſſimo corſo di ſoli venticinque giorni fu fatta egregiamente eſeguire dal Cavalier Gio. Batiſta Contini celebre Architetto di queſta Città, e anch'egli Arcade, ed eccone il proſpetto.

Ma ſiccome le coſe del Mondo non anno mai ſtabil permanenza; così allorchè credevano gli Arcadi d'avere a godere appreſſo queſto magnanimo Principe imperturbabil quiete, ſi videro proſſimi a cadere nel maggior diſturbo, che abbiano mai ſofferto. Imperciocchè nel 1711. diſunitiſi alcuni di eſſi dal rimanente del corpo, non ſolamente fecero ſciſma, ma ſi avanzarono, quantunque non eccedenti il numero di venti, a pretendere di coſtituire l'intera Arcadia, e che tutta l'autorità appo loro riſedeſſe. Queſta controverſia durò lungo tempo, e fu molto ſtrepitoſa nella Corte di Roma; e perche in varie guiſe ne andarono allora in giro le relazioni, ſecondo che da' partegiani venivano ſcritte a loro capriccio; però in grazia della verità, non ſenza ſperanza, che ſia per gradirſi da' lettori, la riferirem quì nella maniera, che dalle ſcritture originali eſiſtenti nel Serbatoio degli Arcadi (a) ſi raccoglie. La cagione adunque princi-

(a) Tom. 5. p. 1. e 2. delle Scrit. origin.

R 2

FIGURE 13.1 Girolamo Odam, 'Bosco Parrasio on the Aventine', in Giovanni Mario Crescimbeni, *Stato della Basilica Diaconale, Collegiata e Parrocchiale di s. Maria in Cosmedin di Roma nel Presente Anno 1719* (Rome: 1719) 131

In the course of its first decades, the academy had moved from the Orsini Villa on the Esquiline to the gardens of the Palazzo Riario, followed by the Orti Farnesiani on the Palatine, the Salviati Garden, the Giustiniani Garden at Porta Flaminia, the Ruspoli garden near Santi Pietro e Marcellino and the Ginnasi gardens on Aventine before permanently moving back to the Janiculum in 1726.[27] These histories document the evolution of the ideal vision of the garden, or *bosco*, as well as the gradually increasing size of the meeting places and addition of more substantial permanent structures. The egalitarian nature of the early settings began to shrink in favour of more visible hierarchies with special seats for cardinals. What had begun as an idea that could be transferred to any small piece of nature, had been given a material form in the span of several decades. Although the regular moves of the Arcadian Academy to new sites made these settings ephemeral in some sense, they were accompanied by clear moves to make tangible the symbols of the academy's identity: the theatres, the *Leges Arcadum* engraved in marble and the *lapide* or memorial stones dedicated to deceased members. The Arcadians had invented a type of academic garden, or, if not invented, as many elements can be found in earlier gardens used by academies, then at least codified it. As its physical form evolved from the first meetings on the rough earth, the mythology of an 'Arcadia reborn' and 'layered' upon modern Rome continued to be developed in the writings of the academicians.

3 The Imaginative Geography of Crescimbeni's Arcadia

Crescimbeni's *L'Arcadia* (1708) [Fig. 13.2] functions as a history of the Arcadian Academy written in the form of a pastoral story. Recent studies have described the text as a founding myth for the Arcadians.[28] Crescimbeni explains it as follows:

> Che questa non è una Storia, che contenga minutamente, e per cronologia tutto ciò, che è accaduto dalla fondazione dell'Adunanza fino al preferite; ma bene un racconto di tutte quelle notizie più Angolari, e riguardevoli, che la tessitura dell'Opera ha potuto abbracciare[29]

27 Morei, *Memorie Istoriche dell'adunanza degli Arcadi* 63–74.

28 Tcharos S. *Opera's Orbit: Musical Drama and the Influence of Opera in Arcadian Rome* (Cambridge: 2010) 178.

29 Crescimbeni G.M., *L'Arcadia* (Rome: 1708) np (12 of front matter).

L'ARCADIA

DEL

CAN. GIO. MARIO CRESCIMBENI

Custode della medesima Arcadia, e Accademico Fiorentino.

A MADAMA

ONDEDEI ALBANI

COGNATA DI N. S. PAPA

CLEMENTE XI.

IN ROMA, MDCCVIII.

Per Antonio de' Rossi alla Piazza di Ceri.

CON LICENZA DE' SUPERIORI.

All'Ill.mo Sig. Antonio Magliabechi

L'Autore

FIGURE 13.2 Title page of Giovanni Mario Crescimbeni, *L'Arcadia del can. Gio. Mario Crescimbeni custode della medesima Arcadia, e accademico fiorentino* (Rome: 1708), Biblioteca Nazionale di Firenze

> This is not a History, which contains in minute detail and chronologically all that has happened since the foundation of the Assembly up to the present time; but rather a narrative of all those more interesting facts that the weaving of the Work has been able to encompass.

The text was published in 1708, almost twenty years after the first meeting of the Arcadians and after several years of moving from garden to garden across Rome. Crescimbeni positions it as a retelling or reimagining of Sannazaro's *Arcadia* (1504) and creates a link between that novel, the ancient history of Arcadia in Greece and the activities of the Academy in early eighteenth-century Rome. Crescimbeni and other founders used Sanazzaro's sixteenth-century text from the beginnings of the Academy to create a structure and identity for the organisation and its members, which Sabrina Norlander Eliasson describes as 'disguise transformed into a sociability of universal character'.[30]

The narrative follows a group of nymphs and poets using the pseudonyms of real members of the Academy as they journey towards Mount Elide to attend the Olympic Games. The images presented in this text allow us to understand how the 'Arcadian gardens' and their activities were understood as metaphors, or even as portals to another place and time. An idea that gardens could act as "virtual" worlds, or as evocations of places that were geographically or temporally distant, was popular in this period. Texts describing gardens often invoked the conceit of the visitor or viewer in the garden transported to new places, or having their imagination triggered and inhabiting a different identity.[31] In *L'Arcadia*, Crescimbeni maps Arcadia onto Rome and the gardens are linked specifically to the Parrhasian Woods. Arcadia had come to play this role in poetic thought, echoing Sanazzaro's achievement in superimposing Arcadia onto the landscape around Naples, or Virgil's in separating the idea of Arcadia from the actual place that existed in Greece.[32] Crescimbeni describes his text as a narrative, not a history, and we might now describe it as an exercise in "imaginative geography", though one in which the ideal is projected onto the familiar rather than the foreign or distant. As Denis Cosgrove observes, 'Arcadia's geography is one of yearning more than finding. It is a nowhere place, a utopia. Arcadia addresses the insistent question of the place

30 Norlander Eliasson, *Portraiture and Social Identity in Eighteenth-Century Rome* 109.

31 Conan M., "Friendship and Imagination in French Baroque Gardens before 1661", in Conan M. (ed.), *Baroque Garden Cultures: Emulation, Sublimation, Subversion* (Washington D.C.: 2005) 353, and Hunt J.D., *The Afterlife of Gardens* (London: 2004) 37.

32 On Virgil's evocation of an Arcadian ideal see Jenkyns R., "Virgil and Arcadia", *Journal of Roman Studies* 79 (1989) 26–39.

that humans occupy, should occupy and, in reverie, perhaps once did occupy, in nature'.[33] Arcadia was a place that had existed, and indeed still existed, but for the members of the academy was symbolic of a state of being required for poetic invention.

In *L'Arcadia*, the pastoral disguises taken on by the academicians are extended to the city of Rome itself. Crescimbeni lays out the rationale for the text in the first pages, explaining the complex relationship between the real and the fictional place as follows:

> Che la necessità ha portato, che alle volte Roma si prenda per l'Arcadia, e alle volte per la stessa Roma; e che le Capanne de'Pastori sieno state situate, nella più parte, ove è tornato in acconcio ; e non già ove, secondo l'ordine della Geografia dell'Antica Arcadia, colla quale lì cammina dall'Adunanza nell'assegnare i luoghi agli Arcadi, avrebbero dovuto essere.[34]

> That necessity has brought it about that sometimes Rome is taken for Arcadia, and sometimes for Rome itself; and that the Shepherds' Huts have been sited, in the most part, where it is fitting; and not where, according to the order of the Geography of Ancient Arcadia, with which the Assembly walks in the assigning of places to the Arcadians, they should have been.

Arcadia is Rome and Rome is Arcadia. In the *Libro Primo* of the book, Crescimbeni describes the nymphs arriving at the Bosco Parrasio. They depart at dawn and pass sites from Arcadia itself: the "selvoso Menalo" (the woods of the Mt Mainalo, sacred to Pan) and Mount Lykaia before coming to the Bosco Parrasio. The descriptions make it clear that the entire city has been reimagined as the Arcadian woods in this story. When Crescimbeni references the changing meeting places over the years, he writes that the academicians moved to another 'part of the Wood'.[35] The book's group of travellers cross the landscape of Arcadian Rome, visiting notable members of the Academy at their *capanne* (shepherds' huts) and engaging in sociable and educational encounters. They discuss poetry, mathematics, philosophy, anatomy and other topics. In Book Seven, the group attends a musical 'festa' presented by the shepherd 'Arcomelo'

33 Cosgrove D., *Geography and Vision: Seeing, Imagining and Representing the World* (London: 2012), 70.

34 Crescimbeni, *L'Arcadia* np (14 of front matter).

35 Crescimbeni, *L'Arcadia* 18.

'con una di quelle bellissime Sinfonie fatte nella nobil Capanna dell'Acclamato Crateo, e poi pubblicate al mondo' ('with one of those beautiful Symphonies made in the Hut of the Acclaimed Crateo, and then published to the world').[36] Arcomelo was the pseudonym of the composer and violinist Arcangelo Corelli and Crateo is his patron Cardinal Pietro Ottoboni.[37]

The meeting place is described in detail as a 'campo spazioso' ('spacious field') with a 'boschereccio Teatro' ('woodland theatre') in the centre and the *capanna* or shepherd's hut (also referred to later as the *serbatoio*) next to this.[38] The theatre is circular in form and its size allows for voices to be heard by the assembled audience. It has two levels of simple seats, which are rustic but beautifully 'vestiti d'odorosa Mortella, e di verde Lauro insieme intrecciati' ('dressed in intertwining fragrant Myrtle and green Laurel').[39] Crescimbeni describes pyramids erected in honour of the deceased Arcadians and links them with those created in 1705 in the Giardino Giustiniani. In a scene that draws on Sannazaro and conjures up the seventeenth-century paintings by Nicolas Poussin and Guercino of "Et in Arcadia Ego", the nymphs stop to inspect these tombs and read their inscriptions.[40] They linger over the Sepulchral Urn dedicated to Queen Christina, who, although deceased prior to the formation of the Academy, was nominated as the Protectress of Arcadians and given the title "Basilissa". Thereafter follows a 'funeral' for Christina and a description of the *Serbatoio*, the space where the Arcadians kept their archive and portraits of members.[41]

The nymphs observe some paintings 'rappresentanti diversi rusticani Teatri' ('representing several rustic Theatres') and they inquire as to why the size and magnificence of the meeting place has increased since the founding of the Academy.[42] The character Alessi, the pseudonym of the poet and Canon Giuseppe Paolucci, explains that the shepherds first met in a small field – the garden of San Pietro in Montorio – and that the first 'theatre' was formed by the 'alti frondose alberi' ('tall leafy trees') and the 'verdi fioriti cespugli'

36 Ibidem.

37 A key to the members and their pseudonyms is published in Giorgetti Vichi A.M., *Gli Arcadi dal 1690 al 1800: Onomasticon* (Rome: 1977).

38 Crescimbeni, *L'Arcadia* 6.

39 Ibidem.

40 Poussin's famous painting was by this time in France, but it is likely that Crescimbeni may have seen or known Bellori's description of it; Bellori Giovanni Pietro, *The Lives of the Modern Painters, Sculptors and Architects*, trans. A. Sedgwick Wohl (Cambridge: 2005) 329.

41 Crescimbeni, *L'Arcadia* 11.

42 Crescimbeni, *L'Arcadia* 15–16.

('green flowering bushes') that were naturally growing.[43] There, the members sat uncomfortably on the bare earth; the simplicity of this first meeting place is linked to the founding of the "repubblica". The second meeting place at the Orsini was larger, but it is still described as simply formed by plants and the seats of bare earth. After this, they moved to the grounds of the Palazzo Riario, once home to Queen Christina, and it is here that 'incominciò ad impiegarsi l'arte, per formare il Teatro' ('art began to be employed to form the Theatre'), though still no more elaborate than a simple circular ditch (*fosso*).[44] From here, the shepherds moved to a site 'che aveva regnato l'antico Arcade Evandro, e Palatino era detto, da Pallante figliuolo di lui' ('where the Arcadian Evander had ruled in ancient times, and was called the Palatine, after his son Pallas').[45] Crescimbeni must have drawn on sources like Livy, who had made the link between Evander, King of Arcadia, and the early settlement of Rome.[46] Despite the historical significance of this site, Crescimbeni's observation features a note of pensive reflection that at this new site the 'pristine simplicity had already vanished from our Republic' and that a much more substantial theatre had been built with special seating for the most eminent of members, weakening the equality between the shepherds.[47]

The transformation of the meeting space from bare earth to grander garden theatres symbolises the tension between the ideal of Arcadia and its physical realisation in the gardens of high-ranking churchmen and noble families. Crescimbeni's text portrays these *prati* and *teatri* as sites of learned discussion, with their own laws and an alternative social structure. Rome must be re-presented as Arcadia to create the conditions under which the *pastori* can pursue their art. Yet even in this narrative of the first decade and a half of the Academy's existence, Crescimbeni depicts the golden age as having already passed.

4 The Arcadian Olympics

In the final book of *L'Arcadia*, Crescimbeni takes his travelling brigade of nymphs and shepherds to the celebration of the *Giuochi Olimpici* (Olympic Games). The events described are based on real proceedings staged by the

43 Ibidem 16.
44 Ibidem 17.
45 Ibidem.
46 Livy, *History of Rome*, trans. Rev. Canon Roberts (New York: 1919) 21.
47 Dixon, *Between the Real and the Ideal* 34.

Arcadians at the Bosco Parrasio. The games were first celebrated in 1693 and remained a semi-regular feature of the Academy's activities in the decades that followed. They were modelled on the Ancient Olympic Games held in the city of Elis west of Arcadia and are described in the classical writings of Pindar and others. The Arcadians reimagined the games as a series of poetic and literary contests rather than physical competitions.[48] The *Giuochi* offer insights into the goals of the Arcadian Academy and how garden settings served to enact their program of reform. The games blended serious poetic contest with imaginative fantasy and sophisticated literary play. In the Academy's *adunanze*, or season of meetings, the games fulfilled a number of important roles: they provided a stage for the performance and judgement of literary compositions that aimed to improve poetry and intellectual discourse, they were used to commemorate lost (*defunti*) academicians and increase the reputation of the Academy, and they became a social occasion for negotiating the status of the academicians and the academy with their patrons in papal Rome.

In his *Memorie istoriche dell'Adunanza degli Arcadi* (1761), Morei explains that the Arcadians decided to celebrate the games according to the same calendar as the Ancient Greeks.[49] In Crescimbeni's *L'Arcadia* he describes how the members maintained their disguises as shepherds and shepherdesses of Arcadia and would visit the nearby site of Elis where the ancient games were held.[50] Morei confirms that the Arcadians revived the games after fifteen centuries not as physical competitions, but as games of intellect, 'sostituì ai giuochi destinati all'esercizio del corpo, altri stabiliti ad esercitare l'ingegno' ('replaced the games intended to exercise the body, with games intended to exercise the mind').[51] Both Morei and Crescimbeni's descriptions of these events underline how the Academy's "modern" games were considered an improvement on the original events, which they present as almost primitive in their emphasis on singularly physical prowess.[52]

Held in 1693, the first games were not published but are described in the first volume of the *Atti d'Arcadi* as follows:

> non solamente nel Bosco Parrasio ma per tutta la nostra Arcadia furon celebrati molti festevoli Giuochi ed in essi molti gentilissimi nostri

48 For a detailed discussion of the Giuochi Olimpici see Silvia Tatti, "I Giuochi Olimpici in Arcadia," *Atti e Memorie Dell'Arcadia* 1 (2012) 63–80.

49 Morei, *Memorie istoriche dell'Adunanza degli Arcadi* 214.

50 Crescimbeni, *L'Arcadia* 296.

51 Morei, *Memorie istoriche dell'Adunanza degli Arcadi* 218.

52 Tatti, "I Giuochi Olimpici in Arcadia" 55–6.

> pastori dieron chiarissimi segni non meno della robustezza del corpo che della vivacità della mente, i quali poi dal gentilissimo Alfesibeo Cario [Crescimbeni] nostro custode furon lodati con nobilissime canzoni secondo il loro merito.[53]

> not only in the Parrasio Wood but throughout our Arcadia many festive Jousts were celebrated and in which many of our most gentle shepherds gave very clear signs not less of the robustness of their bodies than of the vivacity of their minds, who were then praised by the most gentle Alfesibeo Cario [Crescimbeni] our custodian with very noble songs according to their merit.

The games used the ancient Pentathlon format of five events, replacing the sporting competitions with poetic reinterpretations. The first published account of the games from 1701 offers a list of those included in the ancient Pentathlon, naming 'il Corso, il Dardo, il Disco, la Lotta, ed il Salto' ('the Foot Race, the Javelin, the Discus, the Wrestling, and the Long Jump').[54] The subsequent description explains how the physical competitions were adapted to showcase poetry and song. The *Corso* was represented by a game called *L'Oraclo* (the Oracle) and the Javelin or *Dardo* was substituted by the *Giuoco del Conteso* (Game of Dispute), in which the shepherds took part in friendly poetic disagreements. The third game, the Discus, was reimagined as the Game of Wits, in which the poets try to best each other and rise to new heights, with the art of poetry described as a game 'in cui per trasceglier gl'ingegni più elevate, si propone il premio a chi colla mente più alto poggiando, vince i suoi emuli con nuove, e bizzare Canzoni' ('in which, to choose the greatest genius, the prize is offered to the one whose mind rises highest, defeating his rivals with new and bizarre songs').[55] The Game of Transformation followed, in which the poets recall the metamorphosis of the ancient Arcadian King Lycaon who was transformed into a wolf by his vices. The fifth and final game, the Garland, was when 'fogliamo tesser leggiadri ferti di frondi, e fiori Poetici per le nostre Pastorelle' ('we weave graceful garlands of foliage, with poetic flowers for our shepherdesses').[56] The women members of the academy were permitted

53 From the *Atti Arcadici* (held in the Biblioteca Angelica) as transcribed in Tatti, "I Giuochi Olimpici in Arcadia" 66.

54 Crescimbeni Giovanni Mario, *I Giuochi Olimpici celebrati dagli Arcadi nell'Olimpiade DCXX* (Rome: 1701) 19.

55 Ibidem 20.

56 Ibidem.

to participate only in this last game, and in 1702 the poetic 'Garlands' performances included one by the poet Petronilla Paolini Massimo, known in Arcadia as Fidalma Partenide.[57]

These reimagined games focused on poetic composition (and competition), but they also seem to draw on the types of serious play and sociability for which gardens in Italy had long been used.[58] In the Game of Transformation – a replacement for wrestling – the Arcadians: 'fingiamo di trasformare il nostro senso in cose molto inferiori alla nobil condizione dell'Uomo' ('[pretended] to transform our senses into things far inferior to the noble condition of Man').[59] This act of literary transformation mimics earlier representations of metamorphosis that had been popular conceits of the garden experience, from sculptural depictions of the story of Daphne and Apollo to the imitation of natural forces through hydraulic machinery.[60] The goal of this game was for the *pastori* to perform the struggle between nature and reason, and Crescimbeni explains that:

> fingiamo di trasformare il nostro senso in cose molto inferior alla nobil condizione dell'Uomo, come sono gli animali bruti, le piante, e i sassi, perchè riconoscendo la sua deformità, quando non vien regolato dalla ragione, possiamo non essolui contrastare, e vincerlo, e domarlo.[61]

> we pretend to transform our mind into things far inferior to Man's noble condition, such as brute animals, plants, and stones, so that by recognising his deformity, when he is not regulated by reason, we may with it counteract him, overcome him, and tame him.

In the 1710 account, this game is described as allowing the shepherds to gain 'il conoscimento della deformità de' vizj, adombrati ne' corpi inferiori a' nostri, ne' quali ci trasformiamo, per abborrirli, e fuggirli' ('the knowledge of the deformity of vices, which are hidden in the bodies inferior to our own, into

57 Crescimbeni, *I Giuochi Olimpici celebrati dagli Arcadi nell'Olimpiade* DCXX 74.

58 Ribouillault D., "Hermeneutics and the Early Modern Garden: Ingenuity, Sociability, Education", in Enenkel K.A.E. – Melion W. (eds.), *Landscape and the Visual Hermeneutics of Place, 1500–1700* (Leiden – Boston: 2020).

59 Crescimbeni, *I Giuochi Olimpici celebrati dagli Arcadi nell'Olimpiade* DCXX 19.

60 Morgan L., *The Monster in the Garden: The Grotesque and the Gigantic in Renaissance Landscape Design* (Philadelphia: 2016) 78.

61 Crescimbeni, *I Giuochi Olimpici celebrati dagli Arcadi nell'Olimpiade* DCXX 20.

which we transform ourselves in order to abhor and flee them').[62] Gardens had long been associated with the enactment of the struggle between animal nature and human reason, serving as a place where people were free to engage in some earthly – albeit contained – pleasures.

The structure of the games remained much the same for the events held in 1705, 1710, 1721, and 1726. The 1705 games were dedicated to the Arcadians who had recently died, described as *defunti* [Fig. 13.3]. They were held near the Giustiniani garden near the Porta del Popolo and featured the addition of memorials, or 'lapidi', mounted on temporary pyramids around the theatre; Crescimbeni describes these in several texts, including his 1705 account of the games.[63] One of the 'Garlands', the 'Ghirlando di Lauro', was also dedicated to the 'Arcadi Defunti' in that year.[64] The memorialising aspect of the games indicates the balanced path between playful entertainment and serious reform that the Academy, led by Crescimbeni, attempted to walk in its early decades.

The *Giuochi* clearly illustrate the role that the garden played as a stage for serious play, one enabling a 'mimesis' of social confrontations, as described by Amedeo Quondam.[65] Sociability and interaction had always been important aspects of early modern academic activity, with debates and discussions often conceived as entertainment. To this end, the use of metaphor and wit enabled academicians to push against the social and intellectual limits of courtly society. In other words, encounters and disagreements could be dressed in the costume of "play" to manage emotions and issues of decorum. In the Arcadian Academy, this practice was most clearly expressed in the game of Dispute (*le Contese*), described as 'amichevolmente contender tra loro, e co' versi pungersi, e faettarsi, per isfogare ogni amarezza, che tenesse occupati gli animi loro' ('friendly quarrels amongst themselves, and to sting and prick each other with verses, in order to dispel any bitterness that may have occupied their minds').[66] Here, the game provides a means first to reveal and then to dispel disagreements. This rhetorical play was not just poetic conceit but seems to

62 Crescimbeni Giovanni Mario, *I Giuochi Olimpici celebrati in Arcadi nell'Olimpiade DCXXII* (Rome: 1710) 13.

63 Crescimbeni Giovanni Mario, *I Giuochi Olimpici celebrati in Arcadia nell'Olimpiade DCXXI. in lode degli Arcadi defunti* (Rome: 1705) 8; see my discussion of the garden as a site of mourning in Grant K., "The Bosco Parrasio as a Site of Pleasure and of Sadness", *Histoire culturelle de l'Europe* (*Jardin et mélancolie en Europe entre le XVIII*ᵉ *siècle et l'époque contemporaine*) (2019).

64 Crescimbeni, *I Giuochi Olimpici celebrati in Arcadia nell'Olimpiade DCXXI* 113.

65 Quondam A., "L'Istituzione Arcadia. Sociologia e ideologia di un'Accademia", *Quaderni Storici* 8.23 (1973) 423.

66 Crescimbeni, *I giuochi Olimpici celebrati dagli Arcadi nell'Olimpiade DCXX* 20.

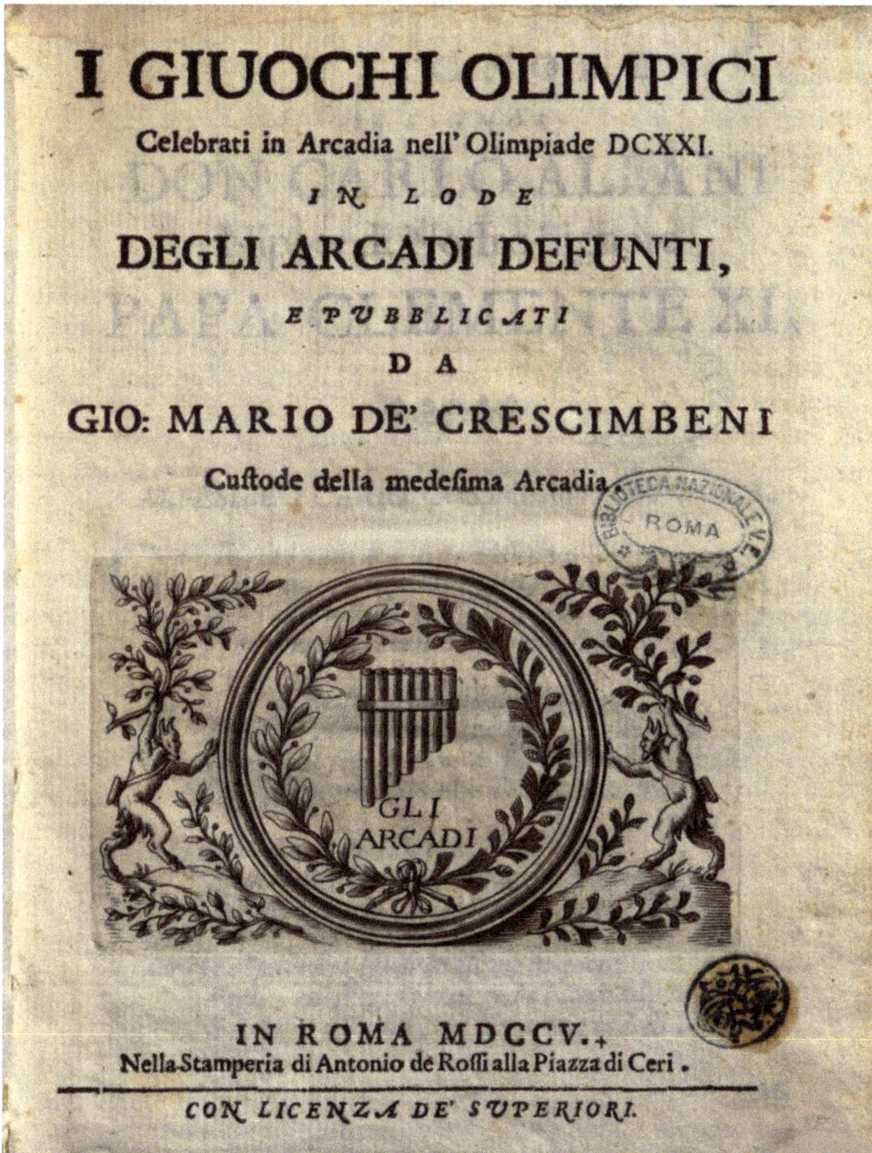
I GIUOCHI OLIMPICI
Celebrati in Arcadia nell' Olimpiade DCXXI.
IN LODE
DEGLI ARCADI DEFUNTI,
E PUBBLICATI
DA
GIO: MARIO DE' CRESCIMBENI
Custode della medesima Arcadia.

IN ROMA MDCCV.
Nella Stamperia di Antonio de Rossi alla Piazza di Ceri.
CON LICENZA DE' SUPERIORI.

FIGURE 13.3 Title page of Giovanni Mario Crescimbeni, *I Giuochi Olimpici celebrati in Arcadia nell'Olimpiade* DCXXI *in Lode degli Arcadi Defunti* (Rome, Antonio de'Rossi: 1705), Biblioteca Nazionale di Roma
IMAGE © INTERNET ARCHIVE

have been a central concern in managing the role of the Academy within the broader Roman community. Tensions between the artistic and literary worlds and those of courtly and ecclesiastical politics were very real. For example, in 1699 the Academy had been forced to move on from the Orti Farnesiani after several satirical verses delivered at a meeting were understood as attacks.[67]

As Silvia Tatti has observed, in 1710, the year before the schism that would split the Academy, Crescimbeni underlined that the Bosco Parrasio was no place for politics. He decried how the academicians had been split by a debate between Pompey and Caesar rather than one between Petrarch and Dante, saying, in other words, that members were debating power factions and not literary style.[68] In the *Direzione* that introduced the published account of the 1710 games, he reiterates the serious side of the event, the different games, and how they shape the objectives of the Academy to reform poetry and support intellectual inquiry. He reminds the reader that the modern games 'surpass' the ancient ones as they are oriented towards training the mind, and that the victors will win immortality through their works and achievements. He also notes the reason for commemorating the dead, 'ma richiamando dalle tombe i compagni defunti della passata Olimpiade, e le loro cospicue azioni' ('recalling from the tombs the deceased companions of the past Olympiad, and their deeds') and highlights how the Arcadians might be led to everlasting life.[69] Finally, Crescimbeni alludes to the egalitarian nature of the academy, reminding readers that the higher-ranked members such as Cleandro (Cardinal Carlo Albani) and Crateo (Cardinal Pietro Ottoboni) have 'scender dall'eminenza del suo altissimo grado, e farsi vostro compagno' ('descended from the eminence of their higher ranks and become your companions'). The games are described as a 'noble' endeavour that shape and regulate the 'Pastoral Republic'.[70] As the academy grew in both size and reputation, clear tensions emerged between the ideal of a democratic republic of writers and poets and the desire from some, like Crescimbeni, for more established patronage and status in Rome and Italy. The garden setting and the Arcadian identity assumed by those within it seems to have played a part in managing this friction.

67 Dixon, *Between the Real and the Ideal* 129 n. 42.

68 Tatti, "I Giuochi Olimpici in Arcadia" 75.

69 Crescimbeni, *I giuochi olimpici celebrati in Arcadia nell'Olimpiade* DCXXII, 13.

70 Ibidem.

5 Conclusion

The tradition of the garden as a liminal space between the real and imagined worlds made it an ideal, and perhaps comfortable, setting in which to adopt a pseudonym and identity as a shepherd or shepherdess and to socialise across a social hierarchy. Although the garden was not the only setting for Arcadian events, it was the primary space where the Arcadian Academy "performed" its ideals. Naturalistic settings seem to have been intentionally chosen for the type of sociability they offered. Elisabetta Graziosi has observed in her study of the role of women in Arcadia that the garden of the Bosco Parrasio and its accompanying *serbatoio* or meeting room was a unique meeting place, more formal than a private salon but less formal than an official embassy or courtly meeting place.[71] The members of the Arcadian Academy in Rome built on a long tradition of using the garden as a setting for sociable debate and education. Although the Academy's earliest garden spaces are largely lost, their writings offer us insight into the extent to which they developed rich narratives and histories to explain the intimate relationship between the intellectual endeavours of an academy and the natural setting of a garden.

References

Åkerman S., "The Forms of Queen Christina's Academies", in Kelley D.R. – Popkin R.H. (eds.), *The Shapes of Knowledge from the Renaissance to the Enlightenment* (Dordrecht: 1991) 165–188.

Alhaique Pettinelli R., *L'Accademia dell'Arcadia e il Bosco Parrasio* (Rome: 2016).

Anderson James, *The Bee, Or Literary Intelligencer* (London: 1792).

Bellori Giovanni Pietro, *The Lives of the Modern Painters, Sculptors and Architects*, trans. A. Sedgwick Wohl (Cambridge: 2005).

Brown A., "Defining the Place of Academies in Florentine Culture and Politics", in Everson J.E. – Reidy D.V. – Sampson L. (eds.), *The Italian Academies 1525–1700: Networks of Culture, Innovation and Dissent* (Abingdon – New York: 2016).

Christian K., "Landscapes of Ruin and the Imagination in the Antiquarian Gardens of Renaissance Rome", in Conan M., *Gardens and Imagination: Cultural History and Agency* 117–37.

71 Graziosi E., "Presenze Femminili: Fuori e Dentro l'Arcadia", in Betri M.L. – Brambilla E. (eds.), *Salotti e Ruolo Femminile in Italia: Tra Fine Seicento e Primo Novecento* (Venice: 2004) 86.

Coffin D., *The Villa in the Life of Renaissance Rome* (Princeton: 1979; Washington D.C.: 2006).

Conan M., "Friendship and Imagination in French Baroque Gardens before 1661", in Conan M. (ed.), *Baroque Garden Cultures: Emulation, Sublimation, Subversion* (Washington D.C.: 2005) 323–384.

Cosgrove D., *Geography and Vision: Seeing, Imagining and Representing the World* (London: 2012).

Crescimbeni Giovanni Mario, *I giuochi Olimpici celebrati dagli Arcadi nell'Olimpiade DCXX* (Rome, Giuseppe Monaldi: 1701).

Crescimbeni Giovanni Mario, *I giuochi Olimpici celebrati in Arcadia nell'Olimpiade DCXXI. in lode degli Arcadi defunti* (Rome, Antonio de' Rossi: 1705).

Crescimbeni Giovanni Mario, *L'Arcadia del can. Gio. Mario Crescimbeni custode della medesima Arcadia, e accademico fiorentino* (Rome, Antonio de' Rossi: 1708).

Crescimbeni Giovanni Mario, *Le vite degli Arcadi illustri, scritte da diversi autori, e pubblicate d'ordine*, vol. 1 (Rome, Antonio de' Rossi: 1708).

Crescimbeni Giovanni Mario, *I giuochi olimpici celebrati in Arcadia nell'Olimpiade DCXXII* (Rome, Antonio de' Rossi: 1710).

Crescimbeni Giovanni Mario, *Prose degli Arcadi* (Rome, Antonio de' Rossi: 1718).

Crescimbeni Giovanni Mario, *Stato della basilica diaconale, collegiata e parrocchiale di S. Maria in Cosmedin di Roma nel presente anno 1719* (Rome, Antonio de' Rossi: 1719).

Crescimbeni Giovanni Mario, *Storia dell'Accademia degli Arcadi istituita in Roma l'anno 1690 per la coltivazione delle scienze, delle lettere umane e della poesia* (London, T. Becket: 1804).

Crescimbeni Giovanni Mario, *L'istoria della volgar poesia* (Venice, Lorenzo Basegio: 1730).

Daston L. – Park K., *Wonders and the Order of Nature 1150–1750* (New York: 2001).

Dixon S.M., *Between the Real and the Ideal: The Accademia Degli Arcadi and Its Garden in Eighteenth-Century Rome* (Newark: 2006).

Dooley B.D., *Italy in the Baroque: Selected Readings* (New York: 1995).

Ehrlich T.L., "Pastoral Landscape and Social Politics in Baroque Rome", in Conan M. (ed.), *Baroque Garden Cultures: Emulation, Sublimation, Subversion* (Washington, D.C.: 2005) 162–163.

Giorgetti Vichi A.M., *Gli Arcadi dal 1690 al 1800: Onomasticon* (Rome: 1977).

Giuli P., "Prospero Lambertini and the Accademia Degli Arcadi (1694–1708)", in Messbarger R. – Johns C.M.S. – Gavitt P. (eds.), *Benedict XIV and the Enlightenment* (Toronto: 2016) 315–340.

Grant K., "Hedge Theatres of Lucca", in Marshall D.R. (ed.), *Art, Site and Spectacle* (Melbourne: 2007).

Grant K., "The Bosco Parrasio as a Site of Pleasure and of Sadness", *Histoire culturelle de l'Europe* (*Jardin et mélancolie en Europe entre le XVIII*e *siècle et l'époque contemporaine*) (2019) https://mrsh.unicaen.fr/hce/index.php_id_1254.html.

Graziosi E., "Presenze femminili: Fuori e dentro l'Arcadia", in Betri M.L. – Brambilla E. (eds.), *Salotti e ruolo femminile in Italia: Tra fine Seicento e primo Novecento* (Venice: 2004) 67–96.

Hunt J.D., *The Afterlife of Gardens* (London: 2004).

Jenkyns R., "Virgil and Arcadia", *Journal of Roman Studies* 79 (1989) 26–39.

Livy, *History of Rome*, trans Rev. Canon Roberts (New York: 1919).

MacDougall E., "The Sleeping Nymph: Origins of a Humanist Fountain Type", *The Art Bulletin* 57 (1975) 357–365.

Malfatti S., "Boschi e mecenati: Nuovi documenti sull'Arcadia delle origini (1690–1707)", in Campanelli M. – Petteruti Pellegrino P. – Russo E. (eds.), *Le accademie a Roma nel Seicento* (Rome: 2020) 239–257.

Minor V.H., "Ideology and Interpretation in Rome's Parrhasian Grove: The Arcadian Garden and Taste", *Memoirs of the American Academy in Rome* 46 (2001) 183–228.

Morei Michele Giuseppe, *Poesie* (Rome, Antonio de' Rossi: 1745).

Morei Michele Giuseppe, *Memorie istoriche dell'adunanza degli Arcadi* (Rome, Antonio de' Rossi: 1761).

Morgan L., *The Monster in the Garden: The Grotesque and the Gigantic in Renaissance Landscape Design* (Philadelphia: 2016).

Norlander Eliasson S., *Portraiture and Social Identity in Eighteenth-Century Rome* (Manchester: 2009).

Predieri D., *Bosco Parrasio: Un Giardino per l'Arcadia* (Modena: 1990).

Quondam A., "L'Istituzione Arcadia. Sociologia e ideologia di un'accademia", *Quaderni Storici* 8.23 (1973) 389–438.

Ribouillault D., "Hermeneutics and the Early Modern Garden: Ingenuity, Sociability, Education", in Enenkel K.A.E. – Melion W. (eds.), *Landscape and the Visual Hermeneutics of Place, 1500–1700* (Leiden and Boston: 2020) 291–325.

Rinaldi A., "Viaggio al Bosco Parrasio. Arcadia Attraverso i Giardini Romani", in Morganti G. (ed.), *Gli Orti Farnesiani sul Palatino* (Rome: 1990) 485–504.

Tatti S., "I Giuochi Olimpici in Arcadia", *Atti e Memorie Dell'Arcadia* 1 (2012) 63–80.

Tcharos S., *Opera's Orbit: Musical Drama and the Influence of Opera in Arcadian Rome* (Cambridge: 2010).

Temple N., *Architecture and the Language Debate: Artistic and Linguistic Exchanges in Early Modern Italy* (Abingdon – New York, 2020).

CHAPTER 14

The Milanese Colony of the Arcadian Academy

Maria Cristina Loi

Abstract

The Milanese colony of the Arcadian Academy, established in the city in 1704 at the behest of Father Gian Antonio Mezzabarba, found the ideal place for its meetings in the garden of Count Carlo Pertusati's residence. Serviliano Latuada describes its layout, with a central path lined by tall trees, a system of symmetrical flower beds, fountains and sculptures. The statue of Hercules in the act of killing the lion of Nemea gave the place the name "Herculean Garden". During the warm months, dances, literary meetings and banquets were held there. Today the Arcadian Garden, the most important example of this typology in eighteenth-century Milan, still bears traces of its past, albeit profoundly modified. Through the documentation in the Milanese archives, it is possible to piece together the life of the Academy and its headquarters, and the role it played in the Milan of that time.

Keywords

Milan – Arcadia Colony – Gardens – Carlo Pertusati – Gian Antonio Mezzabarba

> [...] una Casa Nobile, che a ragione chiameremo Albergo delle Muse, radunandosi ivi l'Accademia degli Arcadi, ed abitandovi co'l suo Padrone le Muse stesse, potendosi ciò dire propriamente di Sua Eccellenza il Sig. Conte Presidente e Gran Cancelliero Don Carlo Pertusati [...] Qui il medesimo Signor Conte Presidente ha fatto disporre un vaghissimo Giardino, ornato de' più odorosi e rari fiori, con alte piante di Cedri ed Agrumi, per la conservazione de' quali dalle ingiurie del Verno, vi si fabbrica a posticcio una Casa di legno, sì bene architettata e connessa, che si concilia l'ammirazione di chiunque la vede. Questo Giardino da una Statua di Ercole in atteggiamento di uccidere il Lione, si denomina Erculeo, del quale se ne aggiunge delinato il disegno, e qui si uniscono in tempo di State, per recitare i loro dotti Poetici Componimenti, gli Arcadi Pastori della Colonia Milanese stabilita nell'anno 1704.

 | DOI:10.1163/9789004517547_015

> [...] A noble house, which we shall rightly call the 'Albergo delle Muse', where the Accademia degli Arcadi meets, and where the Muses themselves live there with their master, this being properly said of His Excellency the Count President and Grand Chancellor Don Carlo Pertusati [...] Here the same Count President has laid out a very beautiful garden, adorned with the most fragrant and rare flowers, with tall cedar and citrus trees, to which, to protect it from the harshness of Winter, a wooden house has been added, so well designed and built that it wins the admiration of everyone who sees it. This garden is called Herculean because of a statue of Hercules in the attitude of killing the lion, of which we show a drawing, and the Arcadian shepherds of the Milanese colony, founded in 1704, meet here in summer to recite their learned poetic compositions.[1]

In his famous *Description of Milan*, Serviliano Latuada devotes a few pages and one of the rare images in his work to the seat of the Milanese colony of the Arcadi, in the garden and palace of Count Carlo Pertusati. The plate depicts a garden delimited by a rectangular perimeter and organised according to a regular design. A central axis serves as the dominant element that orders the composition, directing the symmetrical arrangement of other elements. A first alley flanked by two rectangular flowerbeds with trees on each corner leads to what appears to be the entrance to the garden, which is marked by two small pillars surmounted by large vases. From here, a wide central avenue develops continuing from the axis of the first area, punctuated by a series of regularly spaced trees on either side. Beyond the rows of trees, rectangular beds of varying sizes present a geometric pattern defined by the flowers and plants. The tree-lined avenue culminates in a semi-circular niche that houses the statue of Hercules in the act of killing the Nemean lion mentioned in the text [Fig. 14.1].

1 Latuada Serviliano, *Descrizione di Milano ornata con molti disegni in rame delle fabbriche più cospicue, che si trovano in questa metropoli. Raccolta e ordinata da Serviliano Latuada, sacerdote milanese*, vol. 2 (Milan: Cairoli, 1737) 332–333. The original edition consists of five volumes published between 1737 and 1738 in Milan with forty-eight separate plates by Girolamo Ferroni and Johan Georg Seiller.

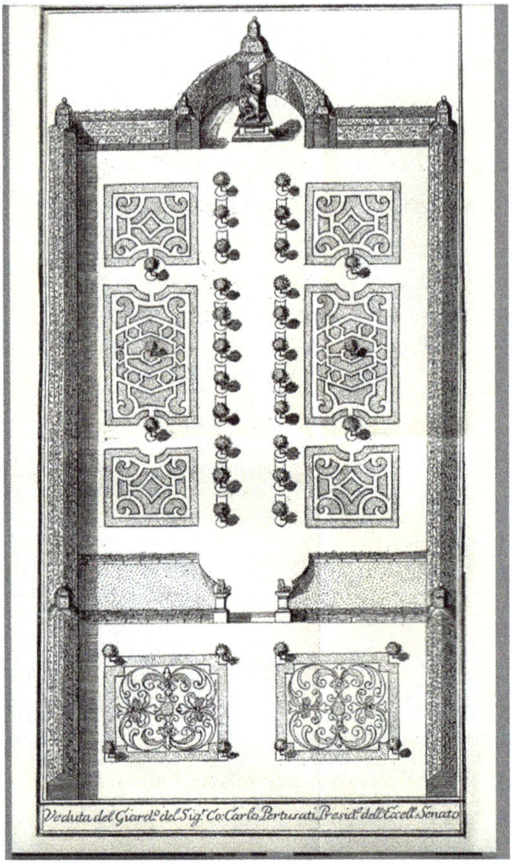

FIGURE 14.1
Serviliano Latuada, *Veduta del Giard°* *del Sig.r Co Caro Pertusati.Presid.e dell'Eccell. Senato*, 1737–1738 (*Descrizione di Milano* […] II, MDCCLI)
IMAGE © MARIA CRISTINA LOI

Latuada's brief but clear text allows us to imagine the garden maintained by the Milanese Arcadi colony and to appreciate its distinguishing features. It is impossible to know with certainty the level of accuracy of this description,[2] whether the image is an effective rendering of the actual garden, or if both are idealised images.[3]

2 Latuada, *Descrizione* II, 332–333. The philologist and prefect of the Ambrosiana, Giuseppe Antonio Sassi, wrote a detailed description of the Milanese Arcadia. See Sassi Giuseppe Antonio, "Arcadum Mediolanensium Colonia", in Argelati Filippo, *Philippi Argelati Bononiensis Bibliotheca scriptorum Mediolanensium, seu acta et elogia virorum omnigena eruditione illustrium* […] *Preamittitur Clarissimi Viri Josephi Antonii Saxii* […] (Milan, In aedibus Palatinis: 1745) XIV, 2, LXII–LXIII.

3 However, it should be remembered that this same description has been handed down over time and that subsequent studies on the Milanese Arcadia, even the most recent ones, refer to this plate. It is also possible to state that Latuada's plates, in comparison with others

The comparison between this plate and Antonio Canevari and Nicola Salvi's famous drawings for the Bosco Parrasio on the slopes of the Gianicolo offers immediate evidence of the profound differences that characterized the two sites of the academy both in terms of physical appearance and use.[4] A simplified environment, the Milanese space represents a reduced scale version of the magniloquent project conceived for the Roman seat. However, many of the characteristics typical of the Arcadian seats are still present and here appear to have been adapted according to the specifics of the Milanese site: a flat, green area full of water, gardens, and "ortaglie" (vegetable gardens). Numerous interventions transformed the area during the nineteenth and the twentieth centuries; a large and irregular trapezoidal lot, private garden, and parcel of buildings constructed in the 1920s and after the Second World War are what is left of the eighteenth-century settlement of the area. It seems that, despite these interventions, the intention was to protect this historical green island and its name, an objective that has been achieved, even if only partially.[5]

The Milanese colony (or "Insubrica") of the Accademia dell'Arcadia settled in Milan in 1704. Father Giovanni Antonio Mezzabarba (Vitanio Gateate) (1670–1705), a member of a noble decurion family from Pavia and regular cleric of the Somaschi congregation, celebrated its inauguration on May 2nd with the ode *All'Arcadia* dedicated to Cardinal Benedetto Pamphili, a member known

dedicated to documented or still existing buildings, are fairly detailed and truthful. This is the case, for example, of the representation of the Palazzo Durini or the Casa degli Omenoni.

4 For an overview of the central themes related to the establishment of the Arcadia Academy and the debates developed among its protagonists, see the studies of Amedeo Quondam. Quondam A., "L'istituzione Arcadia. Sociologia e ideologia di un'Accademia", *Quaderni Storici* 8.23 (1973) 389–438; Quondam A, "Problemi di critica arcadica", in *Critica e storia letteraria. Studi offerti a Mario Fubini*, vol. 1 (Padua: 1970) 515–523. For further and updated information on specific aspects of this academy, see the historic publication of the Academy, "Atti e Memorie dell'Arcadia". It is impossible here to account for the extensive and in-depth bibliography on the Arcadia in Rome. On the architectural project of the Bosco Parrasio on the Gianicolo Hill, see Dixon S.M., *Between the Real and the Ideal. The Accademia degli Arcadi and its Gardens in Eighteenth century Rome* (Newark: 2006) and bibliography. See also Katrina Grant's essay in this book. On the Milanese colony see Seregni G., "La cultura milanese del Settecento", in *Storia di Milano. L'età delle riforme (1706–1796)* vol. 12 (Milan: 1959) 569–586 and the more recent Baragetti S., "Colonie d'Arcadia nella Lombardia austriaca", in Cofano D. – Valerio S., *La letteratura degli Italiani. 1. Centri e periferie* (Foggia: 2011) 1–8.

5 The current limits of the block – corso di Porta Romana, corso di Porta Vigentina, via Quadronno and via San Calimero – do not exactly coincide with the original ones. Nevertheless, they remain the main references to follow the evolution of the area. Further research is being carried out in the archives of the Ospedale Maggiore and the Ospedale Gaetano Pini in Milan in order to define more precisely all the changes of ownership – and the consequent changes in the size and boundaries of each property – that have taken place over the last two centuries.

since 1695 by the name Fenicio Larisseo.[6] The colony, one of the five founded in Lombardy between 1704 and 1722, included the most important representatives of Milan's cultural life, some of whom were already affiliated with the Roman Arcadia. In addition to its founder, members included Carlo Ambrogio Mezzabarba (Eriphilus), patriarch of Antioch and apostolic legate in China, Giuseppe Maria Stampa (Euristeo Parebasio), the future cardinal Giuseppe Pozzobonelli (Vitalgo Pitiuseo), the count Gabriele Verri (Elsindo Sferiano); the Jesuit Tommaso Ceva (Callimaco Meridio), and the scholar and writer Francesco Saverio Quadrio (Ornisco Ippocreneo).[7] Many were members of the Pertusati family, most notably Count Carlo Pertusati (1674–1755) (Pelindo Anigreo, or Anigreio?), President of the Milanese Senate since 1733 and the key figure to whom the history of the colony is inextricably linked. The cenacle in fact met at his residence in Porta Romana in the parish of San Calimero.[8]

At the foot of the statue of Hercules, on the western side of today's garden, an inscribed plaque quotes from a manuscript by Abbot Minola, kept in the Ambrosiana library: 'Nel MDCCIIII per opera di Gio Antonio Mezzabarba fu istituita la prima sede degli Arcadi Milanesi nella casa e giardino detto Orto Erculeo dell'Ecc.ma Famiglia Pertusati' ('In 1704, the first seat of the Milanese Arcadians was established by Gio Antonio Mezzabarba in the house and garden known as the "Orto Erculeo" of the Ecc. Pertusati family').[9] According to the testimony of Giuseppe Antonio Sassi, the Arcadians previously had met in two other seats: 'in Trivultii Principis palatio primum excepta, dein ad Comitios Montii' ('First they stayed at Prince Trivulzio's palace, then at the Counts of

6 On Giovanni Antonio Mezzabarba, see the profile in Crescimbeni, *Notizie Istoriche* II, 291–295. See also Mezzabarba Giovanni Antonio, *Discorso di Vitanio Gateatico pastore d'Arcadia in difesa dell'"Endimione" favola pastorale di Arezio Gateatico [Francesco de Lemene]; indirizzato a Cromiro Dianio [Pietro Antonio Bernardoni] suo Compastore* (Turin, Zappata: 1699).

7 For a list of Milanese Arcadi in the early years, see Ceva Tommaso, *Memorie d'alcune Virtu' del Signor Conte Francesco de Lemene* [...] (Milan, Per Domenico Bellagatta: 1718). The complete list of the Arcadi, updated to 1977, is in Giorgetti Vichi A.M. (ed.), *Gli Arcadi dal 1690 al 1800. Onomasticon*, Arcadia – Italian Literary Academy (Rome: 1977).

8 Cremonini C., "Luca Pertusati", in *Dizionario biografico degli italiani* 82, 2015.

9 Minola Diego Antonio, *Diario Storico-Politico, di alcuni avvenimenti nel secolo XVIII, tomo I, dal 1701 al 1729 inclusivo*, Milan, Biblioteca Ambrosiana Ms. G.123. Minola devotes a few more lines to the establishment of the Academy and the complete citation reads: 'In quest'anno (1704) fu istituita la prima sede degli Arcadi Milanesi nell'eccellentissima casa Pertusati, da Padre Gian Antonio Mezzabarba, chierico regolare somasco, [composta] da vari eruditi personaggi che in tempo di estate recitavano nel giardino li componimenti'. ('In this year (1704) the first seat of the Milanese Arcadi was established in the most excellent Pertusati house, by Father Gian Antonio Mezzabarba, a regular Somascan cleric, [composed] of various learned people who in summertime recited their poems in the garden').

FIGURE 14.2 Marcantonio Dal Re, *Città di Milano*, c.1734. The site of Palazzo Trivulzio and Palazzo Monti; the site of the Giardino d'Arcadia (Milano, Civica Raccolta delle Stampe Achille Bertarelli)
IMAGE © MARIA CRISTINA LOI

Monti's').[10] This citation references two palaces overlooking the Naviglio on opposite banks, both clearly visible in eighteenth-century maps of the city; one was owned by Prince Antonio Gaetano Gallio Trivulzio and the other by Cardinal Monti's heir, Count Cesare Monti [Fig. 14.2].[11] However, the exact chronology of changes to the various seats is uncertain. An extended period of silenced activity followed the death of Mezzabarba in 1705 until April 1715, at which point the cenacle resumed meeting at the palazzo Monti on the

10 Sassi, "Arcadum Mediolanensium Colonia" LXIII.

11 In the eighteenth century, the older structure of Palazzo Monti was modified with two interventions. The first was a redesign of the main façade, carried out by the architect Francesco Croce (1736), and the second concerned the façade overlooking the garden, based on a project by Benedetto Alfieri.

occasion of the end of Lent.[12] This was the second temporary headquarters before their settlement in the Pertusati palace, approximately datable to the 1720s. At the time, the colony had about thirty members.[13]

An examination of the close relationships that linked the three host families, all of whom were in many ways participants, if not protagonists, of the Milanese cultural scene, benefits our understanding of this sequence of changes in meeting places. At the turn of the century, the owners belonged to the intellectual milieu involved in the project of renewing and reforming the arts and sciences; it should also be noted that the Pertusati and Trivulzio were united by marriage alliances.[14] Furthermore, Carlo Pertusati's desire to host Arcadia meetings at his residence seems to have been part of a larger political strategy to consolidate the Alessandrian family's position in Milanese aristocracy. This influence was further secured by his membership in the Palatine Society and, together with other illustrious Arcadians and men of letters, to the group of supporters known as "cavalieri amanti delle lettere", who participated in Ludovico Muratori's (1672–1750) project for a new 'Repubblica delle Lettere' ('Republic of Letters').[15] Although the Milanese colony followed the programme of the Roman seat and its members belonged to or were aware of other contemporary cultural institutions, its activity was primarily focused on and dedicated to the Milanese milieu.[16]

The palazzo in Porta Romana, which was owned by the Ospedale Maggiore, had been rented by Carlo's father, Luca Pertusati, since 1697. In the first half of the century, probably beginning in the 1720s, Carlo Pertusati carried out important transformations to the family residence and adjacent garden. Both

12 Cremonini, "Luca Pertusati", does not specify the date.

13 Baragetti, "Colonie d'Arcadia".

14 Squizzato A., "Carlo e Alessandro Teodoro Trivulzio fra i cultori del buon gusto e della scienza dell'antico", in Bianchi E. – Rovetta A. – Squizzato A. (eds.), *Le arti nella Lombardia asburgica durante il Settecento. Novità e aperture*, Atti del convegno, Milano Università Cattolica del Sacro Cuore – Pinacoteca di Brera, 5–6 giugno 2014 (Milan: 2017) 409–423.

15 On the crisis that arose within the Arcadian community and the schism of 1711 and on the Palatine Society see Quondam, "L'istituzione Arcadia"; id, "Problemi di critica arcadica" and Vischi L., "La Società Palatina di Milano. Studio storico di L. Vischi", *Archivio Storico Lombardo* 7.3 (1880) 391–542; Cremonini C., "L.A. Muratori e la Società Palatina. Considerazioni su cultura e politica a Milano tra Sei e Settecento", in Bona Castellotti M. – Bressan E. – Vismara P. (eds.), *Politica, vita religiosa, carità. Milano nel primo Settecento* (Milan: 1997) 185–212. See also Kapp V., "Muratori e l'idea della Repubblica letteraria d'Italia", *Romanische Forschungen* 114.2 (2002) 191–205 and related bibliography.

16 Moreover, in the *Rime degli Arcadi*, the official collection of works by the Arcadi, only a few are by members of the Milanese colony. See Baragetti, "Colonie d'Arcadia" and the bibliography there indicated.

were purchased in 1721 upon the death of his father, at which time he could draw on his wife Lucrezia Garuffi's dotal assets.[17] Little confirmed information pertaining to the development of this vast green area exists beyond Latuada's description, nor do we know the names of the designers involved in the project. However, historical cartography is suggestive to this end. In the 1700s and stretching through the nineteenth century, this portion of the city was characterised by a low density of constructions, the majority of which were religious buildings and gardens that flourished due to the abundance of city water. The irregularly shaped block was defined by the churches of San Calimero and Santa Maria al Paradiso and by the architecture facing the historic axis that connected the Porta Romana arch built at the end of the sixteenth century with the monumental centre of the city.

More secluded than the two prior seats of the colony, the building overlooked the street with an austere front; the regular rhythm of openings on the main floor were interrupted only by three simple balconies, the central one aligning with the portal [Fig. 14.3]. Little documentation remains on the history of the building, which was demolished after the Second World War. In the *Nuova guida di Milano*, Carlo Bianconi includes it among the palaces of the "sestiere" of Porta Romana, referring only to the 'bella raccolta d'opere di insigni Pittori meritevole della osservazione degl'intelligenti' ('A fine collection of works by eminent painters worthy of the attention of the learned') without providing further information.[18] To this account, we can also add the useful, albeit concise, description of the palace by Paolo Mezzanotte and Giacomo Bascapè published immediately before its demolition.[19] The access vestibule was the most characteristic element of the structure: a rectangular colonnade leading to the garden [Fig. 14.4].[20] While we do not know the details of the internal distribution, old photographs prove the existence of a large room with a fireplace and a large number of paintings on its walls [Fig. 14.5]. One photograph speaks

17 Archivio dell'Ospedale Maggiore, Patrimonio Attivo, cart. 554, Case e Poderi, Milan, Corso Roma. Act dated 30.06.1721. The Palazzo, donated by Girolamo Carcano, was owned by the Ospedale Maggiore of Milan.

18 In the few lines dedicated to the building, Bianconi also mentions an architectural intervention by the Count but offers no further evidence. Bianconi Carlo, *Nuova guida di Milano, per gli amanti delle belle arti e delle sacre e profane antichità milanesi* (Milan, nella Stamperia Sirtori: 1787) 133. See also Bossi L., *Guida di Milano, ossia Descrizione delle città e dei luoghi più osservabili ai quali da Milano recansi i forestieri*, vol. 1 (Milan: 1818) 72.

19 Mezzanotte P. – Bascapè G., *Milano nell'arte e nella storia* (Milan: 1948) 508.

20 The demolition of the building, opposed by the scholar Giacomo Bascapè, who called for its recovery, was completed between 1949 and 1950. See the letter by Bascapè in SbeapMi, Archivio Monumenti, cart. FF/16280/2, 10.19.1949.

FIGURE 14.3 Milan, *Palazzo Pertusati*, façade, early 20th century, SbeapMi. Archivio Fotografico, cart. 850
IMAGE © MARIA CRISTINA LOI

FIGURE 14.4 Milan, *Palazzo Pertusati*, vestibule [1945–1949], SbeapMi. Archivio Fotografico, cart. 850
IMAGE © MARIA CRISTINA LOI

FIGURE 14.5 Milan, *Palazzo Pertusati*, main salon [early 20th century?], unknown source

to Pertusati's passion for the arts, as highlighted by Bianconi, and his proclivity for collecting paintings, coins and books. Indeed, the presence of a large library in the palace is also documented. Latuada states that: 'Ne' solamente pregievole è questo Palazzo pe' giardini, e le rare Pitture, che lo adornano; il più nobile vanto gli reca la unione copiosa di sceltissimi Libri [...]' ('Not only is this palace valuable for its gardens and the rare paintings that adorn it, but its noblest treasure is the abundant collection of precious books [...]').[21]

Carlo Pertusati's collection of books numbered over twenty-four thousand volumes; he organised them into distinctive themed rooms: Ecclesiastical, Historical, Human Letters and the so-called Legal Room.[22] Such a library

21 Latuada, *Descrizione* II, 333.

22 A manuscript catalogue of the Pertusati library (incomplete after the letter F) is in the Braidense Library: *Librorum Bibliothece Regie Mediolanensis, alias Pertusate, Catalogus Auctorum Alphabeticus* (ms. Arm. 55–65). On the library of Carlo Pertusati see Turchetti Grossi M.L., "Brevi note sul Fondo Pertusati della Braidense", *Accademie e Biblioteche d'Italia* 47, 5, (1979) 370–386; Turchetti Grossi M.L. – Gaspari G., "Il fondo Pertusati", in *La Braidense. La cultura del libro e delle biblioteche nella società dell'immagine* (Florence: 1991) 88–89; Zumkeller L., "Da biblioteca di carattere generale a biblioteca storico-umanistica: la Braidense attraverso una rilettura della storia dei fondi", *Accademie e Biblioteche d'Italia* 67. 2 (1999) 42–43. Grossi Turchetti M.L. (ed.), *Biblioteca Nazionale Braidense, I manoscritti datati della Biblioteca Nazionale Braidense di Milano* (Florence: 2004)

containing the most recent and up-to-date studies, a large number of manuscripts and incunabula, and rare editions of the Bible, was among the most important in Milan at the time and attracted many visitors on the *Grand Tour*. It contributed to its owner's fame and augmented the reputation of the palace as an important centre for literates, artists, and clergymen. This rich collection also included classical literature, studies on the history of Milan and Lombardy and volumes on natural history, including Pietro de Crescenzi's *Commoda ruralia*.[23] Following the count's death, his son Luca Pertusati sold the entire collection. About eighteen thousand of the collected volumes were incorporated into the Braidense Library, which, together with the Jesuit legacy, formed its original nucleus, and into the Library of the University of Pavia.[24]

Returning to the subject of the garden, we know, based on Latuada's studies and through an analysis of historical maps, that the garden represented a not particularly large portion of a sizeable green area reaching and even surpassing the limit of the Spanish walls [Fig. 14.6]. Giovanni Filippini's *Iconography of the City and Castle of Milan* offers a general understanding of the state of this area [Fig. 14.7a]. An area lapped by water with orderly gardens and without many buildings is already visible in the document. Despite its vagueness, the Filippini map allows the identification of a series of flowerbeds, neatly arranged around an axis and off-centre from the entrance vestibule to the palace. There may also have been an exedra, although we cannot be more specific. In his map of Milan published in 1737, Latuada did not indicate a building or garden on the site. Marcantonio Dal Re's 1734 plan, although more detailed and informative, does not offer any additional elements that might allow us to follow the progress of works on the site. It does enable us, however, to identify more clearly the relationship between the built areas, the voids (courtyards) and underdeveloped green areas. Only at the beginning of the nineteenth century in Giacomo Pinchetti's and the Astronomi di Brera's maps does a more defined and detailed picture of the site emerge [Fig. 14.7b]. However, the garden had already undergone drastic changes by this time. Acquired by the Melzi, it was transformed into an English garden informed by the most recent

---

3–4. Fossati F., "La catalogazione del fondo degli incunaboli della Biblioteca Nazionale Braidense", *La Bibliofilia* (2020), 167–174, See also Baretta G., "La Biblioteca del conte Carlo Pertusati", in Sciardelli F. (ed.), *Tra i fondi della Biblioteca Braidense* (Milan: 1993) 11–15, http://www.braidense.it/risorse/tra-i-fondi-della-biblioteca-braidense-baretta.pdf.

23 Pietro de' Crescenzi, *Commoda ruralia* (Speyer, Peter Drach: 1495). See Grossi Turchetti M.L., *Gli incunaboli del Fondo Pertusati in Braidense*, http://www.braidense.it/risorse/turchetti_pertusati.pdf.

24 On the Pertusati Fund of the Braidense Library see above, note 22. For Pavia, see the catalogue of the Pertusati Fund at the University Library of Pavia.

FIGURE 14.6 Milan, *Palazzo Pertusati, view toward the palazzo from the garden* [1920s?] SbeapMi. Archivio Fotografico, cart. 850
IMAGE © MARIA CRISTINA LOI

developments in the art of gardens and inspired by the studies of Ercole Silva. At this juncture, its story as the meeting place for the Arcadi of the Milanese colony ends.[25] The Milanese colony ceased its activities in the early years of the third decade of the century, and many of its members joined the Accademia dei Trasformati. The last years were marked by an 'Adunanza' that took place in

25 On the subsequent modifications of the garden, see the report *Il Giardino d'Arcadia in Milan. Protection, Urban Planning, and Reconstruction*, Ministero dei Beni e delle Attività Culturali e del Turismo, Soprintendenza Belle Arti e Paesaggio per le provincie di Milano Bergamo Como Lecco Lodi Monza e Brianza Pavia Sondrio Varese [Now Soprintendenza Archeologia, Belle Arti e Paesaggio per la città metropolitana di Milano, hereafter SABAPMi], Cimmino E. – Marandola S. – Perversi E. – Loi M.C. (eds.) (Milan: 2017) and related archival and bibliographic references: https://www.architettonicimilano.lombardia.beniculturali.it/wp-content/uploads/2016/07/GIARDINO-DARCADIA.pdf. See also Loi M.C., "Dentro il giardino. Il Progetto per il condominio a ville sovrapposte in Corso di Porta Romana a Milano", in Loi M.C. – Sumi C. – Viati Navone A. (eds.), *Giulio Minoletti. Lo spettacolo dell'architettura* (Mendrisio: 2017) 470–479.

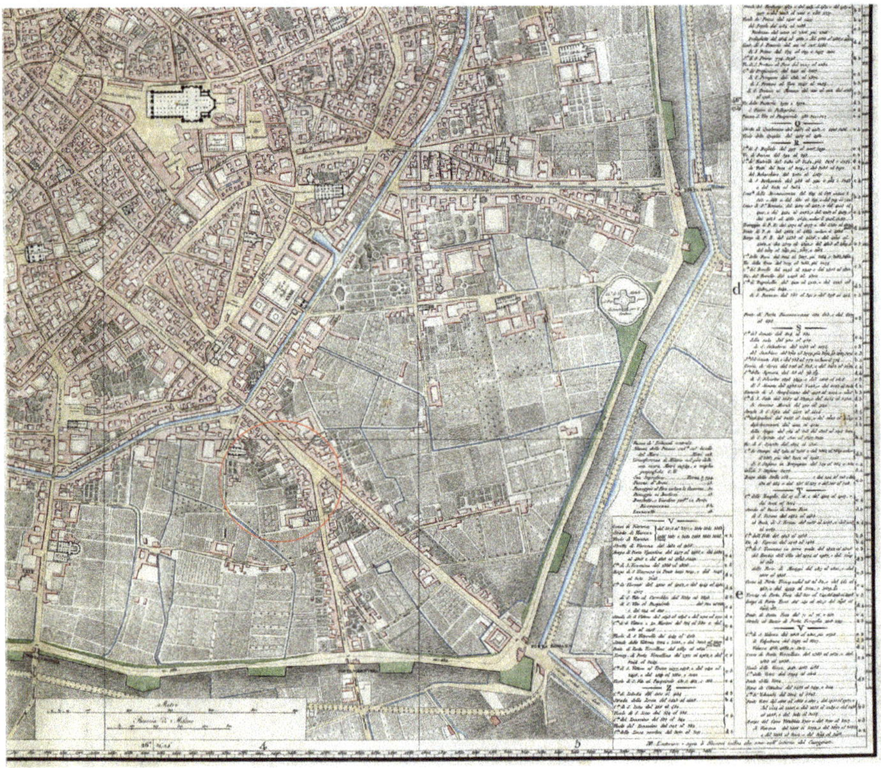

FIGURES 14.7A–B Top: Giovanni Filippini, *Iconografia della città e castello di Milano*, 1722; Bottom: Corpo degli Astronomi di Brera, *Milano Capitale del Regno d'Italia*, 1807–1810. Detail of the area of the Giardino d'Arcadia

IMAGE © MARIA CRISTINA LOI

1729, as reported in the pages of the *Chracas*: 'Jeri sera nella gran Sala di questo Sig. Conte D. Giuseppe Imbonati, Vicecustode della Colonia Milanese degli Arcadi, si tenne una bellissima adunanza [...]'. ('Yesterday evening in the great hall of this Sig. Count D. Giuseppe Imbonati, Vice-Custodian of the Milanese Colony of the Arcadians, a beautiful meeting took place [...]').[26]

In spite of these complex developments, the Arcadian garden seemed to have come back to life in the 1920s, following the restoration of the buildings and the enlargement of the garden, which the Melzi family entrusted to the architect Giuseppe De Finetti in 1924. The events connected with this phase have been reconstructed elsewhere and are beyond the scope of this essay.[27] However, it remains pertinent to mention the influence that the historic seat of the Arcadian colony continued to have in the centuries following its establishment. The twentieth century restoration and enlargement of the garden reinforces its importance and identifies its significant role in the development of Milan. De Finetti's program aimed to preserve the area of the palace and the garden; he began by considering its green area [Figs. 14.8a–b].[28] He later referred to this constraint in his drawings as the driving force behind the project described in his subsequent report. In this document, he studies the relationship between the palace and the Arcadian garden, which he then re-designed exactly according to Latuada's description. He ensured an alignment of the vestibule of the building with the historic axis of Porta Romana in an almost ideological operation, the accuracy of which, however, we are not able to verify [Fig. 14.9].[29]

---

26 See Restante R., *Cronache di vita arcadica tra le pagine del Chracas* (*1719–1772*) (MA diss., Università La Sapienza, Rome: 2021–2022). I would like to thank Stefania Baragetti for this information.

27 See above, note 25.

28 Two legislative measures were taken to preserve the palace and its surroundings: the first in 1912 (SBEAPMi, Archivio monumenti, cart. FF/16280/2, *Palazzo e giardino Melzi, già Pertusati* (*Giardino d'Arcadia*) 20.08.1912), and the second, specifically for the garden, in 1936 (SBEAPMi, Archivio monumenti, cart. FF/16280/2, 06.07.1936).

29 In his report, De Finetti refers to the changes in the extension ('more and less') of the area after its purchase by the Melzi family. By referring to the original "Giardino d'Arcadia" when drawing up his project, De Finetti expressed his desire to recognise the historical value of the garden and relate it to the modern city by means of the "optical telescope". De Finetti's strong idea has survived the postwar transformations that saw the demolition of the building and the construction of new residential buildings. His project is documented by three drawings, a detailed written report, and some photographs that have been preserved in SbeapMi, Archivio monumenti, cart. FF/16280/2 and SBEAPMi, Photo Archive, cart. 850.

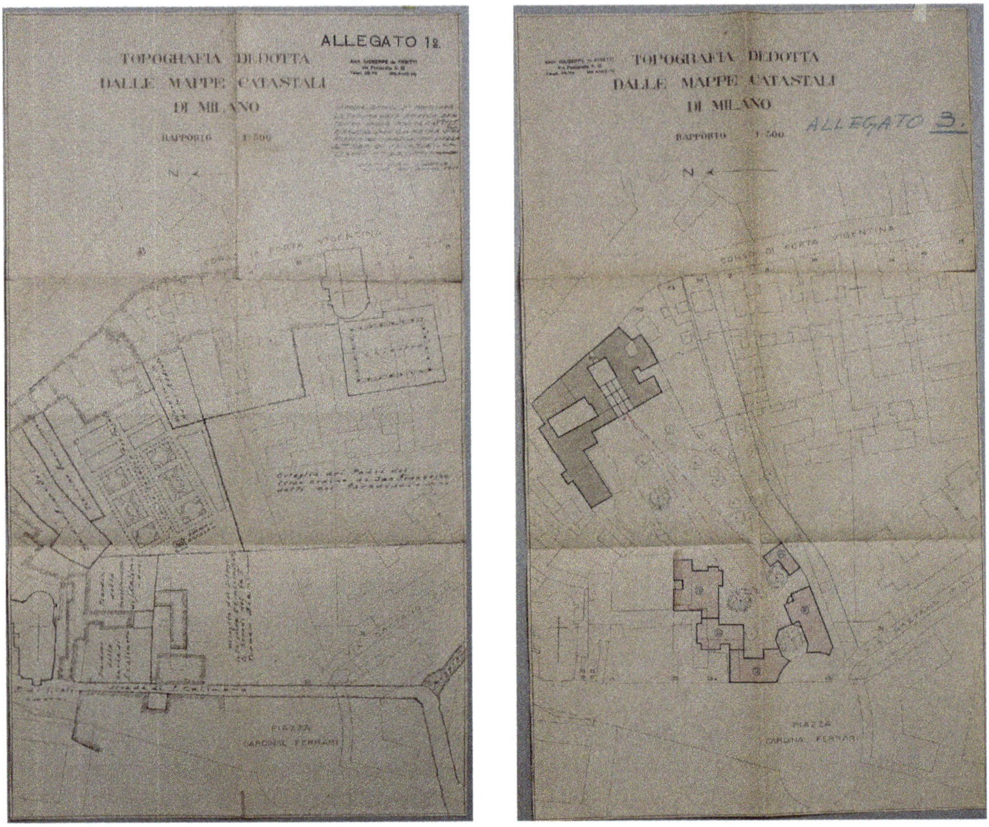

FIGURES 14.8A–B Giuseppe De Finetti, Studies for the redevelopment project of the Giardino d'Arcadia area (proposal to modify the plan of 1912), 1924, SbeapMi. Archivio Monumenti, cart. FF/16280/2
IMAGE © MARIA CRISTINA LOI

As previously mentioned, Count Pertusati had reorganised the large green area into a 'giardino di delizia' ('delightful garden') with a rectangular shape in the Italian style, extending over an almost flat parcel of land divided by paths, flower beds and rows of box and hornbeam trees.[30] The plan presented a structure symmetrical to the main avenue and connected the entrance of the building overlooking the garden to its opposite end. This perspective, delineated by tall trees, culminated in the exedra that housed the Hercules statuary, today located on the northeast side of the garden in the shade of a

30 Crescimbeni, *Notizie istoriche*.

FIGURE 14.9 The relation between the Giardino d'Arcadia and the Palazzo Pertusati and contemporary cartography. Elaboration by Maria Cristina Loi and architect Francesca Critelli (Politecnico di Milano)
IMAGE © MARIA CRISTINA LOI

1950's residential building.[31] The architect responsible for this design remains unknown, but we can ascertain that in the first half of the eighteenth century the architects Francesco Croce and Carlo Croce were engaged in works at the Pertusati villa in Comazzo, near Lodi. This property still exists today, albeit with considerable modifications to the original structure and especially the garden.[32] A series of etchings by Marcantonio Dal Re depicting the villa-palace provide more information than the surviving structures about the original building project. Set on an inverted U-shaped plan, it was surrounded by a garden and dominated by an imposing staircase and large fountain, elements reminiscent of the distinctive features of the Bosco Parrasio on the Janiculum Hill.[33]

The Arcadian garden in Porta Romana missed one fundamental element typical of the Academy's seats, which had found its optimal conditions for construction in Rome: a theatre.[34] As specified, the Pertusati garden was not located on a hill, which compromised not only the possibility of constructing monumental stairways, but also simple, open-air terraces where the 'pastori' could gather and recite songs and poems on warmer days.[35] Even in the maps, there is no trace or indication of a theatre. We are thus led to believe that most meetings in the winter season took place in the halls of the palace, as was suggested by Sassi: '[...] Arcadum Colonia se se in id domicilium recepit, & hyeme quidem in aula superiore domus, celebraorum Pictorum tabulis, circumquaque ornata' ('The Arcadian colony retreated to this residence and, during the winter months, to the main hall of the house, which was decorated with works by famous painters').[36] Moreover, the large greenhouses described by

---

31 The other statues present in the garden today have not been identified, but it is probable that they belonged to the Giardino d'Arcadia.

32 On the villa, owned by the Comazzo branch of the Pertusati family, see Tosetti S., "Villa di delizia o sia palagio campareccio. La Villa Pertusati a Comazzo", in Tartari C.M. (ed.), *Una terra fra civiltà e tradizioni* (Comazzo: 2003) 277–284.

33 Dal Re Marco Antonio, *Ville di delizia o siano palagi camparecci nello Stato di Milano* (Milan, Alla piazza de' mercanti nel portico superiore delle scuole Palatine: 1743). It is possible to hypothesize an intervention of the Croce family, or at least of Francesco Croce, in the Milanese property, considering that he is documented in the plans for the modification of the façade of Palazzo Monti in Milan, the second seat of the Milanese colony.

34 On the tradition of theatres in the Arcadian seats, see for example Pietraroia P., "Il Bosco Parrasio", in *Il Teatro a Roma nel Settecento* (Rome: 1989) 173–198 and Dixon, *Between the Real and the Ideal*, and their respective bibliographies. See also Katrina Grant's essay in this book.

35 For a complete overview of the typical characteristics of the Arcadian seats, see the detailed description by Crescimbeni in *Storia dell'Accademia degli Arcadi istituita in Roma l'Anno 1690 per la coltivazione delle scienze, delle lettere umane e della Poesia, scritta da Gio. Mario Crescimbeni primo custode generale* (London: 1804).

36 Sassi, "Arcadum Mediolanensium Colonia".

Latuada would have offered another alternative shelter and presented guests with a spectacular view of the perpetually blooming oasis. In the summer, the garden was the privileged venue for meetings and a substitute for the living room where parties, dances, and banquets were hosted. It was the place where, without the wooded character of the Roman Bosco Parrasio, nature was transformed into a delimited and ordered environment.

Today the garden still contains a wide variety and wealth of plants and trees, but it is not possible to be certain how many of these are from the original planting. We can ascertain, however, that numerous cedar trees, a symbol of eternity, and other citrus fruits and exotic plants had been planted in Pertusati's time.[37] Many of these varieties found shelter in the winter months in the light, wooden structures mentioned by Latuada that have since been lost, likely located in the northern part of the property. The proliferation of paths, symmetrical flowerbeds, glades, rare plants, fountains, water features and treillages reinforced the symbolic-allegorical role of plants and water in Arcadian meeting places.

The Arcadian garden acted as a precedent for eighteenth-century Milanese gardens. It was the first Lombard example of a winter garden, a typology that became widespread in the second half of the eighteenth century. It was also home to one of the most exclusive institutions in the region and its establishment was the prerogative of a cultural élite that dominated and directed transformations in Milan at the time. This garden, like many others, was not destined to survive the changes that accompanied the end of the century, although its memory has nonetheless persisted in the contemporary city. Its traces include not only libraries, archives and the remaining garden, but also the rich legacy of printed texts and manuscripts mentioned in this essay. The complex evolution of the site and its survival today, despite successive changes of ownership, the devastation of the Second World War and the threat of destruction posed by the post-war "building fever", can only be understood by taking into account the history of the Milanese colony of the Arcadians.

References

Arrigoni P. – Bertarelli A., *Piante e vedute della Lombardia conservate nella Raccolta delle stampe e dei disegni* (Milan: 1931).

Baragetti S., "Colonie d'Arcadia nella Lombardia austriaca", in Cofano D. – Valerio S. (eds.), *La letteratura degli Italiani. 1. Centri e periferie* (Foggia: 2011) 1–8.

37 Ibid.

Baretta G., *La Biblioteca del conte Carlo Pertusati*, in Sciardelli F. (ed.), *Tra i fondi della Biblioteca Braidense* (Milan: 1993) 11–15.

Bianconi Carlo, *Nuova guida di Milano, per gli amanti delle belle arti e delle sacre e profane antichità milanesi* (Milan, nella Stamperia Sirtori: 1787).

Bona Castellotti M. – Bressan E. – Vismara P. (eds.), *Politica, vita religiosa, carità: Milano nel primo Settecento* (Milan: 1977).

Bossi L., *Guida di Milano, ossia Descrizione delle città e dei luoghi più osservabili ai quali da Milano recansi i forestieri*, vol. I (Milan: 1818).

Ceva Tommaso, *Memorie d'alcune virtù del signor conte Francesco de Lemene* [...] (Milan, Per Domenico Bellagatta: 1718).

Cremonini C., *Luca Pertusati*, ad vocem, *Dizionario biografico degli italiani* 82 (2015).

Cremonini C., "L.A. Muratori e la Società Palatina. Considerazioni su cultura e politica a Milano tra Sei e Settecento", in Bona Castellotti M. – Bressan E. – Vismara P. (eds.), *Politica, vita religiosa, carità. Milano nel primo Settecento* (Milan: 1997) 185–212.

Crescimbeni Giovanni Mario, *Notizie Istoriche degli Arcadi Morti*, vol. II (Rome, Antonio de' Rossi: 1720).

Dal Re Marco Antonio, *Ville di delizia o siano Palaggi camparecci nello Stato di Milano* (Milan, Alla piazza de' Mercanti nel portico superiore delle scuole Palatine: 1743).

de' Crescenzi Pietro, *Commoda ruralia* (Spira, Peter Drach: *c.*1495).

Dixon S.M., *Between the Real and the Ideal. The Accademia degli Arcadi and its Gardens in Eighteenth century Rome* (Newark: 2006).

Giorgetti Vichi A.M. (ed.), *Gli Arcadi dal 1690 al 1800. Onomasticon*, Arcadia – Accademia Letteraria Italiana (Rome: 1977).

Grossi Turchetti M.L., *Gli incunaboli del Fondo Pertusati in Braidense*, http://www.braidense.it/risorse/turchetti_pertusati.pdf.

Grossi Turchetti M.L. (ed.), *Biblioteca Nazionale Braidense, I manoscritti datati della Biblioteca Nazionale Braidense di Milano*. Manoscritti datati d'Italia, 10 (Florence: 2004).

Kapp V., "Muratori e l'idea della Repubblica letteraria d'Italia", *Romanische Forschungen* 114.2 (2002) 191–205.

Il Giardino d'Arcadia a Milano. Tutela, pianificazione urbana e ricostruzione, Soprintendenza belle arti e paesaggio per le provincie di Milano Bergamo Como Lecco Lodi Monza e Brianza Pavia Sondrio Varese, 2017, URL: https://www.architettonicimilano.lombardia.beniculturali.it/wp-content/uploads/2016/07/GIARDINO-DARCADIA.pdf.

Latuada Serviliano, *Descrizione di Milano ornata con molti disegni in rame delle fabbriche più cospicue, che si trovano in questa metropoli. Raccolta e ordinata da Serviliano Latuada, sacerdote milanese*, vol. 2 (Milan: Cairoli, 1737).

Loi M.C., "Dentro il giardino. Il Progetto per il condominio a ville sovrapposte in Corso di Porta Romana a Milano", in Loi M.C. – Sumi C. – Viati Navone A. (eds.), *Giulio Minoletti. Lo spettacolo dell'architettura* (Mendrisio: 2017) 470–479.

Mezzabarba Giovanni Antonio, *Discorso di Vitanio Gateatico pastore d'Arcadia in difesa dell'"Endimione" favola pastorale di Arezio Gateatico [Francesco de Lemene]; indirizzato a Cromiro Dianio [Pietro Antonio Bernardoni] suo Compastore* (Turin, Zappata: 1699).

Mezzanotte P. – Bascapè G., *Milano nell'arte e nella storia* (Milan: 1948).

Pietraroia P., "Il Bosco Parrasio", in *Il teatro a Roma nel Settecento* (Rome: 1989) 173–198.

Minola Diego Antonio, *Diario Storico-Politico, di alcuni avvenimenti nel secolo XVIII, tomo I, dal 1701 al 1729 inclusivo*, Milan, Biblioteca Ambrosiana Ms. G.123.

Quondam A., "L'istituzione Arcadia. Sociologia e ideologia di un'Accademia", *Quaderni Storici* 8.23 (1973) 389–438.

Quondam A., "Problemi di critica arcadica", in *Critica e storia letteraria. Studi offerti a Mario Fubini*, vol. 1 (Padua: 1970) 515–523.

Restante R., *Cronache di vita arcadica tra le pagine del Chracas (1719–1772)* (MA diss., Università La Sapienza, Rome: 2021–2022).

Sassi Giuseppe Antonio, *De studiis litterariis Mediolanensium antiquis et novis prodromus* (Milan, J.R. Malatestam: 1729).

Sassi Giuseppe Antonio, "Arcadum Mediolanensium Colonia", in Argelati Filippo, *Philippi Argelati Bononiensis Bibliotheca scriptorum Mediolanensium, seu acta et elogia virorum omnigena eruditione illustrium* […] *Preamittitur Clarissimi Viri Josephi Antonii Saxii* […] (Milan, In aedibus Palatinis: 1745) XIV, 2, LXII–LXIII.

Seregni G., "La cultura milanese del Settecento", in *Storia di Milano. L'età delle riforme (1706–1796)*, vol. 12 (Milan: 1959) 569–586.

Squizzato A., "Carlo e Alessandro Teodoro Trivulzio fra i cultori del buon gusto e della scienza dell'antico", in Bianchi E. – Rovetta A. – Squizzato A. (eds.), *Le arti nella Lombardia asburgica durante il Settecento. Novità e aperture*, Proceedings of the international conference. 5–6 giugno 2014 (Milan: 2017) 409–423.

Storia dell'Accademia degli Arcadi istituita in Roma l'anno 1690 per la coltivazione delle scienze, delle lettere umane e della poesia, scritta da Gio. Mario Crescimbeni primo custode generale (London: 1804).

Tosetti S., "*Villa di delizia o sia palagio campareccio*. La Villa Pertusati a Comazzo", in Tartari C.M. (ed.), *Una terra fra civiltà e tradizioni* (Comazzo: 2003) 277–284.

Turchetti Grossi M.L., "Brevi note sul fondo Pertusati della Braidense", *Accademie e Biblioteche d'Italia* 47, 5 (1979) 370–386.

Turchetti Grossi M.L. – Gaspari G., "Il fondo Pertusati", in *La Braidense. La cultura del libro e delle biblioteche nella società dell'immagine* (Florence: 1991) 88–89.

Vischi L., "La Società Palatina di Milano. Studio storico di L. Vischi", *Archivio Storico Lombardo* 7.3 (1880) 391–542.

Zumkeller L., "Da biblioteca di carattere generale a biblioteca storico-umanistica: la Braidense attraverso una rilettura della storia dei fondi", *Accademie e Biblioteche d'Italia* 67.2 (1999) 41–52.

Index Nominum

www.ingramcontent.com/pod-product-compliance
Lightning Source LLC
Chambersburg PA
CBHW071336230926
51849CB00002B/409